2012

Information and Communications for Development

2012
Information and Communications for Development

Maximizing Mobile

THE WORLD BANK
Washington, D.C.

infoDev

1818 H Street NW, Washington DC 20433
Telephone: 202-473-1000; Internet: www.worldbank.org

1 2 3 4 15 14 13 12

ISBN (paper): 978-0-8213-8991-1
ISBN (electronic): 978-0-8213-9587-5
DOI: 10.1596/978-0-8213-8991-1

Cover photographs: Top and bottom: G. M. B. Akash / Panos; center: Mr. Pierre C. Sibiry Traore, ICRISAT, AgCommons, a program executed by the Consultative Group on International Agricultural Research (CGIAR); right: The Commonwealth of Learning

Cover design: Naylor Design

Table of Contents

Part II

FIGURES

Part I

Part II

TABLES

Part I

Foreword

Mobile phones, a rarity in many developing countries at the turn of the century, now seem to be everywhere. Between 2000 and 2012, the number of mobile phones in use worldwide grew from fewer than 1 billion to around 6 billion. The mobile revolution is transforming livelihoods, helping to create new businesses, and changing the way we communicate. The mobile phone network is already the biggest "machine" the world has ever seen, and now that machine is being used to deliver development opportunities on a scale never before imagined. During this second decade of the new millennium, maximizing the potential of mobile phones is a challenge that will engage governments, the private sector, and the development community alike.

Information and Communications for Development 2012: Maximizing Mobile is the third report in the World Bank Group's series on Information and Communication Technologies (ICTs) for Development, originally launched in 2006. This edition focuses on mobile applications and their use in promoting development, especially in agriculture, health, financial services, and government. Cross-cutting chapters present an overview of emerging trends in mobile applications, the ways they are affecting employment and entrepreneurship opportunities, and the policy challenges presented by the ongoing shift from narrowband to broadband mobile networks. The report features at-a-glance tables for 152 economies showing the latest available data and indicators for the mobile sector (year-end 2011, where possible). The report also introduces an analytical tool for examining the relevant performance indicators for each country's mobile sector, so policy-makers can assess their capacities relative to other countries. A more complete range of ICT indicators is available in the *Little Data Book on Information and Communication Technology 2012*, published alongside this report.

It is our hope that this new report will provide some emerging good-practice principles for policy-makers, regulators, and investors in this complex and constantly changing sector. The World Bank Group already supports a wide range of investment lending programs with an ICT component. According to the report of the Independent Evaluation Group, *Capturing Technology for Development* (2011), around three-quarters of all investment lending projects from the World Bank Group have an ICT component; in addition, more than $4 billion has been invested directly in the ICT sector between 2003 and 2010.

This report marks a shift from the World Bank Group's traditional focus on ICT *connectivity* to a new focus on *applications* and on the ways ICTs, especially mobile phones, are being used to transform different sectors of the global economy. This change of focus reflects how the value created by the industry is shifting from networks and hardware to software and services. The World Bank Group expects that the theme of *transformation* will increasingly guide its investment lending, and this report is aligned with that new direction. Ultimately, the mission of the World Bank is to work for a world free of poverty—a goal that is likely to be achieved more efficiently when ICT investment is integrated effectively alongside investment in sectors such as agriculture, health, and government.

Marianne Fay
Chief Economist, Sustainable Development Network
The World Bank

Preface

The World Bank's new strategy for engagement in the Information and Communication Technologies (ICTs) sector, which comes into force in 2012, is built around three strategic themes: *Innovate*—ICT for innovation and ICT-based services industries; *Connect*—affordable access to voice, high-speed internet, information and media; and *Transform*—ICT applications to transform services for enhanced development outcomes.

This new flagship report on *Information and Communications for Development* builds on these three themes. In particular, the report shows how innovation in the manufacture of mobile handsets—giving them more memory, faster processing power, and easier-to-use touchscreen interfaces—married with higher performance and more affordable broadband networks and services produces transformation throughout economies and societies. Increasingly, that transformation is coming from developing countries, which are "more mobile" than developed countries in the sense that they are following a "mobile first" development trajectory. Many mobile innovations (including multi-SIM card phones, low-cost recharges, and mobile payments) increasingly originate in poorer countries and spread from there.

Since the last *Information and Communications for Development* report was published, almost 2 billion new mobile phone subscriptions have been added worldwide, and the majority of these are in the developing world. This rapid growth does not show the whole picture, however. Alongside the process of enlarging the network is an equally important process of improving the quality and depth of the network as narrowband networks are upgraded to broadband and as basic phones and feature-phones are upgraded to smartphones and tablets. The full range of innovative mobile applications described in this report is not yet available in all countries and to all subscribers, but it soon will be. And the expectation is that developing countries will invent and adapt their own mobile applications, suited to local circumstances and needs. For that reason more research is needed on how mobile applications are used in base of the pyramid households.

This report, like its predecessors, was researched and written jointly by the ICT Sector Unit and by *info*Dev, a global partnership program of the World Bank Group. It has been reviewed by a broad range of experts working in the field, both within and outside the Bank, whose contributions are gratefully acknowledged. Funding is provided by the World Bank as well as *info*Dev's donors, notably the Ministry for Foreign Affairs of the Government of Finland, the Korean Trust Fund for ICT4D, and UKaid. The World Bank Group is committed to continuing its analytical and lending operations to support progress and the sharing of best practices and knowledge, as well as expanding its investments in private ICT companies to further growth in the sector, competitiveness, and the availability of better-quality, affordable ICT services to all the world's inhabitants.

Juan Navas-Sabater
Acting Sector Manager, ICT Sector Unit
The World Bank

Valerie D'Costa
Program Manager, info*Dev*
The World Bank

Acknowledgments

This report was prepared by a team from the ICT Sector Unit (TWICT), *info*Dev, and the Development Economics Data Group (DECDG) of the World Bank Group. The editorial team was led by Tim Kelly and comprised Nicolas Friederici, Michael Minges, and Masatake Yamamichi. Their work was overseen by a peer-review team, led by Marianne Fay, that included Jose Luis Irigoyen, Valerie D'Costa, Philippe Dongier, Phillippa Biggs (ITU), and Christine Zhenwei Qiang.

The principal authors of the chapters in Part I of the report are:

- Tim Kelly and Michael Minges (Executive Summary)
- Michael Minges (Chapter 1)
- Naomi J. Halewood and Priya Surya (Chapter 2)
- Nicolas Friederici, Carol Hullin, and Masatake Yamamichi (Chapter 3)
- Kevin Donovan (Chapter 4)
- Maja Andjelkovic and Saori Imaizumi (Chapter 5)
- Siddhartha Raja and Samia Melhem, with Matthew Cruse, Joshua Goldstein, Katherine Maher, Michael Minges, and Priya Surya (Chapter 6)
- Victor Mulas (Chapter 7)

The principal authors of Part II were Michael Minges and Kaoru Kimura, and the editorial team for the statistical tables comprised Neil Fantom, Buyant Erdene Khaltarkhuu, Kaoru Kimura, Soong Sup Lee, Michael Minges, and William Prince.

Inputs, comments, guidance, and support at various stages of the report's preparation were received from the following World Bank Group colleagues: Maria Amelina, Edward Anderson, Elizabeth J. Ashbourne, Seth Ayers, Alan Carroll, Vikas Choudhary, Toni Eliasz, Tina George, Joshua Goldstein, Aparajita Goyal, Siou Chew Kuek, Katherine Maher, Wonki Min, Fernando Montenegro Torres, Arata Onoguchi, Tiago Peixoto, Mark Pickens, Carlo Maria Rossotto, Leila Search, and Randeep Sudan, as well as from the principal authors.

External reviewers, to whom special thanks are owed, included Phillippa Biggs (ITU), Steve Esselaar (Intelecon), Shaun Ferris (Catholic Relief Services), Vicky Hausmann (Dalberg), Janet Hernandez (Telecommunications Management Group), Jake Kendall (Gates Foundation), Vili Lehdonvirta (London School of Economics), Daniel Leza (Telecommunications Management Group), Bill Maurer (University of California, Irvine), Sascha Meinrath (New America Foundation), Marcha Neethling (Praekelt Foundation), Brooke Partridge (Vital Wave Consulting), Ganesh Ramanathan (Tiger Party), Michael Riggs (Food and

Agriculture Organization), Stephen Rudgard (Food and Agriculture Organization), Brendan Smith (Vital Wave Consulting), Scott Stefanski (Bazaar Strategies), Heather Thorne (Grameen Foundation), Katrin Verclas (Mobile Active), and Anthony Youngblood (New America Foundation).

Special thanks are owed to Phillippa Biggs (ITU), who provided a thorough and dedicated peer review of all chapters, as well as to Denis Largeron and Marta Priftis from TWICT, and to Denise Bergeron, Jose De Buerba, Aziz Gökdemir, Stephen McGroarty, and Santiago Pombo-Bejarano, from the World Bank Office of the Publisher for oversight of the editorial production, design, printing, and dissemination of the book. The infographic in the Executive Summary was prepared by Zack Brisson and Mollie Ruskin of Reboot (www.thereboot.org), with guidance from the editorial team.

A report of this nature would be impossible without the support of our development partners. For this edition of the report, special thanks are due to:

- The Ministry for Foreign Affairs of the Government of Finland for its support for the Finland / *info*Dev / Nokia program on *Creating Sustainable Businesses in the Knowledge Economy*, which supported the production of the report as well as research for chapters 1, 2, 4, and 5.
- The Korean Trust Fund (KTF) on Information and Communication Technology for Development (ICT4D), which supported background research for chapters 2, 3, 4, and 5.
- UKaid, which supported background research for chapter 7 through its support for *info*Dev's analytical work program.

The team would also like to thank the many other individuals, firms, and organizations that have contributed through their continuing support and guidance to the work of the World Bank Group over the three years since the last report in this series was published.

Abbreviations

2G	second generation (mobile communications)
3G	third generation (mobile communications)
4G	fourth generation (mobile communications)
apps	applications
ATM	automated teller machine
CDMA	Code Division Multiple Access (cellular mobile standard)
CGAP	Consultative Group to Assist the Poor
e-payment	electronic payment
e-services	electronic services
ebook	electronic book
eCommerce	electronic commerce
EDGE	Enhanced Data Rates for GSM Evolution (cellular mobile standard)
eGovernment	electronic government
eHealth	electronic health
EV-DO	Evolution–Data Optimized (cellular mobile standard)
GB	gigabyte
GDP	gross domestic product
GNI	gross national income
GPS	Global Positioning System
GSM	Global System for Mobile communications
GTUGS	Google Technology User Groups
HSPA	High-Speed Packet Access (cellular mobile standard)
HTML	hypertext mark-up language
IC4D	*Information and Communications for Development*
ICT	information and communication technology
IM	instant messaging
IMF	International Monetary Fund
ISP	internet service provider
ITU	International Telecommunication Union
kbit/s	kilobits per second
LTE	Long Term Evolution (cellular mobile standard)
MB	megabyte
Mbit/s	Megabits per second
MDGs	Millennium Development Goals
mGovernment	mobile government
mHealth	mobile health
mLab	mobile applications laboratory

NFC	near field communications
NGO	nongovernmental organization
OECD	Organisation for Economic Co-operation and Development
PC	personal computer
PDA	personal digital assistant
PPP	public-private partnership
RFID	radio frequency identification
SAR	special autonomous region
SIM	subscriber identity module
SME	small and medium enterprise
SMS	short message service
TCO	total cost of ownership
TD-SCDMA	Time Division Synchronous Code Division Multiple Access (cellular mobile standard)
UNCTAD	United Nations Conference on Trade and Development
UNDP	United Nations Development Programme
UNESCO	United Nations Educational, Scientific and Cultural Organization
UNICEF	United Nations Children's Fund
USB	universal serial bus
USSD	Unstructured Supplementary Service Data
W-CDMA	Wideband Code Division Multiple Access (cellular mobile standard)
WHO	World Health Organization
WiMAX	Worldwide Interoperability for Microwave Access (wireless standard)

All dollar amounts are U.S. dollars unless otherwise indicated.

Part I

Executive Summary

Tim Kelly and Michael Minges

Main messages

With some 6 billion mobile subscriptions in use worldwide, around three-quarters of the world's inhabitants now have access to a mobile phone. Mobiles are arguably the most ubiquitous modern technology: in some developing countries, more people have access to a mobile phone than to a bank account, electricity, or even clean water. Mobile communications now offer major opportunities to advance human development—from providing basic access to education or health information to making cash payments to stimulating citizen involvement in democratic processes.

The developing world is "more mobile" than the developed world. In the developed world, mobile communications have added value to legacy communication systems and have supplemented and expanded existing information flows. However, the developing world is following a different, "mobile first" development trajectory. Many mobile innovations—such as multi-SIM card phones, low-value recharges, and mobile payments—have originated in poorer countries and are spreading from there. New mobile applications that are designed locally and rooted in the realities of the developing world will be much better suited to addressing development challenges than applications transplanted from elsewhere. In particular, locally developed applications can address developing-country concerns such as digital literacy and affordability.

Mobile applications not only empower individual users, they enrich their lifestyles and livelihoods, and boost the economy as a whole. Indeed, mobile applications now make phones immensely powerful as portals to the online world. A new wave of "apps," or smartphone applications, and "mash-ups" of services, driven by high-speed networks, social networking, online crowdsourcing, and innovation, is helping mobile phones transform the lives of people in developed and developing countries alike. The report finds that mobile applications not only empower individuals but have important cascade effects stimulating growth, entrepreneurship, and productivity throughout the economy as a whole. Mobile communications promise to do more than just give the developing world a voice. By unlocking the genie in the phone, they empower people to make their own choices and decisions.

Near ubiquity brings new opportunities. This 2012 edition of the World Bank's *Information and Communications for Development* report analyzes the growth and evolution of mobile telephony, and the rise of data-based services delivered to handheld devices, including apps. The report explores the consequences for development of the emerging "app economy." It summarizes current thinking and seeks to inform the debate on the use of mobile phones for development. This report looks at key ecosystem-based applications in agriculture, health, financial services, employment, and government, with chapters devoted to each. The story is no longer about the phone itself, but about how it is used, and the content and applications to which mobile phones provide access.

Engaging mobile applications for development requires an enabling "ecosystem." Apps are software "kernels" that sit on a mobile device (typically a smartphone or tablet) and that can often interact with internet-based services to, for instance, access updates. Most apps are used by individual users, but the applications that may prove most useful for development are those usually developed within an ecosystem that involves many different players, including software developers, content providers, network operators, device manufacturers, governments, and users. Although the private sector is driving the market, social intermediaries, such as nongovernmental organizations (NGOs) play an important role in customizing applications to meet the needs of local communities. In many countries, a ready-made community of developers has already developed services based around short message service (SMS) or instant messaging (IM) and is now developing applications for more sophisticated devices. Policy-makers need to create an environment in which players can collaborate as well as compete. That will require rethinking regulations governing specific sectors such as financial services, health, or education. Governments also play a fundamental role in establishing necessary conditions in which mobile communications can thrive through the allocation of wireless spectrum, enactment of vital legislation, and leadership in mobile government, or mGovernment.

The mobile revolution is right at the start of its growth curve. Devices are becoming more powerful and cheaper. But the app economy requires economies of scale to become viable. The report argues that now is the time to evaluate what works and to move toward the commercialization, replication, and scaling up of those mobile apps that drive development. Until recently, most services using mobiles for development were based on text messaging. Now, the development of inexpensive smartphones and the spread of mobile broadband networks are transforming the range of possible applications. Several challenges lie ahead, notably, the fragmentation that arises from multiple operating systems and platforms. It is already clear, however, that the key to unleashing the power of the internet for the developing world lies in the palm of our hands.

Why are mobile phones now considered indispensable?

The report's opening chapter provides an overview of the key trends shaping and transforming the mobile industry as well as their impact on development. The chapter examines the evolution of the mobile phone from a simple channel for voice to one for exchanging text, data, audio, and video through the internet. Given technological convergence, mobile handsets can now function as a wallet, camera, television, alarm clock, calculator, address book, calendar, newspaper, gyroscope, and navigational device combined. The latest smartphones are not just invading the computer space, they are reinventing it by offering so much more in both voice and nonvoice services.

Developing countries are increasingly well placed to exploit the benefits of mobile communications, with levels of access rising around the world. Chapter 1 explores the implications of the emergence of high-speed broadband networks in developing countries, and how the bond between mobile operators and users is loosening, as computer and internet companies invade the mobile space, with a growing number of handset models now offering Wi-Fi capability.

The chapter also examines the size and nature of the mobile economy and the emergence of new players in the mobile ecosystem. The emergence of apps, or special software on handheld devices that interacts with internet-based data services, means that the major issue for the development community today is no longer basic access to mobile phones but about what can be done with phones. More than 30 billion apps had been downloaded worldwide by early 2012, and they make for an innovative and diverse mobile landscape with a potentially large impact on the lives of people in developed and developing countries alike. Growing opportunities for small-scale software developers and local information aggregators are allowing them to develop, invent, and adapt apps to suit their individual needs. Users themselves are becoming content providers on a global scale.

Indeed, the latest generations of mobile telephony are sowing social and political as well as economic transformation. Farmers in Africa are accessing pricing information through text messages, mothers can receive medical reports on the progression of their pregnancy by phone, migrant workers can send remittances without banks. Elections are monitored and unpopular regimes toppled with the help of mobile phones. Texting and tweeting have become part of modern vocabulary.

Mobiles are now creating unprecedented opportunities for employment, education, and entertainment in developing countries. This chapter looks beyond specific examples to identify the broader trends shaping and redefining our understanding of the word "mobile."

A mobile green revolution

Given the dominance of primary commodities in the economies of many developing countries, chapter 2 explores the all-important area of mobile applications designed to improve incomes, productivity, and yields within the agricultural sector, which accounts for about 40 percent of the workforce and an even greater proportion of exports in many developing countries.

To date, voice calls and SMS text messages have proven invaluable in increasing efficiency in smallholder agriculture. They can, for example, provide real-time price information and improve the flow of information along the entire value chain, from producers to processors to wholesalers to retailers to consumers. The basic functions of the mobile phone will continue to remain important for reaching the widest number of people, but the focus of applications development is shifting as the underlying technologies evolve.

Today, increasingly specialized mobile services are fulfilling specific agricultural functions, while multimedia imagery is being used to overcome illiteracy and provide complex information regarding weather and climate, pest control, cultivation practices, and agricultural extension services to potentially less tech-savvy farmers. This chapter also examines the emerging uses of remote and satellite technologies that are assisting in food traceability, sensory detection, real-time reporting, and status updates from the field. It further reviews examples of mobile services in agriculture to draw key learning points and provide direction on how to capitalize on successful examples.

Mobile applications for agriculture and rural development have generally not followed any generic blueprint. They are usually designed locally and for specific target markets, with localized content specific to the languages, crop types, and farming methods. Local design offers exciting opportunities for local content and applications development but may limit the economies of scale realizable from expanding from pilot programs into mass markets, potentially hindering the spread of new and promising applications and services.

The full scope and scale of smartphones and tablets for providing services to agricultural stakeholders have yet to emerge. An enabling environment that can promote the development and use of applications in developing countries must be prioritized to meet the information needs of the agricultural sector.

Keep using the tablets—how mobile devices are changing health care

Chapter 3 examines some of the key principles and characteristics of mobile for health (mHealth), and how mobiles are helping transform and enhance the delivery of primary and secondary health care services in developing countries. Mobile health can save money and deliver more effective health care with relatively limited resources; increasingly, it is associated with a focus on prevention of diseases and promotion of healthy lifestyles.

This chapter reviews on-the-ground implementations of medical health care apps to draw key conclusions about how mHealth can best be implemented to serve the needs of people in the developing world, as well as identifying barriers that must be overcome. It considers some of the unique features of the health care sector and the implications for medical apps in areas such as patient privacy and confidentiality, public and private provision of care, and real-time reporting requirements in crisis or emergency situations.

Modern health care systems are at a tipping point, as consumers take on greater responsibility for managing their own health care choices, and mobile phones could enable a shift in the locus of decision-making away from the state and health institutions to individual patients.

The most substantial challenge for mHealth, however, is the establishment of sustainable business models that can be replicated and scaled up. One step toward addressing this challenge might be a clearer delineation of roles within the health ecosystem between public and private health care providers. Another significant challenge is the effective monitoring and evaluation of mobiles in health, as pilot programs continue to proliferate.

Mobile money

This chapter examines the all-important topic of mobile money as a general platform and critical infrastructure underpinning other economic sectors. Mobile money has transformed the Kenyan economy, where mobile-facilitated payments now equate to a fifth of the country's gross domestic product (GDP). The impact of mobile money is widening elsewhere too, as it is adopted across commerce, health insurance, agricultural banking, and other sectors. Today, the potential of mobile payment systems to "bank the unbanked" and empower the poor through improved access to finance and lower transaction costs is generating growing excitement. Where they exist, mature mobile money systems have often spun off innovative products and services in insurance, credit, and savings.

When connected on a large scale, evidence suggests that the poor are able to use mobile money to improve their livelihoods. Observers remain divided, however, about whether mobile money systems are fulfilling their true growth potential. Innovative offerings, old and new, can succeed only if there is sufficient demand from consumers and firms—a variable missing in many contexts.

The mobile money industry exists at the intersection of banking and telecommunications, embracing a diverse set of stakeholders, including mobile operators, financial services companies, and new entrants (such as payment card firms). In some countries, mobile money systems may be subject to different regulatory practices and interoperability issues, not to mention clashes in culture between banks and mobile operators, so developing the necessary cross-sectoral partnerships can prove difficult. In other countries, well-developed alternative legacy systems are strong competitors to the development of mobile money systems.

This chapter evaluates the benefits and potential impact of mobile money, especially for promoting financial inclusion in the developing world. It provides an overview of the key factors driving the growth of mobile money services, while considering some of the barriers and obstacles hindering their deployment. Finally, it identifies emerging issues that the industry will face over the coming years.

Get a phone, get a job, start a business

The global mobile industry is today a major source of employment opportunities, on both the supply and demand side. Employment opportunities in the mobile industry can be categorized as direct jobs, indirect jobs, and jobs on the demand side. The contribution of the mobile communication sector to employment and entrepreneurship to date is difficult to assess, however, because the seemingly simple mobile phone can generate—and occasionally eliminate—employment opportunities by creating efficiencies and lowering transaction and information costs.

The recent rapid innovation in the mobile sector has generated significant disruptive technological change and uncertainty. This turmoil is also lowering barriers to entry, however, and generating fresh opportunities for small and young firms and entrepreneurs to displace legacy systems, innovate, and grow.

Chapter 5 showcases some of the mechanisms by which the mobile sector supports entrepreneurship and job creation. Some share similarities with traditional donor initiatives, but many are novel ideas, for which the "proof of concept" has been demonstrated only recently or has yet to be demonstrated. This chapter considers the use of specialized business incubators or mobile labs (mLabs) for supporting entrepreneurial activity in the mobile industry, as well as new opportunities that are offered in areas such as the virtual economy (trading goods and services that exist only online) or mobile microwork (work carried out remotely on a mobile device, on micro-tasks, such as tagging images).

It also provides suggestions on how to support entrepreneurship and job creation in the mobile industry. In an industry evolving as quickly as the mobile sector is today, it is vital to tailor support to local circumstances and to evaluate impact regularly.

Using phones to bring governments and citizens closer

In the public sphere, mobiles now serve as vehicles for improved service delivery and greater transparency and accountability. Today, governments are beginning to embrace the potential for mobile phones to put public services literally into the pocket of each citizen, create interactive services, and promote accountable and transparent governance.

Chapter 6 identifies a range of uses for mobiles in government (mGovernment) that supplement existing public services, expand their user base, and generate spin-off services. The revolutionary aspect to mGovernment lies in making government available, anytime and anywhere, to anyone. The chapter also provides a range of examples of mGovernment from around the world as well as a range of best practices and recommendations. It demonstrates how countries can play a constructive role in enhancing sustainability and enabling scale, while maximizing the impact of mGovernment programs.

An important conclusion is that bottom-up ad hoc approaches to mGovernment may endanger economies of scale. Top-down coordinated approaches may be preferable, since they can cut costs in designing, deploying, and operating apps; consolidate demand for communication services across government, thereby eliminating duplication; and include focused actions to build capacity and skills.

Emerging best practices suggest that any government considering the opportunities inherent in mGovernment should focus on enabling technological transformation and building the institutional capacity needed to respond to citizens' demands. Governments looking to adopt mobile tools to become responsive, accountable, and transparent should bear in mind that this process will prove successful and truly transform the government-citizen relationship only when governments take into account both elements—"mobile" and "government."

Onward and upward to mobile broadband

Chapter 7 distinguishes between supply-side policies (which seek to promote the expansion of wireless broadband networks) and demand-side policies (which seek to boost adoption of wireless broadband services) in the mobile broadband ecosystem.

Supply-side policies seek to address bottlenecks and market failures that constrain network expansion and provide incentives for broader wireless broadband coverage. The chapter reviews the following supply-side policy recommendations:

- Boosting the availability of quality spectrum to deploy cost-effective wireless broadband networks
- Eliminating technological or service restrictions on spectrum
- Focusing on expanding network coverage rather than on profiting from spectrum auctions
- Requiring transparency in traffic management and safeguarding competition
- Limiting spectrum hoarding, which could distort competitive conditions in the market
- Fostering the development of national backbone broadband networks
- Encouraging infrastructure and spectrum sharing

Demand-side policies aim at boosting growth in the adoption of wireless broadband services by addressing barriers to adoption and fostering the development of innovative broadband services and applications pulling users' demand toward mobile broadband. The chapter reviews the following demand-side policy recommendations:

- Improving the availability and affordability of broadband-enabled devices
- Boosting the affordability of broadband services
- Fostering the development of broadband services and applications

The chapter concludes that appropriate policy action requires addressing both the supply- and demand-sides of the mobile broadband ecosystem. Policy-makers must evaluate local market conditions before applying specific policies addressing bottlenecks or market failures. The most common breakdowns on the supply side are lack of available spectrum and inadequate backbone networks; on the demand side, the main constraints are lack of affordable mobile devices and broadband services, as well as limited local applications and content. Ultimately, policy-makers must determine which policies to adopt, and how to implement them, based on domestic circumstances and the likely effectiveness of the policy for broadband diffusion in the context of each country.

Appendixes

The *Country Tables* in the appendix to this report provide comparative data for some 152 economies with populations of more than 1 million and summary data for others, with at-a-glance tables focusing on the mobile sector. The report is complemented by the World Bank's annual *Little Data Book on Information and Communication Technology*, which presents a wider range of ICT data.

The Statistical Appendix reviews the main trends shaping the sector and introduces a new analytical tool for tracking the progress of economies at different levels of economic development in widening access, improving supply, and stimulating demand for mobile services.

MAXIMIZING MOBILE FOR DEVELOPMENT

THE PACE AT WHICH MOBILE PHONES SPREAD GLOBALLY IS UNMATCHED IN THE HISTORY OF TECHNOLOGY

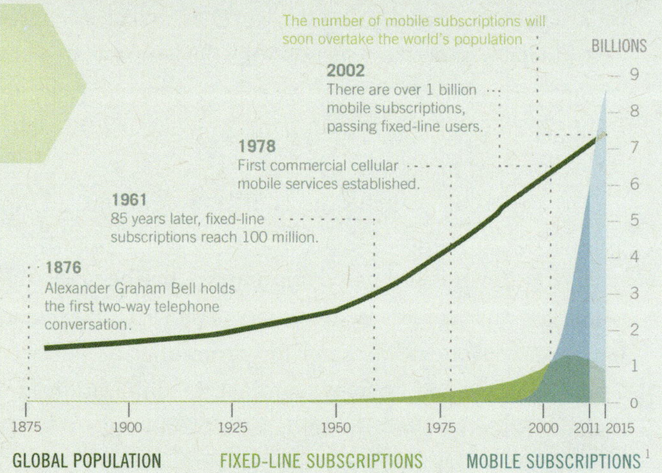

PERCENT OF THE WORLD'S POPULATION WITH MOBILE CELL SIGNAL[2]

OVER 6 BILLION MOBILE SUBSCRIPTIONS WORLDWIDE

75% of the WORLD NOW HAS ACCESS to a MOBILE PHONE[3]

THE DEVELOPING WORLD IS NOW **MORE MOBILE** THAN THE DEVELOPED WORLD

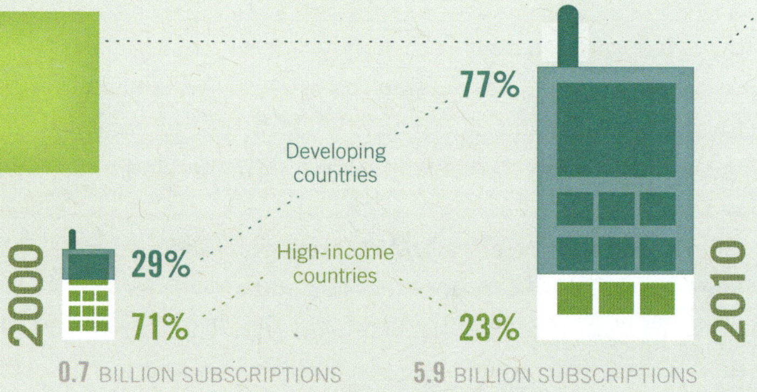

GROWTH OF GLOBAL MOBILE SUBSCRIPTIONS[4]

MOST PHONES ARE OWNED BY PEOPLE **LIVING IN LOW-INCOME REGIONS**

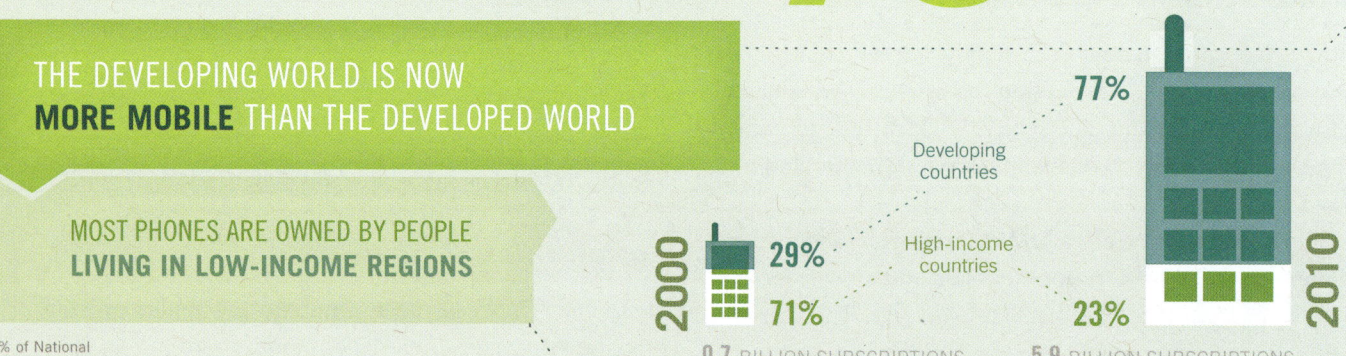

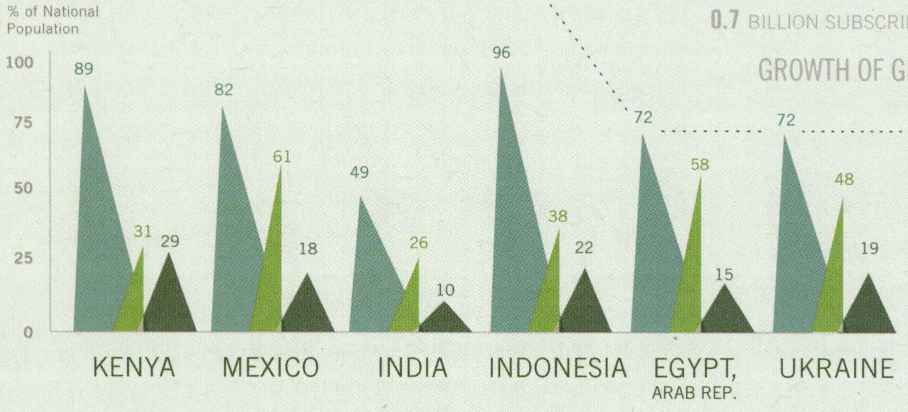

RISE OF NON-VOICE MOBILE USAGE in 2011[5]

ACCESS TO A RANGE OF MOBILE APPLICATIONS HAS **INCREASED DRAMATICALLY THROUGHOUT THE LAST DECADE**

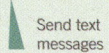

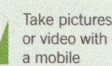

Send text messages | Take pictures or video with a mobile | Use mobile internet

NEAR UBIQUITY BRINGS NEW OPPORTUNITIES

FROM SMS TO SMARTPHONE APPS, **VIRTUALLY ENDLESS APPLICATIONS** ARE NOW AVAILABLE TO USERS IN DEVELOPING COUNTRIES.

MOBILE MONEY

Number of countries using at least one mobile money application[6]

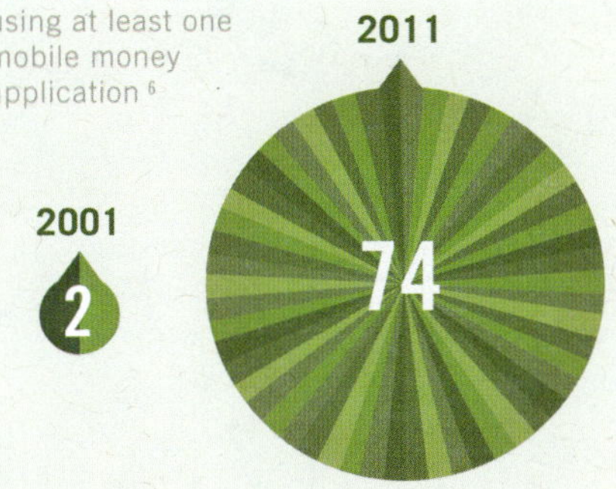

MOBILE HEALTH

Number of countries using at least one mHealth application[7]

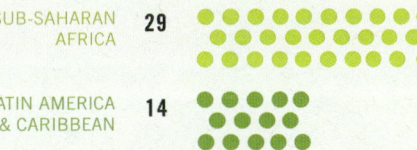

REGION	NUMBER OF COUNTRIES WITH 1+ MHEALTH APPLICATION
SUB-SAHARAN AFRICA	29
LATIN AMERICA & CARIBBEAN	14
MIDDLE EAST & NORTH AFRICA	9
EAST ASIA & PACIFIC	7
SOUTH ASIA	6
EUROPE & CENTRAL ASIA	3

MOBILE APPLICATIONS NOT ONLY EMPOWER INDIVIDUAL USERS, **THEY ENRICH THEIR LIFESTYLES AND LIVELIHOODS, AND BOOST THE ECONOMY AS A WHOLE.**

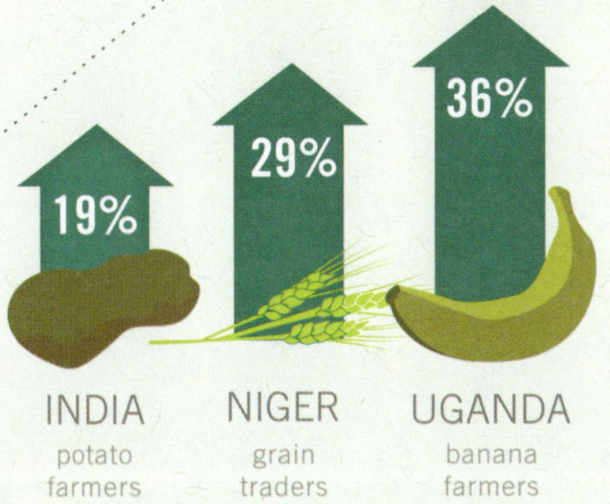

INCOME GROWTH OF FARMERS + TRADERS WITH MOBILE APPLICATION USAGE[8]

Making a development impact requires collaboration.

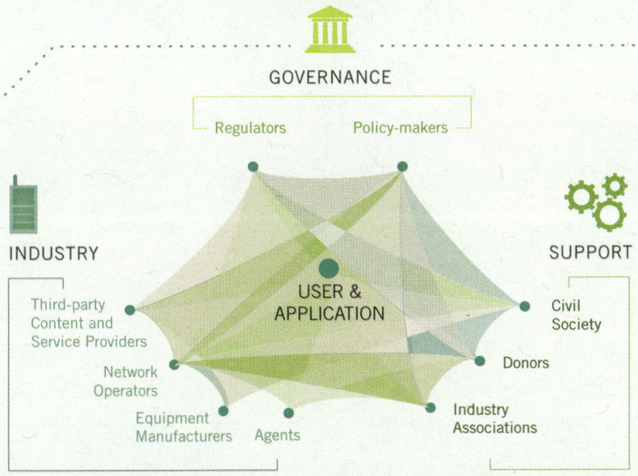

ECOSYSTEM FOR MOBILE APPLICATION DEVELOPMENT

ENGAGING MOBILE APPLICATIONS FOR DEVELOPMENT REQUIRES **AN ENABLING ECOSYSTEM**

Mobile devices are becoming cheaper and more powerful, while networks are doubling in bandwidth roughly every 18 months and expanding into rural areas.

THE MOBILE REVOLUTION IS RIGHT AT THE START OF ITS GROWTH CURVE

GLOBAL FORECAST FOR MOBILE DATA TRAFFIC[9]

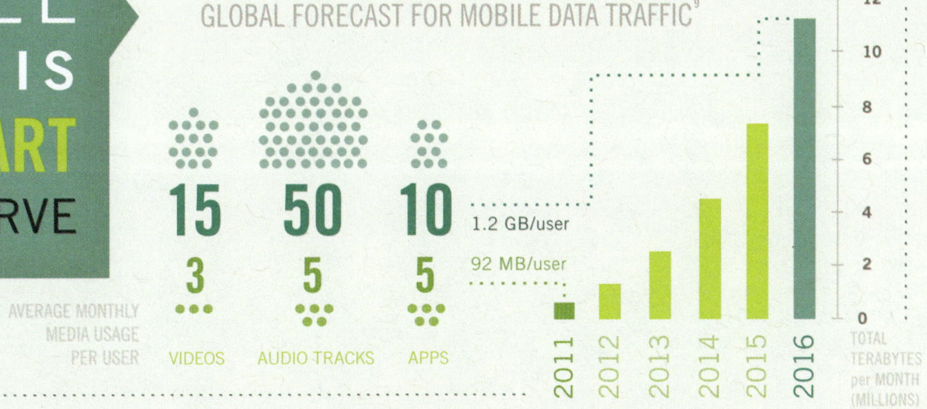

Sources

1. ITU estimates; UN, 2010.
2. ITU, 2012.
3. World Bank estimate.
4. ITU estimates.
5. Pew Research Center, 2011.
6. GSMA Mobile Money Tracker, 2012.
7. Adapted from GSMA mHealth Tracker, 2012.
8. Dixie and Jayaraman, 2011.
9. Cisco, 2012.

FULL REFERENCES AVAILABLE AT www.worldbank.org/ict/IC4D2012

Chapter 1

Overview

Michael Minges

Mobile communication has arguably had a bigger impact on humankind in a shorter period of time than any other invention in human history. As noted by Jeffrey Sachs (2008), who directed the United Nations Millennium Project: "Mobile phones and wireless internet end isolation, and will therefore prove to be the most transformative technology of economic development of our time."

The mobile phone has evolved from a simple voice device to a multimedia communications tool capable of downloading and uploading text, data, audio, and video—from text messages to social network updates to breaking news, the latest hit song, or the latest viral video. A mobile handset can be used as a wallet, a compass, or a television, as well as an alarm clock, calculator, address book, newspaper, and camera.

Mobiles are also contributing to social, economic, and political transformation. Farmers in Africa obtain pricing information via text messages, saving time and travel and making them better informed about where to sell their products, thereby raising their incomes (World Bank 2011a, 353). In India barbers who do not have a bank account can use mobiles to send money to relatives in villages, saving costs and increasing security (Adler and Uppal 2008, 25). Elections are monitored and unpopular regimes toppled with the help of mobile phones (Brisson and Krontiris 2012, 75). Texting and tweeting have become part of the vocabulary (Glotz, Bertschi, and Locke 2005, 199).

Developing countries are increasingly well situated to exploit the benefits of mobile communications. First and foremost, levels of access are high and rising. The number of mobile subscriptions in low- and middle-income countries increased by more than 1,500 percent between 2000 and 2010, from 4 to 72 per 100 inhabitants (figure 1.1a). Second, the age profile of developing nations is younger than in developed countries, an important advantage in the mobile world where new trends are first taken up by youth.[1] Those under age 15 make up 29 percent of the population in low- and middle-income economies but just 17 percent in high-income nations (figure 1.1b). Third, developing countries are growing richer, so more consumers can afford to use mobile handsets for more than just essential voice calls. Between 2000 and 2010 incomes in low- and middle-income nations tripled (figure 1.1c). Fourth, the mobile sector has become a significant economic force in developing economies. Mobile revenues as a proportion of gross national income (GNI) rose from 0.9 percent in 2000 to 1.5 percent in 2010 (figure 1.1d).

These changes are creating unprecedented opportunities for employment, education, and empowerment in developing countries. Local content portals are springing up to satisfy the hunger for news and other information that previously had been difficult to access. The nature of the mobile industry itself is changing dramatically, opening new opportunities for developing nations in designing mobile

Figure 1.1 The developing world: young and mobile

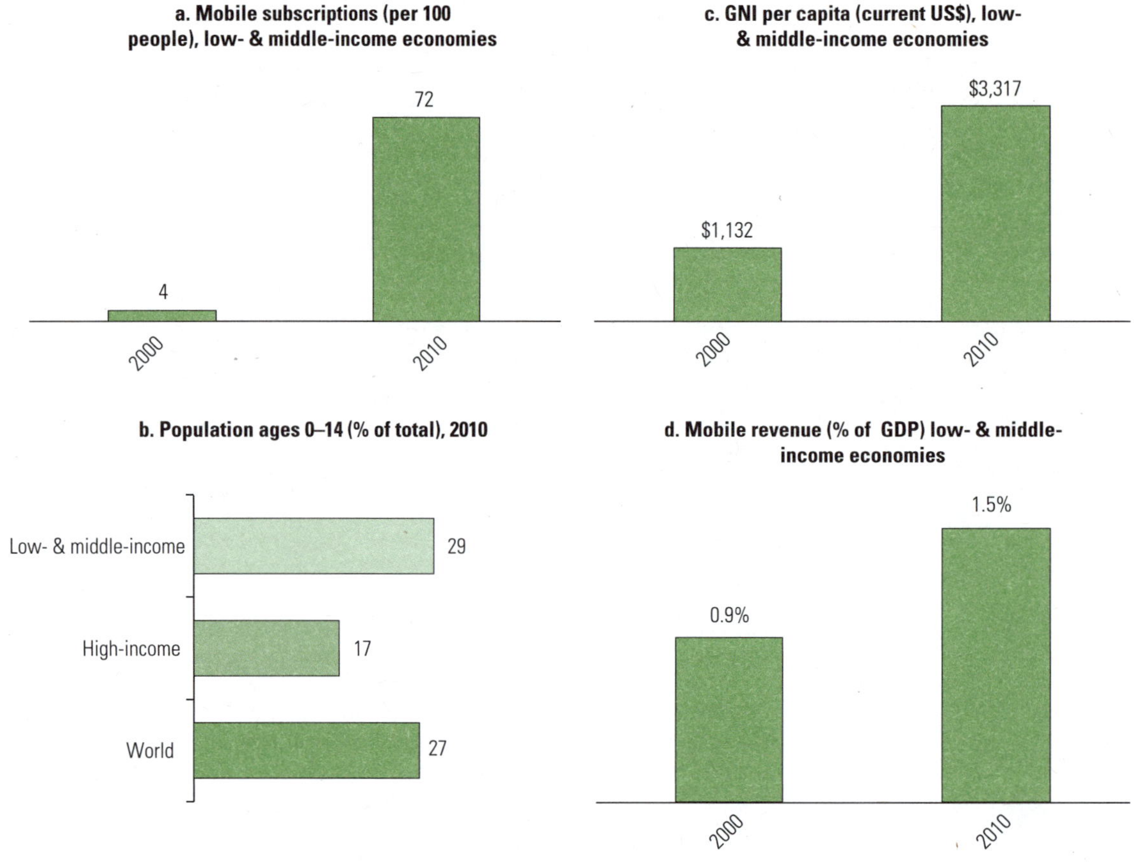

Sources: Adapted from World Bank 2011b and author's own estimates.

applications and developing content, piloting products and services, and becoming innovation hubs. Trendy mobile products and services may be launched in Silicon Valley or Helsinki, but mobile manufacturing usually takes place elsewhere, creating huge opportunities to service, support, and develop applications locally. While key mobile trends are generally adopted around the world, regions such as East Asia are forging their own path for content and applications. New mobile innovation centers are springing up in Beijing, Seoul, and Tokyo, with expertise in specific markets such as mobile gaming and contactless banking.

The emergence of mobile broadband networks, coupled with computer-like handsets, is causing rapid shifts in the ecosystem of the sector. The bond between mobile operators and users is loosening as computer and internet companies invade the mobile space and handsets increasingly offer Wi-Fi capability. Online stores have created a new way for consumers to add content and applications to their mobile phones. Mobile operators are struggling to keep pace with an explosion of data, while networks are converging toward Internet Protocol (IP) technologies and relying on content and data to substitute for declining voice revenues. An increasingly hybrid wireless communications ecosystem will evolve over the coming years.

Although mobile communication is rapidly advancing in most parts of the world, a significant segment of the world's population remains unable to use the latest mobile technologies. Mobile broadband coverage is often limited to urban areas, and current smartphone prices are not affordable for many. Nonetheless, developing-country users are using what they have. Text messaging, mobile money, and simple internet access work on many low-end phones. An emerging ecosystem of local developers is supporting narrowband mobile communicating through scaled-down web browsers, text messaging, social networking, and pay-as-you-go mobile data access. For many users, especially in rural areas, these changes are happening where finding the electricity to recharge a phone is more difficult than purchasing prepaid airtime.

These developments have major implications for the state of access to information and communication technologies (ICTs) in the 21st century. Rich countries have the luxury of both wired and wireless technology, of both personal computers (PCs) and smartphones. Developing countries tend to rely mainly on mobile networks, and phones already vastly outnumber PCs. Applications have to be different to work on small screens and virtual keyboards, while convergence is happening apace. The developed world is also now becoming "more mobile," with average screen size shrinking; while the developing world is now becoming, "more connected," forging ahead with the shift from narrowband to broadband networks on a mobile rather than a fixed platform. Demography is on the side of the developing world, and the economies of scale gained from serving these expanding markets may push the ICT industry as a whole in the direction of a post-PC, untethered world.

One of the challenges facing a report of this nature is that the industry is evolving so rapidly. What is written today is often outdated tomorrow. In addition, given the novelty of many developments and a lack of stable definitions and concepts, official data are scarce or fail to address important market trends. Information from secondary sources is often contradictory, inconsistent, or self-serving. Information about mobile culture is particularly scarce in developing countries. Nevertheless, certain trends are visible, and this opening chapter explores key trends shaping and redefining our understanding of the word "mobile" as an entrée to the review of different sectors in the chapters that follow.

How mobile phones are used

Voice

With all the attention given to mobile broadband, smartphones, and mobile applications, it is sometimes easy to forget that voice communication is still the most significant function and the primary source of revenue for mobile operators.

Voice usage varies considerably both across and within countries. For example, the average Chinese user talks on a mobile phone more than seven times longer per month than the average Moroccan (figure 1.2a). Price is a major factor in calling patterns, with a clear relation between monthly minutes of use and the price per minute. Interconnection fees between operators are a main determinant of price. In some countries these wholesale rates do not reflect underlying costs that drive up the price of mobile calls. A second factor relates to whether the subscriptions are paid in advance (prepaid) or paid on the basis of a contract (postpaid). Prepaid subscriptions are much more popular in developing economies, where incomes may be less stable, but postpaid contracts tend to generate higher usage per subscriber (figure 1.2b).

As with fixed networks, a growing proportion of traffic from mobile devices is moving to Voice over Internet Protocol (VoIP), often routed over Wi-Fi rather than the cellular network, thereby avoiding per-minute usage charges. According to CISCO, a major supplier of IP networking equipment, mobile VoIP traffic is forecast to grow 42 percent between 2010 and 2015.[2] Although mobile VoIP accounts for a tiny share of total mobile data traffic, its value impact on mobile operators is much greater. Skype, a leading VoIP provider, has reported over 19 million downloads of its iPhone application since its launch in 2009. In addition to voice and video, Skype processed 84 million SMS text messages during the first half of 2010.[3] One study forecasts 288 million mobile VoIP users by 2013 (van Buskirk 2010).

Not just for voice anymore

Although voice is still the main revenue generator, its growth has slowed (TeleGeography 2012) as data and text-based applications have grown in popularity, their use made possible by advances in cell phone technology (box 1.1). Mobile applications are the main theme of this book. For many people, a mobile phone is one of the most used and useful appliances they own. Built-in features are indispensable to many for checking the time, setting an alarm, taking photos, performing calculations, and a variety of other daily tasks. Downloadable applications can extend functionalities.

A number of nonvoice applications use wireless networks on a one-off basis (to download, for example); other applications (such as incoming email notifications) are always on. Stand-alone features mean that users do not necessarily need to use a mobile network. For example, downloading of content or applications can be carried out from a PC and then transferred to a mobile phone, or such tasks can be

Figure 1.2 **Talking and paying: mobile voice use and price for selected countries, 2010**

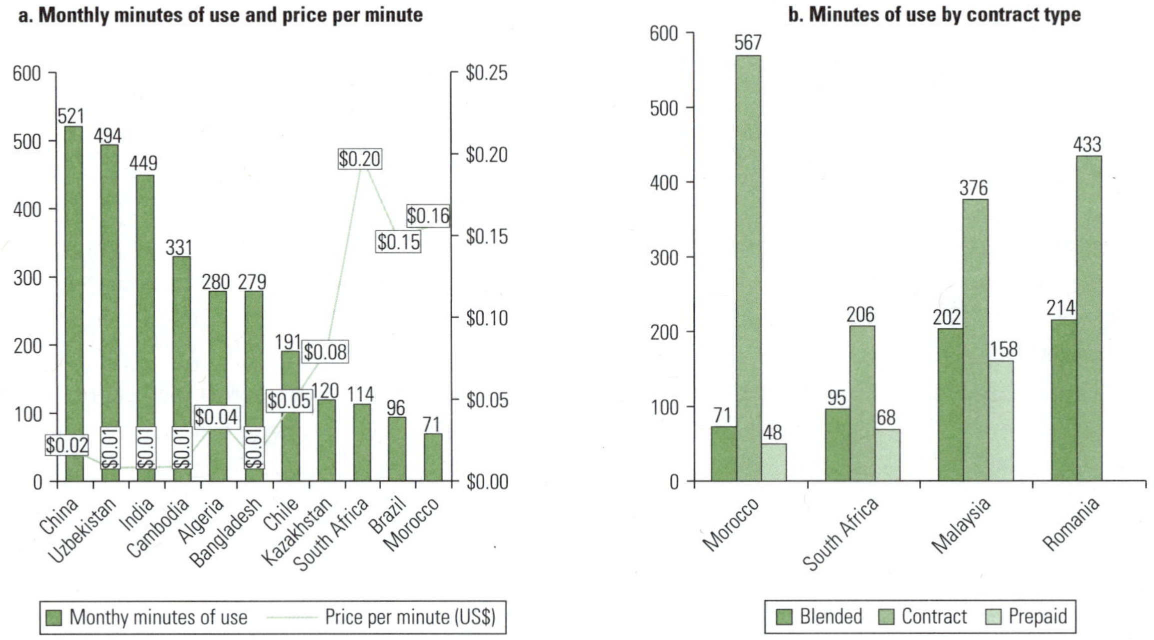

Source: Mobile operator reports.

Note: Data refer to largest mobile operator (by subscriptions). Price per minute is calculated by dividing minutes of use by average revenue per user.

Box 1.1 Mobile phones and applications

The use of mobile phones has evolved dramatically over time and will continue to do so at an ever faster pace, so it is important to define some terms that are used throughout this report, while noting that these definitions are not necessarily stable. Many mobile handsets, particularly in the developing world are so-called **basic phones,** based on the second-generation (2G) GSM (Global System for Mobile communications) standard, first introduced in 1991. GSM offers a number of different services embedded in the standard and therefore available on all GSM-compatible devices, however basic. These include short message service (SMS) text messages of up to 160 characters, and instant messaging using the USSD (Unstructured Supplementary Service Data) protocol. Many of the older "mobile applications," particularly in the developing world, are based on SMS or USSD, because they do not require additional data services or user downloads and are available on virtually any device. Strictly speaking, however, these should be considered **network services** rather than applications (box table 1.1.1). Internet-enabled handsets, or **feature phones,** were introduced with the launching of data services over mobile networks in the early 2000s. These phones supported transmission of picture messages and the downloading of music and often included a built-in camera. **Smartphones** appeared in the late 2000s. They typically feature graphical interfaces and touchscreen capability, built-in Wi-Fi, and GPS (global positioning system) capability.

Smartphones with memories and internet access are also able to download applications, or "**apps**," pieces of software that sit on the phone's memory and carry out specific functions,

(continued next page)

Box 1.1 *(continued)*

Box Table 1.1.1 Mobile devices and their capabilities

Device	Capabilities	Device	Capabilities
Basic mobile phone	*Network services, including:* Voice telephony and voice mail SMS (short message service) USSD (unstructured supplementary service data) SMS-based services, such as mobile money USSD services, such as instant messaging	Smartphone	*As Featurephone plus:* Video camera Web browser GPS (global positioning system) 3G+ internet access Mobile operating "platform" (such as iOS, Android, Blackberry) Ability to download and manage applications VoIP (Voice over Internet Protocol) Mobile TV (if available) Removable memory card
Featurephone	*As basic mobile phone plus:* Multimedia Messaging Service (MMS) Still picture camera MP3 music player 2.5G data access	Tablet	*As smartphone plus:* Front and rear-facing video cameras (for video calls) Larger screen and memory capability Faster processor, enabling video playback Touchscreen with virtual keyboard USB (universal serial bus) port

Note: The list of capabilities is not exhaustive, and not all devices have all features.

like accessing websites or reporting the phone's location and status. In this report, the term "apps" is used to denote such applications that may be downloaded and used on the device, either with or without a fee, in a stand-alone mode. The most popular apps are games. More than 30 billion apps had been downloaded as of early 2012 (Gartner 2012; Paul 2012). Using mobile applications for development usually requires more than simply downloading an app to a user device, however. Specifically, the most useful mobile applications, such as those discussed in this report, typically require an ecosystem of content providers (for instance, reporting price data for agricultural produce, discussed in chapter 2) or agents (such as those providing cash upload facilities for mobile financial services, discussed in chapter 4). These kinds of "**ecosystem-based mobile applications**" are the main topic of this report.

However, technological change continues apace. Newer generations of mobile application may be "**cloud based,**" in the sense that data is stored by servers on the internet rather than locally on the device. Applications that use HTML5 (the current generation of hypertext markup language), for instance, may not require any software to be downloaded. Such applications may have the advantage that they can be used independently of the network or mobile device that the user is currently using. For instance, a music track stored on the "cloud" might be accessed from a user's tablet, smartphone, or PC, and even when the user is roaming abroad. But such a shift depends on much lower prices, without monthly caps, for mobile data transmission.

carried out over Wi-Fi. Indeed, the "mobile" in "mobile applications" refers as much to the type of device as the manner of usage.

A survey (Pew Research Center 2011) carried out across a range of countries at varying economic levels and in different regions illustrates the varied uses of mobile phones (figure 1.3). After voice usage, text messaging is the most widely used: in more than half the countries surveyed, three-quarters of mobile phone owners sent text messages; in Indonesia virtually all mobile users sent text. Although usage rates vary, mobile devices were used to access the internet in all surveyed countries, with almost a quarter of cell phone owners using this feature on average.

Messaging

Despite the attention focused on more glamorous mobile applications, text messaging (or SMS) is a popular and profitable nonvoice application in many countries. Close to 5 trillion text messages were sent worldwide in 2010 (figure 1.4a) accounting for 80 percent of operator revenue from value-added-services, or $106 billion (Informa 2011). This is an attractive revenue source for operators because the cost of transmitting text messages is so low. Although its use in some countries is now starting to decline in favor of instant messaging and phone-based email, SMS remains an alternative for costly voice calls in some countries or suffices for users who do not have access to the internet on their mobiles (or do not know how to use it). Messaging has become popular as a feedback mechanism for voting on TV reality shows and a way of providing value-added services such as banking or pricing information. As a form of asynchronous (that is, non-real-time) communication, it is particularly useful for coordinating meetings or reaching correspondents who are not available to talk (Ling and Donner 2009). Text messaging is also important for applications in the mobile-for-development arena. Many agricultural pricing and health programs for rural dwellers revolve around

Figure 1.3 Mobile phone usage around the world, 2011

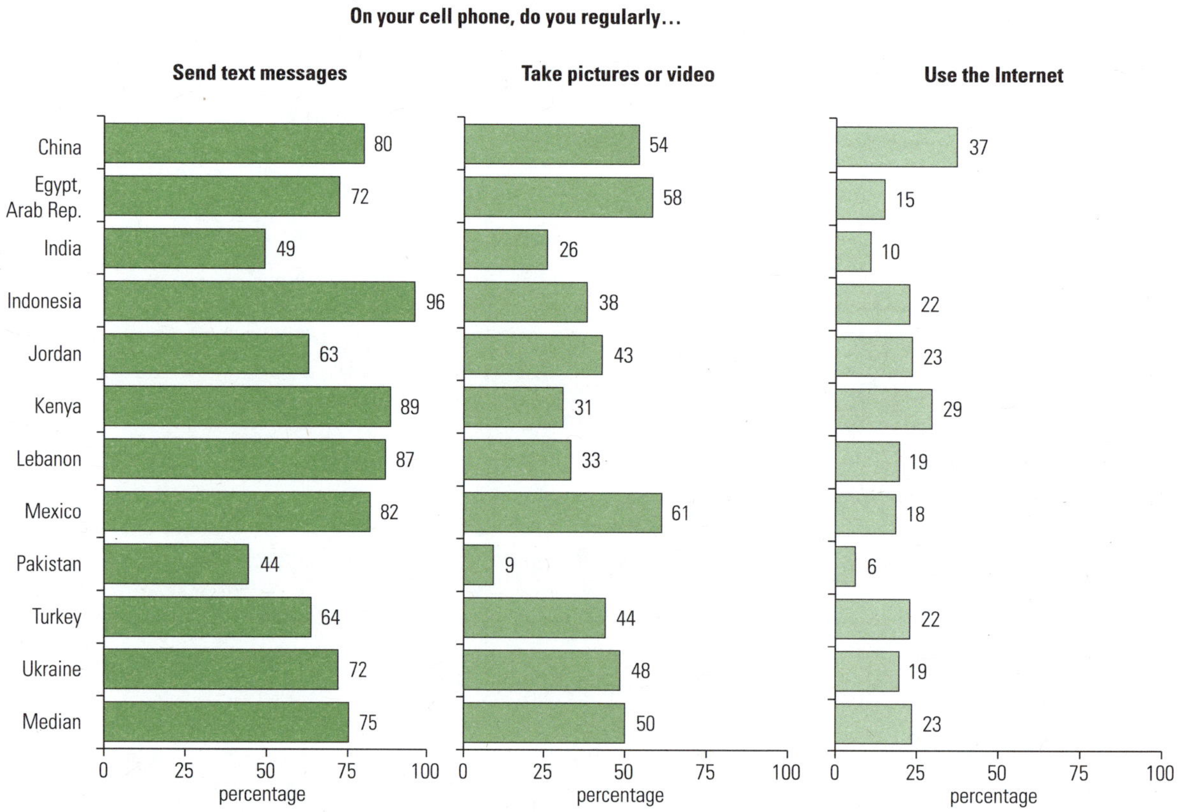

Source: Pew Research Center 2011.
Note: Survey carried out in March–May 2011.

Figure 1.4 Worldwide SMS and Twitter traffic

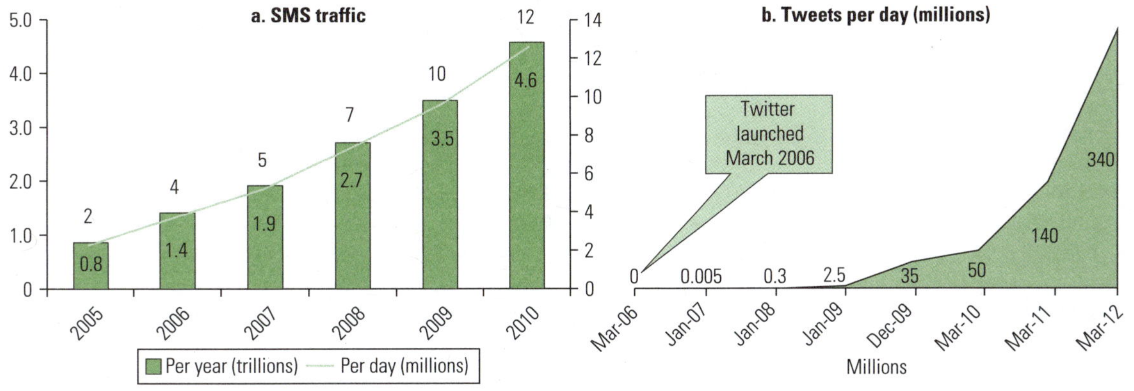

Sources: World Bank estimates (panel a); Twitter 2010, 2011 (panel b).

SMS, and text messaging is used by several governments for citizen alerts.

Twitter, a social networking "microblog" launched in 2006, is also based on short messages, or "tweets," which are intentionally similar to the length of a text message and therefore a good fit for mobile phone use.[4] Around 40 million people (some 37 percent of all Twitter users) were "tweeting" from their mobile devices in April 2010; a year later that number exceeded 100 million (Watters 2010).[5] By March 2012 Twitter users were sending 340 million tweets a day (figure 1.4b).[6] Twitter is integrated with SMS, so tweets can be sent and received as text messages. Twitter short codes have been implemented for several countries so that most SMS tweets are charged at domestic rates. Twitter is working with mobile operators to lower the cost of sending tweets through SMS or USSD or even to make them free. Twitter has rapidly emerged as a tool for social activism and citizen engagement ranging from the Delhi police tweeting traffic updates[7] to tweeting the revolution in the Arab Republic of Egypt.[8]

Web browsing

Access to the internet via a web browser on a mobile device varies across countries depending on costs, education, speeds, and content. Overall, usage is growing, however, with an estimated 10 percent of global internet access coming from mobile phones in 2010, up from 4 percent in 2005. Most popular websites have special versions adapted to mobile devices, although customized mobile browsers, such as Opera, are suited to featurephones.[9] On most smartphones, users are encouraged to download applications from special app stores, sometimes belonging to the operator but increasingly owned by the device platform (such as Apple, Android, Windows, and Blackberry). That arrangement has the convenience of ensuring that the application is suitable for the smaller screen size of mobile devices, although the full range of internet content is still available through a web browser.

Social networking is popular, ranking in the top 10 among mobile internet use in practically every country. Facebook is predominant except in countries such as China and the Russian Federation, where local social networking sites are used. More than 425 million people accessed Facebook through their mobile devices in December 2011.[10]

East Asia in particular is bucking the trend toward use of global applications. The main reason is large domestic markets (such as China, Japan, Republic of Korea), which use non-western alphabets and create huge demand for local content and applications. China Mobile, the world's largest mobile operator, has developed its own applications that mimic global trends in areas such as mobile money, ebooks, video, music, and gaming. But these application are basically closed systems, unfathomable to users that do not speak Chinese and not easily exportable to other countries.

The most downloaded applications for smartphone portals include utilities for tools such as mapping, social networking, chatting, and messaging (table 1.1).

One genre in every list of top downloads across all application portals and all regions is games. The popularity of

Table 1.1 Top mobile applications, June 2011

	Android		Apple		Blackberry	
	Paid	Free	Paid	Free	Paid	Free
1	Beautiful Widgets ($2.85)	Google Maps	Sonic/Sega All-Star Racing ($4.99)	Turtle Fly	One Touch Flashlight ($0.99)	BlackBerry Messenger
2	ROM Manager ($5.86)	Facebook	Angry Birds ($0.99)	Line Jumper	Super Color LED ($1.99)	UberSocial
3	Fruit Ninja ($1.25)	Pandora	Fruit Ninja ($0.99)	Tiny Tower	MegaHorn ($0.99)	Copter
4	Robo Defense ($2.99)	Angry Birds	Tiny Wings ($0.99)	Cars 2 Lite	Tetris ($0.99)	Facebook
5	Root Explorer ($3.83)	YouTube	Angry Birds Rio ($0.99)	Hanging with Friends	Photo Editor Ultimate ($1.99)	WhatsApp Messenger
6	PowerAMP ($5.17)	Words With Friends	Cars 2 ($0.99)	Racing Penguin	Angry Farm ($0.99)	foursquare
7	WeatherBug ($1.99)	Advanced Task Killer	Cut the Rope ($0.99)	Sea Battles Lite	Chat for Facebook ($0.99)	Twitter
8	Better Keyboard ($2.99)	Angry Birds Rio	Hanging with Friends ($1.99)	Dream Bride	BeAlert ($0.99)	Pixelated
9	DocumentsToGo ($14.99)	music download	Camera+ ($1.99)	Super World Adventure	A+ Chat ($0.99)	Free Chat for Facebook
10	Titanium Backup ($6.05)	Yahoo! Mail	Angry Birds Seasons ($0.99)	Facebook	Next Dual Pack ($0.99)	Windows Live Messenger

Source: Respective application stores, June 30, 2011.

games has made millionaires of some application developers (box 1.2) and attests to the significant financial impact the gaming sector is having on the mobile industry.

Games are particularly big in East Asia, accounting for almost half of the estimated global mobile gaming revenue of $5.5 billion in 2008 (Portio Research 2009). In Korea the mobile games sector was worth 424.2 billion won ($390 million) in 2010 even though games downloaded from smartphone application stores operated by Apple and Android were considered illegal because of the government ratings system.[11] That ratings system is set to be loosened, which will likely lead to further market growth. In Japan the mobile games market was estimated to be worth 88.4 billion yen ($1 billion) in 2009 (Toto 2011). China Mobile reported that it had 4.6 million paying users of its online library of 3,000 games in 2010.[12]

The popularity of mobile games and the size of the sector holds opportunities in the areas of software development, virtual cash, and local customization (Lehdonvirta 2011). The traits of game playing, such as acquiring points, leveling, and solving challenges are also entering other fields where applications are used, such as education or social media, in a process called "gamification." The thinking is that users who have become accustomed to using games on their mobile devices would then be more comfortable using similar thought processes in areas that are not entertainment-oriented, including health or business.

Data traffic

Growing mobile data usage is triggering explosive growth in traffic. Social networking entails considerable photo and video exchange and is the leading generator of traffic in many countries (Opera Software 2011). YouTube, the video portal, ranks among the top 10 web applications in most countries. According to CISCO (2012), video is expected to account for more than two-thirds of all mobile traffic in 2016, and mobile data traffic will increase 18-fold between 2011 and 2016.

Mobile operators are struggling to handle all this data and control the traffic. They are adding as much capacity as they can to their networks within investment and spectrum constraints. They are also off-loading traffic to Wi-Fi wherever possible. The most common method for controlling, or "shaping," traffic is through data caps on mobile data plans. Few operators offer truly unlimited mobile data plans, and the cost of exceeding caps can be steep, with users facing a loss or severe disruption of service and dramatically reduced speeds. The case of Hong Kong SAR, China, illustrates well

Box 1.2 How to make a million from Angry Birds

Angry Birds has been a worldwide game sensation. It was the number one Apple iPhone download in countries ranging from Pakistan to Peru and the Philippines to Portugal. Rovio Mobile, a Finnish firm founded in 2003, developed Angry Birds.[a]

In 2009 Rovio released Angry Birds for the iPhone. The company's development of Angry Birds outlines the relationships between game developers, publishers, and giant gaming companies. Rovio initially worked with publisher Chillingo to develop the iPhone version of Angry Birds, keeping the rights for versions on other platforms. Following the sale of Chillingo to gaming company Electronic Arts in October 2010, Rovio developed its own Angry Birds versions for other mobile systems such as Android and Nokia. It is also leveraging its Angry Birds success by expanding into merchandizing with T-shirts and other products.

According to one source, Angry Birds had over 5 million downloads from the Apple app store during the first six months of 2010 alone (Parker 2010). At $0.99 a download, the game generated at least $5 million in revenue during that period.

a. http://www.rovio.com.

the impending wave of data usage that will soon be hitting other countries (figure 1.5a). During 2011 average monthly mobile data usage increased by more than 70 percent to over 500 megabytes (MB) per 2.5G or 3G user. Although Hong Kong is an advanced economy, and therefore well ahead of most developing nations, the same trends can be expected elsewhere at a later date. CISCO (2012) forecasts monthly usage to reach more than 10 exabytes (that is, 1 billion gigabytes) in 2016, with smartphones, laptops, tablets, and mobile broadband networks leading the charge (figure 1.5b). This subject is developed further in chapter 7.

The changing mobile ecosystem

Before the emergence of smartphones, network operators had historically controlled the mobile ecosystem. They were the main point of interface for users regarding devices and applications. Although users were free to purchase their own handsets, operators typically subsidized them where regulation allowed them to do so, at least for the postpaid segment. Users who wanted to talk, send a message, or access the internet did so over the mobile operator's network. Access was often through an operator's "walled garden"—a portal where content providers paid operators to feature their applications. If users went outside the walled garden, they typically had to pay extra. Developments such as value-added text messages and mobile payments widened this ecosystem, but operators essentially remained the gatekeepers.

The app revolution

Operator control started to break down with the emergence of smartphones and other devices that run specific mobile operating systems, incorporate built-in Wi-Fi, and allow users to purchase content and applications through special online stores. The first kink in the direct relationship between operators and users was the BlackBerry, introduced by Canadian company Research in Motion (RIM) in January 1999. Marketed as "wearable wireless email,"[13] the BlackBerry could arguably be called the world's first smartphone. Revolutionary at the time, it allowed subscribers to receive email using RIM's proprietary Enterprise Server. The BlackBerry was a big hit within the corporate world because it ensured that key personnel could receive emails anytime, anywhere. RIM later expanded BlackBerry distribution to reach mass markets, earning $20 billion in revenue in its 2010 fiscal year. RIM has moved into emerging markets and into social networking through its BlackBerry Messenger. The company shipped 52 million devices in its 2010 fiscal year and had some 55 million subscribers in November 2010 (figure 1.6a).[14] BlackBerry App World launched in 2009, but having been an early trendsetter, it is now struggling to keep up with developments elsewhere.

Figure 1.5 Data, data everywhere

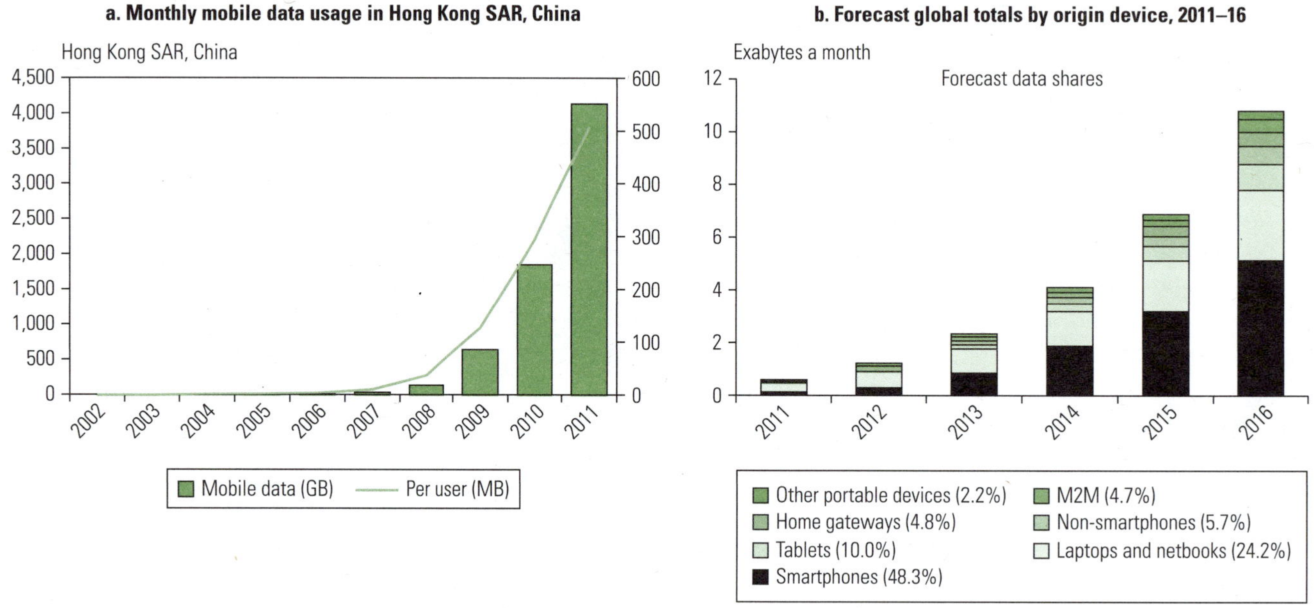

Sources: OFTA 2012 (panel a); CISCO 2012 (panel b).
Note: The compounded annual growth rate for mobile data usage is projected to be 78 percent between 2011 and 2016.

Figure 1.6 Apples and Berries: iPhone sales and Blackberry subscriptions

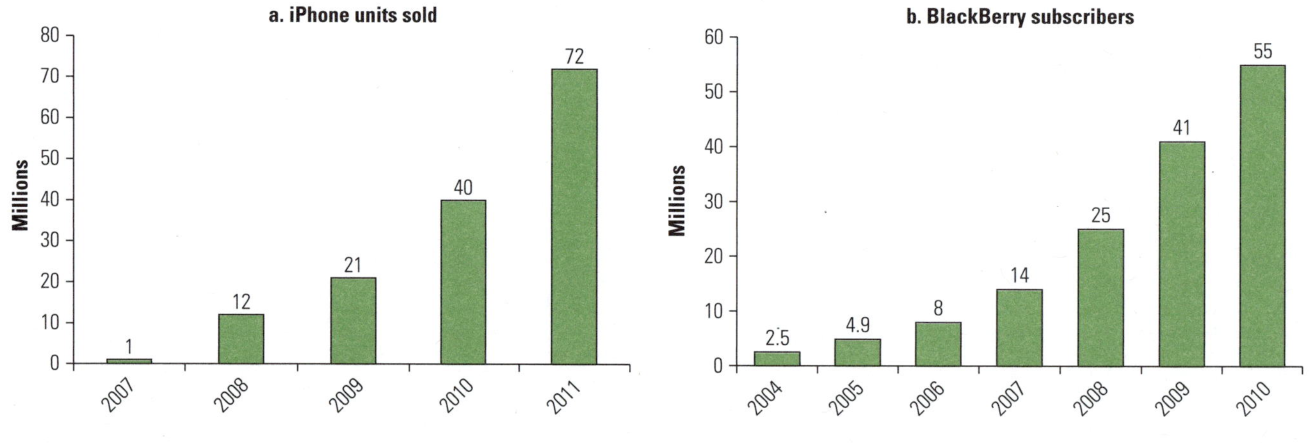

Sources: Apple and RIM operating reports.
Note: Data for Apple refer to fiscal years ending September 25. Data for Blackberry refer to fiscal years ending March.

The industry changed dramatically with the introduction of Apple's touchscreen iPhone in June 2007, followed by the launch of its App Store in July 2008.[15] The exclusive agreements that Apple initially made with mobile operators have now largely ended. In January 2010 the company crossed another milestone, introducing the iPad, its tablet computer. All Apple mobile devices (such as iPhone, iPad, and the iPod music player) are powered by the iOS mobile operating system. The iPhone is distributed through Apple's retail and online stores and also by mobile carriers. In addition to the App Store, iPhone users can download music and video from the iTunes store and ebooks from the iBookstore.

By simplifying and taking ownership of the application platform, handset vendors were able to exert control over the quality of applications on offer and also to create a market for purchasing them. Although the majority of downloaded applications are still free, users are urged to upgrade to paid content or subscriptions, if only to get rid of advertising. By February

2011 Apple had downloaded more than 25 billion applications from the App Store. Sales of the iPhone grew from 1.4 million in 2007 to 72 million in 2011 (figure 1.6b). Revenues from the iPhone and related products and services grew to $47 billion in 2011, accounting for 44 percent of Apple's total sales.[16] An equipment-selling business is rapidly becoming a software-and-services industry, with operators scrambling to provide the spectrum bandwidth to carry the heavy volumes of data traffic while plotting their own applications portals.

Android, Inc., was founded in 2003 to develop mobile phone operating systems and then purchased by search giant Google in 2005. Google made the Android software open source to encourage programmers and handset manufacturers to develop applications and products. The first Android handset, the HTC Dream, was launched in October 2008. Google itself has self-branded several Android phones and developed Android Market (now called Google Play), a portal for obtaining Android applications. By the fourth quarter of 2011 Android had captured just over half the market for smartphone operating systems (Gartner 2012). Google Play offers more than 400,000 applications with over 10 billion downloaded by January 2012 (Paul 2012).

Another significant player is mobile equipment manufacturer Nokia. It has traditionally had a large market share of the handset market, especially in the developing world (figure 1.7). Nokia's mobile operating system, Symbian, is installed on most of these handsets. Thus far, however, Nokia has failed to capture a large share of the smartphone market. In 2011 it forged an agreement with Microsoft to begin offering the Windows operating system on its smartphones.[17]

The rise of smartphones thus sparked tremendous shifts in the mobile ecosystem. A user can now bypass mobile networks completely by downloading content and programs through application stores using Wi-Fi. One survey reported that half the respondents used Wi-Fi to download applications to their mobile phones (In-Stat 2011). Second, users can use VoIP or other applications to communicate instead of the operator's mobile voice service. Third, most handset manufacturers are essentially constrained to using the Android or Windows mobile operating systems for their handsets because RIM and Apple brand their own devices.

As a result of the rise of the smartphone, operators have much less control over the mobile ecosystem. They risk being "genericized," where users do not care about the mobile network brand but instead whether it has the fastest speed, best coverage, cheapest prices, highest quality, or biggest subsidy for popular handsets. Prepaid users, in particular, have little brand loyalty, with high rates of churn in markets where mobile number portability is a regulatory obligation. In some ways, this process is a repeat of the one that occurred in the early 2000s when the rise of the internet threatened to commoditize the "dumb pipes" of telecom operators, only now it is the mobile operators that are under pressure. At the same time, the emergence of HTML5 could cause another disruption in the industry. With the HTML5 standard, apps can be run directly from web browsers, freeing users from

Figure 1.7 Changing market share of mobile handset sales by operating system

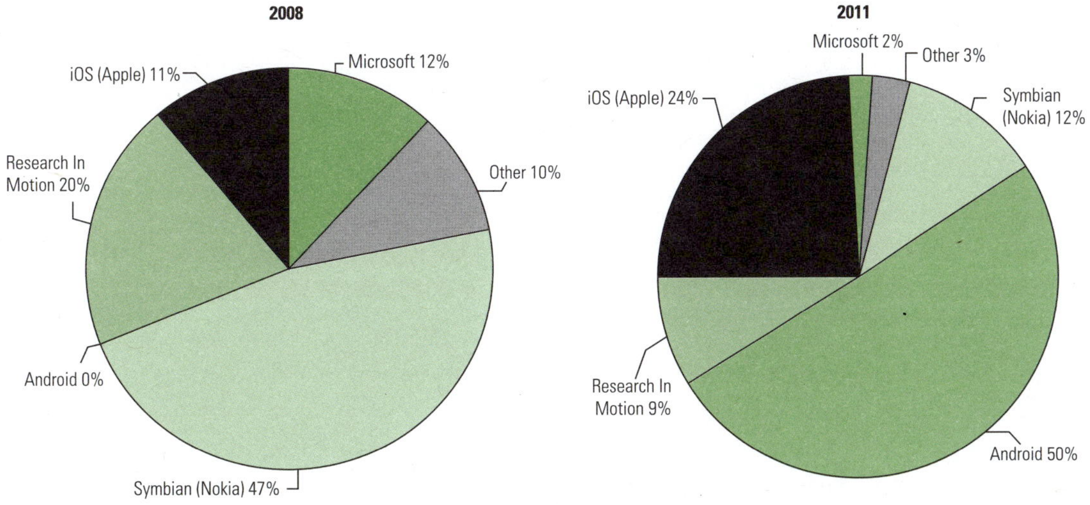

Sources: Adapted from Gartner 2012.

being locked in to a proprietary operating system and creating a new distribution channel for application developers (A.T. Kearney 2011).

Mobile content

The evolution of handsets has driven content providers and aggregators to the mobile industry. In the early days, content largely consisted of ringtones and screen pictures downloaded to customize simple mobile phones. As handsets become more sophisticated and included internet access, more and more of the "big" internet can be reformatted to mobile content, making the "third" screen (after television and PCs) a desirable outlet for the content industry. Content providers have also been aided by the rise of application stores, which allow users to navigate easily to online supermarkets to satisfy their content cravings.

While big technology and media companies dominate content distribution and to some extent content creation, there are opportunities for small software developers and local information aggregators. Examples of these aggregators include:

- Seven out of ten Brazilian internet users visit Brazil's UOL internet portal, formerly Universo Online. It created a mobile version, UOL Celular, with more than 1,000 daily news, weather, and traffic reports. It ranks as the 10th most visited site by Brazilian Opera mobile browser users and the second-leading local site.

- Detikcom is the third most visited site by Indonesian Opera users. It was launched in 1998 and introduced a mobile version in 2002, significantly contributing to growth. It envisions itself as a new media company with partnerships for content and relationships with the country's mobile operators to ensure distribution across the country's mobile networks.

- In South Africa, News24 is a leading portal with updated breaking news. It has a dedicated WAP (wireless access protocol) version for mobile phones. It had more than 500,000 unique visitors to its mobile site in December 2010, up 200 percent over the previous year.[18]

The emergence of cloud computing and multiple types of devices (PCs, tablets, mobile handsets) is creating different distribution markets. On the one hand, companies like Apple produce content only for their own brand. Apple's iBooks, for example, can be read only on Apple devices. This approach ties users to the brand because they cannot use the content they have purchased if they switch brands. On the other hand, companies like Amazon, which makes the Kindle ebook reader, sell software applications that allow Kindle ebooks to be read on multiple platforms. Similarly, Netflix movie streaming is available across a number of platforms. As cloud computing invades the mobile space, it will be possible to run applications remotely instead of having to purchase and download them to the device. This development will create more subscription-like services rather than single downloads. This is good news for developing nations because it lowers the cost of applications and content. But to take advantage of the cloud, users will need good mobile broadband connectivity.

Mobile-enabled social and economic trends

Research shows that mobile networks are having a growing impact on the economy. One of the earliest and frequently cited studies on the subject was carried out by three consultants from the Law and Economics Consulting Group. Using data from 92 countries between 1980 and 2003, they found that an increase of 10 mobile subscriptions per 100 people raised GDP growth by 0.6 percent (Waverman, Meschi, and Fuss 2005). A similar study using data through 2006 found that a 10 percent increase in mobile penetration in developing countries was correlated to a 0.8 percent increase in economic growth (Qiang and Rossotto 2009). Several studies also find that growth in mobile networks is positively correlated to foreign direct investment (Lane et al. 2006; Williams 2005).

Mounting evidence also shows the microeconomic impact of mobile in specific countries and industries. The benefits typically accrue from better access to information brought about through mobile and are typically related to lower transactions costs, savings in travel costs and time spent traveling, better market information, and opportunities to improve one's livelihood (Jensen 2007; Salahuddin et al. 2003; Aker 2008; see also tables 1.2 and 2.1 and box 1.3).

Mobile for development

As noted by the United Nations Development Programme "Mobile phones can enhance pro-poor development . . .

Table 1.2 Mobile and the Millennium Development Goals

MDG	Example
Poverty and hunger	A study on grain traders in Niger found that cell phones improved consumer welfare (Aker 2008). Access to cell phones allowed traders to obtain better information about grain prices across the country without incurring the high cost of having to travel to different markets. On average grain traders with cell phones had 29 percent higher profits than those without cell phones. In the Niger example, demand sprang up organically rather than through a specific program.
Universal education	According to a survey of teachers in villages in four African countries, one-quarter reported that the use of mobile phones helped increase student attendance. A main factor was that teachers could contact parents to enquire about their child's whereabouts (Puri et al, n.d.). Mobile phones have also been used in Uganda to track school attendance so that school administrators can see patterns in attendance, for instance by village, by day of the week, and by season. Tracking attendance for pupils indirectly also tracks absenteeism among teachers (Twaweza 2010)
Gender equality	A study looking at gender differences in the availability and use of mobile phones in developing countries reported that 93 percent of the women who had mobiles felt safer because of the phone, 85 percent felt more independent, and 41 percent had increased income or professional opportunities (GSM Association 2011). The report found that closing the mobile gender gap would increase revenues for mobile operators by $13 billion.
Child health	A program using text messaging to identify malnutrition among rural children in Malawi is notable for its impact on the speed and quality of the data flows.[a] Using a system called RapidSMS, health workers in rural areas were able to transmit weight and height information in two minutes instead of the two months needed under the previous system. The data entry error rate was significantly improved to just 2.8 percent from 14.2 percent in the old system. The improved information flow enabled experts to analyze data more quickly and accurately, identify children at risk, and provide treatment information to the health staff in the field.
Maternal health	One of the earliest uses of mobile technology to improve maternal health took place in rural districts of Uganda in the late 1990s. Traditional birth attendants were provided walkie-talkies, allowing them to stay in contact with health centers and obtain advice. An assessment of the program found that it led to roughly a 50 percent reduction in the maternal mortality rate (Musoke 2002).
HIV/AIDS	In Kenya weekly text messages were sent to AIDS patients to remind them to take their antiretroviral drugs (Lester et al. 2010). Those who received the text messages had significantly higher rates of taking the drugs than those who did not receive them. The study noted that SMS intervention was less expensive than in-person community adherence interventions on the basis of travel costs alone and could theoretically translate into huge health and economic benefits if scaled up.
Environment	According to one forecast, mobile technology could lower greenhouse gas emissions 2 percent by the year 2020 (GSM Association 2009). This reduction can be met through, among other things, widespread adoption of various mobile-enabled technologies such as smart transportation and logistics, smart grids and meters, smart buildings, and "dematerialization" (replacing the physical movement of goods and services with online transmission). Mobile phones can also be used as tools for environmental monitoring. In Ghana, for example, cab drivers in Accra were outfitted with mobile phones with GPS and a tube containing a carbon monoxide sensor to test pollution levels.[b]
Partnership?	MDG target 8F states: "In cooperation with the private sector, make available benefits of new technologies, especially information and communications." Mobile phone penetration in low-income economies has grown from less than one per 100 people in 2000 to almost one per every three by 2010—largely as a result of private sector investment. Of some 800 telecom projects in developing countries with private sector participation between 1990 and 2009, almost three-quarters involved greenfield operations primarily in mobile telephony.[c]

a. "Malawi – Nutritional Surveillance" on the RapidSMS web site: http://www.rapidsms.org/case-studies/malawi-nutritional-surviellence/.

b. http://www.globalproblems-globalsolutions-files.org/unf_website/PDF/vodafone/tech_social_change/Environmental_Conservation_case3.pdf

c. World Bank and PPIAF, PPI Project Database. http://ppi.worldbank.org.

in sectors such as health, education, agriculture, employment, crisis prevention and the environment . . . that are helping to improve human development efforts around the world" (UNDP 2012). The Millennium Development Goals (MDGs) provide a useful framework for assessing the development impact of mobile phones. The MDGs highlight eight priority areas. Examples of the ways mobile phones are being used to address each of the MDGs are given in table 1.2 and throughout this report.

Box 1.3 Smartphones and tablets for development

Box figure 1.3.1 Annotated screenshot of Bangladesh's Amadeyr Tablet

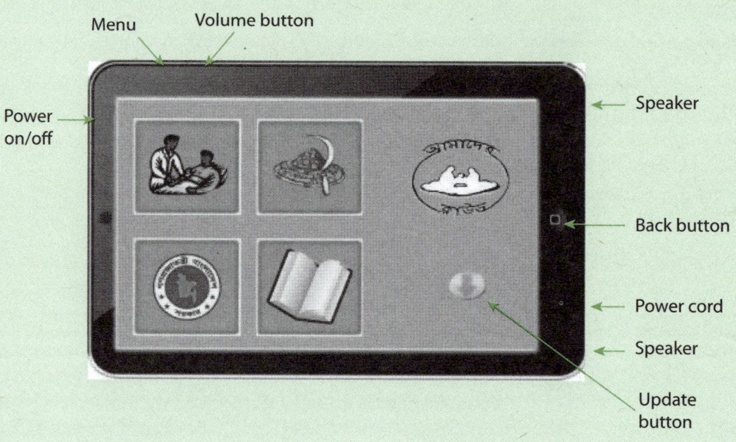

Source: http://amadeyr.org/en/content/amadeyr-tablets.

The introduction of smartphones and lightweight tablet computers has revolutionized the way people access the internet from mobile devices. These powerful touchscreen devices have popularized downloadable apps that can do anything from recognize a song to turn the device into a flashlight. Scaled-down versions of popular office applications for word processing, spreadsheets, and presentations are available for smartphones and tablets as well as ebook software. These devices support internet access over cellular broadband networks and Wi-Fi and often include built-in GPS and still and video cameras.

The graphical user interfaces and touchscreens make them ideal for many developing nations particularly those with non-western alphabets and sizable illiterate populations. Smartphone and tablet penetration is rising rapidly in urban areas of developing countries.

Several initiatives are under way that feature low-cost tablets and investigate the feasibility of devices for rural areas:

- **In Bangladesh**, the Digits to All (DTA) project distributed custom developed tablets (see screenshot) to over 100 households in a rural village to test their feasibility. The $100 Amadeyr tablet uses the Android operating system with software specifically designed and customized for use by semiliterate, illiterate, and bottom-of-the-pyramid users. The tablet uses a touchscreen operated by seeing pictures and hearing instructions given in Bengali, making it user-friendly for illiterate villagers. The project found that villagers who had never used PCs, let alone the internet, were able to use the tablets within a few days and noted: "It is not the rural population who needs to be trained to have access to information but it is the next generation communication technologies that can be tailored to meet the local needs and be made easily accessible by rural communities" (Quadri et al. 2011).

- **India** launched its locally manufactured Aakash tablet in October 2011(Tuli 2011). Priced at around $35 the tablet is aimed for widespread distribution in schools. Apart from its low cost, the Aakash tablet has other features suitable for the Indian environment including data compression techniques that lower consumption and hence reduce Internet access charges. One of the organizations involved in the project forecasts that some 5 million of the tablets will be shipped in 2012, around half of the equivalent PC figure.

(continued next page)

Box 1.3 *(continued)*

- A project in **Tanzania** has been familiarizing farmers with smartphones to introduce them to the features and potential uses (Banks 2011). Although most farmers already had cell phones, they had never used the internet. The smartphones have been used for geotagging climate information and to make videos of farmers offering advice on techniques. The information is uploaded to the internet to share with other farmers. The visually oriented information helped one maize grower to learn from planting mistakes and a few months later he had his first successful harvest.

Governments, the private sector, academia, and the development community all have a role to play in promoting smartphones and tablets for development. Governments in particular can be encouraged by the potential of these devices to take ICT for development to another level through easy-to-use graphical interfaces with Internet connectivity over wireless networks. Just as the One Laptop per Child program helped trigger a reduction in low-end computers, a "One Smartphone/Tablet per Citizen" initiative could help generate mass availability.

Social networking and democracy

Electronic communication has increasingly become two-way: examples include participation through feedback in comments, discussions in forums, and active contribution to applications such as Wikipedia or Mozilla. In addition, the tools for users to generate content have been simplified—not only can most people master text messaging and tweeting but a growing number can also create social networking pages and blogs. Often driven by youth, participation is reaching up the age ladder as these tools and their impact become publicized and popularized.

The increasing availability of these tools and applications on mobile phones is enhancing their popularity. Operators in developing countries are working around the limitations of low-end handsets that do not have internet capabilities by providing ways of interacting with social networking applications through instant messaging, such as MXit in South Africa.[19] Safaricom in Kenya offers special SMS functions allowing users to send and receive Twitter tweets and to update their status and send messages to Facebook.[20]

The diffusion of mobile phones coupled with social networking creates a new space for citizens around the globe to engage in political action concerning democracy, freedom, and human rights. There is disagreement about the extent to which these tools affect appeals for freedom and democracy. Some observers argue that social networking tools empower people to defend freedom and that Twitter should be nominated for a Nobel Peace Prize (Gladwell 2010). Others argue that, while these applications make it easier for people to express themselves, it is "harder for that expression to have any impact." In other words, applications like Facebook and Twitter make it possible for large numbers of people to voice their opinion, but they do so virtually, and these tools are not substitutes for physical participation.

Regardless, recent history has demonstrated that social media along with messages, videos, and pictures sent from mobile phones are useful tools for organizing protests and monitoring democracy and freedom. Examples include:

- One of the first uses of text messaging for social change took place in the Philippines in January 2001. Political activists sent SMS text messages urging Filipinos to assemble at Epifanio de los Santos Avenue (EDSA) in Manila to demonstrate for the impeachment of then-president Joseph Estrada. The message, typically reforwarded by recipients, read: "Go 2 EDSA. Wear blk." During the next few days more than a million people showed up and some 7 million SMS were sent. It is argued that this giant outburst concerned legislators, who allowed evidence in the impeachment trial to be presented. By January 20 Estrada had resigned, blaming his exit on the "the text-messaging generation" (Shirky 2011).

- Thousands of Moldovans demonstrated against the government in the spring of 2009. It was dubbed the "Twitter Revolution," because that application was a main method used to organize the demonstrators. One of Twitter's "Trending Topics" at the time was the tag "#pman" an abbreviation for Piata Marii Adunari Nationale, the main square in downtown Chisinau, the nation's capital and location of the demonstrations. Protestors used the local mobile data network to post tweets from their mobile phones (Morozov 2009).
- In Côte d'Ivoire a so-called "web mash-up" site called Wonzomai ("sentential" in the Ivorian Bété dialect) was created to monitor the 2010 presidential election. Users were provided with telephone number short codes to which they could send free SMS and tweets to report abnormalities that they had witnessed during and immediately after the election. The reports were visualized on a website, which showed the locations where incidents had taken place as well as trends plotted over the duration of the election.[21]

In the Middle East, mobile has unsettled the region's social and political traditions since the mid-2000s (Ibahrine 2009). Its greatest impact to date may have come between 2010 and 2012 when social media played a role in the "Arab Spring" uprisings in Bahrain, Egypt, Libya, the Syrian Arab Republic, Tunisia, the Republic of Yemen, and other countries in the region. As one Egyptian protestor put it: "We use Facebook to schedule the protests, Twitter to coordinate, and YouTube to tell the world."[22] Surges in social networking and demonstrations in these countries appear to be connected. All but one demonstration reportedly took place following the initial call to protest on a Facebook page (figure 1.8). The number of Facebook users in these countries also grew significantly during the demonstrations.

Similarly, Twitter use increased during the Arab Spring. The #jan25 tag, created to organize the first big protest in Egypt falling on that day, remained in active use for several weeks afterward and tag accounted for a majority of Twitter traffic in Egypt through the resignation of President Hosni Mubarak on February 11, 2011. Although there were only around 130,000 active tweeters in Egypt at the time, the #jan25 tag had over 1.2 million mentions, illustrating the viral effect of social networking where a tweet can be retweeted by many other users. The day Mubarak left office, the number of tweets in Egypt reached its zenith at 35,000. During a five-day internet blackout, tweets were sent using proxy servers or through contacts in other countries (Zirulnick 2011).

It is difficult to pinpoint the exact role mobile played in the uprisings because social networking applications can also be used on PCs. In most of the non-Gulf Arab nations, however, mobile ownership far outnumbers computer possession (figure 1.9a). Further the portability and ease of concealment of mobiles are ideally suited to street protests. In addition, camera phones are well integrated with mobile social networking applications, making it relatively simple to record and dispatch images and videos over the high-speed wireless networks available in most Arab nations. In Egypt, almost 60 percent of mobile owners use their phone to take photos or video (figure 1.9b). About 1,000 videos were sent

Figure 1.8 Mapping calls for protest on Facebook to actual "Arab Spring" demonstrations, 2011

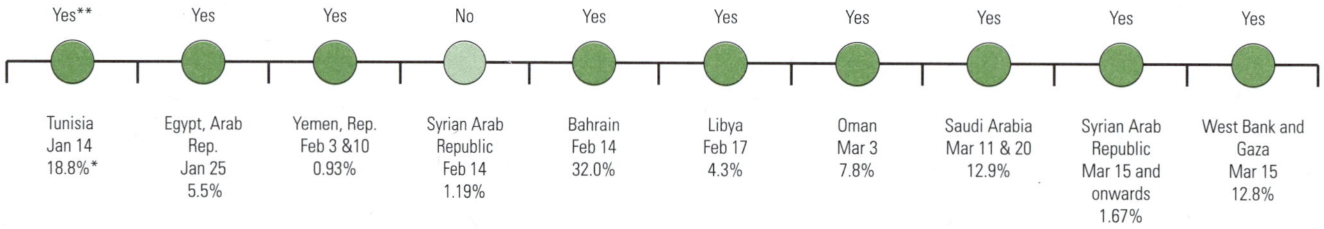

Source: Dubai School of Government, Arab Social Media Report, May 2011.
Note: The percentages underneath each county show Facebook penetration rates at the start of protests.
* Facebook penetration rates at the start of protests in each country.
** Initial protest was not organized on Facebook, although further protests were.

Figure 1.9 Mobile phone versus internet access household availability

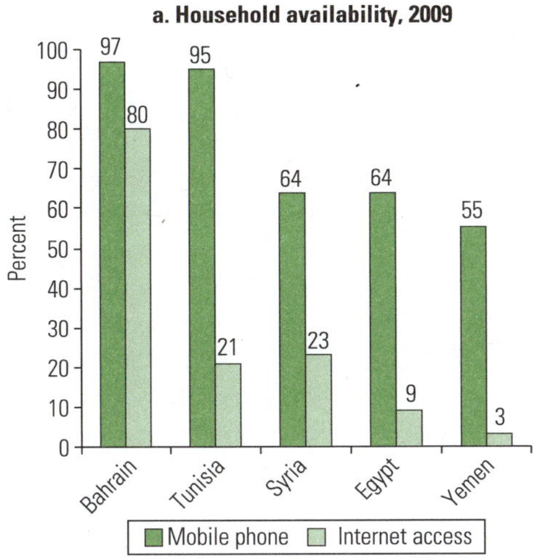

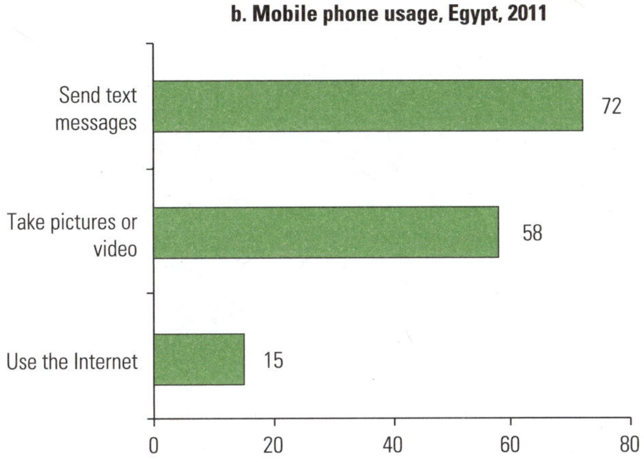

Sources: Gallup 2009; Pew Research Center 2011.

from cell phones to the Al-Jazeera news organization during the Egyptian protests.[23]

Although governments can try to restrict access to the internet and mobile networks, they may pay a heavy price. The Organisation for Economic Co-operation and Development (OECD 2011) estimated that the direct costs to the Egyptian government of shutting down the internet and mobile phone networks during demonstrations was $18 million a day, with a much wider economic impact when factoring in industries such as eCommerce, tourism, and business process outsourcing. Restricting access also tends to have a reverse effect: according to a survey of Egyptian and Tunisian citizens, blocking networks causes "people to be more active, [and] decisive and to find ways to be more creative about communicating and organizing even more" (Dubai School of Government 2011). Short of a complete shutdown, users can find workarounds to blocked applications by using proxies; if close enough, they can also pick up cellular signals from neighboring countries. Intriguingly, some of the countries identified as having the heaviest internet restrictions were also those where social-media-driven demonstrations have taken place (Reporters Without Borders 2009).

Structure of the report

The rest of this report explores these themes in more detail. The report distills work carried out by the World Bank Group and its development partners since the last edition of this report, in 2009, with a particular focus on mobile applications for development. Chapters 2, 3, and 4 have a sectoral focus on the use of mobile applications in agriculture and rural development, health, and financial services respectively. Chapters 5 and 6 are cross-cutting, looking at how mobile communications are contributing to entrepreneurship and employment and how they are being used to bring citizens and government closer together. Finally, chapter 7 looks at the shift from narrowband to broadband mobile networks and the policy implications involved. The Statistical Appendix provides an overview of recent trends in the mobile sector and introduces a new analytical tool. The Country Tables at the end of the report provide an at-a-glance view of the status of mobile communications in World Bank member countries.

Notes

1. "[Y]oung people around the world are more immersed in mobile technology than any previous generation." See Nielsen 2010.
2. http://www.cisco.com/en/US/solutions/collateral/ns341/ns525/ns537/ns705/ns827/white_paper_c11-520862.html.
3. http://www.sec.gov/Archives/edgar/data/1498209/000119312510182561/ds1.htm.
4. A "tweet" is 140 characters (compared to 160 characters for an SMS).
5. For mobile users of Twitter growth in 2010, see http://blog.twitter.com/2011/03/numbers.html.

6. http://blog.twitter.com/2012/03/twitter-turns-six.html.
7. http://trak.in/tags/business/2010/05/24/facebook-twitter-delhi-police/.
8. http://globalvoicesonline.org/2011/01/25/egypt-tweeting-the-day-of-revolution/.
9. http://www.opera.com.
10. "Statistics," http://newsroom.fb.com/content/default.aspx?NewsAreaId=22.
11. In Korea, games must be reviewed and rated by the Games Ratings Board before they can come on the market. See "'Big Bang' of Mobile Games." *JoongAng Daily,* May 17, 2011. http://koreajoongangdaily.joinsmsn.com/news/article/article.aspx?aid=2936279
12. http://www.chinamobileltd.com.
13. RIM (Research in Motion). 1999. Annual Report. p. 2.
14. "Research in Motion Reports Third Quarter Results." Press release. December 16, 2010. http://press.rim.com/financial/.
15. Information on the iPhone is adapted from Apple annual operating reports at http://investor.apple.com/sec.cfm#filings.
16. Apple Inc, 2011 10-K Annual Report, filed Oct. 26, 2011, at: http://files.shareholder.com/downloads/AAPL/1664072048x0xS1193125-11-282113/320193/filing.pdf.
17. "Nokia and Microsoft Announce Plans for a Broad Strategic Partnership to Build a New Global Ecosystem." *Nokia Stock Exchange Release*, February 11, 2011.
18. "News24's Mobile Site Hits the Half-a-Million Unique Users Mark." Press release, January 20, 2011. http://www.news24.com/xArchive/PressReleases/News24-mobile-hits-500-000-users-20110120.
19. http://www.mxit.com.
20. http://www.safaricom.co.ke/index.php?id=1265.
21. "Wonzomai: plateforme d'alertes citoyennes pour les élections présidentielles en Côte d'Ivoire." *Internet Sans Frontières*, October 29, 2010. http://www.internetsansfrontieres.com/Wonzomai-plateforme-d-alertes-citoyennes-pour-les-elections-presidentielles-en-Cote-d-Ivoire_a243.html.
22. http://www.miller-mccune.com/politics/the-cascading-effects-of-the-arab-spring-28575/.
23. http://www.guardian.co.uk/world/2011/dec/29/arab-spring-captured-on-cameraphones.

References

A. T. Kearney. 2011. "The App Frenzy—Just a Short-Lived Fad?" http://www.atkearney.com/index.php/Publications/the-app-frenzyjust-a-short-lived-fad.html.

Adler, R., and M. Uppal, eds. 2008. "mPowering India: Mobile Communications for Inclusive Growth." Aspen Institute. http://www.aspeninstitute.org/sites/default/files/content/docs/pubs/M-Powering_India.pdf.

Aker, J. 2008. "Does Digital Divide or Provide? The Impact of Cell Phones on Grain Markets in Niger." http://www.cgdev.org/doc/experts/Aker%20Cell%20Phone.pdf.

Anderson, J., and Kupp, M. 2008. "Serving the Poor: Drivers of Business Model Innovation in Mobile." *info.* 10: 5–12. http://www.emeraldinsight.com/journals.htm?articleid=1650888&show=abstract.

Banks, K. 2011. "From Smart Phones to Smart Farming: Indigenous Knowledge Sharing in Tanzania." *National Geographic News Watch,* Nov. 30, 2011. http://newswatch.nationalgeographic.com/2011/11/30/smart-phones-meet-smart-farming-indigenous-knowledge-sharing-in-tanzania/.

Brisson, Z., and K. Krontiris. 2012. "Tunisia: From Revolutions to Institutions." *info*Dev. http://www.infodev.org/en/Article.814.html.

CISCO. 2012. "Cisco Visual Networking Index: Global Mobile Data Traffic Forecast Update, 2011–2016." http://www.cisco.com/en/US/solutions/collateral/ns341/ns525/ns537/ns705/ns827/white_paper_c11-520862.html.

Dubai School of Government. 2011. "Arab Social Media Report" (May). http://www.dsg.ae/en/ASMR3/.

Gallup. 2009. "Cell Phones Outpace Internet Access in Middle East." http://www.gallup.com/poll/121652/cell-phones-outpace-internet-access-middle-east.aspx.

Gartner Inc. 2012. "Gartner Says Worldwide Smartphone Sales Soared in Fourth Quarter of 2011 with 47 Per Cent Growth." http://www.gartner.com/it/page.jsp?id=1924314.

Gladwell, M. 2010. "Small Change: Why the Revolution Will Not Be Tweeted." *New Yorker* (October 4). http://www.newyorker.com/reporting/2010/10/04/101004fa_fact_gladwell?currentPage=all.

Glotz, P., S. Bertschi, and C. Locke, eds. 2005. *Thumb Culture: The Meaning of Mobile Phones for Society*. Bielefeld. http://thumbculture.loginb.com/.

GSM Association. 2009. "Mobile's Green Manifesto" (November). http://www.gsmworld.com/our-work/mobile_planet/mobile_environment/green_manifesto.htm.

———. 2011. "Women & Mobile: A Global Opportunity." http://www.vitalwaveconsulting.com/pdf/Women-Mobile.pdf.

Ibahrine, M. 2009. "Mobile Communication and Sociopolitical Change in the Arab World." *Quaderns de la Mediterrània* no. 11. http://www.iemed.org/publicacions-en/historic-de-publicacions/quaderns-de-la-mediterrania/sumaris/sumari-quaderns-de-la-mediterrania-11?set_language=en.

Informa. 2011. "Global SMS Traffic to Reach 8.7 Trillion in 2015." Press release, January 26. http://www.informatm.com/itmg-content/icoms/whats-new/20017843617.html.

In-Stat. 2011. "Mobile Application Downloads to Approach 48 Billion in 2015." Press Release, June 7. http://www.instat.com/press.asp?ID=3155&sku=IN1104930MCM.

Jensen, R. 2007. "The Digital Provide: Information (Technology), Market Performance, and Welfare in the South Indian Fisheries Sector." *Quarterly Journal of Economics* 122 (3): 879–924. doi:10.1162/qjec.122.3.879. http://qje.oxfordjournals.org/content/122/3/879.abstract.

Lane, B., S. Sweet, D. Lewin, J. Sephton, and I. Petini. 2006. *The Economic and Social Benefits of Mobile Services in Bangladesh.*

Lehdonvirta, V. 2011. "Knowledge Map of the Virtual Economy." *info*Dev. http://www.infodev.org/en/Publication.1056.html.

Lester, R., P. Ritvo, E. Mills, A. Kariri, S. Karanja, M. Chung, J. William, et al. 2010. "Effects of a Mobile Phone Short Message Service on Antiretroviral Treatment Adherence in Kenya (WelTel Kenya1): A Randomised Trial." *The Lancet* 376, no. 9755 (November): 1838–45. doi:10.1016/S0140-6736(10)61997-6. http://linkinghub.elsevier.com/retrieve/pii/S0140673610619976.

Ling, R., and J. Donner, J. 2009. "Mobile Communication." http://www.polity.co.uk/book.asp?ref=9780745644134.

Morozov, E. 2009. "Moldova's Twitter Revolution." *Foreign Policy*, April 7. http://neteffect.foreignpolicy.com/posts/2009/04/07/moldovas_twitter_revolution.

Musoke, M. 2002. "Maternal Health Care in Uganda: Leveraging Traditional and Modern Knowledge Systems." *IK Notes*, January.

Nielsen. 2010. *Mobile Youth around the World.*

OECD (Organisation for Economic Co-operation and Development). 2011. "The Economic Impact of Shutting Down Internet and Mobile Phone Services in Egypt" (February 4). http://www.oecd.org/document/19/0,3746,en_2649_201185_47056659_1_1_1_1,00.html.

OFTA (Office of the Telecommunications Authority, Hong Kong SAR, China). 2012. "Key Statistics for Telecommunications in Hong Kong: Wireless Services." http://www.ofta.gov.hk/en/datastat/eng_wireless.pdf.

Opera Software. 2011. "State of the Mobile Web" (April). http://www.opera.com/smw/2011/04/.

Parker, J. 2010. "Rovio: Angry Birds at 60,000 Downloads a Day." *CNET* (August 11). http://reviews.cnet.com/8301-19512_7-20013385-233.html.

Paul, I. 2012. "Android Market Tops 400,000 Apps." *PC World* (January 4). http://www.pcworld.com/article/247247/android_market_tops_400000_apps.html.

Pew Research Center. 2011. "Global Digital Communication: Texting, Social Networking Popular Worldwide." http://www.pewglobal.org/2011/12/20/global-digital-communication-texting-social-networking-popular-worldwide/.

Portio Research. 2009. "Market Notes: Mobile Games in South Korea." http://www.portioresearch.com/Market%20Notes%20Mobile%20Games%20In%20South%20Korea.pdf.

Puri, J., et al. n.d. "A Study of Connectivity in Millennium Villages in Africa." http://www.mobileactive.org/files/file_uploads/ICTD2010%20Puri%20et%20al.pdf.

Qiang C., and C. Rossotto. 2009. "Economic Impacts of Broadband." In *Information and Communication for Development Report: Extending Reach and Increasing Impact*, ch. 3. Washington, DC: World Bank. www.worldbank.org/ic4d.

Quadri, A., K. M. Hasan, M. Farhan, E. A. Ali, and A. Ahmed. 2011. "Next Generation Communication Technologies: Wireless Mesh Network For Rural Connectivity." IEEE Globecom 2011 Workshop on Rural Communications-Technologies, Applications, Strategies and Policies (RuralComm 2011). http://ieeexplore.ieee.org/xpl/login.jsp?tp=&arnumber=6162331&url=http%3A%2F%2Fieeexplore.ieee.org%2Fxpls%2Fabs_all.jsp%3Farnumber%3D6162331.

Reporters Without Borders. 2009. "Internet Enemies." http://www.rsf.org/IMG/pdf/Internet_enemies_2009_2_.pdf.

Sachs, J. 2008. "The Digital War on Poverty." *The Guardian*, August 21. http://www.guardian.co.uk/commentisfree/2008/aug/21/digitalmedia.mobilephones.

Salahuddin, A., H. Baldersheim, and I. Jamil. 2003. "Talking Back! Empowerment and Mobile Phones in Rural Bangladesh: A Study of the Village Phone Scheme of Grameen Bank." *Contemporary South Asia* 12, no. 3 (September): 327–48. doi:10.1080/0958493032000175879. http://www.tandfonline.com/doi/abs/10.1080/0958493032000175879#preview.

Samarajiva, R. 2011. "Challenges of Broadband for Small Pacific Nations." Presentation made at launch of Pacific Islands Regulatory Research Centre, Nov. 10–11. http://www.pirrc.org/home/index.php?option=com_edocman&task=document.viewdoc&id=6&lang=en.

Shirky, C. 2011. "The Political Power of Social Media." *Foreign Affairs* (February). http://www.foreignaffairs.com/articles/67038/clay-shirky/the-political-power-of-social-media.

TeleGeography Inc. 2012. "International Market Trends." Presentation at PTC, January 15, 2012. http://www.telegeography.com/page_attachments/products/website/telecom-resources/telegeography-presentations/0002/7639/PTC_2012_Workshop.pdf.

Toto, S. 2011. "How Big Is Japan's Social Gaming Market?" February 20. http://www.serkantoto.com/2011/02/20/japan-social-gaming-market-stats/.

Tuli, S. 2011. "The Internet Revolution: Act 2." Presentation given at the World Bank on the Askash tablet, December 8. http://go.worldbank.org/0RIXUMDMU0.

Twaweza. 2010. "CU Tracking School Attendance in Uganda." http://twaweza.org/index.php?i=221.

Twitter. 2010. "Measuring Tweets." Twitter Blog (February 20). http://blog.twitter.com/2010/02/measuring-tweets.html.

———. 2011. "200 million Tweets per day." Twitter Blog (June 30). http://blog.twitter.com/2011/06/200-million-tweets-per-day.html.

UNDP (United Nations Development Programme). 2012. "Mobile Technologies and Empowerment: Enhancing Human Development through Participation and Innovation." http://www.undpegov.org/mgov-primer.html.

Van Buskirk, E. 2010. "Five Reasons Cellphones and Mobile VoIP Are Forging an Unlikely Truce." *Wired* (April 23). http://www.wired.com/epicenter/2010/04/mobile-voip-truce/.

Watters, Audrey. 2010. "Just the Facts: Statistics from Twitter Chirp." *ReadWriteWeb* (April 14). http://www.readwriteweb.com/archives/just_the_facts_statistics_from_twitter_chirp.php.

Waverman, L., M. Meschi, and M. Fuss. 2005. "The Impact of Telecoms on Economic Growth in Developing Countries." Vodafone Policy Paper Series 2 (March). http://info.worldbank.org/etools/docs/library/152872/Vodafone%20Survey.pdf.

Williams, M. 2005. "Mobile Networks and Foreign Direct Investment in Developing Countries." Vodafone Policy Paper Series.

World Bank. 2011a. *ICT in Agriculture eSourcebook.* www.ictinagriculture.org.

———. 2011b. *World Development Indicators, 2011.* http://data.worldbank.org/data-catalog/world-development-indicators.

Zirulnick, A. 2011. "Egypt's Protests, Told by #Jan25." http://www.csmonitor.com/World/Global-News/2011/0125/Egypt-s-protests-told-by-Jan25.

Chapter 2

Mobilizing the Agricultural Value Chain

Naomi J. Halewood and Priya Surya

In many developing countries the agricultural sector plays a significant role in the national economy. The sector employs about 40 percent of the total labor force in countries with annual per capita incomes ranging from $400 to $1,800 (World Bank 2008). Developing countries will continue to rely heavily on the agricultural sector to ensure employment for the rural poor and food security for growing populations as well as to meet challenges brought on by climate change and spikes in global food prices.

Improving efficiencies in the agricultural value chain is central to addressing these challenges. Increasing productivity in agriculture is also critical to reducing poverty. Greater productivity can boost farmers' income, especially for smallholder farmers and fishers, who have limited resources to leverage in growing and marketing their produce. Creating a more efficient value chain also requires engaging many stakeholders, from farmers growing crops and raising cattle to input suppliers to distributors.

The potential benefits of using mobile phones to connect these diverse stakeholders along the agricultural value chain speak for themselves. For rural populations, geographically dispersed and isolated from knowledge centers, the information and communication capabilities of the mobile phone can be even more valuable. Close to 6 billion phones are in use today and are accessible to the 70 percent or so of the world's poor whose main source of income and employment comes from the agricultural sector (World Bank 2012).

The mobile revolution in agriculture is not driven by mobile phones alone. Other mobile devices such as smartphones and tablets have already begun to have an impact as information delivery channels. These devices can carry applications that are much more sophisticated than those available in the basic mobile phone. As the cost of these devices declines, they will increasingly be adopted in developing contexts.

This chapter examines how services provided on mobile phones and other mobile devices have begun to change the way stakeholders across the agricultural value chain make decisions regarding inputs, production, marketing, processing, and distribution—decisions that can potentially lead to greater efficiencies, reduced transaction costs, and increased incomes. The chapter also examines the key challenges mobile service providers are facing in scaling up their operations to reach critical mass and to ensure sustainability for the development of a whole ecosystem of different stakeholders. Based on this analysis, the chapter concludes by drawing key policy considerations.

Making information mobile

Among the numerous technological developments in the information and communication technology (ICT) sector, mobile phones have had the most pronounced impact in developing countries. As detailed in chapter 1, adoption has been driven by improved accessibility and affordability made possible through the expansion of mobile networks that are cheaper to deploy than fiber-optic cable infrastructure. The capacity or bandwidth available on mobile networks continues to increase as the technology evolves, enabling more data-intensive services to be delivered through sophisticated devices such as smartphones and tablets.

The most common device in developing countries is still the basic mobile phone, and hence most of the examples cited in this chapter are for mobile services provided through the text-based SMS (short message service) (see table 1.1). An SMS of up to 160 characters can be sent from one phone to another. SMS messages can be used to communicate, inform, and share knowledge on various aspects of agricultural and rural life. The SMS function is generally bundled into the price of a subscription or prepaid package; in many, but not all, developing countries, SMS costs a small fraction of the price of a voice call and can be sent asynchronously, that is, without the caller and the called party having to be online at the same time. Messages sent using USSD (Unstructured Supplementary Service Data) have a functionality similar to instant messaging and can be used when both parties are online, for instance, to access information from a database; USSD messages are sometimes cheaper than SMS messages.

As prices continue to decline, data-enabled devices such as feature phones, smartphones, and tablet computers are expected to become more accessible to more people. These devices include an operating system, which means they have computing capabilities and can carry software applications, referred to as mobile applications. In the past year tablet computers have started to revolutionize various entertainment and knowledge-based industries such as music, videos, books, newspapers, and magazines. Combining the operational potential of a computer, the communications capabilities of a phone, and the versatility of a notepad, companies have already started selling no-frills tablets for less than the cost of some mobile phones ($50–$150).

These data-enabled devices, along with their increasing affordability, can have a range of implications for the development of mobile applications, including ease of use, richer multimedia that can transform agricultural extension services, and the ability to access relevant information on demand in local languages. While cost may still be a barrier for smallholder farmers,[1] community knowledge workers, and local entrepreneurs, users are increasingly able to afford these mobile devices, incorporating them in their work to collect and disseminate information. Devices targeted for this market increasingly use offline technology such as USB (universal serial bus) media to overcome connectivity issues.[2]

Mobile and remote wireless sensors and identification technologies also have an important role to play in gathering data and information relevant to agricultural production, such as temperature, soil composition, and water levels. Illustrative examples of emerging uses of these non-cellular technologies in developing countries are given throughout this chapter.

Increasingly, specialized mobile services targeted to specific agricultural functions are becoming more available (table 2.1). The basic functions of a mobile phone—sending and receiving voice calls and text messages—are invaluable in increasing efficiency in smallholder agriculture by improving the flow of information along and between

Table 2.1 Mobile-enabled solutions for food and agriculture

Improving access to financial services*	Mobile payment platform Micro-insurance system Microlending platform	Increasing access and affordability of financial services tailored for agricultural purposes
Provision of agricultural information	Mobile information platform Farmer helpline	Delivering information relevant to farmers, such as agricultural techniques, commodity prices, and weather forecasts
Improving data visibility for supply-chain efficiency	Smart logistics Traceability and tracking system Mobile management of supplier networks Mobile management of distribution networks	Optimizing supply-chain management across the sector, and delivering efficiency improvements for transportation logistics
Enhancing access to markets	Agricultural trading platform Agricultural tendering platform Agricultural bartering platform	Enhancing the link between commodity exchanges traders, buyers, and sellers of agricultural produce

Source: Vodafone 2011.

* The role of mobiles in finance is discussed in chapter 4.

various stakeholders in the value chain from producer to processor to wholesaler to retailer to consumer. Furthermore, mobile phones also enable smallholder farmers to close the feedback loop by sending information *to* markets, not just consuming information *from* markets.

Improved access to agricultural information

The expansion of mobile networks provides a unique and unparalleled opportunity to give rural smallholders access to information that could transform their livelihoods. This section explores the role of mobile applications in mitigating some of the informational costs that producers in developing countries face in obtaining better yields, increasing their income, and managing uncertainty. The most common uses of SMS and USSD in the context of agriculture include access to price information, disease and meteorological information, and information on growing and marketing practices (extension services).

Price information

The prevailing market price signals the aggregated demand and value on any given day and fluctuates over time. Before the expansion of mobile networks, agricultural producers were often unaware of these prices and had to rely on information from traders and agents to determine whether, when, where, or for how much to sell their crops. Delays in obtaining this data or misinterpretation of second-hand pricing information has serious consequences for agricultural producers, who may end up underselling their products, delivering too little or too much of the product, or having their products wither away. Further, reliance on traders or agents creates rent-seeking opportunities, adding to the agricultural workers' cost of business.

This "information asymmetry" often results in price dispersion—drastically different prices for the same products in markets only short distances apart—and thus lost income for some farmers and higher prices for consumers. Numerous studies have shown the benefits of ICT in promoting access to price information, including increases of up to 24 percent in incomes for farmers and up to 57 percent for traders and price reductions of around 4 percent for consumers depending on the crop, country, and year of study (table 2.2).

A study (Aker 2010) conducted in Niger from 2001 to 2006 found that the introduction of mobile phones had reduced grain price dispersion by 6.4 percent and reduced price variation by 12 percent over the course of one year. Further, the study notes that the impact (or benefits) of mobile phones tends to be greater in markets that are more remote. Pricing for the agricultural sector requires village-level information and generating relevant localized information can be costly and time-consuming. To address this challenge, and to improve local livelihoods, Grameen AppLab in Uganda and Reuters Market Light in India (box 2.1) have collaborated with the government agencies and nongovernmental organizations (NGOs) to employ farmers and extension service providers to collect information.

Feature-enabled phones with camera and GPS (global positioning system), and smartphones have already begun to emerge in rural areas, where they are being used by field workers responsible for collecting data. At volume, the cost of data can be much cheaper than SMS in some countries. For example, through the Grameen Foundation's partnership with a telecommunications operator in Uganda, data is dramatically less expensive than SMS for the volumes their Community Knowledge Workers use. A worker can earn $20 a month from disseminating and collecting information and another $20–$30 from charging farmers' phones from their solar charger.

Disease and meteorological information

Disease and meteorological information is also required by farmers on a frequent basis. Without such information, farmers may be unable to use timely measures to stem losses from climate shocks and poor yields caused by crop diseases. Mobile phones can serve as the backbone for early warning systems to mitigate these risks and safeguard incomes.

For example, a publicly funded pilot project in Turkey provides locally relevant information to farmers in Kastamonu province, where producers maintain orchards susceptible to frost and pests (Donovan 2011). Initially, nationally aggregated weather data collected in urban areas was used but proved to be inaccurate and of limited use to farmers in the provinces, because of differing microclimates from farm to farm in temperature, humidity, precipitation, and soil fertility. Five small meteorological stations and 14 small reference farms were then established to collect data on these variables, enabling accurate pest monitoring. Given the wide use of mobile phones with SMS capability,

Table 2.2 Impact of ICT on farmers, traders, and consumers

Location, product, medium (study authors)	Farmer income (%)	Trader income (%)	Consumer savings (%)	Comments
Uganda, maize, radio (Svensson and Yanagizawa 2009)	+15			Increase in price paid to farmers attributed to farmers' improved bargaining power
Peru, range of enterprises, public phones (Chong, Galdo, and Torero 2005)	+13			Farm incomes increased, but incomes for nonfarm enterprises increased more
India (West Bengal), potatoes, SMS (M. Torero, IFPRI, pers. comm.)	+19			Yet to be published, but both information through SMS and price ticker boards in markets shown to be important
Philippines, range of crops, mobile phones (Labonne and Chase 2009)	+11–17			Commercial farmers, but not subsistence farmers, showed income gains; perceived increase in producers' trust of traders was also reported
India (Madhya Pradesh), soybeans, web-based e-Choupal (Goyal 2008)	+1–5 (average: 1.6)			Transfer of margin from traders to farmers, effect seen shortly after e-Choupal established
Sri Lanka, vegetables, SMS (Lokanathan and de Silva, pers. comm.)	+23.4			Appreciable price advantage over control group over time, plus benefits such as increased interaction with traders and exploring alternative crop options
India (Maharashtra), range of products, SMS (Fafchamps and Minten n.d.)	No significant effect			In this one-year study, quantitative analysis did not show any overall price benefit, but auction sales in state were thought to affect this finding; price benefits of 9 percent were observed at farm gate sales and among younger farmers
Morocco, range of crops, mobile phone (Ilahiane 2007)	+21			Small sample showed usual behavioral changes; higher-value enterprises took a more proactive approach to marketing via mobile phone
India (Kerala), fisheries, mobile phones (Jensen 2007)	+8		–4	Outlier in the sense that fish catches are highly variable and fishermen have their own boat transport
Uganda, range of crops, SMS and radio (Ferris, Engoru, and Kaganzi 2008)	Bananas +36; beans +16.5; maize +17; coffee +19			Awareness of market conditions and prices offers more active farmers opportunities for economic gain
Niger, grains, mobile phones (Aker 2008)		+29	–3 to –4.5	Traders increased margin by securing higher prices through greater capacity to search out better opportunities
Ghana, traders, mobile phones (Egyir, Al-Hassan, and Abakah 2010)		+36		Traders using mobile phones tended to sell at higher prices but also tended to be larger-scale traders than nonusers
Kenya, wholesale traders, mobile phones (Okello 2010)		+ 7		Improved trader margin through combination of cheaper buying prices and higher sale price
Ghana, maize, groundnut, and cassava, SMS (Subervie 2011)	+10			Half of those surveyed receiving market prices via SMS saw increase in incomes

Source: Updated from Dixie and Jayaraman 2011.

the project supplies timely information so that producers can apply pesticides as and when needed, resulting in lower production costs and improved crop yields. Savings amounted to about $2 a tree, with overall savings estimated to be as much as $1 million a year. Considering the cost required to set up this service (around $40,000), this project may be viewed as a success.

Information on growing and marketing practices

Information shortfalls exist in many areas throughout the agricultural production cycle. Whether for growing crops, fishing, or raising livestock, the producer must make decisions on cultivating certain crops or livestock, crop inputs, pest management, harvest, postharvest, marketing, and sale.

Box 2.1 How Reuters Market Light generates hyperlocalized information

An international news giant launched Reuters Market Light (RML) in 2007 to provide market prices and weather and crop advisory services to farmers in India. Invented by a Reuters employee, this service offers highly customizable market information to farmers through text messages delivered to mobile phones.

To subscribe, farmers call a toll-free number to activate the service in the local language and specify the crops and markets in which they have an interest. Farmers receive four to five SMS alerts with relevant information each day. Initial studies show that farmers who receive the service typically gain 5–10 percent more income.

RML is one of India's largest market information services, serving 250,000 customers across tens of thousands of villages. It delivers customized information to India's farming sector covering over 250 crops, 1,000 markets, and 3,000 weather locations across 13 Indian states in 8 local languages.

The company employs over 300 office staff in eight states to process localized agricultural information. The teams, organized according to content type, scour media sources for agricultural news (including market prices, pest and disease reports, government programs, weather reports, and local news). This information is sorted by geography and sent to the appropriate subscribers. RML's growth shows that embracing a wide network of people—including, in this case, price collectors, agricultural institutes, and other information providers—is a vital success factor for mobile applications ecosystems.

Such detailed processing can involve large sunk costs with relatively high monthly operating costs of $4 a customer. There is a trade-off between the provision of local information and scalability. Local teams are needed to collect data, and expansion into new areas may involve additional content provision costs, limiting economies of scale. Costs therefore climb in parallel with new subscribers. Because it relies solely on income from this single service, RML's market remains relatively small and is not yet profitable.

RML has sought to reach as many customers as possible through a number of strategies, including sales offices in postal offices, local shops, input suppliers, and banks. Customers obtain RML through basic SMS using prepaid scratch cards that give access to the service for a given amount of time.

RML competes with traditional information services (radio, market intermediaries, newspapers) and other services that use mobile phones. IFFCO Kisan Sanchar Limited (IKSL) offers similar market information for rural farmers but uses voice messages so illiterate farmers can use the service. Achieving economies of scale is essential for profitability. In 2009 RML reportedly crossed the $1 million sales mark.

Sources: Adapted from Donovan 2011 and Qiang et al. 2012.

Farming organizations and cooperatives provide farmers with a broad range of information, as well as institutional links to large-scale suppliers and distributors. These organizations give farmers a collective voice and more visibility in the agricultural value chain. Many of these organizations started out by providing information and services through leaflets, radio, and internet sites, but they are increasingly using the mobile platform to provide tailored information to farmers (box 2.2). These organizations are being used to supplement and support existing face-to-face trainings for farmers and livestock owners.

Smallholder farms are often disadvantaged compared with larger enterprises because of their inability to leverage economies of scale in procuring inputs, marketing their goods, and sharing machinery and knowledge. Successful agricultural cooperatives and farmer groups have solved this problem by enabling small farmers to pool their resources and improve their bargaining power vis-à-vis large producers and traders. Cooperatives can also be ideal networks to launch and manage mobile information services, because they can provide highly relevant and localized information, and drive farmer adoption through existing social networks. Coopeumo, a Chilean farming cooperative with fewer than 400 members, uses text messages to help small-scale farmers increase productivity. Through the Mobile Information Project (MIP), nearly 200 farmers receive daily messages including market prices and weather forecasts directly from the internet to their mobile phones. The MIP provides two different services—DatAgro and Yo Agricultor. DatAgro provides targeted weather updates that are particularly useful to farmers at critical points such as planting and harvest. Yo Agricultor is a sophisticated web portal for farmers supported by the Chilean government that uses MIP to send messages to further its outreach to groups that have more limited internet access. The MIP software works on the basic phones (costing around $15–$20) that farmers tend to use and is effective over slow networks.

While many farmer groups have seen success in forming long-standing cooperatives in Latin America, such cooperatives are less prevalent in Sub-Saharan Africa. Organizations serving them, and companies operating in the value chain, thus face different needs and opportunities. In areas where farmers are less networked, the interventions may need to be more robust—building up social networks to reach the poorest—and to ensure the information is relevant and actionable in order to drive farmer adoption of new technology services.

A recent addition to the kind of information available to farmers is digital images of agricultural land. The Seeing Is Believing West Africa (SIBWA) project—started by scientists at the ICRISAT (International Crops Research Institute for the Semi-Arid Tropics)—involves local extension service providers and farmers in Burkina Faso, Ghana, Mali, and Niger, who interpret information from very high resolution imagery (VHRI) taken from satellites. The images are used to gauge the relative fertility of the soil (through light reflectivity) and to measure the size and shape of fields. Many farmers may not know the precise size of their land, so the SIBWA team works with the farmers to determine the optimal amounts of fertilizer, pesticide, and seeds needed to cover their land evenly. Knowing the size and shape of fields can help rural communities plan for future developments, including investments in irrigation, for example. The SIBWA team also worked with local NGOs with expertise in specialized technologies and extension services to complement their efforts (Deloitte 2012).

Box 2.2 A pregnant pause for Sri Lanka's cows

The Information and Communication Technology Agency (ICTA) of Sri Lanka discovered that between 2003 and 2008, more than half of the country's 560,000 milk cows were not in fact pregnant at any given time, resulting in a loss of 30–35 days' worth of milk. Low pregnancy rates resulted from a lack of timely access to artificial insemination and breeding services. The eDairy program was introduced in 2009 to enable farmers to request veterinary and extension services (related to issues such as animal health, artificial insemination, milk prices, and construction of dairy stalls) through a simple SMS interface or on touchscreen tablets. Farmers type in their personal identification code and the code of the service they need. The request is then sent to all registered suppliers, so they can contact the farmers directly. Farmers usually obtain feedback within a few hours. So far, 300 farmers have registered for the service. According to Sri Lanka's Department of Dairy Foods, milk production could be increased by 30 percent if artificial insemination services were requested and supplied in a timely manner. Moreover, the ICTA estimates that farmers could earn an additional $262 per calf each year.

Source: Adapted from Qiang et al. 2012.

Improving data visibility for value-chain efficiency

In addition to improved information services for producers, mobile services can also enable better access to markets and other value-chain stakeholders such as traders, input suppliers, and end users. Mobiles can help agribusiness companies and wholesale buyers connect with geographically dispersed producers. This section explores how mobiles and mobile applications create value in the value chain by linking producers to distributors and retailers through better record-keeping and traceability.

Improving logistics

Transporting produce requires coordination between producers, truckers, and, at times, warehouse owners and aggregate traders. Many producers, especially in remote and rural areas, must carry their produce themselves, often by foot, to the nearest collection point. Coordinating transportation is also key to larger traders who aggregate produce for sale in urban areas or for export. Studies show that so far traders are using their websites to relay information on transport and logistics. Some of these services, however, could also be provided on a mobile phone.

The Zambia National Farmers Union operates an SMS-based information service that provides information on commodity prices to farmers. To complement the service, the union has also launched an electronic transport system that allows registered transporters to publicize the arrival and delivery times of loads or cargo.[3] They have three main services, one through which producers can publicize the size of their load and where it is located for pickup, the second for transporters on the way back from the market with an empty truck that could potentially be used to haul products from the market to the village, and the third a directory of transporters that allows producers to contact a transporter directly. This service is being provided through a website in Zambia, but in Morocco, a similar service is using mobile phones. Through the use of voice and SMS, farmers coordinated with local truckers to improve product transport and identify where to deliver their products. Some farmers developed a two-way trade, bringing products back from the market to sell in their own rural communities (Dixie and Jayaraman 2011).

Another example is M-Farm Ltd,[4] an agribusiness company established by a group of women developers, that emerged from the IPO48 competition, a 48-hour boot-camp event aimed at giving mobile and web developers a platform to launch their applications. Besides the staple text-based service for obtaining price information, M-Farm enables suppliers to publicize information on special offers to farmers. This format follows a global trend in deal-of-the-day websites that feature discounted offers at local retailers, such as the Groupon service in the United States.

Tracing products from farm gate to market

The growing globalized and interdependent nature of food production and distribution, combined with raised awareness of food-borne diseases, has shed light on the need to ensure food safety in the global food supply chain.[5] These trends have catalyzed effective technological innovation to trace the food supply from point of origin to the consumer (Karippacheril, Rios, and Srivastava 2011)

The International Organization for Standardization (ISO) defines traceability as the ability to trace the history or location of the item or product under consideration. Traceability is therefore a common element of both public and private systems for monitoring compliance (with regulations on quality environmental, or other product or process attributes related to food). Traceability is becoming increasingly relevant to developing countries that want to gain or expand into new export markets. Smallholder farms, which often lack resources to keep up with strict and changing food safety standards on their own, are now increasingly turning to cooperatives and aggregators who are leveraging ICTs to improve traceability. By opening up new specialized market opportunities, the use of ICTs has led to improved consumer protection and food safety on the one hand, and better livelihood outcomes for farmers on the other (box 2.3).

For this challenge, radio frequency identification (RFID) chips are emerging as a solution for traceability. Placed on a crate of apples or in the ear of a cow, the chip can collect data such as motion, temperature, spoilage, density, light, and other environmental variables though an interface with wireless sensor networks. Traceability systems for bulk products have been implemented in developing countries, even among small farmers.

Representing more than 500,000 small farmers, the National Coffee Growers association in Colombia has leveraged RFID technology to improve traceability and recordkeeping on coffee quality standards. At a cost of

Box 2.3 Tracking specialty coffee

Lack of traceability during the growing and procurement process is a major constraint for producers growing for high-value export markets, such as specialty coffee. For the cooperatives and companies that manage the exports, emerging mobile technology—smartphones and tablets—can play a major role in capturing, tracking, and accessing valuable information from growing practices to crop quality.

Sustainable Harvest is a coffee importer that works with 200,000 farmers in Latin America and East Africa. Extending its relationship-based procurement model to the digital platform, the organization and its farmer training offices have introduced a new coffee traceability program—called the Relationship Information Tracking System, or RITS—to help coffee growers become more efficient, reliable, and quality-focused through a new mobile or tablet-based information tracking system.

RITS provides farmer cooperatives with the ability to trace each step of the value chain. Using a cloud-based application, the cooperative managers can record deliveries of coffee from each member including details of coffee varieties and quality scores for each lot of coffee received. The application also tracks the certification status of each delivery, processes farmer payment, and generates reports on farmer productivity, payments, and samples.

Roaster clients can access videos, photos, quality, and lot information from their supplier cooperatives. The application has been designed for Apple's iPad and iPhone, but it can be used in any smartphone through the web browser. Devices with large touchscreens allow for easier input of a large variety of information. The application can record information offline, and then upload to the online database when connectivity is restored.

In 2011 Sustainable Harvest also launched RITS Ed, an iPad app that delivers agricultural training videos on organic coffee production and quality control that co-op managers can use to assist their members. Sustainable Harvest also plans to expedite the application process for third-party certification (organic, for example) through the launch of a new module, RITS Metrics, that will enable more robust, and customizable reports.

RITS is currently testing the program with two cooperatives in Peru with 500 members and one cooperative in East Africa with 1,840 members.

Sources: USAID 2011; http://www.sustainableharvest.com/; Annerose 2010.

$0.25 a tag, encased wear-resistant tags with unique farm identification numbers are distributed to farmers. These tags are read at each step to market, thus helping to maintain the stringent standards required for this high-value specialty coffee.[6]

RFID chips are also commonly used to trace animal movements, enabling the monitoring of animals from cradle to grave. The Namibian Livestock Identification and Traceability System (NamLITS) (Collins 2004), implemented in 2005, focuses on nurturing livestock production for export markets. More than 85 percent of agricultural land in Namibia is used to raise livestock, and beef production constitutes 87 percent of agricultural revenue. The objective of NamLITS is to implement a traceability system to help in the control, risk management, and eradication of bovine diseases such as foot-and-mouth disease. The use of RFIDs to replace traditional paper-based recording, has increased the accuracy of the data and the speed with which it is disseminated. It has also contributed to a more vigorous market: the Namibian livestock market increased approximately $83 million in 2010 (Deloitte 2012).

Mali is a landlocked country with 80 percent of employment in subsistence agriculture and fishing. In the 1990s the government identified mangoes as having potential for diversifying the country's exports. It faced a number of challenges, however, including meeting increasingly stringent criteria regarding the origin of products, the way they are grown, the fertilizers and pesticides used, and how they are packed. With the support of donors and NGOs, Fruit et Legumes du Mali (Fruilema), an association representing 790 small producers and five exporting companies, launched a web- and mobile-enabled platform through which potential buyers can track and monitor their mangoes (Annerose 2010). The consumer can type the number shown on a tag attached to the fruit into a website to get the exact details of where the mango came from, its producer, and the methods used to cultivate the mango. To leverage the mobile phone platform, Fruilema partnered with a Senegalese mobile operator, Manobi, to pay farmers an additional 9 cents a pound when they entered data on their produce on the Manobi website. One of the key challenges Fruilema faces is to make sure farmers send in all the necessary information to meet the criteria for exporting (Deloitte 2012).

Enhancing access to markets

Mobile phones, although owned and used by individuals, can nevertheless have an important impact in linking markets and key stages of the value chain. A recent study of farmers conducted in Bangladesh, China, India, and Vietnam found that 80 percent of farmers in these countries owned a mobile phone and used them to connect with agents and traders to estimate market demand and the selling price (Minten, Reardon, and Chen n.d.). More than 50 percent of these farmers would make arrangements for sale over the phone. Another study (Muto and Yamano 2009) found that as remote communities in Uganda were provided with access to a mobile network, the share of bananas sold rose from 50 to 69 percent of the crop. This effect, however, was not observed for maize, which is a less perishable crop.

Improved understanding of real-time market dynamics can help farmers deal with external demand, such as switching to high-demand but riskier (perishable) products (Sen and Choudhary 2011). Risky products include crops that are easily ruined if the rainy season arrives too early, for example. The growing sophistication and knowledge of value chains also means that farmers can work directly with larger intermediaries, capturing more of the product's value. Farmers are able to expand their networks and establish contacts directly with other buyers in other areas (Shaffril et al. 2009). Aside from the overall impact of mobile phones on marketing and market linkages, certain mobile applications can help aggregate information between buyers and sellers (box 2.4).

As mobile service and applications providers in agriculture become more knowledgeable about the needs of the farmers as well as their behavior, they are developing increasingly sophisticated applications. In 2000 ITC (Indian Tobacco Company), a large conglomerate in India, broke new ground by establishing e-Choupal—kiosks with computers—in rural villages, where farmers are able to access price, planting, and weather information. Since then, the company has been working to provide its services over mobile phones. ITC has been piloting a new virtual

Box 2.4 DrumNet, the value chain on your mobile phone

More than two-thirds of Africans rely on agriculture for a living, yet because of the lack of complete information, high transaction costs, and inefficient value chains, farmers, intermediaries, and buyers are unable to effectively collaborate in the fragmented market. Pride Africa's DrumNet project is an integrated platform that uses various ICTs, including mobile phones, to provide producers, traders, and financial service providers with an end-to-end solution to procuring inputs, linking to buyers, and finalizing credit and payments.

Starting with fast-growing horticulture and oilseed industries in Kenya, DrumNet ran a series of pilots that delivered services to agro-buyers, banks, farm input retailers, and farmers. The pilots were implemented in five different Kenyan provinces and are reported to have involved over 4,000 small-scale farmers.

Before farmers plant crops, DrumNet's network of entrepreneurs negotiates contractual arrangements between buyers and farmers. These agreements allow farmers to access credit

(continued next page)

Box 2.4 *(continued)*

from partner institutions such as Equity Bank and to purchase inputs from certified retailers. At harvest, DrumNet franchise representatives coordinate produce aggregation, grading, and transportation through agreements with local field agents and transporters. DrumNet tracks and facilitates the entire process through the use of complimentary manual and SMS applications.

Benefits to the stakeholders include:

- Farmers reduce transaction costs by accessing both credit and markets through DrumNet and are able to pay off their loans with their farm produce proceeds. Farmer income is reported to have risen by an average of 32 percent.
- Large-scale buyers are freed from the requirement of managing cumbersome transactions to ensure reliable supplies of produce from multiple smallholders.
- Input sellers can access new customers without having to sell products on credit.
- Banks and microfinancial institutions are able to tap into a currently inaccessible market for savings and credit while avoiding high transaction costs.

The process creates an enabling environment for agricultural finance in a number of ways:

- Banks are assured at the time of lending that farmers have a market for their produce and the means to adequately serve that market, which indicates a healthy revenue stream.
- Banks offer in-kind credit to farmers for inputs.
- Cashless payment transfers reduce strategic default, since farmers cannot obtain revenue until their outstanding loans are fully paid.

The DrumNet project employs tested value-chain approaches to promote agricultural lending. Its operating cost of about $6.80 a user is high, and DrumNet is facing difficulties because it has not yet reached a critical mass that would allow it to stand alone without donor funding. Farmers' inability to attain sufficient crop yields, because of irregular and insufficient rain and other factors, has also threatened the success of the project.

Sources: Adapted from Deloitte 2012, Qiang et al. 2012; and http://www.prideafrica.com.

commodity exchange, Tradersnet, that enables the direct purchase and sale of coffee by producers and wholesale purchasers over an internet-based trading platform. SMS messages are sent to users' mobile phones every morning with the offers and grades available for purchase on that day. At the end of the day, users receive a text message with details of what actually took place (Vodafone 2009).

In Ghana, TradeNet established Esoko to serve as a central repository of price information to be run by a centralized agency such as the government. The people who set up Esoko soon realized that the agricultural sector consists of many decentralized markets where a single price cannot suffice. Therefore, Esoko became a mobile and web-enabled repository of current market prices and a platform to enable buyers and sellers to make offers and connect to one another. Using a bronze/silver/gold/platinum subscription model, Esoko has also been able to offer differentiated service to a diverse customer base. In a recent study of 600 smallholder farmers in northern Ghana, the French National Institute for National Research (INRA) found that farmers have seen a 10 percent revenue increase since they began receiving and using Esoko SMS market prices (Egyir, al-Hassan, and Abakah 2010).[7]

Policy considerations

The examples provided in this chapter demonstrate that food producers and intermediaries are already able to do more with their mobile phones to raise farm incomes and the efficiency of the value chain. Governments have a role to play in ensuring that innovation in this area continues. An enabling environment for mobile services, applications, and other devices, such as RFIDs and remote sensors, includes three support pillars:

- *Business models.* Many of the services described in this chapter rely on public funding and are in pilot stages. DrumNet and RML, while they provide robust business models, are still figuring out how to address high per-user costs, by either scaling up or adding new services to increase the number of subscribers. Public funding, applied through pull mechanisms and results-based financial incentives such as challenge funds, can provide grants and soft loans to innovators who are experimenting with new technologies and business models until they can become financially viable. The public sector can also innovate in its own agricultural programs to create more client-oriented information and knowledge services that leverage mobile technology. Finally, governments can play a catalytic role in facilitating collaboration and dialogue between various private sector players, public sector service providers, and academia and knowledge centers.
- *ICT skills.* Information needs in developing countries are highly localized; therefore, nurturing a domestic ICT skills base in the workforce is crucial to the development of mobile applications and services in the agricultural space. Several of the examples cited in this chapter are from India and Kenya, where the strong presence of skilled software professionals and entrepreneurs has significantly helped these countries lead in producing relevant and high-quality development-focused application services. Governments have a critical role to play in ensuring that the education curricula at the secondary, tertiary, and vocational levels properly reflect the needs of the emerging digital economy. In addition to the pull-based mechanisms and challenge funds described above, technology hubs and technology incubation programs can have a crucial role in encouraging entrepreneurship and emergence of an industry in this space.
- *Supporting infrastructure.* To make the more powerful mobile devices, such as smartphones and tablets, more accessible and affordable, governments will need to ensure that the private sector is capable of offering mobile broadband services at affordable prices. That requires an enabling environment where competition between telecommunications providers is robust.

In addition to supporting the emergence and growth of the mobile services industry, governments could also benefit from the data generated through mobile phone networks and remote sensors. For example, information on price, weather, and diseases could potentially be aggregated so that research institutions and relevant government agencies can analyze and monitor trends. The highly relevant and up-to-date information generated from this type of analysis can inform higher-level policy dialogue on topics such as commodity pricing, subsidy effectiveness, climate change, and trade. Further, by disclosing the aggregated data and analysis to the public, people who initially provided the data, such as farmers, input suppliers, and distributors, would benefit from the analysis—an important component of the Open Data Initiative that many developing countries are implementing.

Conclusions

As information becomes more accessible through the use of mobile devices for stakeholders throughout the agriculture value chain, people are gradually moving toward more efficient ways of producing agricultural products, increasing incomes, and capturing more value by linking fragmented markets. Key benefits include increases in productivity and income for farmers and efficiency improvements in aggregating and transporting products. Although elements of the mobile agriculture platform are emerging in developing countries, the full potential has yet to be realized. The mobile services cited here are simply tools, and without the proper supporting pillars such as those described above, the key challenges that hamper their sustainability will be difficult to overcome.

Looking forward, governments will need to examine their role in creating an enabling environment for innovators seeking ways to meet the needs of this information-intensive sector. Specific ICT strategies for the agriculture sector would help guide both the public and private sector in creating this enabling environment. These policies should take into account the need for new business models in specific country contexts and facilitate inputs such as the supporting infrastructure (broadband services) and the IT industry (IT skills). Technologists, governments, NGOs, private businesses, and donor agencies are just starting to work together to leverage mobile technologies for greater inclusion of rural and poor communities into their spheres of activity.

Notes

1. The definition of smallholder varies across countries and regions but generally refers to farmers with limited volumes of yield and low or uncertain income. According to the Food and Agriculture Organization (FAO 2004), smallholder farmers often cultivate less than one hectare of land in favorable areas, whereas they may cultivate 10 hectares or more in semi-arid areas, or manage 10 head of livestock.
2. Examples are the new tablets from the Canadian firm Datawind, which have been much in demand in emerging markets such as India, Turkey, and Thailand. http://www.bbc.co.uk/news/technology-17218655.
3. http://www.znfu.org.zm/index.php?option=com_wrapper&view=wrapper&Itemid=89.
4. http://afrinnovator.com/blog/2010/11/02/video-pitch-of-ipo48-winner-m-farm.
5. The main source for this section is Karippacheril, Rios, and Srivastava 2011.
6. Colombia Coffee: "Finalists Unveiled for the Fourth Annual RFID Journal Awards," *RFID Journal,* March 18, 2010, http://www.rfidjournal.com/article/view/7467.
7. http://www.esoko.com/about/news.htm.

References

Aker, J. C. 2008. "Does Digital Divide or Provide? The Impact of Mobile phones on Grain Markets in Niger." Working Paper 154. Center for Global Development, Washington, DC. http://www.cgdev.org/content/publications/detail/894410.

———. 2010. "Information from Markets Near and Far: Mobile Phones and Agricultural Markets in Niger." *American Economic Journal: Applied Economics* 2 (3): 46–59. http://ideas.repec.org/a/aea/aejapp/v2y2010i3p46-59.html.

Annerose, D. 2010. "Manobi: ICT for Social and Economic Development." Presentation to the World Bank, Washington, DC, August 12.

Chong, A., V. Galdo, and M. Torero. 2005. "Does Privatization Deliver? Access to Telephone Services and Household Income in Poor Rural Areas Using a Quasi-Natural Experiment in Peru." Working Paper 535. Inter-American Development Bank, Washington, DC. http://www.iadb.org/res/publications/pubfiles/pubwp-535.pdf.

Collins, J. 2004. "African Beef Gets Tracked: Namibia Beef Tracking by Savi Technologies." *RFID Journal,* December 10. http://www.rfidjournal.com/article/articleprint/1281/-1/1/.

Deloitte. 2012. "Agriculture Sector Report." In *Transformation-Ready: The Strategic Application of Information and Communication Technologies in Africa.* World Bank and African Development Bank. http://www.etransformafrica.org/sector/agriculture.

Dixie, G., and N. Jayaraman. 2011. "Strengthening Agricultural Marketing with ICT." Module 9 in *ICT in Agriculture e-Sourcebook.* World Bank, Washington, DC. http://www.ictinagriculture.org/ictinag/sourcebook/module-9-strengthening-agricultural-marketing.

Donovan, K. 2011. "Anytime, Anywhere: Mobile Devices and Services and Their Impact on Agriculture and Rural Development." Module 3 in *ICT in Agriculture e-Sourcebook.* http://www.ictinagriculture.org/ictinag/sites/ictinagriculture.org/files/final_Module3.pdf.

Egyir, I. S., R. al-Hassan, and J. K. Abakah. 2010. "The Effect of ICT-Based Market Information Services on the Performance of Agricultural Markets: Experiences from Ghana." Unpublished draft report, University of Ghana, Legon.

———. 2011. "ICT-based Market Information Services Show Modest Gains in Ghana's Food Commodity Markets." Paper presented at a conference on Development on the Margin, University of Bonn, October 5–7.

Fafchamps, M., and B. Minten. n.d. "Impact of SMS-Based Agricultural Information on Indian Farmers." Unpublished draft report.

FAO (Food and Agriculture Organization). 2004. "Framework for Analyzing Impacts of Globalization on Smallholders." Rome. http://www.fao.org/docrep/007/y5784e/y5784e02.htm.

Ferris, S., P. Engoru, and E. Kaganzi. 2008. "Making Market Information Services Work Better for the Poor in Uganda." CAPRi Working Paper 77. World Bank, CGIAR Systemwide Program on Collective Action and Property Rights, Washington, DC. http://www.capri.cgiar.org/pdf/capriwp77.pdf.

Goyal, A. 2008. "Information Technology and Rural Markets: Theory and Evidence from a Unique Intervention in Central India." University of Maryland.

Ilahiane, H. 2007. "Impacts of Information and Communication Technologies in Agriculture: Farmers and Mobile Phones in Morocco." Paper presented at the Annual Meetings of the American Anthropological Association, December 1, Washington, DC.

Jensen, R. 2007. "The Digital Provide: Information (Technology), Market Performance, and Welfare in the South Indian Fisheries Sector." *Quarterly Journal of Economics* 122 (3): 879–924. http://qje.oxfordjournals.org/content/122/3/879.abstract.

Karippacheril, T. G., L. D. Rios, and L. Srivastava. 2011. "Global markets, Global Challenges: Improving Food Safety and Traceability While Empowering Smallholders through ICT." Module 12 in *ICT in Agriculture e-Sourcebook*. World Bank, Washington, DC. http://www.ictinagriculture.org/ictinag/sites/ictinagriculture.org/files/final_Module12.pdf.

Labonne, J., and R. S. Chase. 2009. "The Power of Information: The Impact of Mobile Phones on Farmers' Welfare in the Philippines." Policy Research Working Paper 4996, World Bank, Washington, DC. http://papers.ssrn.com/sol3/papers.cfm?abstract_id=1435202##.

Minten B., T. Reardon, and K. Chen. n.d. "The Quiet Revolution of 'Traditional' Agricultural Value Chains in Asia: Evidence from Staple Food Value to Four Mega-cities." Unpublished draft, International Food Policy Research Institute, Washington, DC.

Muto, M., and T. Yamano. 2009. "The Impact of Mobile Phone Coverage Expansion on Market Participation: Panel Data Evidence from Uganda." *World Development* 37 (12): 1887–96. http://www.sciencedirect.com/science/article/pii/S0305750X09000965.

Okello, J. 2010. "Effect of ICT-based MIS Projects and the Use of ICT Tools and Services on Transaction Costs and Market Performance: The Case of Kenya." Unpublished draft.

Qiang, C. Z., S. C. Kuek, A. Dymond, and S. Esselaar. 2012. *Mobile Applications for Agriculture and Rural Development*. World Bank. http://go.worldbank.org/YJPDV8U9L0.

Sen, S., and V. Choudhary. 2011. "ICT Applications in Agricultural Risk Management." Module 11 in *ICT in Agriculture e-Sourcebook*. World Bank, Washington, DC. www.ICTinAgriculture.org.

Shaffril, H. A. M, M. S. Hassan, M. A. Hassan, and J. L. D'Silva. 2009. "Agro-Based Industry, Mobile Phone and Youth: A Recipe for Success." *European Journal of Scientific Research* 36 (1): 41–48. http://www.eurojournals.com/ejsr_36_1_05.pdf.

Subervie, J. 2011. "Evaluation of the Impact of a Ghanian-based MIS on the First Few Users Using a Quasi-Experimental Design." INRA (French National Institute for National Research). http://www.esoko.com/about/news/pressreleases/2011_15_12_Esoko_INRA.pdf.

Svensson, J., and D. Yanagizawa. 2009. "Getting Prices Right: The Impact of the Market Information Service in Uganda." *Journal of the European Economic Association* 7 (2–3): 435–45. http://onlinelibrary.wiley.com/doi/10.1162/JEEA.2009.7.2-3.435/abstract.

USAID (U.S. Agency for International Development). 2011. *Sustainable Harvest*. ICT and Agriculture Profile. http://microlinks.kdid.org/sites/microlinks/files/resource/files/SustainableHarvestProfile.pdf.

Vodafone. 2009. "India: The Impact of Mobile Phones." http://www.icrier.org/pdf/public_policy19jan09.pdf.

———. 2011. "Connected Agriculture: The Role of Mobile in Driving Efficiency and Sustainability in the Food and Agriculture Value Chain." http://www.vodafone.com/content/dam/vodafone/about/sustainability/2011/pdf/connected_agriculture.pdf.

World Bank. 2008. *World Development Report 2008: Agriculture in Development*. Washington, DC. http://siteresources.worldbank.org/INTWDR2008/Resources/WDR_00_book.pdf.

———. 2011. *ICT in Agriculture eSourcebook*. www.ictinagriculture.org.

World Bank. 2012. "World Development Indicators on Agriculture and Rural Development." http://data.worldbank.org/topic/agriculture-and-rural-development.

Chapter 3

mHealth

Nicolas Friederici, Carol Hullin, and Masatake Yamamichi

Calling a doctor is a natural response to getting sick in most of the developed world, but that is not always an option in many developing countries. The spread of mobile phones in developing nations promises to change that, however, by enabling health professionals to speak directly with their patients, to arrange health care services such as appointments, and to monitor symptoms.

This chapter is concerned with what happens once basic communications are widely available. How can mobile devices be used to enhance health care? How can mobile devices improve the efficiency and effectiveness of health care interactions between patients and immediate health care providers (such as doctors and hospitals), as well as between patients, providers, and other institutions involved with health (such as health information portals, insurance companies, and government agencies)?

Early on, the term mHealth was narrowly defined to mean wireless telemedicine involving the use of mobile telecommunications and multimedia technologies and their integration with mobile health care delivery systems (Istepanian and Lacal 2003).[1] However, this definition does not do justice to the wide variety of stakeholders and types of uses that mHealth spans today. In this report, a broader definition is adopted: "mHealth encompasses any use of mobile technology to address health care challenges such as access, quality, affordability, matching of resources, and behavioral norms [through] the exchange of information" (Qiang et al. 2012). It is a dynamic field for innovative new services that move health care away from pure public service delivery toward seeing the patient as a consumer. Mobile health software and services have proved to be versatile tools for collecting data at the point of action, potentially resulting in more accountable management of information in health care delivery, increasingly going beyond telemedicine.[2] Table 3.1 summarizes some of the more important mHealth categories.

Why mHealth? Opportunities and challenges

How can mobile communications help to achieve public and private health sector objectives, and what policies can help facilitate mHealth deployments? On the supply side, mobile communications can help provide health care services more quickly and cheaply in many cases, mainly by focusing on primary, preventive, and self-empowered approaches to health care. From the demand perspective, mobile phones can make it easier and more convenient not only to find relevant information quickly but also to enter health data and engage in interactive services, such as symptom tracking and online communities of patients.

For mHealth to deliver, mHealth application developers should ideally consult with medical or health informaticians trained to understand the information flows involved in health care processes.[3] At the same time, to reach a wider market and to achieve sustainability, many

Table 3.1 Major categories of mHealth services and applications

mHealth category	Typical fields of application	Description/use cases	Examples (and sources)
Improving management and decision-making by health care professionals	• Treatment of medical conditions • Prescriptions • Targeted provision of information and marketing about health care products[a]	• Remote patient tracking • Updating and verification of digital medical records, accessible to health care providers and pharmacists • Delivery of health insurance and savings products	• Moca (Celi et al. 2009) • 104 Mobile (see box 3.3)
Real-time and location-based data gathering	• Health care delivery and logistics • Crisis mapping • Resource allocation	• Monitoring and surveillance of disease outbreaks for more timely reporting of symptoms and containment of epidemics • Crisis mapping after natural disasters • Reporting of urgent health needs • Real-time provision with information on available health facilities and resources • Supply chain management • Access to health emergency services and rapid response systems	• RapidSMS in Somalia (Vital Wave Consulting 2011) and Ethiopia (see box 3.2) • Trilogy/International Federation in Haiti (Qiang et al. 2012) • Desert Medicine Research Centre Interactive Voice Response System in India (Chalga et al. 2011)
Provision of health care to remote and difficult-to-serve locations	• Remote provision of health care services • Extending the reach of health care • Complementing traditional face-to-face health care services	• Medical advice, reminders counseling, monitoring, simple diagnoses • Focusing on areas where only limited physical infrastructure is available, such as remote and rural areas, including telenursing, teleradiology, telepsychiatry, and tele-education.	• Remote mobile health care for rural communities in Sri Lanka (Perera 2009) • Moca (Celi et al. 2009)

Fostering learning and knowledge exchange among health professionals	• Medical knowledge repositories • Virtual communities • Event and conference organization	• Retrieving best practices, international standards, and patient histories from other health care professionals • Local communities • Expert crowdsourcing for health information wikis • Virtual classrooms, webinars, and the like	• RAFT network in Burkina Faso, Côte d'Ivoire, Mali, and Senegal (Vital Wave Consulting 2011)[b] • Moca (Celi et al. 2009)
Promoting public health	• Delivery of health information • Awareness building and campaigning • (Mass-oriented) tele-education	• Games, quizzes, and other nontraditional mechanisms • Conventional mHealth prevention and education campaigns • Medication reminders	• Text to Change in Uganda (Vital Wave Consulting 2011)
Improving accountability	• Transparency for usage of funds • Feedback systems	• Public health fund flow tracking • Interactive portals for comments and complaints	• Kgonfalo (see table 3.2), Botswana—UPenn Partnership (2012) • Transparency International in Northern Uganda[c]
Self-management of patient health	• Enabling better self-help and limiting transactions • Lowering health care costs (shifting tasks to the patient) • Patient empowerment	• Patients obtain accurate information • Patients can better understand their diagnoses, for example, by checking medical records • Mainly focused on noncommunicable diseases and may deal with health indicators such as weight and blood pressure	• MEDAfrica (see box 3.1) • Calorie Counter (popular downloadable app)[d]

a. See Patil (2011) for suggestions on how to integrate traditional product marketing concepts, as well as social marketing, into mHealth for developing countries.
b. RAFT (Réseau en Afrique Francophone pour la Télémédecine) http://www.who.int/workforcealliance/members_partners/member_list/hugraft/en/index.html.
c. http://www.ict4democracy.org/about/partnerproject-briefs/ti/.
d. The "Calorie Counter: Diet and Activities" was one of the 10 most popular Apple iPhone/iPad apps in 2011; see http://mobihealthnews.com/15229/top-10-iphone-medical-apps-for-2011/7/.

mHealth applications need to offer standardized services that can be delivered and accessed by nonexperts. Finally, mHealth should be integrated with larger eHealth programs and aligned with the delivery of offline health care services (box 3.1). So far, short message service (SMS) texting has probably been the most prominent mode of delivery (see table 3.1 and box 3.2), perhaps, according to some research, because texting is already an integral part of mobile usage culture (Gombachika and Monawe 2011). Increasingly, however, mHealth services are also offered as

Box 3.1 Kenya: A breeding ground for mHealth applications

Box figure 3.1.1 MedAfrica app

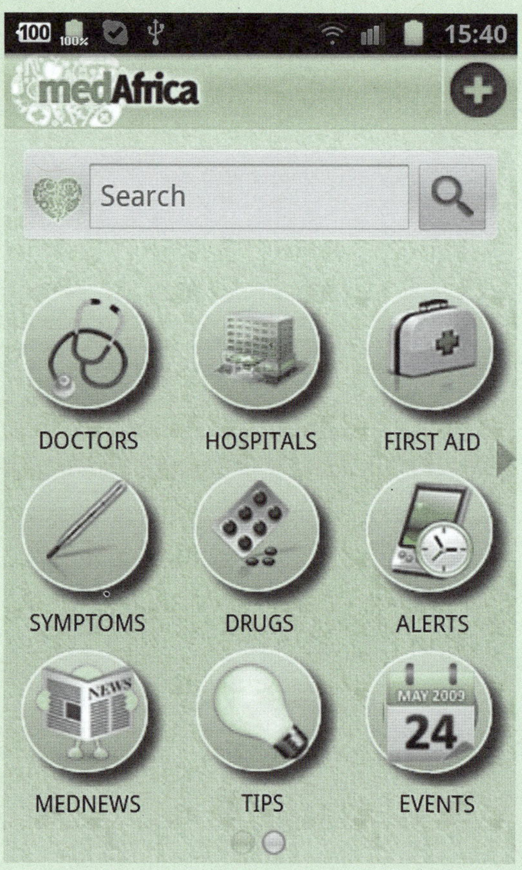

Kenya has emerged as a leading player in mobile for development, largely because of the success of the mPesa mobile payment ecosystem, based on local application developers, projects mounted by local nongovernmental organizations (NGOs), favorable governmental policies, foreign investment, and stable economic conditions. This active ecosystem has benefited the health sector, with many mHealth applications being piloted in Kenya.

Unfortunately, the proliferation of pilot programs, with diverse goals and stakeholders, has fragmented the Kenyan mHealth landscape: standardized platforms that are well-integrated with the local health care system are lacking; few projects have been endowed with long-term funding; and systematic evaluation and impact studies are scarce.

Only recently have more streamlined coordination and division of responsibilities started to emerge. Increasingly, the government is taking over mHealth implementation and ensuring that it complements national policy, while NGOs undertake research, monitoring, and evaluation. Kenya certainly offers an insightful repertoire of mHealth applications.

A recent notable effort is MEDAfrica.org, a company launched in November 2011 that is currently being incubated in the *info*Dev mobile applications lab in Nairobi (see chapter 5). The application integrates symptom checkers, first-aid information, doctor and hospital directories, and alert services into a single, customizable mobile information platform (see screenshot in this box). MEDAfrica won the East African application contest Pivot 25.

MEDAfrica is also pioneering a viable business model, which has attracted worldwide media and investor attention. Other Kenyan mHealth applications are based on remote monitoring or supply chain management through simple SMS technology. Examples include systems for HIV medication reminders, children's health monitoring, early-infant HIV diagnosis, and medicine validation through scratch codes.

Sources: Adapted from Qiang et al. 2012 and www.medafrica.org.

Box 3.2 Ethiopia: SMS helps in monitoring UNICEF's food supply chain

Box figure 3.2.1 RapidSMS in Ethiopia

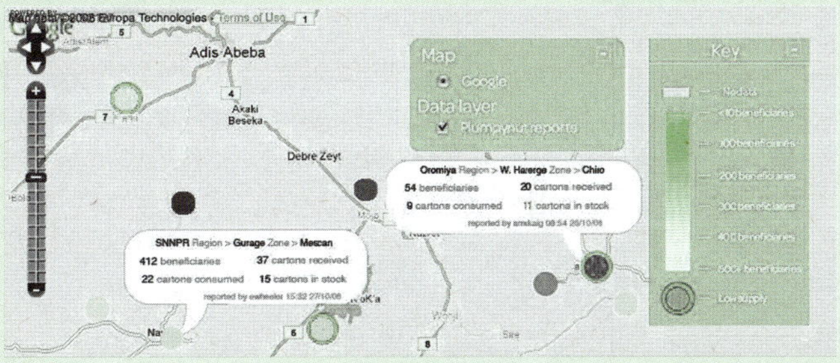

Ethiopia faces significant challenges to effective performance in its health sector. The country is struggling to reduce maternal and child mortality, preventable communicable diseases, and malnutrition. While some policy initiatives have achieved notable success, Ethiopia is likely to meet only one of the six health-related targets under the Millennium Development Goals (MDGs) by 2015. Health care coordination, monitoring, and supply chain management are largely deficient. Funding is limited and well-trained health staff scarce.

In 2008 Ethiopia was hit by severe droughts, leading UNICEF to administer a large-scale food distribution program. Because of the country's poor telecommunication infrastructure and low technical literacy, however, inventory management at local distribution centers was arduous: teams of monitors had to travel back and forth to the centers to deliver handwritten reports of inventory. Inventory analysis could be started only after the data had been delivered, which often took several weeks. In response, UNICEF worked with RapidSMS, which helped cut delays in data transmission related to paper-based collection. Transmission times were reduced from about two months to approximately two minutes. Additionally, data quality discrepancies decreased from 14.2 percent to 2.8 percent while generating significant cost savings. In developing RapidSMS, UNICEF has shown that mHealth applications can represent a feasible low-tech response to challenging conditions such as those in Ethiopia.

RapidSMS was designed to be a simple supply chain management tool, which automatically integrates inventory information sent by SMS into a central database in real time. SMS technology is easily accessible and robust, and minimal training is needed to use the application. RapidSMS allows for stock taking, new admissions, precise location of distribution centers, and analysis of the quantities of food distributed and consumed. This analysis was sped up by immediate visualizations in graphs and maps, accessible by offices in all locations. Monitors still have to travel to distribution centers, but from there they can immediately send stock information to the server. Saving days in travel literally saves lives: UNICEF can respond to shortages and deliver new food resources more promptly, rapidly, and efficiently.

The RapidSMS system is a success story, but several issues arose, including a lack of standards for coding distribution centers, poor allocation of responsibilities, and slow resolution of technical issues. This experience underlines the need for ICT systems to be integrated into existing health care systems as well as the need for capacity building to use ICT effectively.

Source: Adapted from Vital Wave Consulting 2011.

voice-based systems (Chalga et al. 2011) or as specific applications that can be downloaded to a mobile device, as the MEDAfrica example in box 3.1 illustrates.

Optimism about the potential of mHealth is growing; indeed, its potential to be a cost-effective solution for health care in developing countries has led to a growing influx of funds, mainly from public sector and civil society donors (Vital Wave Consulting 2011). In turn, the funding is scattered and mHealth implementations are too often standalone pilot programs. Further, mHealth can help consumers and communities in the developing world keep themselves informed and take more control of their health choices.[4] The opportunities that mobile phones offer for health monitoring could mean that people will start thinking of their phones as personal digital assistants (PDAs) to take care of their health. In parallel, entry barriers for the supply of applications are often lower for mHealth than for other eHealth services or conventional delivery of health care, because small start-ups and local developers can develop mobile software with relatively few resources and can address a much wider potential user base. The shift from eHealth to mHealth can also create an opportunity for a shift from top-down to bottom-up approaches, from government to consumer initiatives, and from centralized to decentralized spending, if mHealth initiatives are effectively implemented.

However, the health sector remains both complex and challenging. The most relevant challenges to the greater uptake of mHealth include:

- *Insufficient financial resources.* Obstacles to comprehensive mHealth solutions are often financial, especially in the developing world. In particular, if no payment structures have been established, it is unclear who should cover the costs for mHealth in private health care (consumers, governments, insurance companies?). This is critical, since the largest part of the cost is often related not to the development of the mHealth application but to the integration of mHealth services with other health care infrastructure.

- *Lack of sustainable business models.* The roll-out of mHealth and other eHealth products and services needs sustainable business models and revenues. Besides a lack of public and private investments in developing such products and services, low-income countries often lack human resources and purchasing power on the demand side. Thus, business models cannot simply be adapted from the developed world but must be designed to match the scarcity of resources both on the demand and supply side.

- *Privacy and security concerns.* Typically, mHealth faces significant privacy and security concerns, with limitations on access to patient data that can complicate interactions between different systems such as primary care, emergency care, and insurance.

- *Limited evidence.* Reliable assessments on the impact of mHealth services are scarce, making it difficult to justify adoption and implementation.

- *Difficult coordination of stakeholders.* Orchestrating diverse private, public, and development sector interests for mHealth can be challenging. Clear roles have yet to be defined, and role models are lacking. The different stakeholders have different goals and strategies that often overlap and conflict, leading to frictions and inefficiencies.

- *Interoperability issues.* Piecemeal implementation of mHealth products and services has led to a lack of interoperability between applications that run on different devices and platforms.

The potential of mHealth

Despite the growing popularity of mHealth, in both usage and commercial terms, there is a disappointing lack of comprehensive studies evaluating its impact. Overall, mHealth is often associated with lower costs and improvements in the quality of health care, but also with a focus on the prevention of diseases and promotion of healthy lifestyles. In line with this assertion, a recent study estimates that mHealth reduces data collection costs by approximately 24 percent, costs of elderly care by 25 percent, and maternal and perinatal mortality by 30 percent (Telenor Group 2012). The same study finds that mHealth can improve compliance with tuberculosis treatment by 30–70 percent.

Given consumers' higher purchasing power and their shown willingness to pay for mHealth applications and devices (IBM Institute for Business Value 2010; MobiHealthNews 2009), huge business opportunities have been identified, mainly in the developed world. Of note, in 2011 the third convention of the mHealth Summit attracted more than 3,600 visitors, up from fewer than 800 in 2009 (mHealth Alliance 2011), and the mHealth market went up from $718 million in 2011 to an estimated $1.3 billion in 2012 in the United States alone (Telecoms Tech 2012). In particular, mHealth applications aimed at individuals are growing in popularity. The number of health-related applications in Apple's App Store grew from just over 4,000 in February 2010 to more than 15,000 by September 2011; roughly 60 percent of these were aimed directly at consumers, with the most popular applications relating to cardio workouts and diet (figure 3.1). The most popular health-related search in 2011 was for information on chlamydia, a sexually transmitted disease, suggesting that the privacy offered by mobile access to health information is important to users. Also the use of "wellness apps" is seen to be growing: an estimated 30 percent of smartphone users are likely to use them by 2015 (Telenor Group 2012). Currently, applications focusing on individuals are mainly geared to developed countries, where purchasing power and education are higher.

Yet, mHealth arguably offers even greater potential in the developing world, where mobile phones serve not only as communication tools but also as key means for accessing health information, obtaining medical insurance, and making payments. As long as macroeconomic conditions are at least somewhat favorable, a lack of existing structures can translate into a greater scope for mHealth solutions; it is expected that major emerging economies, finding themselves in rapid transition to new health care structures, will see the strongest uptake of mHealth in the years to come (Freng et al. 2011). Because of the diversity of mHealth applications and the limited potential of mHealth commercialization, however, the larger economic or development impact of mHealth is difficult to assess, and there is a lack of systematic data for the developing world that would justify higher-level investments (Qiang et al. 2012).

Nonetheless, more than 500 mHealth projects have been deployed around the world (Telenor Group 2012). According to the World Health Organization's Global Observatory for eHealth (GOe), some 83 percent of 112 surveyed countries had at least one mHealth program in operation, with the majority reporting at least four types of application (WHO 2011). The same survey showed that the mHealth adoption gap between low- and high-income countries is fairly small: 77 percent of the former and 87 percent of the latter reported they had implemented at least one mHealth program. In absolute terms, Africa is still the region with the most countries with mHealth deployments, while the developed world and other developing regions have seen stronger adoption growth in recent years (figure 3.2).

Figure 3.1 Relative popularity of consumer health applications in Apple's App Store, 2011

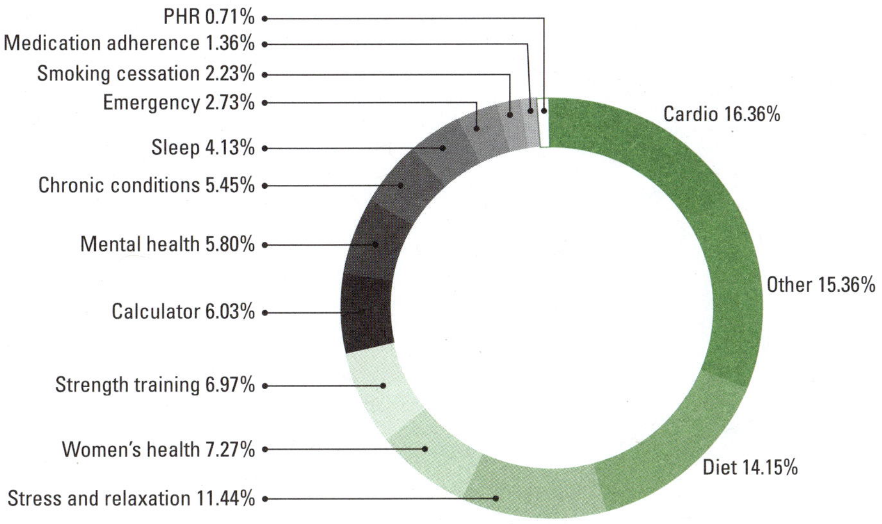

Source: MobiHealthNews 2011.
Note: PHR = personal health records.

Figure 3.2 Number of countries with at least one mHealth deployment, by World Bank region

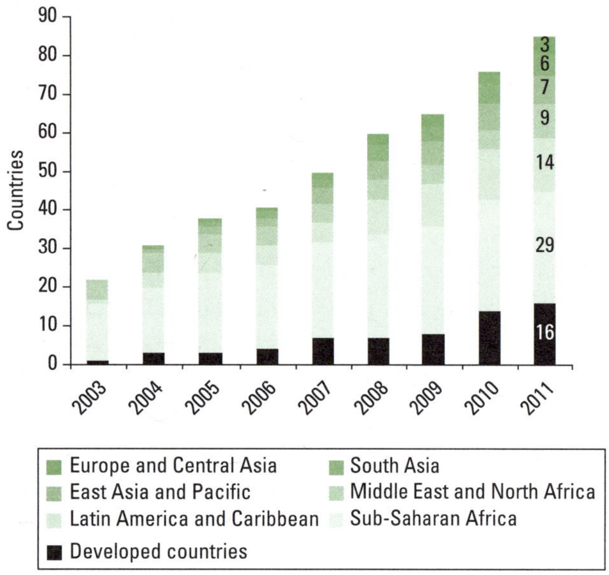

Source: Adapted from GSMA mHealth Tracker 2012.

The mHealth ecosystem

The emergence of mHealth initiatives in many parts of the world can make it difficult to assess their impact in a coherent manner. Increasingly, mHealth stakeholders are realizing the need to arrive at a more holistic understanding of the subject not only to base implementation on best practices but also to factor in local circumstances. Moreover, the large number of different stakeholder groups requires that their different roles and responsibilities be clarified as well. Because mHealth always exists within and interacts with a country's larger health care system, it will be affected by public policy, private sector influence, diverse patient needs, and the interests of several other participants.

A useful framework for the mHealth ecosystem is provided in a World Bank report on mobiles in health (Qiang et al. 2012), which positions mHealth at the nexus of health, technology, and financial services, with government influencing all three of these spheres (figure 3.3). This positioning is in line with a common argument that mobile financial services can enhance the impact of mHealth initiatives (mHealth Alliance and WEF 2011).

Figure 3.3 mHealth ecosystem

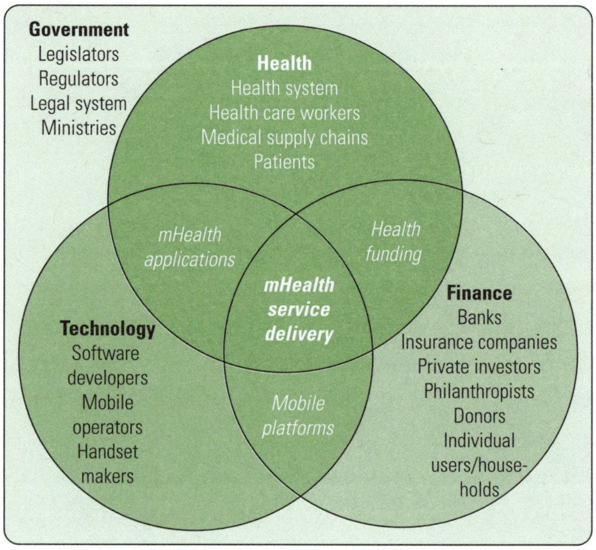

Source: Qiang et al. 2012.

Business models for mHealth

Basing mHealth services on a sustainable business model is vital for implementing mHealth. The first decision that an mHealth organization has to make is what financing model to adopt. Broadly, the options are nonprofit, for-profit, or hybrid.

- Nonprofit organizations may rely less on investments from the private sector and more on large blocks of funding from ministries, multilaterals, and other major donors. Often, a nonprofit mHealth organization's goal is not revenue maximization, but maximum development impact and improvement of patients' health outcomes.
- In contrast, for-profit organizations focus on developing services and products that generate revenues to be distributed to investors and owners, although they may also include a philanthropic element, for example, in probing the opportunities in new markets.
- Whereas health care almost always implies strong public sector involvement, there is certainly potential in mHealth for for-profit projects as well, suggesting that hybrid models may be an appropriate option. For instance, a subscription to mDhil's medical information service in India, costs 1 rupee ($0.02) a day, which is in line with the purchasing power of its target consumers—young Indians between 18 and 25 (Qiang et al. 2012).

For both nonprofits and for-profits, clear value propositions for the still emerging mHealth industry have yet to be established. Value chains in mHealth can be complicated even for simple applications (Vital Wave Consulting 2009), so capturing and monetizing value for only one among many stakeholders in the chain can be difficult (Freng et al. 2011). Because demand for mobile applications for health care delivery is high, however, consumers might be willing to pay for mHealth, for instance, if the service can help them avoid disease and the high opportunity costs of suffering from a medical condition. Cumulative losses in global economic output from noncommunicable diseases are expected to amount to $47 trillion, or 5 percent of GDP, by 2030, so governments also have an interest in promoting better access to health information (Chand 2012).

For now, however, the main challenge for the mHealth industry in low-income countries has been to continue to deliver services once initial funding of pilot projects ends and to scale up or replicate effective models in large-scale implementations. This challenge results in part from a lack of long-term feasible business models. In developed countries, many mobile apps are offered free of charge, with revenues derived from advertising. Fee-for-services is a secondary model (for example, some health apps in Apple's App Store cost up to $79). To obtain sustainable investment in this emerging industry, the private sector needs to demonstrate effective and robust mobile apps that address both local and national health needs, especially for low- and middle-income countries, where average per-user revenues are lower. In cases where incentives for the private sector are not strong enough (that is, where the market prospects are too uncertain, or consumers lack purchasing power), the public sector will have to fill the gap, for example, by directly subsidizing mHealth services, limiting administrative cost for licensing, or engaging in public-private partnerships.

The business models for mHealth must follow the actual health care needs of individuals and the public to be sustainable. As health is considered a public good, the business models should also be aligned with public policy interventions. Investment in mobile applications for public health issues such as noncommunicable diseases should help reduce the costs of health care services and guarantee a healthier population and workforce for developing economies.

Principles for implementing mHealth applications

Clearly, the field of mHealth is just beginning to demonstrate its full potential. So far, there have been many pilot projects and scattered initiatives, with dramatically varying levels of success (table 3.2). This section briefly outlines some of the principles that hold across contexts from these early mHealth initiatives.[5]

Avoid a one-size-fits-all approach

Mobile health applications must be designed to respond to people's needs and suited to their local context. A common pitfall is that, once an application is working well technically and is seen to have high potential, there is an immediate enthusiasm to implement it everywhere regardless of context. Conducting readiness assessments with users can help avoid such overextension. Close involvement of health practitioners in the design and development of mobile app content can also ensure accuracy for public health programs. A good example of successful adaptation to various contexts and sectors is RapidSMS—this system has been used for food supply chain management in Ethiopia (see box 3.2), as a citizen outreach system in Senegal (see table 3.2), and as an emergency response tool in Somalia (Vital Wave Consulting 2011). The success of RapidSMS has also sparked the development of other mHealth tools, such as ChildCount+ (Vital Wave Consulting 2011).

Maintain flexibility

Policy-makers must be careful not to overregulate mHealth nor to prescribe, from the top down, how applications are to be implemented. Because mHealth technology is cheap to implement and change, it can be a tool for achieving efficiencies and improved flexibility in the health system. Mobile health can also be combined with other mobile services to enhance its impact. The industry may also evolve freely, including in ways that the health sector may not anticipate. For example, mHealth and mMoney can be integrated in a variety of useful ways. A patient might receive a prescription through an mHealth application and pay for it using an mMoney transfer or bank account, all using the same mobile phone. Health care workers who spend most of their time in the field and transfer information to health systems through mobile phones could receive their wages in the same way.

Table 3.2 Selected examples of mHealth projects and lessons learned

Country	Application	Services provided	Scale/location	Key success factors and lessons learned
Botswana Botswana-UPenn Partnership (2012)	**Kgonafalo:** A remote diagnosis facility using camera-equipped mobile handsets and tablets to send photos of patients for treatment advice.	Initially focused on oral health, it now covers radiology, cervical cancer, and dermatology. Photos sent from rural areas are used to determine whether to transport patients for treatment in Gaborone.	Initially piloted in 6 locations, it is now ready to be scaled up to 25 locations, before going nationwide, with funding from the government. The pilot program assessed 230 cases over a 6-month period.	The initial implementation, using technology from a Bangladeshi company, was not robust, and was replaced by an application developed locally by PING (Positive Innovation for the Next Generation). In addition, handsets were replaced with Android Tablet PCs preloaded with medical databases and treatment guidelines.
Kenya Changamka (2012)	**Changamka MicroHealth:** A smart card that enables payments to be made for medical services via mPesa, linking health care providers with medical insurance companies.	Smart cards are available (from supermarkets and the like) to outpatients and pregnant mothers. Patients can buy health care packages that include consultation, lab test, and drugs, for example, or top up credit with amounts as low as K Sh 100 ($1.20).	Launched in 2008, it now covers 18 service distributors and 29 health care facilitators across Kenya but is mainly concentrated in Nairobi. Available to all citizens.	This health application is designed to be user friendly and to respond to selected high-demand services. Improved interoperability has created value, especially in linking many different service providers and insurance companies.
Peru WawaRed (2012)	**WaWaRed:** A mobile application providing timely messages for pregnant women.	SMS messages and basic health information delivered at different stages during pregnancy. It also includes a symptom checker, which can be accessed and used via SMS.	Launched in 2010, WaWaRed is available in Ventanilla Distrito, a vulnerable community of Lima, and serves the Callao community of 5,000 people. It is now being scaled up nationally.	Involvement of pregnant mothers has been facilitated by a high adoption rate of mobile phones. This case has also highlighted that fathers need to be involved in maternal health education for effective communication with health care providers.
Chile Centro de Informatica en Salud (2012)	**Centro de Informatica en Salud:** Provision of health care to the elderly at home, in a project called Cuidado Domiciliario.	Providing health information to and from mobile devices, such as phones and tablets, Devices store electronic health records to facilitate care at home.	Launched in 2011 to serve 3,000 elderly people in the Pedro Aguire Cedra district, it is now being scaled up nationally.	An assessment of digital literacy among the elderly was carried out before the project launch. Awareness raising and training is essential to engage effectively with health care services when using mobiles for health care at home.
Senegal RapidSMS (2012)	**RapidSMS:** Implementation in Senegal through the Jokko initiative, with UNICEF and Tostan, an NGO.	Citizen engagement with health care providers through an SMS aggregation service, allowing short texts to be distributed to a large network of users.	Launched in mid-2009, the Jokko initiative now serves 800 communities in 8 African countries.	Significant costs have been saved by using SMS aggregation to broadcast text messages to multiple recipients for the price of a single message. The messaging process may take up to 8 hours depending on the technology used, so it may not be effective for emergency alerts.
South Africa Cell-Life (2012)	**Cell-Life Aftercare:** an SMS alert service for patients following HIV retroviral therapy.	Patients receive SMS alerts when medication is due, along with other health tips. Mobile phones are also used for data capture by nurses following patients.	Begun in 2001 as a research project at the University of Cape Town, this initiative became a company in 2005. It is currently working in partnership with over 50 organizations.	In South Africa, the prevalence of HIV and AIDS in adults is close to 20 percent. Cell-Life has developed a philosophy of "Dispense, Communicate, Capture."

Sources: Assembled from diverse sources (see References).

Take standards and interoperability into account

Although apps should be adapted to local context, designing a separate and incompatible application for every stakeholder group or every locale frequently leads to large inefficiencies. Applications often benefit from economies of scale and reach—the power of singular mHealth services can be multiplied by their ability to work together, operate on common platforms, and share information. Making interoperability a prerequisite for new mHealth applications could help reduce inefficiency or duplication. Accordingly, a lack of standards is seen as preventing the scaling up of applications and, thus, to be a key obstacle to achieving cost savings through mHealth (Telenor Group 2012). The perspective should go beyond the health sector: seamless integration with other mobile platforms, such as mobile money, can enhance the value of mHealth applications even more.

A push for more universal platforms can come from the top (for instance, as part of a national eHealth strategy that encompasses mHealth) or from the bottom (especially at the point of care through mobile phones). The greatest value will be realized when both strategies are used and complement each other. International standards for hardware and software platforms can ensure interoperability among mHealth applications and other mobile tools, while also enabling the development of locally relevant applications. International bodies such as the mHealth Alliance, the Health Metrics Network, and the Continua Health Alliance are helping to forge cooperation in the development of globally recognized standards and metrics. For example, to achieve seamless exchange of data elements, HL7 and ISO standards have been widely used for electronic health records. Standards and interoperability must be addressed early on—consolidating many fragmented or incompatible services is hard, as cases like Kenya (see box 3.1) or Ethiopia (see box 3.2) have shown.

Evaluate existing information systems

Multiple health information systems exist and data are gathered with or without mobile applications. Reliable assessments of these systems are useful to identify where mHealth is needed and how it can best be implemented. Evaluation of delivery flows of health services should also be taken into consideration. Mobile apps may prove complementary to existing solutions, especially for remote data collection and telemedicine—for example, in the cases of the Health Management and Research Institute in India (box 3.3), the EpiHandy program in Uganda, the IHISM system in Botswana, and the Dokoza system in South Africa (Vital Wave Consulting 2011). Accordingly, it is estimated that mHealth can double the number of rural patients reached by a physician (Telenor Group 2012).

Poor evaluation of current information systems before entering the digital arena may result in fragmented or inappropriate health care applications. For example, it is vital for mHealth applications to be interoperational with eGovernment applications in other sectors. A success story in this context is Rwanda, which has implemented an overarching eHealth initiative combining patient record tracking, transmissible disease monitoring, and supply chain management, as well as mHealth telemedicine apps for health professionals in remote areas (Vital Wave Consulting 2011).

Track key success indicators for monitoring and evaluation

The need for evaluation does not end once an mHealth application has been implemented. To move from pilot projects to full-scale implementation, evidence is needed on the impact of mHealth applications, along with identification of operational efficiencies and detailed estimates of cost savings. In short, monitoring and evaluation (including tracking project-specific success indicators) are necessary right from the beginning of an mHealth implementation. However, only 7 percent of low-and lower-middle income countries report that they evaluate their mHealth initiatives (WHO 2011), and only a few systematic analyses of nongovernmental projects exist. A rare exception is WelTel (Lester et al. 2010), an SMS-based tool for tracking compliance with antiretroviral therapy. Peer-reviewed evidence confirmed its positive influence on health outcomes beyond the initial stages, which, in turn, led to the continuation of funding for the project and its increased sustainability.

Ensure quality and content of health information

The content and quality of health information must be tailored to end users and decision-makers. Lack of trust is a major resistance factor against the use of mobile applications in health care provision. Similarly, local languages and cultures often represent major barriers to adoption. One notable example of the relevance of trust in health information is the Indian mobile platform mDhil. While it received significant private sector investment and

Box 3.3 India: Health Management and Research Institute—104 Mobile

India's Health Management and Research Institute (HMRI) is a public-private partnership between the state government of Andhra Pradesh (which bears 95 percent of costs) and the Satyam Foundation (which bears 5 percent of costs) based in Hyderabad, Andhra Pradesh. HMRI launched "104 Mobile" in 2008[a] to improve local health services by replacing the traditional health care system with mHealth applications for disease surveillance, prevention counseling, telemedicine, and supply chain management.

104 Mobile sends medical units (MUs) to habitations more than three kilometers away from the nearest public health service provider to provide medical care to rural populations. Each MU circulates on a fixed date every month, ensuring continuity of care. Maternal and child health are prioritized, along with the diagnosis and management of chronic diseases such as diabetes, hypertension, asthma, and epilepsy. 104 Mobile deployed 475 MUs to 22 districts throughout Andhra Pradesh. Generally, treatments at clinics tend to be costly, and more than half of unmet requests for outpatient care could be treated by phone in rural areas—a potential that 104 Mobile can exploit through its hotline for medical consultations.

HMRI has delivered the following major benefits (partly thanks to the integration of mHealth applications):

- Expanded the service area covered by 25 percent
- Services may cost as little as one-tenth of those provided by the government
- Up to 55 percent of 600,000 unmet requests for outpatient treatment could be treated by phone
- 1.26 million pregnant women each received an average of three antenatal care check-ups
- 2.9 million people with chronic diseases were screened, tested, and provided with medication
- Over 10 million unique electronic health records were established, making this one of the largest public electronic health record databases worldwide

Source: Qiang et al. 2012.

a. 104 Mobile has been transitioned back to the government of Andhra Pradesh and the service is currently operated by the Ministry of Health under the government.

was able to attract a fairly large base of paying customers, one of its biggest challenges was to establish credibility and win the trust of its users, given the inaccuracy and lack of clarity of much of the health information it had to draw on (Qiang et al. 2012).

Respect privacy and confidentiality

Although awareness of the issue of data privacy is often low in developing countries, the case can be made that mechanisms guaranteeing some level of privacy and confidentiality are a universal requirement for mHealth. Evidence from developed economies shows that privacy and confidentiality are important success factors in the management of public and personal data, especially in the case of infectious and transmissible diseases. For example, the privacy and confidentiality of personal data of patients is vital to prevent discrimination in the workplace. The dangers of poor privacy requirements are often visible only after the damage is done (for example, once security leaks are exploited by hackers), making this a natural field for government intervention and regulation (Qiang et al. 2012). However, privacy regulation should be limited according to context. For instance, health records on nonstigmatizing infectious diseases (such as dengue fever) should be shared quickly and widely, while a patient's interests in the confidentiality of personal data might triumph, for example, in cases of sexually transmitted diseases.

Enable public-private partnerships

Policy-makers contemplating mHealth should consider bringing private sector stakeholders to the table. If administered wisely, public investment and technology partnerships enriched with competitive incentives (through tenders and challenges, for example) can improve the quality of mHealth apps and services and improve choice. Often, public-private partnerships (PPPs) can benefit from a division of labor based on the respective competencies and resources of the stakeholders. For instance, private mobile operators and software developers might be better situated to provide the technological platform and develop the mHealth applications, while governments can provide a favorable regulatory environment and integration with the existing (public) health care system (Qiang et al. 2012). Governments might also use a PPP approach to spark innovation from a more agile private sector. This approach seems to be very effective; the largest and most scalable mHealth initiatives are mostly supported by PPPs (WHO 2011). One notable project is ChildCount+: the Kenyan government, the Millennium Villages Project, and UNICEF collaborated with Zain and Sony Ericsson as technology partners to develop a monitoring and tracking system with a focus on easily treatable diseases; the new system is expected to dramatically reduce child and maternal mortality (Qiang et al. 2012).

Offer training and take literacy into account

Mobile health services will have a greater impact on health outcomes where their users have high levels of literacy (and for health workers, training) in ICT and health. Proficiency with mobile devices and computers saves time and reduces errors. As a result, during mHealth implementation, the technical literacy of users needs to be factored in, and staff have to be trained to use the necessary technology. In addition, training in technical and organizational skills is often needed to launch, scale, and sustain mHealth interventions. For instance, major barriers to adoption of telemedicine in Uganda were lack of knowledge and skills on the part of health care staff (Isabalija et al. 2011). There are many ways to achieve improvements in these areas: dedicated training institutions, public information campaigns, programs in schools, and even software for mobile devices that trains people in their use and in treatment methods. All of these may ultimately require oversight to ensure that the information being conveyed corresponds to best practices and health system priorities.

Ensure the commitment of leaders

The mHealth industry is today at a pivotal moment in its rapid evolution. To realize the industry's full potential for improving health outcomes, long-term leadership is needed from government and from the health, technology, and financial sectors. Their leadership will help supply the industry with better inputs—both tangible (such as handset technology and financing) and intangible (such as market regulations, standardization of software, and rules for using bandwidth). It will also ensure that mHealth services correspond to health sector priorities. The impact of committed leadership can be magnified by a series of multipliers—improvements in reach, affordability, quality assurance, behavioral norms, and matching of resources—that can boost health outcomes. High-level leadership within government is especially crucial for forging inter-ministry partnerships.

Conclusions

As this chapter has argued, mHealth applications have the potential to transform health care systems in low-income economies: mHealth can generate cost savings and provide more effective health care delivery within relatively limited resources. Modern forms of health care are at a tipping point where consumers are taking on more responsibility for managing their own health choices, and mobile phones could contribute greatly to this shift of decision-making from state and health institutions to the individual. However, the most substantial challenge for mHealth is the establishment of viable and sustainable business models that can be replicated and scaled up. One step forward could be clearer delineation of roles within the health ecosystem between public and private health care providers. Accordingly, for "macro"-focused public health purposes, the World Health Organization (WHO 2005) recommends that mHealth be integrated into a country's broader eHealth strategy. Finally, a missing component is the effective monitoring and evaluation of mHealth, which could inform the design of more successful mHealth applications at this critical stage of their development.

Notes

1. One of the first uses of the term mHealth was in 2008 when the Rockefeller Foundation engaged global eHealth experts at Bellagio, Italy; see Mishra and Singh (2008).
2. For a comprehensive review on the provision of telemedicine in the developing world, see Wooton et al. (2009).
3. Medical informatics professionals are trained in medicine and computer sciences and information theory. See www.imia.org for more details.
4. The role of social intermediaries, including civil society organizations and community-based organizations, should not be overlooked. They can focus on health workers, building their capacity and training them in ICT skills. In addition, they can offer help directly to citizens in poor and isolated communities who do not possess adequate ICT skills, for instance by timely provision of necessary information to minimize information asymmetries, and sometimes by providing training on how to use mobile applications.
5. We also refer readers to a more detailed list of Calls to Action, divided by stakeholders, in Vital Wave Consulting (2009).

References

Botswana-UPenn Partnership. 2012. http://www.med.upenn.edu/botswana/.

Celi, L., L. Sarmenta, J. Rotberg, A. Marcelo, and G. Clifford. 2009. "Mobile Care (Moca) for Remote Diagnosis and Screening." *Journal of Health Informatics in Developing Countries* 3 (1): 17–21.

Cell-Life. 2012. http://www.cell-life.org/home.

Centro de Informatica en Salud. 2012. http://www.centrodeinformaticaensalud.org/?page_id=24.

Chalga, M. S., A. K. Dixit, B. Shah, and A. S. Bhati. 2011. "Real Time Health Informatics System for Early Detection and Monitoring of Malaria in Desert District." Jaisalmer, India. *Journal of Health Informatics in Developing Countries* 5 (2).

Chand, S. 2012. "Silent Killer, Economic Opportunity: Rethinking Non-Communicable Disease." Chatham House briefing paper GH BP 2012/01. London. http://www.chathamhouse.org/sites/default/files/public/Research/Global%20Health/0112bp_chand.pdf.

Changamka. 2012. http://www.changamka.co.ke/html/welcome.html.

Freng, I., S. Sherrington, D. Dicks, N. Gray, and T. Chang. 2011. "Mobile Communications for Medical Care." http://www.csap.cam.ac.uk/media/uploads/files/1/mobile-communications-for-medical-care.pdf.

Gombachika, H., and M. Monawe. 2011. "Correlation Analysis of Attitudes towards SMS Technology and Blood Donation Behavior in Malawi." *Journal of Health Informatics in Developing Countries* 5 (2).

GSMA mHealth Tracker. 2012. http://apps.wirelessintelligence.com/health/tracker/.

Health Management and Research Institute. 2012. "Our Work, Mobile Health Services." Hyderabad, India. http://www.hmri.in/104-Mobile.aspx.

IBM Institute for Business Value. 2010. "The Future of Connected Health Devices." http://public.dhe.ibm.com/common/ssi/ecm/en/gbe03398usen/GBE03398USEN.PDF.

Isabalija, S. R., K. G. Mayoka, A. S. Rwahana, and V. W. Mbarika. 2011. "Factors Affecting Adoption, Implementation and Sustainability of Telemedicine Information Systems in Uganda." *Journal of Health Informatics in Developing Countries* 5 (2).

Istepanian, R. S. H., and J. Lacal. 2003. "Emerging Mobile Communication Technologies for Health: Some Imperative Notes on MHealth." Proceedings of the 25th Institute of Electrical and Electronics Engineers Annual International Conference: Engineering in Medicine and Biology Society, Cancun Mexico.

Lester, R., et al. 2010. "Effects of a Mobile Phone Short Message Service on Antiretroviral Treatment Adherence in Kenya (WelTel Kenya1): A Randomised Trial." *The Lancet* 376 (9755): 1838–45.

mHealth Alliance. 2011. "2011 mHealth Summit Attracted Industry Leaders from Around the Globe." December 13. http://www.mhealthsummit.org/pdf/mhs11_wrap_release.pdf.

mHealth Alliance and WEF (World Economic Forum). 2011. "Amplifying the Impact: Examining the Intersection of Mobile Health and Mobile Finance." http://www.mhealthalliance.org/news/making-connection-between-mobile-health-and-mobile-finance.

Mishra, S., and I. P. Singh. 2008. "mHealth: A Developing Country Perspective." http://www.ehealth-connection.org/files/conf-materials/mHealth_%20A%20Developing%20Country%20Perspective_0.pdf.

MobiHealthNews. 2009. "Wireless Health: State of the Industry 2009 Year End Report." http://mobihealthnews.com/wp-content/Reports/2009StateoftheIndustry.pdf.

———. 2011. "Consumer Health Apps for Apple's iPhone." http://mobihealthnews.com/13368/report-13k-iphone-consumer-health-apps-in-2012/.

Patil, D. A. 2011. "Mobile for Health (mHealth) in Developing Countries: Application of 4 Ps of Social Marketing." *Journal of Health Informatics in Developing Countries* 5 (2): 317–26.

Qiang, C. Z., M. Yamamichi, V. Hausman, R. Miller, and D. Altman. 2012. "Mobile Applications for the Health Sector." ICT Sector Unit, World Bank, Washington, DC. http://siteresources.worldbank.org/INFORMATIONANDCOMMUNICATIONANDTECHNOLOGIES/Resources/mHealth_report_(Apr_2012).pdf.

Perera, I. 2009. "Implementing Healthcare Information in Rural Communities in Sri Lanka: A Novel Approach with Mobile Communication." *Journal of Health Informatics in Developing Countries* 3 (2): 24–29. http://www.jhidc.org/index.php/jhidc/issue/view/8.

RapidSMS. 2012. "The Jokko Initiative." http://www.rapidsms.org/case-studies/senegal-the-jokko-initiative.

Telecoms Tech. 2012. "US$1.3 billion: The Market for mHealth Applications in 2012." http://www.telecomstechnews.com/blog-hub/2012/jan/30/us-13-billion-the-market-for-mhealth-applications-in-2012/.

Telenor Group. 2012. "New Study: The World Is Ready for Mobile Healthcare." Press release. http://telenor.com/news-and-media/press-releases/2012/new-study-the-world-is-ready-for-mobile-healthcare/.

Vital Wave Consulting. 2009. "mHealth for Development: The Opportunity of Mobile Technology for Healthcare in the Developing World." Washington, DC, and Berkshire, UK: UN Foundation-Vodafone Foundation Partnership. http://www.vitalwaveconsulting.com/pdf/mHealth.pdf.

———. 2011. "eTransform Africa Sector Report: Health Sector Study." http://www.etransformafrica.org/sites/default/files/Complete-Report-and-Summary-Health.pdf.

WawaRed. 2012. http://wawared.lamula.pe/.

Wooton, R., N. G. Patil, R. E. Scott, and K. Ho, eds. 2009. *Telehealth in the Developing World*. Royal Society of Medicine Press/IDRC.

WHO (World Health Organization). 2005. "Global eHealth Strategy." World Health Assembly Resolution 58.58, Geneva. http://apps.who.int/gb/ebwha/pdf_files/WHA58/WHA58_28-en.pdf.

———. 2011. "mHealth: New Horizons for Health through Mobile Technologies: Second Global Survey on eHealth." http://whqlibdoc.who.int/publications/2011/9789241564250_eng.pdf.

Chapter 4

Mobile Money for Financial Inclusion

Kevin Donovan

Mobile financial services are among the most promising mobile applications in the developing world. Mobile money could become a general platform that transforms entire economies, as it is adopted across commerce, health care, agriculture, and other sectors. To date, at least 110 money mobile systems have been deployed, with more than 40 million users. The most well-known system, M-PESA, started in Kenya and is now operational in six countries; it has 20 million users who transferred $500 million a month during 2011.[1] While the benefits of mobile money payment systems are clear, observers remain divided over whether mobile money systems are truly fulfilling their growth potential.

This chapter evaluates the benefits and potential impact of mobile money, especially for promoting financial inclusion in the developing world, before providing an overview of the key factors driving the growth of mobile money services. It also considers some of the barriers and obstacles hindering their deployment. Finally, it identifies emerging issues that the industry will face over the coming years.

Mobile money: an ecosystem approach

At the most basic level, mobile money is the provision of financial services through a mobile device (box 4.1).[2] This broad definition encompasses a range of services, including payments (such as peer-to-peer transfers), finance (such as insurance products), and banking (such as account balance inquiries). In practice, a variety of means can be used such as sending text messages to transfer value or accessing bank account details via the mobile internet (figure 4.1). Special "contactless" technologies are available that allow phones to transfer money to contactless cash registers.

Although mobile phones are central to all these uses, mobile money is more than just technology—it needs a cash-in, cash-out infrastructure, usually accomplished through a network of "cash merchants" (or "agents"), who receive a small commission for turning cash into electronic value (and vice versa).

Because the mobile money industry exists at the intersection of finance and telecommunications, it has a diverse set of stakeholders, with players from different fields in competition. Mobile network operators, banks, and increasingly new entrants, such as payment card firms, continue to catalyze the industry with innovative offerings, but to be sustainable, these must be met with sufficient demand from consumers and firms—a variable missing in many contexts. A host of supporting businesses, such as agents and liquidity management firms, are also necessary. In areas where it has proved successful, mobile money has created a platform for start-ups to build upon (Kendall et al. 2011). Finally, all of this must happen in an environment with appropriate government regulations for both finance and the ICT

Box 4.1 One device, many channels

Mobile phones are multifunctional devices that allow for a variety of communication methods. These range from ubiquitous voice and SMS channels to more sophisticated means such as software applications or web browsers. To be a viable solution for mobile money, the channel should ideally be universally available (including the cheapest mobile phones) and must be secure. In practice, this requirement largely limits mobile money to using a standard network service, such as USSD (Unstructured Supplementary Service Data) or SMS (short message system), or an application preloaded on a unique SIM card. Since mobile operators control both of these channels, they remain essential gatekeepers in deploying mobile money.

Sources: http://mmublog.org/blog/on-channels/; http://www.ictinagriculture.org/ictinag/sites/ictinagriculture.org/files/web_Module3.pdf.

Figure 4.1 Different types of mobile financial services

Mobile finance *including* Credit | Insurance | Savings	Mobile banking *including* Transactional | Informational
Mobile payments *including* Person-to-person | Government-to-person | Business-to-business	

Source: Adapted from Gencer 2011.

sector, as well as appropriate safeguards for consumer protection.

The financial inclusion imperative

Poverty is more than just a lack of money. It involves a lack of access to the instruments and means through which the poor could improve their lives. Exclusion from the formal financial system has increasingly been identified as one of the barriers to a world without poverty. In many developing countries, more than half of households lack an account with a financial institution, while small firms frequently cite difficulty in accessing and affording financing as a key constraint on their growth. This exclusion does not necessarily mean that the poor lack active financial lives: in fact, the fragility of their situation has led to the development of sophisticated informal financial instruments. However, the use of only informal instruments means that the poor are limited in their ability to save, repay debts, and manage risk responsibly. On a macroeconomic level, these financial constraints on the poor can slow economic growth and exacerbate inequality (Demirgüç-Kunt, Beck, and Honahan 2008).

Finding innovative models to extend financial services to the poor has now become an urgent challenge. The excitement around mobile money has arisen in part because it is widely seen as an effective way to provide access to finance to millions of people around the globe. According to the Consultative Group to Assist the Poor (CGAP), roughly 1 billion people have a mobile phone but no bank account. Providing them access to mobile financial services will involve difficult implementation that is unlikely to succeed quickly.

In addition to extending financial services to the poor, mobile money is expected to improve productivity by increasing the efficiency and lowering the cost of transactions, improving security, generating new employment opportunities, and creating a platform on which other businesses can grow.

Mobile money could transform financial inclusion. "Where most financial inclusion models have employed either 'credit-led' or 'savings-led' approaches, the M-PESA experience suggests that there may be a third approach—focusing on building the payment 'rails' on which a broader set of financial services can ride," wrote the authors of one report (Mas and Radcliffe 2010). As illustrated in the next section, while benefits from the simple diffusion of an improved infrastructural "rail" are significant, even greater impact arises because mobile money systems can serve as a platform for additional innovations, whether they be bill payment services that avoid lengthy queue times or more striking examples such as efficient conditional cash transfers for drought relief or compensation.[3] In places where no financial infrastructure exists, this type of change is truly transformational.

What is the impact of mobile money?

According to data from the GSM Association, most of the 100-plus deployments of mobile money systems have been in developing countries, with around half in Africa alone (figure 4.2). Mobile money systems can be made available wherever there is wireless phone service, helping to overcome distance, as well as the lack of branch offices in rural areas (box 4.2).

Since mobile money is often linked to financial inclusion, it is vital to understand how and under what conditions mobile money applications can extend financial services to the poor. Support for mobile money initiatives from governments, nongovernmental organizations, and the international development community needs to be justified by assessing the impact on development goals such as financial inclusion, poverty reduction, increased productivity, and risk management.

Although the mobile money industry has achieved significant scale in only a handful of countries, a growing number of studies are establishing its impact in a variety of areas. Its potential advantages include benefits arising from the inherent characteristics of the services; benefits arising organically from widespread usage and network effects; and benefits arising from purposeful and innovative applications, either made by developers or created by people's uses of mobile money services.[4]

Inherent benefits

Mobile money is often successful because it is considerably cheaper than other alternatives to cash. In an international comparison of 26 banks, McKay and Pickens (2010) found that branchless banking (including mobile money) was 19 percent cheaper on average than alternative services. At low transaction amounts or for informal money transfer options, this difference more than doubled.[5] In Kenya M-PESA was routinely one-third to one-half as expensive as alternative systems. Lower costs directly translate into money the poor can keep—in Kenya the amount of money remitted increased when transferred using M-PESA compared with traditional forms of remittances. Conversely, where transaction costs are high, as in Botswana where the cost per transaction is a minimum of 8 pula ($1.07), mobile money has been slow to take root.

Well-supervised mobile money can be safer than alternatives, including cash. Early studies of M-PESA in low-income areas found that the risk of muggings declined, because cash was less evident. Because it is less visible than cash, mobile money also has consequences for privacy and autonomy. Research has found that women are able to have personal savings without seeking permission from their husbands (Morawczynski 2009), but, of course, this autonomy holds true for both genders.[6]

The speed and liquidity of mobile money are also key benefits. The limited assets the poor own often take the form of valuable objects (such as livestock or gold), which are relatively illiquid. In times of crisis, such assets can be difficult to realize quickly, and their value may decline if the

Figure 4.2 Global mobile money deployments

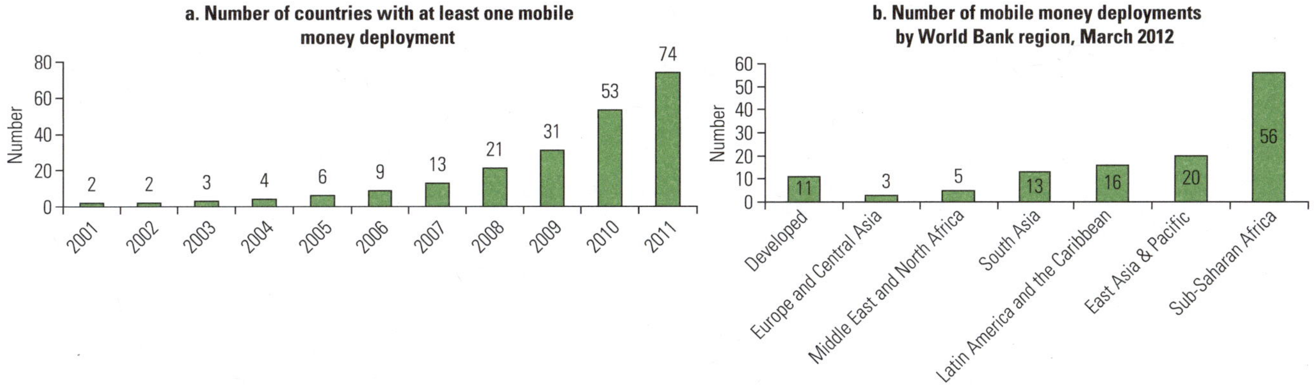

Source: GSMA Mobile Money Tracker 2012.

Box 4.2 Using mobile money

Mobile money applications are typically small pieces of software embedded on a SIM card or available over a mobile network. A customer can use an inexpensive mobile to send value to someone else. To change this digital value into cash, a user simply visits a retail agent who verifies the user's identity and makes the switch. In this way, money can cross enormous distances at the speed of a text message. Consider a young Tanzanian who has moved to Dar es Salaam to find work. With mobile money, he can send regular, small payments to his family at their rural home without needing to pay and trust a courier or take it himself. His family can then exchange the digital value for cash at a local agent.

market floods with other families seeking to convert similar assets to cash at the same time. Moreover, sending gold bracelets or cash to a family or friend in need can be a risky enterprise. Mobile money can be an accessible and convenient medium for the delivery of financial services and more reliable than traditional, informal methods.

Benefits from scale

In some jurisdictions, mobile money has achieved critical mass, so nonusers are encouraged to adopt the systems used by their peers. When the poor are connected on a large scale, they are able to use mobile money to improve their livelihoods. The best data available on this point comes again from Kenya, where households with access to mobile money were better able than those without to manage negative shocks (including job loss, death of livestock, or problems with harvests). Whereas households that did not use M-PESA saw consumption fall by 6–10 percent on average, M-PESA users were often able to fully absorb the shocks, because they received more remittances and lost less to transaction costs (Suri and Jack 2011). Evidence of such "livelihood strategies" was also evident during the violence following Kenya's 2007 election, during which M-PESA "became one of the only means through which [residents of Nairobi's informal Kibera settlement] could access cash" (Morawczynski 2009). Even in less tumultuous times, mobile money at scale can serve to meet the needs of the poor: research in Kenya found that M-PESA was a useful means to access cash. Often the poor lack fungible sources of exchange such as cash, and through the network of cash agents and people's contacts willing to send value, mobile money allows many to get cash when and where they need it (Stuart and Cohen 2011).

Mobile money can also prove commercially significant for service providers, when it reaches scale. Although the transaction fees that mobile money providers charge are individually quite small, in total, they can represent an important revenue source. For example, Safaricom, the mobile operator that offers M-PESA, reported mobile money revenues for the first half of 2011 of K Sh 7.9 billion ($90 million). In addition, cash agents may also gain commercial benefit from the fees they receive.

Benefits from innovation

Improving the ability of the poor to transfer money is certainly beneficial, but in isolation, mobile transfer services do not capture the full potential of mobile money to enhance financial inclusion. Early studies of South African mobile money found that while it had the potential to advance financial inclusion, it had not increased access to banking, especially compared with nontechnological efforts, such as a particular type of bank account designed especially for the poor (Porteous 2007). In Kenya, for example, the predominant use of M-PESA is still sending money, although some people use it for savings (Stuart and Cohen 2011). Access and use of more sophisticated financial services such as savings, credit, and insurance could prove far more beneficial to the poor. To develop these services, businesses, governments, and other institutions must innovate actively on top of the payment services that are being deployed by mobile money operators.[7]

Some organizations are deliberately using mobile money to enhance their traditional offerings. For example, during a recent drought in Niger, a set of randomly selected households received cash transfers via mobile money (Aker et al. 2011). In comparison with physical cash, this trial found lower variable costs for senders, as well as lower costs for recipients. Over the course of the crisis, recipient households also enjoyed better diets and depleted fewer assets.

Insurance, credit, and savings services are now being developed atop mature mobile money systems. Kilimo Salama is a micro-insurance product that uses M-PESA to provide payouts to smallholder farmers whose crops fail. In its second year of operation, 12,000 farmers were insured, and 10 percent of those received payouts of up to 50 percent of their insured inputs (Sen and Choudhary 2011). Likewise, Equity Bank and Safaricom have partnered to offer M-Kesho, a mobile service that offers microsavings accounts, credit, and insurance. As individuals develop financial histories with mobile money, the ability to provide credit can expand because financial institutions will be able to analyze those histories and assign credit scores.

The impact of mobile money is also likely to extend to the public sector through increased efficiency and reach. Government adoption of mobile money is still in its infancy, but a study by McKinsey for the Gates Foundation estimates that connecting poor Indian households to an electronic payment system for cash transfers would have considerable impact through reduced leakages, transaction costs, and overheads (Lochan et al. 2010). It would also improve the government's ability to monitor financial flows, collect tax revenues, and reduce illicit activity. Government use of mobile money—such as salary disbursements—could prove to be an enormous driver of the service throughout the economy on the whole.

Growing mobile money: challenges and success stories

Despite a growing number of successes, the mobile money industry faces a number of challenges. Mobile money deployments in developing countries often target customers who may be poor, dispersed, and remote. Mobile money also spans two distinct industries with different business models. Telecommunications and payments are transaction-based, with fees collected on transactions; conversely, banking is float-based, with money earned through holding deposits. Developing the necessary cross-sectoral partnerships—including bridging cultures and regulations—may therefore prove difficult.

Additionally, mobile money services represent a two-sided market, and new deployments must convince both agents (supply) and customers (demand) to sign up for the service in sufficient quantity to be viable. Building and properly incentivizing the agent network is no small task, and maintaining the necessary cash liquidity at the outlets can prove a constant challenge. Winning and retaining the trust of customers, including those who are poor and new to the technology, is central to success. Commercial viability in this industry requires scale, and operators are faced with the trade-off between higher costs to recoup their investments or lower costs to reach scale and build a mass market (Mas and Radcliffe 2010).

Despite these challenges, mobile money has grown in a variety of markets. Although the International Finance Corporation (IFC) identified more than 50 factors influencing the growth of mobile money, 3 are especially important (IFC 2011): regulation, competition with other instruments of financial access, and user perceptions and skills.

Regulation

Since mobile money straddles finance and telecommunications, it faces regulation originating within two different sectors. For mobile money to develop, regulations must encourage inclusiveness, while minimizing fraud and risk. The uncertainty associated with innovative industries means that regulations must be incremental and proportional. Kenya's initial success with mobile money was arguably based on a virtual absence of formal regulation in favor of industry-government engagement (World Bank 2010). However, since mobile money services manage the limited capital of the poor, caution is essential (USAID 2010).

Successful regulation is usually marked by collaborative exchange between industry, government, and civil society. For example, regulation should allow agents outside of bank branches to handle financial transactions and develop tiered anti-money-laundering and know-your-customer (AML/KYC) requirements. To facilitate more sophisticated service offerings, ongoing regulatory development will be necessary—for example, most mobile money is regulated as "payments," "denying e-money accounts the benefit of interest payments and deposit insurance" (Ehrbeck and Tarazi 2011). In considering these new regulatory issues, protection against fraud and failure, including regular monitoring by financial regulators, is essential. But it is also important to remember the goal is to find ways to provide society's poor with financial services, and often mobile is the most promising way.[8]

Existing status of finance and mobile industries

Mobile money is by no means the only instrument for extending access to finance to the poor; cooperatives, savings and loans groups, and even ATMs (automated teller machines) are popular throughout the developing world. Among the factors that will determine whether mobile money will succeed is the extent to which alternative options are accessible and desirable. In places with sophisticated financial or mobile industries, the commitment of leading firms to mobile money can do much to drive adoption of the service, but already-existing alternatives or a limited market size can limit the economies of scale necessary for mobile money to succeed. On the other hand, too low a volume of existing financial services can be a detriment for mobile money because cash agents need a way to manage their liquidity (such as traveling to bank branches, for example). In short, mobile money is one part of the solution that also requires other forms of infrastructure and resources (box 4. 3).

User perceptions, behavior, and skills

The success of mobile money also rests on various factors relating to end users. There may be considerable distrust of the formal financial services, or people may be uneasy about parting with their cash. Mobile money operations need to create a clear and trustworthy value proposition that fits within social and cultural practices. For example, mobile phones are widely available, but they are not universal, and many people in the developing world share or rent phones. Designing mobile money requires a careful understanding of these diverse interactions.

Emerging issues in mobile money

Mobile money is a fast-moving and wide-ranging industry, but as it matures and evolves, several emerging issues are worth increased attention. This section flags these issues as a first step toward finding longer-term solutions.

Technological issues

It was technological change—in the form of less expensive phones and expanded network coverage—that made mobile money feasible. As mobile telephony continues to evolve toward more sophisticated devices and services, the range of feasible mobile money applications will continue to expand.[9] Over the coming years, three technological developments will have a significant impact on mobile money: the rise of smartphones, near field communications, and biometrics.

Smartphones. Over the coming years, smartphones will become more widespread in developing markets. The relatively well-off and young individuals who will adopt them first will serve as important trendsetters, but adoption will eventually become more widespread. Already, in Kenya, Huawei is offering an Android-powered smartphone for under $100, and when smartphones begin to be sold on the second- and third-hand market, they will be even more widely accessible. The enhanced capabilities of smartphones will mean that mobile money applications will move beyond channels closely controlled by the mobile operators to platforms that are more open to competition (although SMS and USSD functionality will remain important for reaching a broader base of customers). Because smartphones serve as a gateway to the internet, a broader range of applications will become available, enhancing the need for interoperability. These changes will be accompanied by opportunities, such as the chance to use graphical interfaces with illiterate populations, and challenges, such as the growth in data traffic and increased burden on network capacities. Smartphones will also drive home the importance of device-makers to mobile money.

Near field communications. Near field communications (NFC) is a technology that allows devices to communicate through mere proximity, usually by waving a specially equipped phone or card near a receiving device, as opposed to having to physically swipe it. NFC could serve to make transactions more efficient and secure by reducing errors, such as those that arise from mistyped numbers. In the coming years, more phones will be equipped with NFC, which is expected to become more popular for financial transactions. For mobile money, this means that transactions can be completed by waving a phone near a receiver, as opposed to having to text value to a recipient. Since NFC requires a new infrastructure to receive the payments, it may be slow to grow, but as wallets become digitized onto phones, mobile money agents and businesses may start to use their own NFC-enabled smartphones to receive payments. Already

Box 4.3 Business models for mobile money

Although it has received both direct and indirect support from the public sector, to date, mobile money remains a private sector enterprise. To achieve profitability, mobile money providers have pioneered three general business models: mobile-operator-led, bank-led, and collaborative. Because operators control the mobile platform and have significant distribution capacity through their existing retail agent networks, it is logical that mobile money deployments will often be initiated by operators who may partner or collaborate with a bank. In some places, such as Pakistan, where the operator Telenor purchased a 51 percent stake in Tameer Microfinance Bank, the boundaries between the two entities may be blurred.

A variety of business models exist for mobile money. Although M-PESA popularized a model based primarily on peer-to-peer transfers, mobile money systems elsewhere are quite different. For example, in South Africa, WIZZIT is an independent mobile money provider that works over all mobile networks and that has partnered with banks to provide customers with easily accessible accounts. In Thailand, the two relatively successful mobile money operations have partnered with retailers from the start and emphasize bill payment offerings.

According to the International Finance Corporation's *Mobile Money Study*, in a given market, the business case for mobile money will be driven by those players with the strongest incentive to develop mobile money; the primary value proposition for targeted customers; and the regulation, demand, and partnership requirements. Combining these variables, the International Finance Corporation has developed mobile money demand curves that show how mobile money has different appeal in different environments.

Box figure 4.3.1 Mobile money demand curves

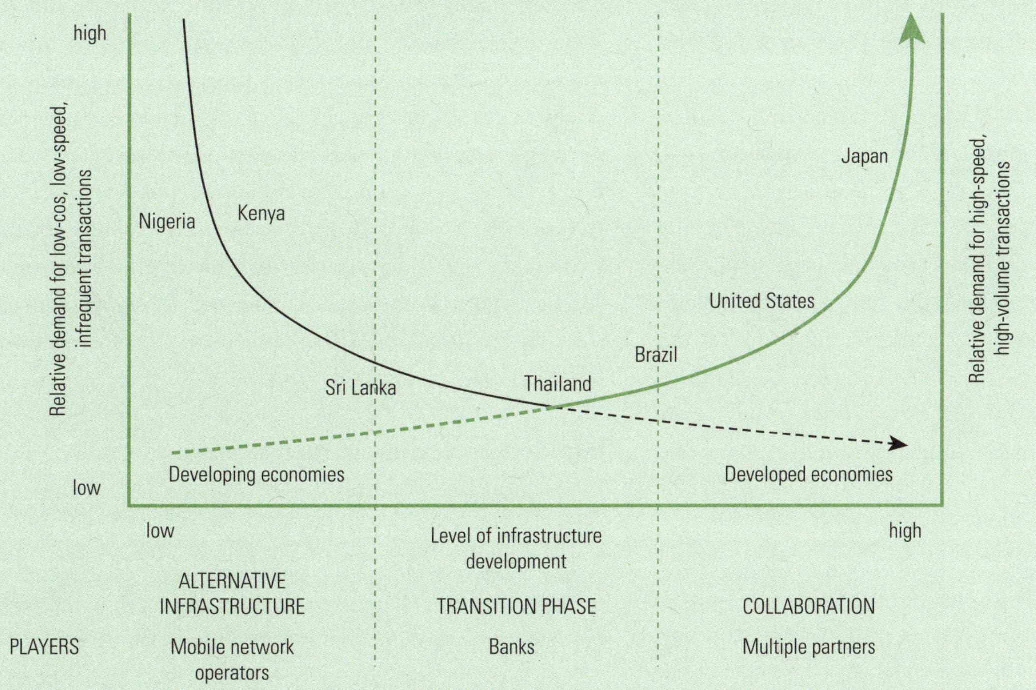

(continued next page)

Box 4.3 *(continued)*

The black curve represents mobile money demand for developing economies. As developing countries progress, financial infrastructure develops, and competition from banks, credit card companies, and other financial institutions increases. The black curve becomes dotted because demand changes from low-cost, low-speed, and infrequent to high-speed and high-volume as represented by the blue curve. The green curve starts off dotted because developed countries already have substantial financial infrastructure, thus demand for low-cost, low-speed, infrequent transactions is low. The continuum is divided into three parts: alternative infrastructure, transition phase, and collaboration. In developing economies mobile money acts as an alternative infrastructure to existing financial services; during the transition phase mobile money moves from an alternative infrastructure to a complementary one. In the collaboration phase mobile money must fully integrate with the financial infrastructure.

Source: IFC 2011.

at the start of 2012, Absa, a large South African bank, was testing NFC deployments for its payments.

Biometrics. The Center for Global Development estimates that over 450 million people in developing countries have had their biometric data recorded, and this number is expected to triple over the next five years (Gelb and Decker 2011). The most ambitious biometric program—India's Project Aadhaar, which is aiming to provide a universal ID system for all citizens, including iris scans, ten fingerprints, and a picture of each face—has been explicitly linked to financial inclusion.[10] These identification schemes are typically associated with security initiatives, but they are also seen as a means of improving delivery of cash by governments and development agencies. Many of these programs are in the early stages, and significant challenges abound. Deploying biometric systems can be very expensive, and ensuring high accuracy is often out of reach. Further, it is likely to raise political concerns given the implications for citizen privacy, so some countries are opting for less intrusive means of identification.

The changing role of agent networks

Understanding the human dynamics of a growth market is essential. Building and incentivizing networks that serve as the cash-in, cash-out point of contact, as well as customers' primary interface with the brand, is difficult and costly. Many operators have found that existing airtime resellers are useful agents, but other intermediaries (such as large-scale retail chains or post offices) are also likely candidates. This development is important because increased competition, not to mention the possibility of digital money lessening the need for cash, could reduce agents' profits: in Kenya, M-PESA agents have already seen daily profits drop from $5 to $4 (Pickens 2011).

As mobile money providers have realized the importance of agents in their business models, four interlinked problems have emerged: profitability, proximity, liquidity, and trust (Maurer, Nelms, and Rea forthcoming). The agent model is founded on the exchange of cash through a franchise model, so the profitability of agents is vital for success. If the agent network grows too quickly and saturates the market, however, mobile money agents may not have sufficient transactions to remain in business. If agents' costs for managing their cash liquidity are too high, they will also suffer. Finally, if the agents behave improperly or fail to develop relationships with their customers, the all-important client trust will not develop.

Internationalization of mobile money

International remittances are one of the largest sources of external financing in developing countries and often serve as a lifeline to the poor.[11] However, the costs of transmitting money from abroad are often large and uncertain. For example, according to World Bank data, the cost of sending money across the Tanzania-Kenya border was nearly 10 times the price of sending money from the United Kingdom to Pakistan in 2011 (figure 4.3). Easing and improving international remittances will have significant development impacts, just

Figure 4.3 The most and least expensive remittance corridors

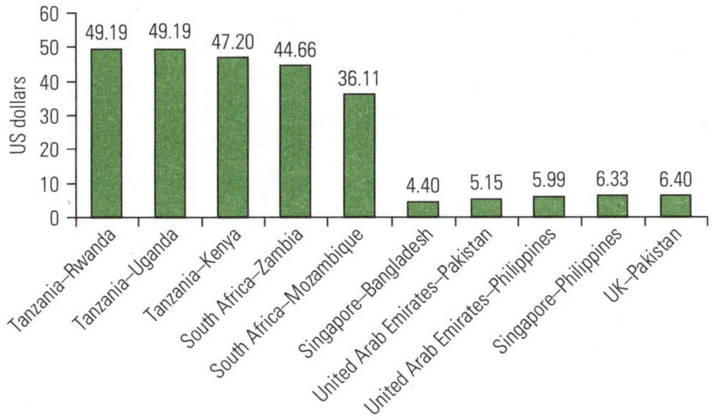

Source: World Bank (http://remittanceprices.worldbank.org).
Note: Data is for Q3 2011.

as easing remittance transactions at the domestic level has done (Maimbo, Saranga, and Strychacz 2011). Prices are high because of underdeveloped payment systems infrastructure, inappropriate legal frameworks, and the difficulty many migrants have obtaining identification in order to access finance; a lack of transparency, competition, and consumer protection has also kept prices high. Mobile money could do much to ease this situation, but regulatory assistance and the creation of the appropriate payment systems infrastructure will be required.

Policy-makers are justifiably concerned about criminal and terrorist financing, as well as the monetary policy issues arising from illegal cross-border remittance flows, but regulators need to give increased attention to easing the policy constraints on internationalization of mobile money. Because multinational negotiations are time-consuming, smaller pilot projects could be implemented to explore how to improve the regulation of international mobile money remittances. Regions with currency unions, such as parts of West Africa, or where existing infrastructure is present, such as between Mexico and the United States, may lead the way here because foreign exchange considerations have been eliminated.

Competition and interoperability

Additional regulatory attention is also needed for issues of competition and interoperability. Like other network industries, economies of scale and high barriers to entry could create uncompetitive market outcomes in the mobile money industry. In cases where a mobile money service is tied to a dominant mobile network operator (as in the case of Kenya's Safaricom, which has 68 percent of the mobile subscribers market; see Communications Commission of Kenya 2011), that operator is at an advantage in dictating the terms of the product.

The appropriate form of regulation is still emerging and will depend on context. Premature competition regulation may even stymie the growth of mobile money. As a recent World Economic Forum report noted, "initial adoption appears to be driven by constrained access to formal financial services, as opposed to well-developed institutions and competitive markets"(WEF 2011). On the other hand, waiting too long to curtail anticompetitive practices may incur social and financial costs to society.

One of the main ways to reduce mobile money market domination is through interoperability (box 4.4). Interoperability can occur at various levels: in Nigeria, where the Central Bank has been keen to avoid a dominant market player, interoperability is required at the level of the bank, the switch, and the payment channel (IFC 2011). In other countries, mobile money occurs in a "walled garden" because interoperability is not technically allowed. Consumers wishing to swap between mobile money services must have multiple SIM cards and use cash to exchange between different digital wallets (incurring time, effort, and extra fees).

Sensing a market opportunity, third-party firms are beginning to offer interoperability between different mobile money services. Because these interoperability systems are often unofficial, however, they remain tenuous. While some observers are of the opinion that consumer demand will ultimately pressure providers to allow interoperability in time (IFC 2011), others detect a potential market failure.

Mobile money operators are often reluctant to allow formal interoperability because, after investing heavily in their product, they do not want to make it easy for customers to move their money to competitors. In fact, in markets where customers frequently change mobile operators to save money, mobile money services are seen as a key way of keeping customers locked into an operator's own network. However, it has been argued that interoperability will benefit operators by expanding the pool of customers, reducing incentives to have multiple SIM cards (and thus to make calls on competing networks), and minimizing the need for retail agents to have cash, which is costly to move around between different agents (Mas 2011). Interoperability

Box 4.4 Interoperability and innovation in mobile money

The excitement surrounding successful mobile money deployments has spurred significant additional innovative activity. Surveying the landscape in Kenya, Kendall et al. (2011) found that M-PESA has emerged as a platform for a wide variety of new applications and services. Businesses have started integrating M-PESA into their activities, often to improve efficiencies and reduce costs. Other entrepreneurial ventures offer entirely new services based on the mobile phone, such as a medical savings plan from Changamka Microhealth Ltd. Finally, an entirely new category of businesses is developing; these businesses serve as intermediary bridge-builders, allowing others to integrate with mobile money. For example, Kopo Kopo is a start-up that offers smaller financial institutions and competing mobile money providers the technical means to integrate with M-PESA.

There is reason to worry that this initial flourishing is tenuous. The lack of seamless interoperability (for example, through an M-PESA application programming interface) is a common complaint, raising the costs of working with M-PESA. Because it is a proprietary service of Vodafone, the businesses building on top of the platform are highly dependent upon the choices of Vodafone and its local affiliate, Safaricom.

Source: http://www.microsave.org/research_paper/analysis-of-financial-institutions-riding-the-m-pesa-rails.

may also benefit mobile money agents who currently have to maintain redundant infrastructure for each mobile money deployment they wish to serve, as well as enhancing overall efficiency gains in the economy. But because premature interoperability may limit the market's development, regulators must approach this issue with caution.

Universal access and service

The populations least likely to feel the benefits of mobile money are societies' poorest citizens because they have the least connectivity, ability to pay, and requisite skills.

Both mobile network operators and financial institutions find it commercially infeasible to operate in remote rural areas. In the realm of telecommunications, this market failure has led to universal access and service funds that aim to connect all citizens, and the rationale for extending those to programs to mobile financial services should be considered.

Because mobile money has been driven by for-profit entities, most transactions incur a fee that many poor find difficult to pay, even if they are willing to do so because of the convenience and speed of transfer. Regulators must ensure that the mobile money industry is competitive to allow well-functioning market forces to drive prices down. As mentioned, interoperability could serve as a primary lever by which to reduce redundant costs and expand access.

Finally, many would-be mobile money users lack the necessary skills—including basic and quantitative literacy—that are necessary to fully realize the benefit of mobile money. Mobile money providers have an incentive to educate consumers about their products, and governments can support this through promoting transparent business practices.

Product innovation for meaningful financial inclusion

Today, concerns about excluding the poorest from mobile money are premature in most developing countries. Despite the runaway success of a few deployments, in the vast majority of cases, mobile money services have struggled to achieve the scale at which they might raise distributional concerns.[12] Surveying the globe, CGAP found that only one in four branchless banking services (a broad category that includes mobile money) had more than 1 million registered customers, and of those launched since 2007, only 1 in 15 has more than 250,000 active customers (Fathallah, Mino, and Pickens 2011). Furthermore, customer use of many mobile money services remains low—often only a couple of transactions a month. In many cases, the transaction fees remain too high to enable mobile money to replace cash for petty purchases.

At the moment, however, a "product gap" exists in most countries between the financial services the poor are being offered and the services they want (Morawczynski and Krepp 2011). The model so successfully pioneered by M-PESA—starting with peer-to-peer transfers—has been widely replicated but may not fit well in other contexts. For example, an extensive ATM network already meets many of the consumer financial needs in Thailand. By definition, mobile money will not have a comparative advantage in every location or for every service, so the business environment must be enabling and open to allow businesses to pioneer new forms of mobile money tailored to local circumstances.

Product innovation is also essential to realize the full potential of mobile money. Currently, only 1 in 8 branchless banking deployments offer functionality beyond basic peer-to-peer transfers and e-wallet services. Indeed, the IFC's study of mobile money in Brazil, Nigeria, Sri Lanka, and Thailand found that the most popular uses for mobile money were essentially moving money over *distance*. However, customers also want the ability to move money over *time* (in the form of savings, insurance, and credit). As argued above, simply formalizing people's finances onto the mobile platform falls short of meaningful financial inclusion—for that, the simple "additive" models of mobile money (where mobile is just another channel) is to move to "transformational" mobile money (where finance is extended to those previously unbanked, excluded populations) (Porteous 2007). While a mobile payment infrastructure is a first step, tailored products and services that enable the poor to better manage and capitalize on their assets must follow.

Mitigating the growing pains

In celebrating mobile money as a disruptive innovation, it is important to remember the second half of that phrase. The introduction of technology into communities can upset existing practices, sometimes causing stress or worse. Although humans are adaptive and generally adopt mobile money willingly, it is worth being on guard for undesirable disruptions from innovation. For example, ethnographic work from Kenya suggests that mobile money users in Nairobi who had previously traveled frequently to family in rural areas did so less often after they began to use mobile money, leading to family troubles arising from worries about their whereabouts, potential infidelity, and financial stress.

Another example might be the use of mobile money in microfinance. Many microfinance supporters believe that the social pressures exerted through face-to-face group meetings are essential for generating the high rates of repayment that make microfinance viable. If they are correct, the disintermediation created by mobile money could prove harmful to microfinance. The Gates Foundation argues that bringing together different models such as banks, co-ops, savings-led groups, and mobile money could leverage their respective strengths, instead of "creating a single synthetic model" (WEF 2011).

Finally, as mobile money matures, people are increasingly discussing the "cashless society." Although that is unlikely, mobile money may displace many uses of cash. Already, the Central Bank of Nigeria is promoting "cashless Lagos" in an effort to reduce the amount of cash circulating in the economy in favor of electronic transactions, including direct credit and debit, payment cards, internet-based services, and mobile money. The U.S. Agency for International Development, too, is arguing for adopting alternatives that are "better than cash" (USAID 2010). If this trend toward replacing cash continues, financial transactions could become uniquely identified and recorded, introducing complexities for consumer privacy. Others have suggested that the provision of money by private companies over private infrastructure risks undermining an important function of the public sector, namely, that the means of value transfer are not "owned" by anyone.

Conclusions

Many of the characteristics that make mobile money so promising—its scale and impact, its varied uses, and the novelty of its role—are also reasons why achieving these hopes is so difficult. While exciting, the success of a few mobile money deployments should not shelter the fact that those examples remain the exception, not the rule. With this caution in mind, governments, donors, and industry have good reason to support the creation of vibrant mobile money services that include the world's poor in financial markets and allow them to manage and use their own money. Although far from the only mechanism, mobile is certainly one of the most powerful means by which to realize this promise.

Notes

1. *Vodafone Annual Report 2011* (http://www.vodafone.com/content/annualreport/annual_report11/business-review/strategy-in-action/focus-on-key-areas-of-growth-potential/emerging-markets.html).
2. Mobile money can be considered a subsector of a wider industry—branchless banking that uses a variety of methods and technology to extend financial access.
3. An example of the latter is the World Bank–funded initiative to use mobile phones to compensate ex-combatants in the Democratic Republic of Congo; see http://www.mdrp.org/PDFs/In_Focus_3.pdf.
4. For additional information on the adoption and impact of mobile money, see Institute for Money, Technology and Financial Inclusion (http://www.imtfi.uci.edu) and the Financial Services Assessment project (http://www.fsassessment.umd.edu/).
5. In reality, the savings are likely even greater for mobile money because this study grouped mobile money with other methods of branchless banking and did not account for the savings arising from the reduced travel.
6. At the same time, anecdotal evidence suggests that the need to go to an agent to cash in or cash out can advertise a person's relative wealth, perhaps increasing risk.
7. Despite the justifiable promise of such approaches, a word of caution is worthwhile. Innovation implies the possibility of failure, and given the precarious situations of the poor, entities wishing to improve the poor's financial situation through mobile money must take every caution to understand the risk involved; see, for example, USAID 2010. As is evident with other industries working with the poor, changing incentives and policies can result in disaster. Furthermore, creating dependencies on private infrastructure can be disastrous in the event of bankruptcy or other disruptions.
8. For additional information, see the regulatory resources from the Consultative Group to Assist the Poor and Chatain et al. (2011).
9. Although device innovation gets the majority of attention, larger developments, such as cloud computing or network standard negotiations, could serve as the underlying infrastructure for mobile money.
10. However, high-profile disputes around the program in December 2011 emphasized the clashes that are likely to emerge with large-scale biometric programs.
11. Although international remittances are expected to become increasingly important to the mobile money landscape, it is essential not to lose sight of the opportunity presented in the market for domestic remittances. Kendall and Maurer (2012) document nationally representative surveys of eight African countries and "a vast and untapped domestic payments market" with 64 million people in the surveyed countries not using any formal remittance instrument.
12. Of course, this is not to say that these subscale examples will not, in the future, raise those concerns. Indeed the very purpose of this section is to consider that possibility.

References

Aker, Jenny, Rachid Boumnijel, Amanda McClelland, and Niall Tierney. 2011. "Zap It to Me: The Short-Term Impacts of a Mobile Cash Transfer Program." Working paper 268, Center for Global Development, Washington, DC.

Chatain, Pierre-Laurent, Andrew Zerzan, Wameek Noor, Najah Dannaoui, and Louis de Koker. 2011. *Protecting Mobile Money against Financial Crimes: Global Policy Challenges and Solutions*. Washington, DC: World Bank.

Communications Commission of Kenya. 2011. "Quarterly Sector Statistics Report: Q3 2011." Nairobi.

Demirgüç-Kunt, Asli, Thorsten Beck, and Patrick Honohan. 2008. *Finance for All?: Policies and Pitfalls in Expanding Access*. Washington, DC: World Bank.

Ehrbeck, Tilman, and Michael Tarazi. 2011. "Putting the Banking in Branchless Banking: The Case for Interest-Bearing and Insured E-Money Savings Accounts." *The Mobile Financial Services Development Report*, 37–42. Washington, DC: World Economic Forum.

Fathallah, Sarah, Toru Mino, and Mark Pickens. 2011. "The Case for More Product Innovation in Mobile Money and Branchless Banking." Consultative Group to Assist the Poor, Web log post. October 14. http://technology.cgap.org/2011/10/14/the-case-for-more-product-innovation-in-mobile-money-and-branchless-banking/.

Gelb, Alan, and Caroline Decker. 2011. "Cash at Your Fingertips: Biometric Technology for Transfers in Developing and Resource-Rich Countries." Working paper 253, Center for Global Development, Washington, DC.

Gencer, Menekse. 2011. "The Mobile Money Movement: Catalyst to Jump-Start Emerging Markets." *Innovations: Technology, Governance, Globalization* 6, no. 1: 101–17.

GSMA Mobile Money Tracker. 2012. "Global Mobile Money Deployment Tracker." Available at http://www.wirelessintelligence.com/mobile-money.

IFC (International Finance Corporation). 2011. *Mobile Money Study 2011*. Washington, DC. http://www.ifc.org/ifcext/globalfm.nsf/Content/Mobile+Money+Study+2011.

Kendall, Jake, and Bill Maurer. 2012. "Understanding Payment Behavior of African Households: A Vast and Untapped Market." http://pymnts.com/commentary/Tips-for-2012-Understanding-Payment-Behavior-of-African-Households-A-Vast-and-Untapped-Market/.

Kendall, Jake, Bill Maurer, Phillip Machoka, and Clara Veniard. 2011. "An Emerging Platform: From Money Transfer System to Mobile Money Ecosystem." *Innovations: Technology, Governance, Globalization* 6, no. 4: 49–65.

Lochan, Rajiv, Ignacio Mas, Daniel Radcliffe, Supriyo Sinha, and Naveen Tahilyani. 2010. "The Benefits to Government of Connecting Low-Income Households to an E-Payment System: An Analysis in India."*Lydian Payments Journal* no. 2 (December). http://ssrn.com/abstract=1725103.

Maimbo, Samuel, Tania Saranga, and Nicholas Strychacz. 2011. "Facilitating Cross-Border Mobile Banking in Southern Africa." Africa Trade Policy Note 1, World Bank, Washington, DC.

Mas, Ignacio. 2011. "Three Enemies and a Silver Bullet." Web log post. *Mobile Money for Unbanked*. GSM Association (March 9). http://mmublog.org/blog/three-enemies-and-a-silver-bullet/.

Mas, Ignacio, and Daniel Radcliffe. 2010. "Mobile Payments Go Viral: M-PESA in Kenya." In the "Yes Africa Can: Success Stories from a Dynamic Continent" series. World Bank, Washington, DC (March). http://ssrn.com/abstract=1593388.

Maurer, Bill, Taylor Nelms, and Stephen Rea. Forthcoming. "Bridges to Cash: Channeling Agency in Mobile Money." *Journal of the Royal Anthropological Institute*.

McKay, Claudia, and Mark Pickens. 2010. "Branchless Banking 2010: Who's Served? At What Price? What's Next?" Focus Note 66, Consultative Group to Assist the Poor, Washington, DC.

Morawczynski, Olga. 2009. "Examining the Usage and Impact of Transformational M-Banking in Kenya." In *Internationalization, Design and Global Development*, ed. Nurgy Aykin, 495–504. Berlin: Springer.

Morawczynski, Olga, and Sean Krepp. 2011. "Saving on the Mobile: Developing Innovative Financial Services to Suit Poor Users." *The Mobile Financial Services Development Report*, 51–58. Washington, DC: World Economic Forum.

Pickens, Mark. 2011. "CGAP Releases Agent Management Toolkit." CGAP Web log post (February). http://technology.cgap.org/2011/02/10/cgap-releases-agent-management-toolkit/.

Porteous, David. 2007. "Just How Tranformational Is M-Banking?" FinMark Trust.

Sen, Soham, and Vikas Choudhary. 2011. "ICT Applications for Agricultural Risk Management." *ICT in Agriculture Sourcebook*, 259–84. Washington, DC: World Bank. http://www.ictinagriculture.org/ictinag/sites/ictinagriculture.org/files/final_Module11.pdf.

Stuart, Guy, and Monique Cohen. 2011. *Cash-In, Cash-Out: The Role of M-PESA in the Lives of Low-Income People*. Financial Services Assessment.

Suri, Tavneet, and Billy Jack. 2011. "Risk Sharing and Transaction Costs: Evidence from Kenya's Mobile Money Revolution." Working paper. Massachusetts Institute of Technology, Cambridge, MA, and Georgetown University, Washington, DC. http://www.mit.edu/~tavneet/Jack_Suri.pdf.

USAID (U.S. Agency for International Development). 2010. *Mobile Financial Services Risk Matrix*. Washington, DC. http://www1.ifc.org/wps/wcm/connect/14d0748049585fbaa0aab519583b6d16/Tool+10.14.+USAID+MFS+Risk+Matrix.pdf?MOD=AJPERES.

WEF (World Economic Forum). 2011. *Mobile Financial Services Development Report*. http://www.weforum.org/issues/mobile-financial-services-development.

World Bank. 2010. "At the Tipping Point? The Implications of Kenya's ICT Revolution." Kenya Economic Update, Edition 3, Washington, DC (December). http://siteresources.worldbank.org/KENYAEXTN/Resources/KEU-Dec_2010_Powerpoint.pdf.

Chapter 5

Mobile Entrepreneurship and Employment

Maja Andjelkovic and Saori Imaizumi

Given its strong recent growth, the global mobile industry is now a major source of employment opportunities on both the supply and demand side. Employment opportunities in the mobile industry can be categorized into direct jobs and indirect jobs, with a diverse labor force supplying each category. Direct jobs are created by mobile operators and manufacturers in professions that range from engineers to managers to sales support staff. The International Telecommunication Union (ITU) estimates that around 1.5 million people are directly employed in the industry worldwide (ITU 2011). The total number of jobs fitting this narrow "direct" description may continue to grow slowly but may begin to decline as the industry becomes commoditized. Indirect jobs, however, show strong potential for new growth, in professions broadly associated with the industry such as application development, content provision, and call center operations. Indirect jobs can be created by mobile operators and manufacturers as well as by third-party content and device producers, including entrepreneurs. In some emerging markets, outsourcing of mobile content development can also create significant numbers of indirect jobs. In India alone, the mobile industry is expected to generate around 7 million indirect jobs during 2012 (COAI 2011).

This report argues that faster mobile networks and more capable smartphones make mobile communications a platform for innovation across different sectors (such as health, agriculture, and financial services), supporting overall employment numbers in an economy. The greatest potential for employment growth therefore derives from demand for services enabled by mobile phones. For many entrepreneurs in developing countries and rural areas, a mobile device is a tool not only for contacting customers and accessing the internet, but also for making financial transactions, establishing a client database, or coordinating just-in-time supply-chain deliveries. Such critical business functions can enable small firms to thrive in locations where accessing markets or selling new products would otherwise be impossible. It is difficult to estimate the number of people establishing new companies or the employment generated as small and microenterprises expand, but mobile phones undoubtedly contribute to this process.

It is also difficult to say with certainty how much the mobile communication sector has contributed to employment and entrepreneurship to date, because no global count exists. It seems clear that the sector is a net generator of jobs, however, even though it can occasionally eliminate employment opportunities. For example:

- In the United States alone, the mobile app industry provided an estimated 466,000 jobs in 2011 with annual growth rates of up to 45 percent from 2010 to 2011 (TechNet 2012).

- In Canada a large proportion of mobile apps are used to deliver games to handheld devices. The gaming sector is expected to expand by 17 percent over the next two years, driven by proliferating mobile broadband access; as a result, mobile games are likely to generate a greater number of employment opportunities. Of the 348 gaming companies in the country, 77 percent expect to hire new graduates in 2013 (Secor Consulting 2011).
- Mobile money schemes have generally proved to be net generators of jobs. Safaricom's M-PESA system supports 23,000 jobs for agents in Kenya alone.[1] Airtel Kenya, the second-biggest mobile operator, plans to recruit some 25,000 agents for its mobile money service, Airtel Money.[2]
- By boosting access to information about market demand and prices, mobile phones can also improve conditions for entrepreneurship.[3] A number of studies have shown that cell phones make entrepreneurial ventures less risky, mainly by reducing information search costs.[4]

This chapter showcases some of the mechanisms by which the mobile sector can support entrepreneurship and job creation, with the aim of informing policy-makers, investors, and entrepreneurs themselves. Some of these approaches share similarities with traditional donor initiatives, but many are novel ideas for which the "proof of concept" has been demonstrated only recently. In an industry evolving as quickly as the mobile sector is today, it is vital to tailor support to the local circumstances and to evaluate impact regularly. As a framework for entrepreneurial activities, the chapter examines open innovation, and considers one particular way of supporting entrepreneurial activity in the mobile industry, namely, specialized business incubators, or mobile labs. The chapter reviews mobile microwork and the potential of the virtual economy, and then reverses the lens to consider mobile phones as a tool for job seekers. Finally, it summarizes suggestions to support entrepreneurship and job creation in the mobile industry.

Open innovation and mobile entrepreneurship

The rapid innovation in the mobile sector is creating uncertainty and disruptive technological change, while lowering barriers to entry and generating opportunities for small and young firms and entrepreneurs.[5] The rise of entrepreneurship in the mobile industry is therefore unsurprising. The lack of vertical integration and direct competition between operators, handset manufacturers, and content providers has resulted in a complex environment of different technological standards and innovation in business models, with ample space for growing new businesses. New information-sharing and collaboration practices that transcend the closed communication channels are characteristic of newly establishing markets. Rapid information flow dynamics were present in the early stages of other high-tech industries, including the semiconductor industry in the 1970s, PC software in the 1980s, and the internet in the 1990s.

In today's open innovation model, partners, customers, researchers, and even competitors are finding new ways to collaborate in the product development process. The paradigm of open innovation assumes that firms can, and should, use external as well as internal ideas and paths to market as they seek to advance their technology.[6] Today, in many sectors there is a need to complement internally oriented, centralized approaches to research and development (R&D) with more open, networked methods, because useful knowledge has become more dispersed (both within and outside firms), while the speed of doing business has increased. Collaborative approaches to innovation also offer new ways to create value, especially in fast-changing industries. To capitalize on fresh opportunities, innovators must find ways to integrate their ideas, expertise, and skills with those of others outside the organization to deliver the result to the marketplace (Chesbrough 2003; Aldrich and Zimmer 1986; Teece and Ballinger 1987).

One of the most promising areas for entrepreneurship is in mobile software applications, where the barriers to market entry for individual developers and small and medium enterprises (SMEs) are generally low. Mobile apps can be written by programmers working for device manufacturers, network operators, content providers, or software development firms, and they can also be created directly by individual freelance professionals. In emerging, as in more developed markets, there is no "natural" place where applications originate; for the most part, network operators and device manufacturers provide their own apps, with other apps supplied to market directly by developers. This room for independence allows developers who also have entrepreneurial ambitions to start their own apps-based businesses. Many SMEs and individual entrepreneurs in the developing world offer their services at competitive rates compared with those in rich countries, but the vast array of choices of platforms and distribution models can be challenging to navigate. For example, most apps for simple, low-end phones are written for SMS (short message service), while apps for midrange devices often rely on mobile internet access and may be written in Java or PHP programming languages. Smartphone applications can be written for the proprietary Apple iOS, BlackBerry, or Windows platforms, or for the open source Android, among other options. According to one survey, in 2011 developers used an average of 3.2 platforms concurrently, which was a 15 percent increase over 2012 (Vision Mobile 2011). While this growth may be interpreted as an indication of low barriers to entry, it is, rather, a sign of the necessity for developers to hone skills in multiple platforms, because no one knows which of these platforms—if any—will become dominant in the future. In other words, developers choose to diversify their skills because the market, at the moment, demands variety and flexibility. Marketing and distributing dilemmas are especially challenging: app stores based on operating systems compete with those managed by handset manufacturers and major global brands, and programmers must decide which store, or stores, will be most effective as a delivery vehicle of apps to their potential customers.

Informal industry networks for mobile entrepreneurship

The lack of formal information channels and uncertainty mean that mobile entrepreneurs must keep up-to-date with changes in standards and industry developments, resulting in frequent socializing and informal networking between mobile entrepreneurs and developers. Informal social networks, consisting of acquaintances, mentors, investors as well as other mobile entrepreneurs, or peers, serve three distinct purposes in the development of new ventures—discovering opportunities, securing new resources, and obtaining legitimacy—all of which are necessary for the survival of a young firm (Elfring and Hulsink 2003). Entrepreneurs may have initiative, an appetite for risk, creative ideas, and business acumen, but they may also need complementary resources to produce and deliver their goods or services. Social networks are important sources of support and knowledge and can provide access to distribution channels, capital, skills, and labor to start new business activities (Greve and Salaff 2003).

One way to support jobs created through entrepreneurship in an era of open innovation is through structured social networking events that can help define business opportunities, identify talent, and draw investment into the mobile sector in emerging markets. Networking events can also graft best practice lessons from the ground back into the development and donor communities. An early example of an informal social networking organization is Mobile Monday (www.mobilemonday.net), an open community platform of mobile entrepreneurs, developers, investors, and industry enthusiasts. Mobile Monday fosters business opportunities through live networking events. It provides a space for entrepreneurs to demonstrate new products, share ideas, and discuss trends from local and global markets. Founded in 2000, in Helsinki, the community has grown to more than 100 city chapters and is managed by 300 volunteers around the world.[7]

More narrowly focused organizations, such as Google Technology User Groups (GTUGs) (www.gtugs.org), cater to participants interested in a particular developer technology. These groups provide training for developers using the open Android mobile platform, followed by minimally structured networking events.[8] GTUGs vary in format, from a dozen people who may get together to watch a corporate video, to large groups involved in product demos, lectures, and competitions dubbed "code sprints" and "hackathons." Smaller, local networks have also been formed in many cities. For instance, Nairobi-based AkiraChix provides networking and training for women and girls unfamiliar with software design. It cultivates the careers of young developers of both genders by providing training in programming and mobile application development (box 5.1). In Nepal Young Innovations, the group behind the Kathmandu-based organization Mobile Nepal, regularly hosts "bar camps"—open conferences where entrepreneurs and developers give presentations and provide feedback.[9] In Georgia the business social network "mTbilisi" promotes corporate partnerships, coordinates online and in-person events focused on incubating mobile start-ups, and provides a space for testing new ideas and designs. This project aims to bridge the gap between online and mobile application concepts—such as eCommerce applications, virtual guides, informational bases, or search engines.[10]

Features and dynamics of informal networks of entrepreneurs

Mobile developers and entrepreneurs interviewed for this report identified both informal gatherings and more

Box 5.1 AkiraChix

AkiraChix hosts informal gatherings, workshops, and competitions for mobile developers and entrepreneurs in Kenya. In 2011 AkiraChix and iHub, a community of technology innovators in Nairobi, partnered to host AppCircus, a traveling showcase of the most creative and innovative apps. Twelve finalists were given the chance to pitch their mobile apps to an audience. The overall winner was Msema Kweli, who developed a mobile app that helps keep track of Community Development Fund projects. It uses data made available by the Kenyan government through its Open Data Initiative to track spending and progress by constituency. Users can then report and comment on different projects. The app can be adapted to increase transparency and accountability for any community or development project. The first runner-up, Martin Kasomo, developed Hewa App, a mobile cloud-based music marketing and distribution platform for Kenyan artists and record labels to sell their music online. The jury highlighted the ease of use of this app and its attempt to tackle the problem of local music distribution in Kenya. Third place was claimed by Bernard Adongo of NikoHapa for a customer engagement and promotions application for businesses, which relies on news-sharing and geolocation tools to build a customer community.

Source: http://akirachix.com.

structured social networking activities (as mentioned above) as helpful to innovation and entrepreneurship in the development of mobile applications. Respondents from Kenya, Nepal, and Uganda indicated that they are initially cautious about sharing ideas and information but that they freely provide lessons and support once they are established and have begun implementing their business ideas.[11] Entrepreneurs may first test options for starting their own business within a circle of carefully selected contacts. As a second step, during the planning stage, entrepreneurs often mobilize a large, informal network of friends, colleagues, mentors, and other acquaintances, since they may not know who exactly can help them (Berglund 2007). Information exchange in informal environments carries risks for fragile new businesses, including the threat of idea theft: promising ideas risk being taken over not only by peers and direct competitors but also by larger companies, which, instead of hiring the idea generator to complete the work, may assign an internal team to develop the project in-house. To mediate such risks, once the project design stage has begun, entrepreneurs choose smaller, trusted groups from a wider social network to form product development teams. Entrepreneurs recognize that without a plan for execution, an idea is irrelevant. Many individuals may recognize demand for a specific product or service; however, it is execution that "makes or breaks" an app. Developers and entrepreneurs tend to rely on their informal networks to identify potential partners, mentors, or peers who can be consulted in confidence and relied on to help move a viable product from mind toward market. Once collaboration is under way, individuals may come back to the network to talk about their example of successful partnership and to share challenges. In other words, the interaction pattern seems to circle from a group setting to one-on-one interaction and back to the wider network.

The rewards of networking usually greatly outweigh the risks. Many mobile entrepreneurs note that collaboration is essential, because few applications can be successfully brought to market by a single developer, let alone expanded to additional platforms and maintained afterward. Market information, idea validation, and partnerships are among the most frequently cited rewards of participation in social networks, according to more than 80 percent of participants in our survey (figure 5.1). Access to finance (including small amounts raised from friends and family) and mentorship opportunities were other important rewards, listed by more than 60 percent of respondents. Finally, marketing support is another benefit of participating in informal peer groups. On the risk side, more than 35 percent of respondents are

Figure 5.1 Rewards and risks from entrepreneur participation in social networks

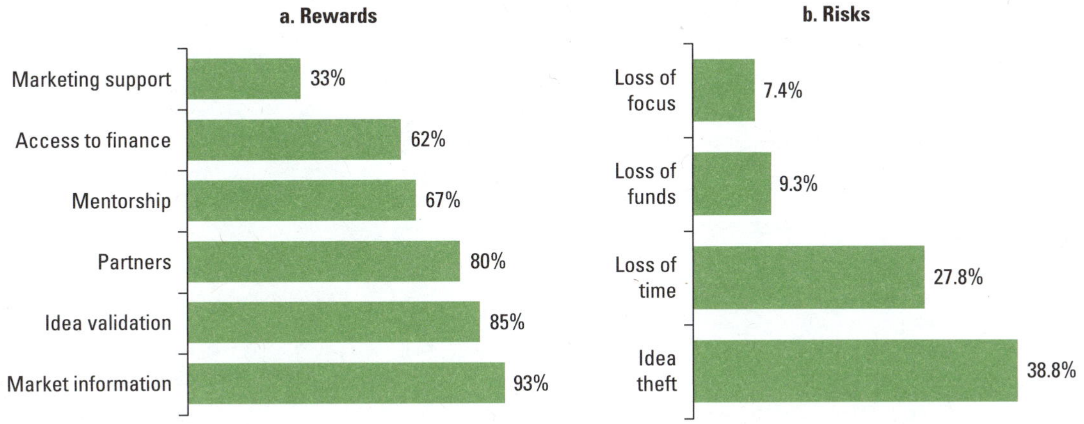

Source: Author interviews.
Note: Risks and rewards as perceived by mobile industry entrepreneurs, based on a sample size of 54, split between Kenya, Nepal, and Uganda.

concerned about idea theft, in particular by more established businesses; however, even these entrepreneurs recognize the necessity of vetting or validating ideas with their peers and consider the risk of idea theft to be tolerable. Loss of time, funds, and focus are concerns for 28 percent, 9 percent, and 7 percent of respondents, respectively.

The marketing of mobile applications is typically the biggest expense and also the activity about which developers are often the least enthusiastic. Developers often rely on partners or enterprise customers for all aspects of marketing, which, if executed poorly, can stall the adoption of an otherwise successful app. For small teams of developers working on "mass market" apps, marketing strategies can include dissemination and awareness-raising through word of mouth, Twitter, Facebook, email, and SMS. Successful incubators, such as iHub Nairobi,[12] act as useful "amplifiers" of marketing efforts, because local media and investors tend to follow their announcements and activities closely.

Participants report that small groups (from 4 or 5 people up to 20) are the most helpful form of networking in discussing ideas and execution. Larger groups can be too impersonal or too strongly driven by formal presentations. As a result, many organizers (including Mobile Monday Kampala[13]) use breakout groups to ensure more meaningful conversations at their events. Network sponsors can help strengthen social networks by attracting well-known figures or VIPs to the meetings, as much as by direct financial support. Attracting respected experts to address attendees can be helpful in drawing out participants and broadening the number and scope of conversations within the network.

Mobile incubators

While the informal networks of mobile entrepreneurs and developers described above can provide many resources, including knowledge and connections to investors, demand for more formal, hands-on learning spaces and supportive office environments is also strong.[14] A typical business incubator may house 5 to 20 start-up companies in a shared space offering common office equipment and conference facilities. Most employ a resident manager who coordinates business assistance, training, and other services, such as business plan development; accounting, legal, and financial advice; coaching and help in approaching investors; marketing; and shared services, such as administrative support. Once a client or resident business is deemed financially viable, it moves its operations outside the incubator, enters the market, hires new staff, and expands its contribution to the economy (Lewis 2001).

A number of incubators, or "labs," focused on mobile entrepreneurs have been established in emerging markets, including Grameen Foundation's AppLabs in Uganda and Indonesia, and *info*Dev's regional mobile applications laboratories, or "mLabs." (figure 5.2; box 5.2). Launched over the past five years, these labs are still in an experimental stage, but they offer several early lessons. Mobile labs facilitate

Figure 5.2 *info*Dev's network of mLabs

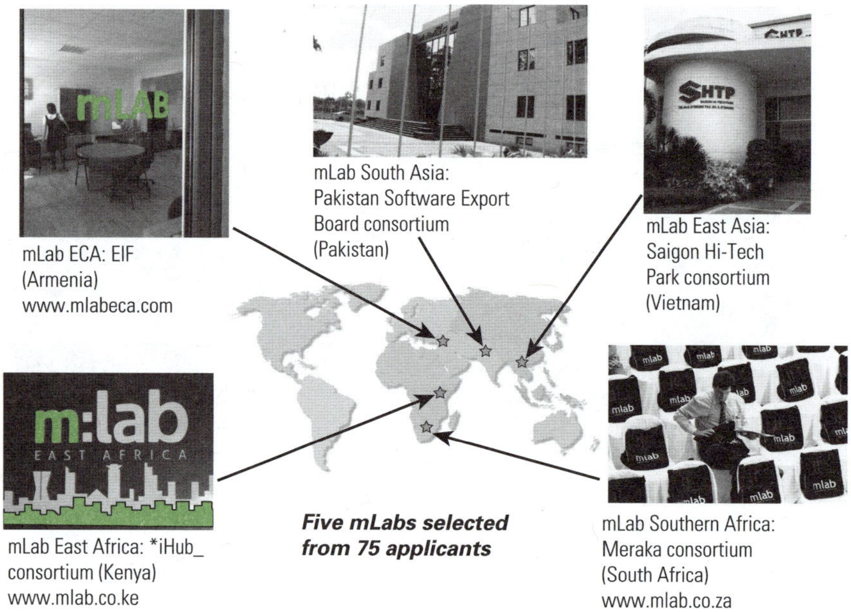

Source: *info*Dev

Box 5.2 *info*Dev's mLabs and mHubs

In response to demand by local mobile entrepreneurs, the World Bank Group's *info*Dev program, in collaboration with the Government of Finland and Nokia, has established a network of five mobile application labs, or mLabs, and eight mobile social networking hubs, or mHubs. In Armenia, Azerbaijan, Georgia, Moldova, Kenya, Tanzania, South Africa, Uganda, Nepal, Pakistan, and Vietnam, mLabs and mHubs facilitate demand-driven innovation by grassroots entrepreneurs, so breakthrough low-cost, high-value applications can be developed. Each mLab is a technology-neutral physical space with testing facilities for developing the technical skills and business sense needed to build scalable mobile solutions into thriving businesses that address social needs. As well as providing state-of-the-art equipment, the labs offer technical training and workshops, and they connect developers and entrepreneurs with potential investors, experts, and public sector leaders. The labs are complemented by eight mHubs, which focus on bringing together various stakeholder communities in the mobile industry and providing advice, mentorship, idea and product development competitions, and access to investors through regular informal events and conferences. Both the mLabs and mHubs are run and used by local communities working to increase the competitiveness of enterprises in mobile content and applications and are part of a wider mobile innovation program, seeking to develop talent and produce successful companies with strong growth potential.

Sources: Examples of mLab and mHub activities can be found on select websites: mlab.co.ke | mlab.co.za | mobilenepal.net | akirachix.com.

demand-driven innovation by grassroots entrepreneurs, so breakthrough low-cost, high-value apps can be brought to market. Although specialized incubators are not unusual, those focusing solely on mobile app businesses are a recent phenomenon. That presents both a challenge and an opportunity, because lessons and best practices can be borrowed from related ventures, but ample opportunity exists to develop new formats tailored to the mobile sector. Ideally, mobile labs should be designed in a way that enables them to remain open and adaptable to their environment, so lessons can be incorporated continuously throughout the lab's existence.

Mobile lab managers identify their members' greatest needs as start-up capital and opportunities to network with mobile ecosystem players and other technology entrepreneurs. In addition, many mobile app entrepreneurs need specialized business training to understand the mobile ecosystem, market demand, or both. Further, because mobile app development needs a special set of technical abilities, many app developers need specialized technical training to continuously update their programming skills. Networking with local business professionals can enhance the incubation experience, providing entrepreneurs with highly customized advice that can accelerate the growth of their business. Mobile labs can offer a wide range of services, including "business accelerators"—intensive training and direct mentoring meant to quickly increase the value of a company and to help management develop a viable growth strategy. In poor or remote areas, virtual incubation—business training, advice, mentorship, and networking over a distance and without a dedicated workspace, as well as links with knowledgeable diaspora members—can be particularly helpful. The service offerings implemented by any given lab or incubator should reflect the environment and characteristics of the region where it is located. These characteristics often dictate the services that can be offered and the most likely mix of revenue streams. Incubators may be instituted as nonprofit organizations, for-profit companies (usually when they do not receive grant funding), or foundations. The business models and legislation of a given country usually dictate the most advantageous status for an incubator. Regardless of the regulatory environment, however, partners are essential to the ultimate success of a mobile incubator through their support of the organizations' development and distribution efforts. That is because, ideally, the incubator sits near or at the center of the value chain for mobile content creation and, in its role as an integrator, brokers essential partnerships with all key mobile ecosystem players (Vital Wave Consulting 2011).

Even in developed countries, mobile incubators are a recent phenomenon. In the United States the prominent mobile incubator Tandem Entrepreneurs was launched in 2011 to enable a group of experienced entrepreneurs to provide resources and mentorship to early-stage mobile start-ups. The incubator also offers each resident company seed funds and a collaborative workspace in Silicon Valley.[15] As mobile services become more sophisticated and widespread, the potential of mobile entrepreneurs to contribute to the economies of both developed and developing countries is likely to grow. Most businesses based around mobile app technology are at an early stage of development but may offer enormous employment and economic potential, similar to that of the software industry in the early 1980s. Supporting networking and incubation of entrepreneurs in this space is an important way to ensure such potential is tapped.

Mobile microwork

New employment opportunities in mobile communications are not restricted to highly skilled developers and entrepreneurs but can also extend to a relatively low-skilled labor force. "Microwork" refers to small digital tasks (such as transcribing hand-written text or determining whether two photos show the same building). Typically, such tasks can be completed in a few seconds by a person without special skills or training, but they cannot be readily automated. Workers are paid small amounts of money for completing each task. For such work to be broadly accessible to workers from developing countries, it should be performed via mobile devices as well as PCs. The mobile microwork market is still very much in its infancy, however (box 5.3).

Currently, microwork employs more than 100,000 people and contributes to a virtual global economy valued at $3 billion a year, according to a recent *info*Dev study (Lehdonvirta and Ernkvist 2011). To understand how a mobile user may be able to tap into additional sources of income, consider, for instance, the growing gaming industry, which enables online gamers to become microworkers compensated in virtual game currency that can often be cashed in for real monetary gains. Today's online game

Box 5.3 Mobile microwork: JANA

JANA, a service developed originally by Nathan Eagle as TxtEagle, relies on SMS to connect users to a wide range of more complex media and communication technologies. It also acts as an aggregator of microwork tasks and assigns workers tasks that can be completed on a mobile phone, including, for example, data entry, translations, and transcriptions. With the help of partnerships with 220 mobile operators in 80 countries, it then compensates workers with mobile money or airtime minutes.

Source: www.jana.com.

market is very competitive, with monthly subscription fees for some games nearing zero. Instead of charging players, leading online game producers can earn revenue by selling virtual currency to players. The players buy virtual goods and value-added services inside the game using virtual currency. Third parties—monetization service providers—facilitate the exchange of real money into virtual funds. Two such monetization services providers, Gambit and TrialPay, allow gamers to pay for purchases by carrying out micro-tasks. After completing assigned microwork, the player is paid in virtual currency, which can be traded for virtual goods or converted to real money.

Because virtual workers come from a global pool, international microwork aggregators must be able to provide compensation in foreign countries. This is complex in any market, but it is especially challenging in developing regions, where traditional financial infrastructure can be limited. Mobile money schemes, which are more advanced in developing than developed countries, provide a viable option for payment for microwork via mobile phones (box 5.4).

Leila Chirayath Janah of Samasource works with refugees in Dadaab, Kenya, who are paid for performing small tasks for Samasource's clients, including Google and CISCO. She suggests that microwork may be a natural complement to microfinance, noting that, whereas microfinance can enable entrepreneurs to operate small businesses serving local needs (such as producing chickens on a small farm), microwork allows them to reach beyond the local market and develop a variety of skills. Samasource now facilitates virtual assistance via microwork, including for clients from the developed world. Janah also notes that, while typical microwork tasks are not necessarily intellectually stimulating, they encourage interaction with technology and access to global online social networks, which is "vital to having a voice in the modern world."[16]

Although third-party gaming services have existed for more than a decade, the general microwork industry remains relatively new and undeveloped, with mobile microwork in an even earlier stage of development. And despite the relative simplicity of tasks required, microwork faces the challenge of breaking down larger business procedures or analytical problems into smaller components that can be executed by microworkers. This is a technical, as well as procedural, problem that warrants further research by the development and business communities alike. A number of new ventures are considering potential solutions, in the hope of entering a market that is likely to grow into billions of dollars a year over the next five years. Easier-to-use interfaces and better distribution channels are also needed, if mobile microwork is to prove a viable employment option for some of the poorest and least educated workers in developing countries (Lehdonvirta and Ernkvist 2011).

Mobiles and recruitment

In many countries, coordination and information failures arise between the demand and supply sides of the labor market. While the demand for employment exists both in the formal and informal sectors, information on recruitment is often limited to those with a strong social network or access to job postings via the internet. The mobile phone can extend this access to those job providers or job seekers for whom PCs are an ineffective or unavailable channel of exchange. A number of emerging business models are using mobile communications for improving coordination and information flows in the labor market. At least four such services are

Box 5.4 Turning ideas into applications: "Mobile To Work" challenge

Idea competitions can be an effective way to encourage creativity and identify the most promising product blueprints in a quickly innovating industry. To harness mobile microwork for development, *info*Dev has organized the Mobile To Work (M2Work) challenge to developers to come up with fresh thinking on ideas for microwork tasks that could be commissioned remotely to create employment opportunities in the developing world. The challenge, published online on February 1, 2012, at www.ideasproject.com/m2work, attracted some 944 proposals by the April 2 deadline. Prizes, sponsored by UKaid and the Department for Foreign Affairs of the Government of Finland and worth up to $40,000, are being awarded to the best ideas. The overall winning proposal, submitted by Aadhar Bhalinge of India, suggests a smart rickshaw network to crowd-source maps at a very low cost in developing nations by employing fleets of rickshaw drivers to feed live traffic updates into a subscription service. The regional winners also will benefit from mentorship and a hackathon designed to turn the best ideas into functional applications.

Source: www.ideasproject.com/m2work.

already up and running: Babajob (India), Assured Labor (Latin America), LabourNet (India), and Souktel (Middle East and North Africa, as profiled in box 5.5). Two others, Pakistan Urban Link and Support (PULS) and Konbit (Haiti), have developed their systems and will soon start operating.

Skilled, educated workers may already have access to existing web-based job-matching services such as Monster.com, but job-matching services that are mobile-based will be even more important for people without access to web-based services. Mobile-to-web technology will be beneficial for people with a certain level of skills and education (that is, basic literacy) but not enough knowledge to create a marketable résumé or access online resources. Employers also find it hard to identify low-skilled workers for entry-level jobs in developing countries, because existing job-matching services mainly target highly skilled candidates. Mobile-to-web technology promises to bridge some of these gaps.

Building trust among users is the most challenging task for the job-matching business. Each of the new organizations mentioned above offers additional and customized services to meet the specific needs of local users, including interview, résumé writing, and networking skills training for job seekers, and access to a special database for employers. Depending on the job seeker's target market and country of operation, mobile phones may be used for different aspects of the job-matching business process. Most of these organizations use mobile phones for registration and job-match notifications for job seekers. The actual job-matching service is conducted mostly via web-based databases.

Of course, such technology cannot fully replace the traditional interview process. Once employers become interested in certain candidates, they can access job seekers' information and then contact them directly for an interview. Use of SMS text messaging can be popular where its cost is significantly lower than that of voice services; however, in multilingual environments with illiterate populations, calls and voicemail remain particularly valuable.

Perhaps the greatest impact of mobile communications on jobs lies not so much on recruitment techniques, but rather on the *structure* of employment. Beyond creating more vacancy notices, mobiles can stimulate entrepreneurial activity, as the demand for mobile industry hubs and mobile incubators has shown, and it can create many more opportunities for self-employment, part-time work, and flexwork. In a mobile-driven economy, second and third jobs will become much more common—and much more important.

Conclusions and considerations for policy-makers

Overall, the rise of mobile technology carries great potential for employment, but with increased reach of powerful and affordable mobile devices, jobs may also be lost. Mobile technology can occasionally eliminate jobs, especially where efficiencies are created or resources made available that replace human input. For example, as more individuals acquire their own mobile phones, the demand for "village

Box 5.5 Business processes for job seekers and employers: Souktel's JobMatch

JobMatch find the perfect match. on your mobile phone.

Job Seekers	Employers
1 Sign Up Right from your mobile phone, by texting "register" to 37191.	**1 Register** Using your mobile phone or secure website.
2 Create Mini-CV Use your phone to create an SMS "mini-CV" and upload it to our main database, so hundreds of employers can find you.	**2 Create Mini-job Ad** Create a simple job ad on your phone or online. Upload to the main Souktel database, so thousands of job-seekers can find it and call you.
3 Browse Jobs Browse thousands of jobs via SMS on your phone, or find the exact job that matches your CV info. Get employer phone numbers for follow-up.	**3 Browse CVs** Browse thousands of CVs by phone or web, or find the exact CV that matches the criteria of your job. Get job-seeker details, along with phone numbers for follow up.

Founded in 2005 by graduate fellows at Harvard University, MIT, and the Arab-American University of Jenin (West Bank and Gaza), Souktel launched a trial service in 2007. Within a year, over 100 of the 400 new college graduates who participated in the pilot found work or internships, and more than three-fifths of employers who used this service cut their recruiting time and costs by up to half. With a $100,000 grant from the World Bank Group, the service has been launched at three more college campuses in partnership with the Ministry of Education, then franchised in Morocco, Somalia, and the United Arab Emirates; and it is expected to launch in the Arab Republic of Egypt and Rwanda.

Leveraging the high penetration rate of mobile phones, Souktel developed a job information software platform to connect job seekers with employers via a mobile device. One of the unique characteristics of Souktel is its franchise business model. Souktel has used this model to achieve a rapid growth in new markets. Each country uses a customized version of the JobMatch platform for a franchise fee and a recurrent annual support fee. In return, per-use revenue from local user fees charged to job seekers and employers accrue to the franchisee, helping to ensure each franchise's long-term cost coverage and sustainability.

As a way of measuring its impact, Souktel uses weekly database tracking of service use (searches, match requests, job alerts); monthly phone surveys of "matched" job seekers and employers; and bi-annual "match retention" phone surveys and institutional partner surveys. Positive outcomes are observed in the reduction of time spent looking for employment (from an average of 12 weeks to 1 week or less), wage increases (64 percent of matched job seekers in the West Bank and Palestine surveyed in 2009 reported a 50 percent increase in average monthly wages, from $500 a month to $750 a month), and a reduction in hiring costs and time (70 percent of West Bank and Gaza employers surveyed in 2009 reported a 50 percent reduction in hiring costs and time, while 75 percent of the same sample confirmed a mean 5 percent increase in annual profits). Challenges have included working with the different mobile carriers. The cost of SMS, which averages about $0.05 a message in the West Bank and Gaza, is also a barrier to wider usage.

Sources: Author interview and http://www.slideshare.net/guest923d97/souktel-jobmatch-overview.

phones," teleshops, and other phone-sharing services may disappear in many countries (matching the demise of public payphones in many countries, following the widespread adoption of mobile phones), taking away with it an important source of jobs. In sum, however, with growing mobile penetration rates, the mobile industry is widely expected to produce a net increase in jobs:

- The direct number of jobs in the mobile industry from 1996 to 2011, as reported by governments to the ITU, shows a clear upward trend in most (although not all) countries (ITU 2011).
- As the adoption of mobile technology increases, new jobs are needed to support sales of prepaid cell phone minutes, mobile money transactions, and other mobile-based services.
- The introduction of mobile broadband is expected to generate significant revenues and jobs, especially in related spin-off industries, including the development of mobile applications.
- Nontraditional business plans (such as those based on microwork) are another source of potential growth in jobs enabled by mobile technologies.
- The labor market can benefit from the ability of mobile apps to improve efficiency and lower costs in matching job candidates and employers.

This chapter has outlined a number of tools for enabling growth of employment opportunities in the mobile ecosystem, including:

- Supporting informal community networks and activities such as business competitions and hackathons to promote open collaboration, mentorship, and introduction of entrepreneurs and investors, and to identify viable new business ideas
- Investing in mobile hubs and incubators, or mobile labs, in order to equip entrepreneurs with updated technical skills, to provide them with tools necessary for product prototyping such as testing facilities, and to identify businesses with growth potential through business evaluation and acceleration programs
- Facilitating creation of micro- and virtual work opportunities
- Investing in better mobile platforms for recruiters and job seekers as well as platforms that extend work beyond traditional work spaces and times

To capitalize on the potential of mobile technologies to support entrepreneurship and employment, policy-makers may consider whether current regulation supports an enabling environment for mobile broadband and entrepreneurship, whether to provide financial support for entrepreneurs and incubation systems, and whether to incorporate some of the aforementioned tools in their public service offerings, such as schools and vocational training institutions, in order to increase employment opportunities in the mobile ecosystem.

Notes

1. These could be considered part-time or supplementary jobs, because M-PESA agent tasks are often combined with other merchant duties. http://www.safaricom.co.ke/index.php?id=252; http://www.bloomberg.com/news/2010-10-14/safaricom-of-kenya-will-boost-access-to-credit-insurance-for-unbanked-.html.
2. Bharti Airtel took over Zain Kenya's network in 2010. Some of the Bharti Airtel agents will also be M-PESA agents, but others will be new.
3. Mobile Entrepreneurs in Ghana. http://www.webfoundation.org/projects/mobile-entrepreneurs/
4. As but one example, see Aker 2008.
5. This environment can be contrasted with one of stability, continuity, and homogeneity of the more established economy. The link between entrepreneurship and economic performance at the individual, firm, and societal levels has been shown in numerous studies that provide a framework of dual causality between a strong period of entrepreneurship and a growing and rapidly innovating economy. See, for example, Audretsch and Thurik 2000, p 26, and Wennekers, Uhlaner, and Thurik 2002.
6. The phenomenon of open innovation is explored, among other things, at Open Innovation Africa Summit, organized jointly by *info*Dev and Nokia. The first two Summits were held in Nairobi in November 2010 and in May 2012; see http://www.infodev.org/en/Article.640.html.
7. www.mobilemonday.net.
8. www.code.google.com.
9. http://www.younginnovations.com.np/.
10. http://www.facebook.com/mTbilisi.
11. Nairobi and Kampala interviews conducted by authors. See also Pfeiffer and Salancik 2003.

12. www.ihub.co.ke.
13. www.momokla.ug/.
14. Globally, the shortage of employees with information technology skills has persisted in recent years. See, for instance, http://us.manpower.com/us/en/multimedia/2011-Talent-Shortage-Survey.pdf.
15. http://techcrunch.com/2011/11/01/mobile-startup-incubator-tandem-opens-new-"mobilehome"-in-silicon-valley-now-accepting-applicants/.
16. http://www.socialedge.org/blogs/samasourcing/archive/2009/08/25/microwork-and-microfinance.

References

Aker, J. 2008. "Does Digital Divide or Provide? The Impact of Cell Phones on Grain Markets in Niger." http://www.cgdev.org/ doc/events/2.12.08/Aker_Job_Market_Paper_15jan08 _2.pdf.

Aldrich, H. E., and C. Zimmer. 1986. "Entrepreneurship through Social Networks." http://papers.ssrn.com/sol3/papers.cfm?abstract_id=1497761.

Audretsch, D. B., and A. R. Thurik. 2000. "Capitalism and Democracy in the 21st Century: From the Managed to the Entrepreneurial Economy. *Journal of Evolutionary Economics* 10 (1): 17–34.

Berglund, H. 2007. "Opportunities as Existing and Created: A Study of Entrepreneurs in the Swedish Mobile Internet Industry." *Journal of Enterprising Culture* 15 (3): 243–73. http://www.henrikberglund.com/Opportunities.pdf.

Chesbrough, H. 2003. "The Era of Open Innovation." *MIT Sloan Management Review* 44 (3): 35–41.

COAI (Cellular Operators Association of India). 2011. "Indian Mobile Services Sector: Struggling to Maintain Sustainable Growth." Study commissioned from Price Waterhouse Coopers. http://www.coai.in/docs/FINAL_03102011.pdf.

Elfring, T., and W. Hulsink. 2003. "Networks in Entrepreneurship: The Case of High-Technology Firms." *Small Business Economics* 21 (4): 409–22. http://www.ingentaconnect.com/content/klu/sbej/2003/00000021/00000004/00403594.

Greve, A., and J. Salaff. 2003. "Social Networks and Entrepreneurship." *Entrepreneurship, Theory & Practice* 28 (1): 1–22. http://homes.chass.utoronto.ca/~agreve/Greve-Salaff_ET&P.pdf.

ITU (International Telecommunication Union). 2011. *Yearbook of Statistics 2011.* http://www.itu.int/ITU-D/ict/publications /yb/index.html.

Lehdonvirta, V., and M. Ernkvist. 2011. "Converting the Virtual Economy into Development Potential." *info*Dev. http://www.infodev.org/en/Publication.1056.html.

Lewis, D. 2001. "Does Technology Incubation Work?" *Reviews of Economic Development Literature and Practice* (Rutgers University) https://umdrive.memphis.edu/jkwalkr1/public/business_incubator/do%20business%20incubators%20work.pdf.

Pfeiffer, J., and G. Salancik. 2003. *The External Control of Organizations: A Resource Dependence Perspective.* Stanford: Stanford Business Books.

Secor Consulting 2011. *Canada's Entertainment Software Undustry in 2011.* A report prepared for the Entertainment Software Association of Canada. http://www.secorgroup.com/files//pdf/SECOR_ESAC_report_eng.pdf.

TechNet. 2012. "Where the Jobs Are: The App Economy." http://www.technet.org/wp-content/uploads/2012/02/TechNet-App-Economy-Jobs-Study.pdf.

Teece, D., and E. Ballinger. 1987. *The Competitive Challenge: Strategies for Industrial Innovation and Renewal.* Cambridge, MA: Harper and Row.

Vision Mobile. 2011. "Developer Economics 2011." http://www.visionmobile.com/rsc/researchreports/VisionMobile-Developer_Economics_2011.pdf.

Vital Wave Consulting. 2011. "Mobile Applications Laboratories Business Plan." *info*Dev. http://www.infodev.org/en/Publication.1087.html.

Wennekers, S., L. Uhlaner, and R. Thurik. 2002. "Entrepreneurship and Its Conditions: A Macro Perspective." *International Journal of Entrepreneurship and Education* 1 (1): 25–68. http://people.few.eur.nl/thurik/Research/Articles/Entrepreneurship%20and%20its%20Conditions_%20a%20Macro%20Perspective.pdf.

Chapter 6

Making Government Mobile

Siddhartha Raja and Samia Melhem with Matthew Cruse, Joshua Goldstein, Katherine Maher, Michael Minges, and Priya Surya

Governments around the world, in varying stages of economic development and with diverse technological and institutional capacities, are adopting or investigating mobile government (mGovernment). Several examples of how civil society, the private sector, and entrepreneurs are delivering service improvements using mobile tools have been discussed in chapters 2–5. This chapter focuses on how mobile tools are helping governments to deliver public services more widely and to improve processes of governance.

Yet, the mere introduction of mobile tools cannot serve as a panacea for structural deficiencies in governments' capacities or processes. Initial experiences suggest that the benefits of mGovernment will likely accrue to those governments that put in place policies and programs that not only enable technological transformation but also promote needed institutional reforms and process redesign. The increased demand for services and governance stimulated by this technological transformation will require an increased capacity to supply those services and improve governance. Recognizing the rapid evolution of the field, this chapter identifies some emerging best practice policies and programs that could support the technological transformation and needed institutional capacity development to unlock the benefits of mGovernment.

A typology of mGovernment

Mobile government involves using mobile tools to change either the interactions between users and government or the processes of government. In 2012 tools in use include mobile networks (such as broadband, Wi-Fi, and voice-centric), mobile devices (tablets, smartphones, featurephones), their associated technologies (voice calling, SMS text messaging, location detection, internet access), and software in the form of network services and applications.

Mobile government matters because it has the potential to liberate users from the physical or location-related constraints inherent in conventional service delivery and traditional electronic government (eGovernment) services. With more than 6 billion mobile telephone subscriptions worldwide in early 2012, and more than four-fifths of the world's population covered by mobile telephone networks, mGovernment can make public services and processes available and accessible just about anywhere, at anytime, to almost anyone.

Table 6.1 summarizes three forms of mGovernment. Typically, governments adopt a combination of these three types to achieve their service delivery and governance objectives, and in so doing, provide accountability, transparency, and responsiveness to their citizens. First, mobile tools can be used to *supplement* existing eGovernment applications

Table 6.1 Three types of mGovernment

mGovernment	Supplement	Expand	Innovate
Definition	Mobile tools add a channel to existing eGovernment services and processes.	Mobile tools allow conventional services to reach previously un- or underserved constituents.	Mobile tools are used to develop new services for service delivery and governance.
Example	The Republic of Korea with widespread e-Government, has added wireless portals and interfaces to e-services (such as transport tickets, renewals, confirmations).[a]	Bangladesh's Health Line provides citizens with medical advice through a telephone hotline, cutting travel time and waiting at health centers.[b]	In the Democratic Republic of Congo, mobile tools allow citizens to participate in budgeting, by voting on how to spend local budgets.[c]
Opportunities	Mobile devices, which are more widespread than traditional computers, connect more citizens to existing e-services.	Widespread mobile tools allow conventional services to reach previously excluded citizens including the poor, rural populations, and people with disabilities.	Combined innovation in technology and government processes creates new opportunities for citizens to engage with and hold government accountable.
Limitations	Full advantage is not taken of unique capabilities of mobile tools (such as location determination, built-in cameras); limited to existing eGovernment services.	Benefits are limited by the design and nature of the conventional service and institution; do not necessarily improve the government-citizen relationship.	Extent of innovation depends on local political, economic, and capacity constraints; might need more time to deploy.
Implications for government	Marginal: related to being able to provide any related "physical" service at the needed location and time.	Moderate to significant: government capacity needs to grow to serve more citizens; may need process re-engineering.	Significant: needs changes to government processes, creating response capacity.

a. http://www.futuregov.asia/articles/2011/mar/21/korean-city-opens-mobile-app-centre/.
b. http://healthmarketinnovations.org/program/healthline-bangladesh.
c. http://wbi.worldbank.org/wbi/news/2012/02/17/mobile-enhanced-participatory-budgeting-drc.

based on traditional personal computers (PCs), adding a new channel to reach citizens or manage processes of governance. Supplementary mGovernment adds the dimension of mobility to existing electronic services.

Second, mobile tools can *expand* the reach of conventional public services or government processes to citizens who are unserved or underserved, often because of their remote location or the nonavailability of PCs and internet access. Broad mobile coverage and widespread access to and familiarity with mobile telephones, give governments the opportunity to reach people who might not otherwise have easy access to these public services and processes. These two types—supplementary and expansionary—are also instrumental, focusing more on the "mobile" in mGovernment.

Third, mGovernment can use the introduction of mobile tools to *innovate* new ways for governments to interact with and involve constituents, creating new types of services and governance processes. Innovative mGovernment programs intend to change not only the technology of interaction but also the nature of service delivery or the process. For example, they allow participatory budgeting[1] and community mapping of infrastructure and services.[2] Experiments in mobile-enabled mapping by urban slum dwellers, for example, suggest that innovative mGovernment could actually transform governments' design process for urban development programs by directly involving beneficiaries.[3] Possibilities like these have profound implications for innovative mGovernment.

Although the specific form of a service will vary depending on the availability or advancement of technology, governments could use these different types of mobile services regardless of the technical base or socioeconomic status. In the case of transformative mGovernment, for example, applications using smartphones or basic devices can allow citizens to report nonemergency municipal problems, track responsiveness, and participate in virtual social spaces to put pressure on municipalities to address community issues.[4]

There are some limits on what might be possible to accomplish on a mobile device with a smaller screen or less powerful computing capability than a traditional personal computer has; more traditional eGovernment services will thus continue to have an important role. Both the design of mobile devices as well as their (and networks') capabilities are constantly evolving, however, and the future might see more powerful mGovernment services working alongside, or as replacements for, traditional eGovernment services. Governments will thus need to consider carefully which services can make the transition to mobile, weighing the capabilities of both users and technologies in the process.

Drivers for mGovernment

Why have local, provincial, and national governments and public agencies around the world become interested in mGovernment? Experience thus far suggests that two sets of factors are driving governments to look at mGovernment: global developments that create the environment for governments to consider mobile tools, and the opportunity mGovernment offers to governments seeking to improve service delivery and promote good governance.

Global developments

Three sets of global developments are creating an environment in which mGovernment has become relevant. These are the creation of the underlying technology base in the form of mobile networks and devices, deepening innovation in mobile applications and services, and shifts in the ability of citizens to voice their demands using these technologies combined with increasing pressure on governments to respond to those demands.

First, as chapter 1 shows, mobile networks are spreading even as devices become ever more capable. Mobile networks now have the capacity to deliver a mix of voice, audio-visual, and data services, creating an opportunity for governments to reach more citizens and offer new services through other than conventional means. And while the vast majority of the world's population now uses basic mobile telephones, more powerful mobile devices such as smartphones and tablets are being increasingly adopted (Hellstrom 2008).

Second, as illustrated in chapters 2–5, there is tremendous growth in innovation in the development of applications and services that use mobile technologies. While initial innovation focused on commercial and entertainment applications, more recently there has been a rapid increase in innovative mobile applications and services for social or economic development (Qiang et al. 2012a, b). A growing list of individuals, cooperatives, not-for-profit and nongovernmental organizations, private firms, and public agencies are experimenting with and using mobile applications and services in interesting ways (OECD and ITU 2011; UNDP 2012). As the frontier of innovation begins to touch many public services, it often compels or encourages governments to experiment with these technologies.

Third, individuals have begun to harness these technologies and applications to voice their demands, mobilize communities, and engage with various levels of governments (box 6.1). Even if the results of such efforts vary,[5] combined with ongoing global political and economic transformations in recent years, this voiced demand for responsive services and good governance by citizens through alternative means has increased pressure on some governments to respond.

Because these developments affect different governments in different ways, the speed with which governments adopt mobile tools is certain to vary. Yet, as the subsequent examples illustrate, few governments at any level anywhere in the world are not interested in going mobile (OECD and ITU 2011, 119–50).

The opportunity of mGovernment

In comparison with the growing volume of evidence on the benefits of eGovernment (*info*Dev 2009; Hanna 2010), the impact of many mGovernment services is still unknown. Even without clear evidence of the benefits, many governments nonetheless have begun to explore the possibility of mGovernment if only in low-risk or limited ways. A small number of governments are undertaking major efforts to mainstream mobile tools in service delivery and governance. This section describes some of the more sector- or function-specific examples first, beginning with a discussion of citizen-facing examples and following with examples of internal process-oriented tools. It then discusses broader and, in some cases, government-wide initiatives.

Sector- or function-specific programs. There are many examples of sector- or function-specific mGovernment programs. The simplest ones use mobile tools as a means for government to reach citizens to provide information or simple services or to coordinate internal processes.

Common examples are emergency notifications for adverse weather events or for changes to water or energy supplies. Moldova's Ministry of Agriculture and Food Industries is working with a local agriculture cooperative to pilot

Box 6.1 The mobile telephone as a tool for citizen voice and empowerment

Mobile devices, especially mobile telephones, have become important tools for citizens to express their opinions, mobilize groups, and report on events as they unfold (UNDP 2012). Although mobile telephones and associated applications cannot substitute for community mobilization and democratic processes, they can and have played a role in organizing citizens, especially through social media such as Facebook and Twitter (Brisson and Krontiris 2012).

Perhaps the best-known example is the Ushahidi platform, which emerged in Kenya in response to the violence that erupted after the 2007 election. Ushahidi has now become an open source platform that anyone may use to create an incident-reporting system, by crowdsourcing information using multiple channels such as SMS, email, Twitter, and the web. The information is used to create a map of events to give users a visual image of event hotspots. It has been applied in circumstances as diverse as election monitoring, disaster recovery, and crime reporting.

More recently, feature- and smartphones have been used widely in the ongoing political changes in the Middle East. Citizens have collected and disseminated information during recent events in Egypt, for example, through mobile-based tools including SMS, and for users with more sophisticated devices, through Twitter and YouTube (see chapter 1).

Box figure 6.1.1 Screenshot of the original Ushahidi mash-up

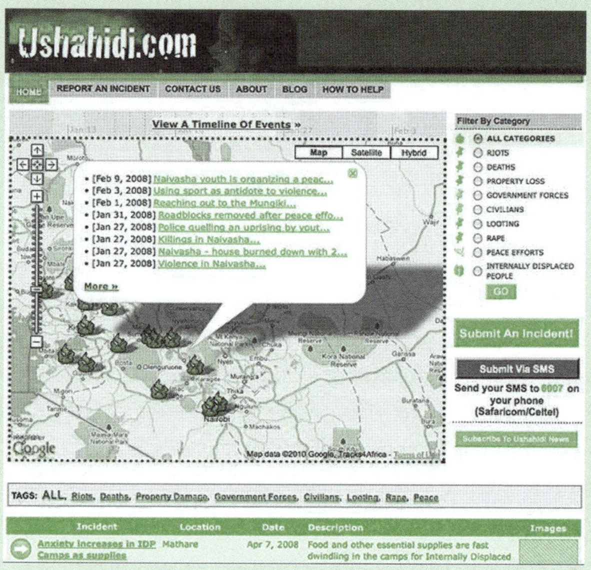

Sources: Stauffacher, Hattotuwa, and Weekes 2012; http://ushahidi.com/about-us; UNDP 2012.

an adverse weather alert service for farmers.[6] Similar examples come from Malaysia and the United States, where SMS is used to alert citizens about limited drinking water supplies or energy blackouts (OECD and ITU 2011). A number of educational systems use SMS to provide students with examination results. The state of Kerala in India has used SMS to send students examination results on request since 2010, reducing the need to wait in queues.[7]

Mobile tools have also shown potential in cutting out intermediaries while improving broader economic outcomes. In Bangladesh, sugarcane farmers now receive an SMS telling them when they should bring their product to sugar mills. In the conventional system, a paper notification might either be misplaced or misdirected by rent-seeking intermediaries. After a successful trial, this system, e-Purjee, was extended to about 200,000 farmers and all 15 of the country's state-owned sugarcane mills, and a feature was added alerting farmers when their payment was ready. Sugar production rose 62 percent following the introduction of e-Purjee, and farmers are benefiting from a more transparent system.[8]

Integration with mobile-based payment systems offers consumers of public resources the opportunity to pay for services anytime and any where and also simplifies revenue collection for governments. Many cities in Europe and the

United States have integrated payment for parking or transport services into mobile applications. In Bangladesh students can also apply for their university entrance examinations through SMS, reducing the need for them to travel to the university to submit an application. Fees are deducted from the applicant's mobile phone account. Following a successful pilot, 28 postsecondary educational institutions implemented the system in 2010.[9] Qatar's Hukoomi service allows citizens to access and pay for a range of services through their smartphone or computer, including utility bills and parking or traffic fines.[10] Complaint reporting through mobile-based SMS has also been expanding throughout the world.

Mobile government efforts have made use of mobile's potential for wider citizen engagement and participation to strengthen accountability and transparency in public services and processes. These efforts are typically innovative, because they often change the delivery or management of a conventional service or process. For instance, the Department of Education in the Philippines worked with the Affiliated Network for Social Accountability in East Asia and the Pacific to set up a website called checkmyschool.org. This is a government-to-citizen online and mobile-based interactive tool that allows citizens to view pertinent statistics on local schools. The site includes budget allocations, teacher and textbook information, and test scores for about one-fifth of the 44,000 schools in the country. It also gives local teachers and parents a public place to post areas of concern that they feel need to be addressed. All users are able to view the government's responses to these posts. Seeking to improve education service delivery through transparent and accountable behavior by school staff, checkmyschool.org has increased community participation and vigilance and improved teacher behavior.[11]

Municipalities and local police departments have begun to use mobile tools to innovate and encourage citizen participation in incident and issue reporting and tracking. Guerrero, Mexico, was able to cut response times to citizen complaints from 72 hours to 24 hours using Citivox.[12] This service provides real-time report management, crowdsourcing reports from people using mobile telephones to register complaints or opinions on everything from simple municipal issues to violent crimes. Follow-up by public agencies has led to wider citizen participation in the service.[13] Similarly, cities across the United States are saving time and money with SeeClickFix, a citizen-reporting tool that allows people to geo-tag nonemergency municipal issues, such as potholes or graffiti, with their mobile phones.[14] With more than 57,000 incidents reported and a 45 percent fix rate between January and October 2010 across multiple cities, this application shows promise for efficient and streamlined citizen-government interactions.

Public agencies are also using mobile tools to support internal functions and to improve resource and program management. For example, electricity companies are beginning to use mobile networks to get real-time consumption data from wireless-equipped smart meters.[15] This will allow electricity networks and consumers to be better informed about consumption patterns, enabling new tariff models.

Governments are beginning to use mobile tools to manage resources more efficiently. Liberia's water resource management plan seeks to improve access to the half of the rural population that does not have access to potable water. The public works ministry deployed 150 data collectors to map all of its roughly 7,500 publicly accessible water points with a mobile geo-tagging and monitoring tool called FLOW (Field Level Operations Watch). The process gave the ministry a visualization of the status of water points, allowing an updated needs assessment and leading to more effective resource allocation.[16]

The possibility of using location sensing, either through global positioning systems (GPS) embedded in devices or by using mobile networks, has also created new service possibilities. In the city of Cebu, in the Philippines, taxi drivers are using GPS-enabled mobile phones to receive traffic data and dispatch information. The data is used to generate maps in real-time that identify areas with traffic congestion and to generate traffic volume estimates.[17]

Cities are also using mobile devices to monitor the status of ongoing programs. Auckland, New Zealand, piloted a project with Municipal Reporter, a GPS-based handheld system that allows the city to monitor its employees and resources. The handheld monitors are saving the city more than over 30 person-hours a week on highway maintenance work. Auckland is currently in the process of shifting all maintenance management to a GPS-based system. Such tools also can help monitor programs in difficult security or climatic conditions. For example, similar technologies, using GPS-enabled smartphones, have been used in Afghanistan to monitor the quality and progress of road construction.[18]

It is also possible to embed unique identifiers in physical objects that mobile phones can recognize (Gartner 2011). Such tools can allow citizens, for instance, to report a broken streetlamp or park bench; officials can then use the same technology to monitor repairs.

Civil society or international agencies have also used platforms to support government service delivery by improving efficiency and reducing waste. For example, UNICEF created a mobile-based data collection tool called Rapid SMS (see box 3.2 in chapter 3).[19] In Hong Kong SAR, China, the Mobile Field Inspection System enables inspectors to use touchscreen PDAs (personal digital assistants) to enter inspection information at the scene, as well as to review the results of past inspections. Inspectors can send their reports through their mobile phones without going to the office. The PDAs were designed for easy use to shorten the training time. Some of the benefits include an approximate 10 percent increase in productivity, a 1.5-hour daily timesaving per inspection team, and elimination of duplicate work.[20]

The wide range of countries and sectors covered in this short list of examples is evidence of the growing interest in and use of mobile tools by governments at different levels and in varying stages of economic development. These examples also display a range of implementation arrangements. In some cases, such as with FLOW in Liberia, projects have been initiated by single agencies. In other cases, multiple partners come together to deploy the tool and respond to citizens' demands. An example is SeeClickFix, where the responsibility of complaint registration, traditionally a government function, is shared between a private organization and the city municipality. Governments adopt these services because they involve and engage citizens in incident and problem reporting through a third party, building trust and credibility. At the same time, such services also build pressure on governments to perform, opening government processes to public scrutiny.

Government-wide initiatives. Apart from the many initiatives coming through bottom-up efforts, a few governments have also begun mainstreaming mGovernment in a larger and more coordinated way, taking a top-down approach in some cases. Some governments, such as that of the state of Kerala in India (box 6.2), have started on such coordination relatively early; others such as the Republic of Korea have evolved to realize the need for such coordination (box 6.3).

Such government-wide initiatives span the range from having an overall mGovernment strategy for mobile services to creating facilities for multiple government agencies to use to deploy services. Countries as diverse as Afghanistan, India,[21] and the United States[22] have been developing mobile-specific strategies that address issues such as how to align activities across agencies, encourage innovation within an overall technical or process framework, and support the development and delivery of services. Other countries have incorporated mobility in their overall ICT strategies. For example, Singapore's government has already deployed more than 300 mGovernment services and has plans, as part of the Singapore eGovernment master plan to create "more feature-rich and innovative mobile services" between 2011 and 2015.[23] Similarly, the U.K. government has identified mobile technologies as an area for attention in its Government ICT Strategy of 2011.[24]

Some governments have also begun to create shared facilities that may be used by multiple agencies. These facilities are similar to those run by private firms that offer news, entertainment, or information services. A number of governments have developed shared services platforms that give citizens access through a common entry point to a range of services. Such platforms allow costs to be shared across multiple agencies, consolidate demand for telecommunications services, and focus human capacity. The governments of Jordan[25] and of the state of Kerala in India (see box 6.2), for example, have implemented shared services platforms that deliver a wide range of SMS, interactive voice response (IVR), or simple text-data services that citizens access using a short code. Among the less developed countries, the government of Afghanistan is also planning to set up a government-wide mobile services delivery platform, which will allow government services to reach the half of all Afghan households that have mobile phones; for many the phone would become the first medium for regular interaction with the government.[26]

In countries where smartphones are common, governments have begun to create points of entry such as mobile sites (the United Kingdom's direct.gov, for example[27]) or even government "app stores." Such facilities allow citizens easy discovery, access, and use of mGovernment applications. In 2010 the U.S. government created such an app store with the intention of making it easy for citizens to access information and services using their smartphones.[28]

Box 6.2 Kerala's mobile government program

The southern Indian state of Kerala has a population of 33 million. Leveraging the wide use of mobile telephones, the Kerala State IT Mission (KSITM) leads a province-level mGovernment program. The objective was to allow equitable access and enable social impact by reaching people with mobile devices, rather than only those who are able to afford and access computer-based internet services.

The centerpiece in Kerala's m-Government architecture is a common service delivery platform (SDP) that integrates various channels such as voice, data, and SMS. The KSITM manages the SDP, supervising a private firm, MobMe, which set up the SDP. All government departments can access the SDP to enable the cost-effective design, development, and deployment of various mGovernment applications. This arrangement avoids duplication of effort and cuts capital spending on stand-alone systems. By integrating with all telecommunication companies, the SDP eliminates the need for individual coordination by government agencies. The KSITM also provides technical assistance to public departments to design and launch mobile applications.

Services include a common "short code" for the government (citizens dial KERALA or 537252 to access services). The service has created an additional incentive for the government to offer services relevant to consumers, including citizen voting on a social reality show where village governments present their successes, posting scores for major exams, and processing movie and bus ticket reservations. The KSITM has also set up an electronic SMS (eSMS) gateway for various government departments to communicate throughout their own units and departments and across institutions. An interactive voice response system supports government customer service call centers and was used to conduct an energy availability survey. A Mobile Crime and Accident Reporting Platform has been used by Keralan police to enhance public safety and law and order. Now, the state is looking to adopt a mobile payments platform, so citizens can pay government fees from their handsets. The state continues to improve and scale up initial mobile applications, such as multimedia messaging service-based accident and crime reporting.

Since its launch in December 2010, the program has involved more than 60 government agencies, facilitated more than 3 million interactions between the government and citizens, deployed at least 20 mGovernment applications, and captured some 200,000 photos for crime and accident reporting purposes. As the KSITM sees it, this is a start to shifting government-citizen interactions from "red tape" to a "red carpet."

Yet, the state faces various challenges in using mobile technology to create transformative change. Successful applications for citizen participatory monitoring and reporting remain elusive. Other key challenges are the low resource and process capacity of public agencies, which limit the ability of the state to respond or improve its accountability.

Source: http://www.expresscomputeronline.com/20110214/technologysabhaspecial24.shtml.

Having such coordinated and broader approaches to mGovernment does not mean that governments should or will need to stop bottom-up or innovative application development. Governments will need to encourage quick deployment of innovative applications when the demand arises. Moreover, as the U.S. government's draft federal mobile strategy indicates, one size does not fit all, and there will be a need to accommodate agency-specific programs. Such coordination should enable innovation by guiding the choice of technical standards and providing facilities where needed.

Challenges for governments

Two key challenges for governments seeking to implement mGovernment are to enable the technology transformation and to respond increased demand for services and good governance.

Box 6.3 Evolving toward coordination: the case of the Republic of Korea

By 2011 government agencies in the Republic of Korea had launched more than 160 mobile applications covering internal processes, access to information, and public service delivery. Problems soon emerged, however, because the applications lacked a common framework. As a result, there was a redundant development of products, mismatch of technical standards across ministries and agencies, and the lack of a clear direction for budget priorities around mGovernment services.

To address these challenges, the government in 2011 launched a five-year, $55 million strategy to integrate mGovernment, focusing on both internal processes and public services. This strategy establishes a common framework for developing simple mobile websites, hybrid websites customizable by operating system, and mobile applications. For each of the five years, the strategy sets priorities ranging from security to quality assurance and authentication, to the establishment of a mobile common data management system. It also provides a detailed guide to the user interfaces and experiences with mobile government websites.

Enabling the technology transformation

Governments that are interested in mGovernment will need to ensure that mobile tools are widely available to citizens, that public agencies are ready and able to adopt these technologies, and that the ecosystem of applications and services developers is in place to deliver needed services.

Simple mobile telephones are now commonplace across the world, and mobile networks are widespread. However, governments will need to ensure that the populations or geographies they wish to target are adequately covered. This issue is especially important if technology choices are more sophisticated—using feature- or smartphones, for example—because mismatches could keep citizens from accessing public services.

Public agencies will also need to have the ability to adopt these technologies. In many countries, that is likely to involve closing gaps in technological or human capacity, ensuring financial sustainability, and overcoming political or bureaucratic resistance. These considerations are similar to those seen for eGovernment services in the past, and indeed, such factors limited the adoption of many of those programs and reduced their long-term impact.

Many developing country governments are not in a position to carry out mobile applications development on their own, so it will be critical for them to work with partners in the private or nonprofit sector. In some cases, countries have local technology companies that could develop and even manage mGovernment applications. Many governments might face a shortage of talent in applications development, however, or might not easily find willing partners. Such constraints might slow down mGovernment efforts or increase costs if nonlocal resources have to be called upon.

Creating institutional capacity

Even if mGovernment gains widespread acceptance, concerns remain about the increased demand for responsive services and good governance. It is thus important to match technological progress with increases in institutional capacity and, depending on the scale of change, possibly to restructure government. Institutional capacity is a greater issue with mGovernment than eGovernment because of the wider reach of mobile tools and consequently the larger number of citizens that likely would use such tools.

True transformation needs governments to pay close attention to re-engineering processes, reforming institutions, and creating an environment for greater accountability and transparency. Such major shifts often need significant political leadership and capital to implement, and they inevitably take time. At the very least, governments should have the institutional capacity in place to respond to citizen demands because the move to mobile exponentially increases the capacity for citizens to demand services and good governance.

As Ben Berkowitz, the co-founder of SeeClickFix, explains, "The most important part of the process is the 'fix.' Without that, the incentive for participation disappears."[29] Echoing a similar sentiment, Lishoy Bhaskar, vice president at MobMe—the implementer of the Kerala shared service delivery platform—finds that many government officials in the developing world understand the benefits of mGovernment but often hesitate to implement it because "there is no one to fix the potholes even if they are reported."[30]

The risk in not responding is that citizens will quickly lose trust and interest in participating in mGovernment programs. This risk extends not only to those programs that propose to make governance transparent and accountable but also to those where technologies are supposed to improve service quality by reducing wait times or simplifying processes. If a government is unable to follow up on the expansion of service—for example, by being unable to serve the increased number of patients that show up at health clinics because of better information on medical conditions—it risks losing credibility.

Emerging best practices for going mobile

How might governments respond to the challenges inherent in going mobile? Emerging best practices—summarized in this section and in table 6.2—suggest a range of actions governments could take to boost technological take-up and improve institutional capacity.

Enabling a sustainable technological transformation

Create a strategy for mGovernment. A holistic mGovernment strategy or strategic framework can help governments identify gaps in technology and human capacity, in financial sustainability, and in the applications development ecosystem. It can also help raise the profile of mGovernment, potentially leading to high-level political support. And mGovernment programs should be aligned with broader national development programs and strategies.[31] Such a strategy could also define needed technology, service, and data standards; identify common facilities and resources to be developed within the government; and look for opportunities for partnering with civil society, the private sector, and entrepreneurs. The strategy could also define ways to make these programs financially sustainable. It will be important, however, to avoid restricting innovation and flexibility. Furthermore, coordination should not imply that some types of sector- or function-specific systems should never exist independently; some services (such as in health or education) will have specific needs and might be justifiably separate in their implementation.

Enable innovation. Much of the development in mobile applications and services worldwide has come from innovation by nongovernment agencies. Governments are often late adopters of this technology. Hence, there is much to gain from allowing such innovation to continue, with governments encouraging innovation and working with partners such as mobile networks, applications

Table 6.2 Policies and programs to promote mGovernment

	Enabling a sustainable technological transformation	Strengthening institutional capacity to respond
Policies	• Create a strategy for mGovernment • Enable innovation • Make mobile technology accessible and affordable • Enable mobile payments • Define standards for technologies and content	• Enable shared responsibility in service delivery • Promote efficiencies in resource allocation and management and in processes • Build trust
Programs	• Create shared facilities • Support content creation and use in local languages • Mobilize and train users • Support public-private partnerships	• Train government officials on strategic uses of mGovernment • Incentivize testing through iterative processes, user-centric design, and risk-reduced innovation programs

developers, and civil society organizations to design and pilot applications. At the least, interested agencies within government should be encouraged to move swiftly toward implementing "quick wins" that demonstrate the validity of the approach and hence secure greater support among other participants. Definition of technology standards and opening of government facilities (such as data centers or data sets) will help direct such innovation and avoid undesirable fragmentation of systems. Governments could also partner with universities and mobile networks to develop skills among potential mobile application developers.

Make mobile technology accessible and affordable. Governments will need to promote universal access and service for specific user groups where mobile networks have yet to reach, especially because these groups also tend to be the unserved for regular government services. Efforts should also focus on improving the affordability of devices and services. Some countries may be able to reduce the price of devices by cutting excessive taxes, duties, or levies. Service prices might be reduced by consolidating demand across government, for example, through purchases of bulk SMS or IVR minutes. The reader is directed to chapter 7 on this topic (see also Kelly and Rossotto 2012; Muente-Kunigami and Navas-Sabater 2010).

Enable mobile payments. Many government transactions involve the transfer of money to citizens or payment of fees by citizens. Enabling mobile payments will allow citizens to make and receive payments securely, even if they do not have bank accounts or cannot securely carry cash, and will encourage them to use mobile-based services. The reader is directed to chapter 5 for further discussion.

Adopt standards for technologies and content. Governments can help to enable innovation by adopting standards for technologies and content. For example, the Open 311 framework is a protocol developed by a combination of government and civil society organizations and adopted by municipalities for location-based collaborative issue tracking.[32] Adopting Open 311 could help standardize complaint or issue management applications across government, making them interoperable. Such standards could also extend to how government agencies open and share the information and data they produce. Such information, when digitized and openly available, could facilitate the creation of mGovernment services (box 6.4).

Create shared facilities. Some governments, such as those in Kerala state and Jordan, are creating shared facilities to develop, deploy, operate, and manage mGovernment services. For citizens, such common facilities would make access simpler and more organized by enabling "single windows" (Hellstrom 2008). For the government, these resources include the hardware and software needed to run applications as well as the communication services to connect with users through mobile telecommunications networks (such as text messages, voice minutes, data services). These facilities could also include commonly used tools to simplify development and deployment of mGovernment services (such as survey tools, peer-to-peer communication tools, short codes). Such shared facilities for mGovernment could also link with efforts to create government cloud-computing facilities.[33]

Support content creation and use in local languages. As with any technology, cultural context, user capability, and local relevance will drive adoption and success. Ensuring that mGovernment services remain focused on beneficiaries is important, especially in the case of service delivery or information provision. Governments will need to engage with a wide range of stakeholders—technologists, communities, users, intermediaries, and public service providers—to design and develop demand driven and user-centric applications and services. Updated content will have to be created or kept in local languages, and the content and the application will need to fit the needs and ICT literacy levels of users.

Mobilize and train users. Users beyond early adopters need to understand the benefits of using mGovernment services. Community-level intermediaries can play a vital role in educating users and driving adoption of applications. Critically, however, evidence of government responsiveness and improved service delivery and governance will be the most effective means to attract citizens to this platform.

Encourage public-private partnerships to support mGovernment. Both private and public sector efforts will

Box 6.4 Open data and mobile access in Kenya

Governments are beginning to open public data sets and make them accessible to the public and civil society. With mobile telephones being more widespread than PCs, it is not surprising to see more open data being made available on mobile platforms with interesting consequences.

In July 2011 Kenya became one of the first African countries to launch an open data initiative, making some 160 government datasets open to the public, with more on the way (www.opendata.go.ke). The aim was to lead to a more responsive and citizen focused-government. Among the initial data sets that were uploaded are poverty surveys by district, budget by government department, and plans for future changes in electoral districts and health facilities. A beta site was launched in June 2009 and the public site a year later.

But in Kenya, as in many other developing nations, mobile ownership far exceeds PC ownership, so to increase transparency and widen access, facilitating mobile access to the data is an important goal. Kenya's experience with Ushahidi (see box 6.1) created a local precedent for this. To support the development of mobile applications that would open up the government data, the Kenya ICT Board launched the Tandaa Digital Content Grant, offering up to 30 awards totaling \$1.5 million. An early success came when exam results were made available on mobile phones.

Providing data to citizens, civil society, and entrepreneurs will support their ability to engage with the government and help develop new ideas and services. As such, open data is part of making the government a platform on which stakeholders and constituents can engage, interact, and create.

Box figure 6.4.1 Screenshot from Open Data Kenya website, showing poverty and pupils per teacher

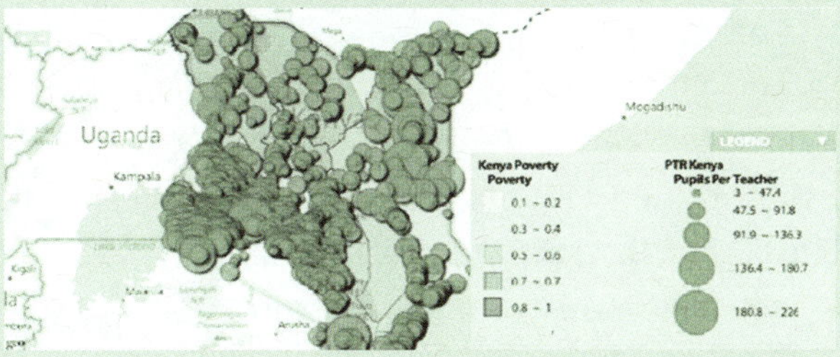

Source: Adapted from Rahemtulla et al. 2011.

complement and strengthen each other. Initially, the private sector will focus on commercially viable applications including media and infotainment, mCommerce, and advertising- or subscription-based information services. With the right incentives and given the opportunity, private entities can supplement state technological capacity and create and deploy applications that serve public needs or support program management. The examples of Kerala (see box 6.2), where the IT Mission has contracted with a private company, and of SeeClickFix, a private group working with municipalities, suggest such new possibilities. Such partnerships could also help close technological or human capacity gaps, with private firms taking on the responsibility of managing the technology and sharing some of the financial or political risks.

Strengthening the institutional capacity to respond

Enable shared responsibility in service delivery. A key consideration is how the nature of service delivery will change as technology and its use evolves. It is difficult to predict the extent of transformation in services. However, governments can begin to prepare by looking for ways to share responsibility, which can also create the possibility of increasing capacity. Three options exist: governments could transfer responsibility for service delivery to the private sector or civil society, share responsibility for serving citizen demands with other actors, or continue to supply improved or enhanced services but with the help of private and civil society actors. These models can exist side by side. For example, many countries have transferred responsibility for infrastructure construction and operation (roads, power, telecommunications) to the private sector while retaining or sharing responsibility in others areas such as education or health services. In any case, governments will need to consider how the re-engineering of processes could open new models for service delivery and remove any unneeded legal or regulatory impediments to transferring or sharing responsibility where such models are valuable.

Promote efficiencies in processes and in resource allocation and management. Governments can encourage the use of mGovernment tools by creating opportunities for greater efficiencies within existing workflows and processes. In an analogous example, the government of Bhutan encouraged civil servants to use electronic communications technology while cutting office stationery budgets.[34] As was the case in Bhutan, adequate training and capacity building will be needed to support the transition to the use of mobile tools.

Build trust. One of the most critical, yet often ignored, aspects in mGovernment is to balance the increased interaction between governments and citizens with the need to ensure privacy and security. There are three aspects to this—the security of private information, avoiding the perception of surveillance, and managing anonymity—which are discussed in box 6.5. Legal and ethical views on privacy vary from government to government and also depend on social context. Yet, in every case, governments must maintain the expected level of trust through a combination of legal and technical actions. Infringements must be dealt with quickly. A related area for consideration is the development of electronic or mobile identification services to protect citizens' identities in their interactions with governments and to prevent data leakage and fraud. The government of Moldova is now developing a system to create a unified way to solve, for any electronic or mobile application, security-related tasks such as identity management, authentication, and transaction authorization.[35]

Train government officials on strategic uses of mGovernment. Governments will need to undertake some capacity-building programs to develop skills of government officials to understand and use mGovernment tools. In Afghanistan, the Ministry of Communications and IT coordinates government training of chief information officers (CIOs) with targeted mGovernment-related training. It is also creating a team of mGovernment advisors—international experts who could advise on strategic interventions—to support the cadre of CIOs and officials keen to deploy mGovernment tools.

Incentivize testing, user-centric design, and innovation. Governments could consider promoting innovative approaches to applications development and operation through innovation challenges or competitions;[36] set up incubators that provide entrepreneurs within and outside government a physical, social, and intellectual space to develop innovative services; or support national innovation policy programs. A forthcoming Innovation Support Program in Afghanistan explicitly targets the development of products for improved public service delivery and adoption. Governments should also borrow from techniques employed in the private sector for the development and adoption of new technology platforms and services, such as iterative, pilot-based service rollout, and user-centric design to ensure relevance and usability.

Conclusions

The ubiquity of the mobile telephone has created an opportunity for governments around the world to improve service delivery and enhance governance. Mobile tools also create the opportunity for citizens to participate directly and engage with governments like never before. Already, examples from a wide range of countries, provinces, and cities are showing that mGovernment is taking hold and helping supplement, expand, and innovate services and governance.

Mobile government is relatively nascent and the potential of mobile devices continues to evolve, so new ideas are certain to emerge to help make governments mobile. Based on experience thus far, however, governments seeking to go mobile will need to create an enabling environment for technology transformation as well as the institutional capacity to respond to citizen requests for service.

In closing, any government seeking to adopt mobile tools should keep in mind that this process will successfully

Box 6.5 Challenges to trust and credibility

As governments find more ways to deliver services using mobile and geo-location technologies, concerns over security and privacy are mounting. If used properly, mGovernment can promote transparency and accountability of service delivery. However, citizens often express concern about the security of their private and confidential information, possible surveillance, and anonymity, among other issues.

It is vital that governments create a legal and technical framework to protect data from corruption or leakage. Without strong protection or the quick resolution of any breach, citizens will be wary of sharing their information with the government, and efforts to connect and interact would quickly be undermined. Internet users already face security problems—for example, so-called "Trojans" or "malware" can compromise personal computers and gather private data from users illegally.

While location- and context-based services offer powerful opportunities, illegal or unwarranted surveillance must also be avoided. Again, citizens need to be assured that installing applications or using services will not compromise their privacy. Governments will need to exercise care in securing their systems and software to avoid any perception of surveillance. For example, the Data Protection Working Party, an independent European Union advisory body on data protection and privacy, has suggested that users of smartphones and other mobile devices give clear and explicit consent and have a clear understanding of how the data will be used, before location data is collected.

Finally, citizens might seek anonymity (or pseudonymity) as they become more vocal to avoid the risk of reprisals due to their views. Governments may need to consider which services require identification and which services (anticorruption hotlines, for example) might be more popular if citizens can remain anonymous when they make a report.

Balancing privacy concerns against the government's need to ensure that it is dealing with legitimate users of the service should not be a barrier to exploring mGovernment. Rather, it should be the catalyst for ongoing conversations regarding the strength of privacy laws and proper auditing alongside the ability to share information.

Source: http://ec.europa.eu/justice/data-protection/article-29/documentation/opinion-recommendation/files/2011/wp185_en.pdf.

transform the government-citizen relationship only when governments enable the transformation of both elements—"mobile" and "government."

Notes

1. "Participatory budgeting" implies citizen involvement in the budgeting and allocation of public resources through direct democracy; see, for instance, http://www.youtube.com/watch?v=hZYm0kEvkAo; http://www.tnpp.org/2011/12/mobile-participatory-budgeting-dr-congo.html.
2. www.mapkibera.org.
3. Initial efforts toward this aim are under way in Dar es Salaam, for example, where citizens are involved in mapping community resources as a first step toward improving resource allocation for urban services. See http://blogs.worldbank.org/ic4d/node/535.
4. See, for example, http://seeclickfix.com/, http://www.fixmystreet.com/, and http://plus1lab.com/about-cityreporter/.
5. See varying opinions and views on the role of social media and ICT in recent political events: http://pitpi.org/index.php/2011/09/11/opening-closed-regimes-what-was-the-role-of-social-media-during-the-arab-spring/; http://www.twq.com/ 11autumn/docs/11autumn_Alterman.pdf; and http://www .time.com/time/world/article/0,8599,2104446,00.html.
6. http://www-wds.worldbank.org/external/default/WDSContentServer/WDSP/IB/2010/07/19/000334955_20100719024447/Rendered/PDF/530500PAD0IDA11B01OFFICIAL0USE01091.pdf.

7. http://www.hindu.com/2010/05/02/stories/2010050255260400.htm.
8. http://www.epurjee.info/Implementation.php.
9. http://www.ictdata.org/2011/10/going-digital-in-bangladesh.html.
10. http://www.ictqatar.qa/en/department/national-programs/e-government/hukoomi.
11. www.checkmyschool.org.
12. http://citivox.com/.
13. http://thanassiscambanis.com/sipa/?p=276; http://www.informationactivism.org/en/citivox.
14. http://seeclickfix.com/.
15. http://www.telenor.com/en/news-and-media/press-releases/2011/telenor-to-measure-your-electricity-consumption.
16. http://www.wsp.org/wsp/sites/wsp.org/files/publications/WSP-FLOW-Liberia-QandA.pdf.
17. http://www.citynet-ap.org/images/uploads/resources/DhakaNov27.pdf (p. 36).
18. http://aidc.af/aidc/.
19. http://www.rapidsms.org/case-studies/malawi-nutritional-surviellence/.
20. http://www.itu.int/ITU-D/asp/CMS/Events/2011/ict-apps/s1_ITU_souheil.pdf.
21. http://www.mit.gov.in/content/framework-mobile-governance.
22. http://mobility-strategy.ideascale.com/a/pages/draft-outline.
23. http://www.egov.gov.sg/c/document_library/get_file?uuid=4f9e71be-fe35-432a-9901-ab3279b92342&groupId=10157 (p. 7).
24. http://www.cabinetoffice.gov.uk/content/government-ict-strategy.
25. http://www.jordan.gov.jo/wps/portal/?New_WCM_Context=/wps/wcm/connect/gov/eGov/Home/e-Government+Program/E-Services/Shared+Services/SMS+Gateway.
26. http://documents.worldbank.org/curated/en/2011/03/13995882/afghanistan-ict-sector-development-project (pp. 24-25).
27. http://www.direct.gov.uk/en/Hl1/Help/YourQuestions/DG_069492.
28. http://apps.usa.gov/.
29. Interview with Mr. Berkowitz, June 2011.
30. Interview with Mr. Bhaskar, December 2011.
31. This is also noted in the draft mGovernment strategy outline for the U.S. federal government; see http://mobility-strategy.ideascale.com/a/pages/draft-outline.
32. http://open311.org/learn/.
33. http://www.cloudbook.net/directories/gov-clouds/government-cloud-computing.php.
34. http://www.bhutanobserver.bt/ministries-try-frugal-stationery-use/.
35. http://egov.md/upload/CN-mobile-eID-eGC-June-2011.pdf.
36. http://whatmatters.mckinseydigital.com/innovation/prizes-a-winning-strategy-for-innovation.

References

Brisson, Z., and K. Krontiris. 2012. *Tunisia: From Revolutions to Institutions*. *info*Dev. http://www.infodev.org/en/Publication.1141.html.

Gartner. 2011. "Executive Advisory: The Untapped Potential of Mobile: Connecting the Physical World to the Online World" (April 27). http://www.gartner.com/id=1656117. Registration required.

Hanna, N. 2010.*Transforming Government and Building the Information Society: Challenges and Opportunities for the Developing World*. New York: Springer.

Hellstrom, J. 2008. "Mobile Phones for Good Governance: Challenges and Way Forward." Draft discussion paper. http://mobileactive.org/research/mobile-phones-good-governance-challenges-and-way-forward.

*info*Dev. 2009. "eGovernment Primer: Using ICT for Public Sector Reform." http://www.infodev.org/en/Project.39.html.

Kelly T., and C. Rossotto, eds. 2012. *Broadband Strategies Handbook*. Washington, DC: World Bank. www.broadband-toolkit.org.

Muente-Kunigami, A., and J. Navas-Sabater. 2010. "Options to Increase Access to Telecommunication Services in Rural and Low-Income Areas." World Bank Working Paper 178, Washington, DC. books.google.com/books?isbn=0821381407.

OECD (Organisation for Economic Co-operation and Development) and ITU (International Telecommunication Union). 2011. "M-Government: Mobile Technologies for Responsive Governments and Connected Societies." http://dx.doi.org/10.1787/9789264118706-en.

Qiang, C. Z., S. C. Kuek, A. Dymond, and S. Esselaar. 2012a. "Mobile Applications for Agriculture and Rural Development." World Bank. http://go.worldbank.org/YJDV8U9L0.

Qiang, C. Z., M. Yamamichi, V. Hausmann, R. Miller, and D. Altman. 2012b. "Mobile Applications for the Health Sector." World Bank. http://siteresources.worldbank.org/INFORMATIONANDCOMMUNICATIONANDTECHNOLOGIES/Resources/mHealth_report_(Apr_2012).pdf.

Rahemtulla, H., J. Kaplan, B-S. Gigler, S. Cluster, J. Kiess, and C. Brigham. 2011. *Open Data Kenya: Case Study of the Underlying Drivers, Principal Objectives and Evolution of One of the First Open Data Initiatives in Africa*. Open Development Technology Alliance. http://www.scribd.com/WorldBankPublications/d/75642393-Open-Data-Kenya-Long-Version.

Stauffacher, D., D. Hattotuwa, and B. Weekes. 2012. *The Potential and Challenges of Open Data for Crisis Information Management and Aid Efficiency: A Preliminary Assessment.* ICT for Peace Foundation. http://ict4peace.org/wp-content/uploads/2012/03/The-potential-and-challenges-of-open-data-for-crisis-information-management-and-aid-efficiency.pdf.

UNDP (United Nations Development Programme). 2012. "Mobile Technologies and Empowerment: Enhancing Human Development through Participation and Innovation." http://www.undpegov.org/sites/undpegov.org/files/undp_mobile_technology_primer.pdf.

Chapter 7

Policies for Mobile Broadband

Victor Mulas

This final chapter looks to the future and provides policy recommendations for expanding the range and uptake of mobile applications for development. In practical terms, that means looking at the shift toward mobile broadband networks. Broadband has a positive impact on growth and development (Qiang and Xu forthcoming). Mobile broadband, in particular, is expected to show an even higher positive effect on economic growth, especially in developing countries. Thus, mobile broadband development and diffusion across the economy is a subject of policy action. Unlike other information and communication technology (ICT) services, such as fixed-line voice telephony, broadband (including mobile broadband) behaves as an ecosystem where the supply and demand sides interact and mutually reinforce each other. Hence, both aspects of the ecosystem—supply and demand—need to be addressed by policy initiatives (Kelly and Rossotto 2011). Supply-side policies aim at promoting and enabling the expansion of mobile broadband networks; demand-side policies seek to increase adoption of mobile broadband services. Policy recommendations for both supply and demand are addressed below.

The mobile broadband opportunity and developing countries

As discussed in chapter 1, broadband has an important effect on economic growth and development. Numerous studies have found a positive relationship between broadband penetration and economic growth, particularly in developing countries (Qiang and Rossotto 2009, 45; Friedrich et al. 2009, 4; Katz et al. 2010, 2; Digits 2011). One of the transmission channels of this growth is linked to the transformational effect of broadband throughout the sectors of the economy, raising productivity and efficiency (Kelly and Rossotto 2011). Mobile broadband has been found to have a higher impact on GDP growth than fixed broadband, through the reduction of inefficiencies (Thomson et al. 2011).

Mobile telephony has already demonstrated that networks that use spectrum, such as mobile networks, are often the most efficient infrastructure for expanding ICT services worldwide, especially in developing countries, which generally suffer from a shortage of fixed infrastructure (see Statistical Appendix). Such is the case for broadband, which is now growing faster in developing countries than in developed ones, with a compound average growth rate of over 200 percent since 2009. In some countries, such as Colombia, Kenya, South Africa, and Vietnam, mobile broadband is already the main platform for broadband access, having surpassed fixed broadband by over 10 times in the two African countries and almost 3 times in Vietnam (figure 7.1).

Even so, the broadband gap between developing and developed countries is increasing.[1] Whereas around half of mobile connections provide broadband access in

Figure 7.1 Broadband subscriptions in selected countries per platform (mobile vs. fixed)

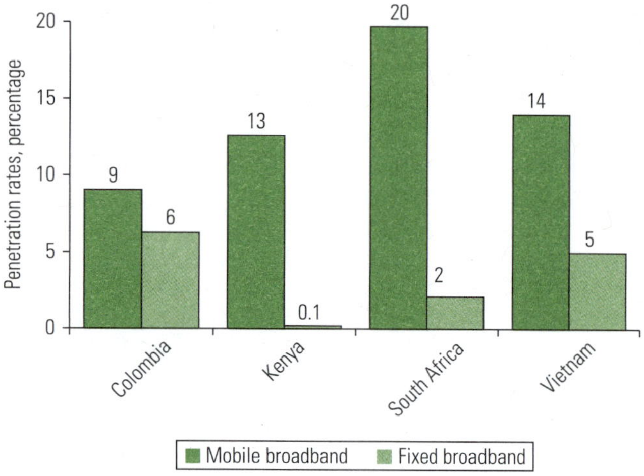

Sources: TeleGeography Inc. database, March 2011, and World Bank database for population data.
Note: Data are for the third quarter, 2011, for Colombia, Kenya, and South Africa; second quarter, 2011, for Vietnam.

Figure 7.2 Broadband as an ecosystem where supply and demand factors interact with each other

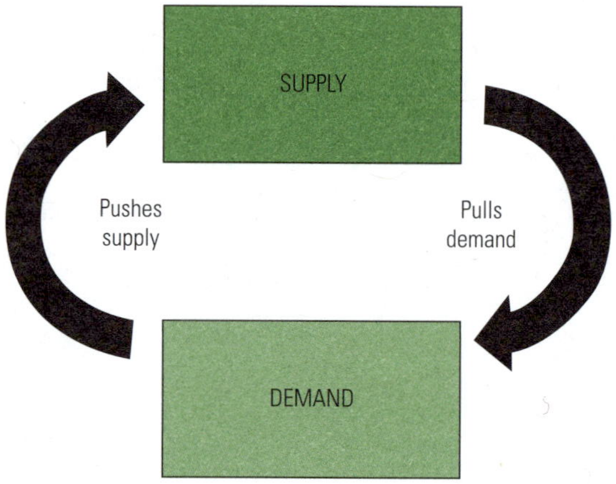

Source: World Bank.

developed countries, in developing countries this percentage is below 10 percent. The different pace of mobile broadband adoption has many causes, one of which has been more aggressive policies in developed countries to enable and foster the implementation of mobile broadband technologies. As shown by examples in Chile, Germany, Sweden, and the United States, to name but a few, policies that foster mobile broadband allow for its faster and wider diffusion.

Policy recommendations for facilitating mobile broadband diffusion

To understand how policy-making can promote and enable broadband, it is useful to understand the various elements that influence broadband diffusion. By contrast with other ICT services, such as voice, broadband works as an ecosystem, where the supply and demand sides interact and reinforce each other (Kelly and Rossotto 2011, 25). Thus, broadband diffusion not only requires the supply of access through network coverage expansion, but also the development and availability of demand-side enablers, such as affordable smart devices and content and applications that respond to user needs (figure 7.2).

With this framework in mind, policies to support and enable broadband diffusion through mobile networks can be categorized as either supply-side or demand-side policies.

Supply-side policies

Supply-side policies aim to expand mobile broadband networks by addressing the bottlenecks and market failures that constrain network expansion and by providing incentives for wider mobile broadband coverage. Bottlenecks and market failures differ among countries, and policy-makers and regulators should assess their specific market conditions, prioritizing those policies that are relevant to their domestic bottlenecks and market failures. However, two main bottlenecks are relatively common worldwide: insufficient availability of spectrum, and inadequate backbone networks.

The following policy recommendations focus on these common bottlenecks, as well as on incentives for expanding the coverage of mobile broadband networks.

Ensure sufficient availability of quality spectrum to deploy cost-effective mobile broadband networks. Availability of spectrum may become a bottleneck to the development of mobile broadband networks for various reasons. First, to facilitate rapid deployment of these networks, operators need spectrum that is technically adapted to the most cost-efficient mobile broadband technologies. Technologies are designed to be more efficient in specific spectrum bands. International harmonization provides the benefits of economies of scale for network equipment. As a result some bands are much more commercially attractive than others. If spectrum is not offered for the bands where the most cost-efficient technologies work, operators have to opt for other less efficient options, which can result in more limited investments or no investments at all.

Second, operators need spectrum in the bands that are most effective for deploying mobile broadband technologies. For instance, a fourth-generation broadband mobile technology such as Long-Term Evolution (LTE) can operate in multiple frequency bands, but the lower bands (such as 700 and 800 megahertz, or MHz) can be more cost-effective, allowing for both wider coverage from fewer radio base stations (an important consideration for rural area deployments) and higher powers to support building penetration (an important consideration in urban areas). Using optimal frequency bands can also assist with the high availability of network equipment and lower prices resulting from global economies of scale. Continuing with the previous example, deployments of LTE networks driven by U.S. and European operators have generally been more successful in the 700 and 800 MHz bands. That has resulted in more affordable network equipment in these two bands.

Third, blocks of spectrum must be sufficiently large to allow cost-efficient provision of mobile broadband, with multiple operators. LTE, for example, allows operations with different-sized blocks of spectrum (from 1.4 to 20 MHz); the size of the spectrum blocks and the pairing of frequencies determines the maximum broadband speed and the cost of deploying mobile broadband networks based on this technology. Because data traffic and bandwidth are growing rapidly, operators may need larger blocks of spectrum to cope with demand and avoid congestion, particularly in urban areas. Use of Wi-Fi networks to offload mobile broadband traffic from cellular networks can also help to offset congestion pressures over these networks. However, these complementary networks will not be able to solve the growing congestion problem by themselves. Although forecast to almost double, Wi-Fi offload traffic is expected to handle only around 20 percent of total mobile broadband data by 2016 (CISCO 2012).

To minimize bottlenecks in the availability of spectrum, policy-makers and regulators should assess spectrum needs and available cost-efficient technologies and release to the market spectrum of suitable and sufficient quality for these technologies. In some case, policy-makers and regulators may need to refarm spectrum (the practice of making spectrum available by moving existing users or organizing band use more efficiently) and reassign legacy users with less valuable uses or less efficient technologies to other bands. Permitting spectrum trading among operators also allows for spectrum refarming for more efficient uses through private sector–led transactions. The digital switchover (the process whereby analog television has been superseded by digital television) has allowed spectrum managers worldwide to liberate spectrum for other uses, particularly mobile broadband. That in turn has allowed policy-makers worldwide to institute spectrum refarming. In the United States, the 700 MHz band, where LTE networks are currently being deployed, was released as a result of the digital switchover. Similarly, in Europe, countries such as Sweden and Germany have taken advantage of the digital switchover to release spectrum in the 800 MHz band for their LTE networks.

Eliminate technological or service restrictions on spectrum. The availability of spectrum is not the only issue. Technical or technological restrictions or mandated uses that require the spectrum to be used for other services could still act as a bar to mobile broadband technologies. Eliminating such restrictions, and making spectrum technologically neutral, allows operators to choose the most efficient technology to deploy on broadband services. Market mechanisms for spectrum allocation, such as auctions or secondary trading, should help to ensure that available spectrum is used efficiently. This is valid not only for current mobile broadband technologies, such as WiMAX, HSPA, or LTE, but also for other technologies that may be developed in the future. Applying the principle of technological neutrality is as relevant for new spectrum being released as for spectrum that has already been allocated, particularly second- and third-generation (2G, 3G) band spectrum. Operators can thus leverage existing network deployments in the 2G- and 3G-bands, such as GSM (Global System for Mobile communications), and Wideband CDMA (Code Division Multiple Access), by turning over part or all of the spectrum they already use for these services to advanced mobile broadband technologies (in-band migration).

This practice has been successfully applied for 3G technologies within the 2G bands in many countries, particularly in Latin America where operators could launch 3G services before 3G licenses where awarded or in bands initially awarded for 2G services. In Mexico operators launched 3G services in 2007 and 2008 using both CDMA and Universal Mobile Telecommunications System (UMTS) technologies, well before 3G spectrum licenses were awarded in 2009. In Brazil operators started launching CDMA-3G services in 2004, before 3G licenses were awarded. In addition, the regulator allowed the use of 2G-awarded spectrum for 3G services as 3G spectrum licenses were awarded in 2007.[2] Allowing the use of existing spectrum for any technology-neutral use (given that these technologies do not result in harmful interferences) also enables operators to follow a phased and scalable approach to transition from 2G/3G technologies to 4G technologies (such as LTE).

Focus on expansion of network coverage rather than on spectrum proceeds. High up-front spectrum costs may limit the capital available for operators to invest in coverage beyond the most affluent areas (EC 2002; Delian 2001; Bauer 2002). There are several methods for awarding spectrum rights, the most common ones being auctions, beauty contests, and hybrid methods of these two. Although auctions are generally considered more efficient than beauty contests, auction designs aimed at increasing up-front revenues for the government do not achieve the highest social welfare benefits (Hazlett and Munoz 2008, 2010). Indeed, auctions that extract high rents from operators may result in delays of investments or in concentration of network coverage in urban and high-income areas, while rural and low-income areas are not served (Patrick 2001). The results of the 3G auctions in Europe, where high proceeds were achieved, but 3G network deployment was delayed for several years and a number of licenses were returned, showed that high up-front costs may result in low or delayed investment (Gruber 2006). To encourage coverage in underserved areas, some governments, such as Chile (box 7.1), Germany (Brugger and Oliver 2010; Wireless Intelligence 2011), and Sweden,[3] have introduced hybrid methods adding specific coverage obligations to mobile broadband spectrum licenses to cover underserved areas, or "white spots."

Require transparency in traffic management and safeguard competition. Demand for mobile broadband is growing exponentially. Mobile data traffic, spurred by mobile broadband growth, is expected to grow more than 26 times in five years (figure 7.3; CISCO 2012). The expansion of data-hungry devices, such as smartphones and tablets, are already resulting in exponential increases of traffic in some countries (see figure 1.5).

Unlike fixed broadband technologies that can make use of the almost unlimited capacity of fiber optics to cope with growing data traffic, mobile broadband networks must work with finite allocations of spectrum. Mobile operators rely on optimization of networks and traffic management to increase efficiency, at least in the short term.[4] However, operators may also use optimization and traffic management techniques to hinder competition through data caps and by blocking or "throttling" access to applications. For instance, mobile network operators may limit the bandwidth available to those applications that threaten to deprive them of revenue, such as Skype used as a substitute for voice calls. To avoid such practices, regulators have been imposing limits on traffic prioritization while permitting optimization of mobile broadband networks, within the network neutrality concept.

Network neutrality generally refers to the notion that an Internet Service Provider (ISP) should treat all traffic equally, including any content, application, or service (Atkinson and Weiser 2006). Based on this principle of nondiscrimination, a growing number of jurisdictions have adopted regulations that range from barring ISPs from managing internet traffic in a way that discriminates among content providers to permitting "best efforts" to deliver content on equal terms. These regulations have generally not been applied to mobile networks, however. In some cases, the justification for the exemption has been to allow mobile broadband networks to develop. Some governments are now beginning to regulate certain practices, for example by requiring full access to certain applications (such as Voice over IP services, like Skype).[5] It is also useful to promote transparency on the part of operators to explain how they are applying traffic management.

Limit spectrum hoarding that could distort competitive conditions in the market. Making spectrum available to the market is critical for developing mobile broadband, but this spectrum also must be used efficiently. Operators should use their spectrum allocations to provide services and not to distort the market or impede other providers from entering the market. To avoid these pernicious effects, governments have introduced limitations in awarding spectrum, such as

Box 7.1 Using reverse auctions to match spectrum allocations with coverage obligations in Chile

In Chile, the government provided spectrum in multiple bands for mobile broadband in underserved rural areas. Chile offered subsidies through a reverse auction (resulting in a government subsidy of more than $100 million) to develop mobile broadband in around 1,500 municipalities in rural areas, where no broadband service was available. Extending coverage to these areas could mean that 90 percent of Chile's population would have broadband coverage. Minimum service conditions for broadband access (such as a 1 Mbit/sdownlink) and a ceiling on prices was established. The winner of the auction, Entel Movil, started deploying mobile broadband in these areas in September 2010.[a] The large expansion of mobile broadband services in the country, has permitted Entel Movil to achieve the largest share of mobile broadband connections in the country, surpassing its other two main competitors (figure 7.1.1).

Box figure 7.1.1 Mobile broadband subscriptions per operator in Chile

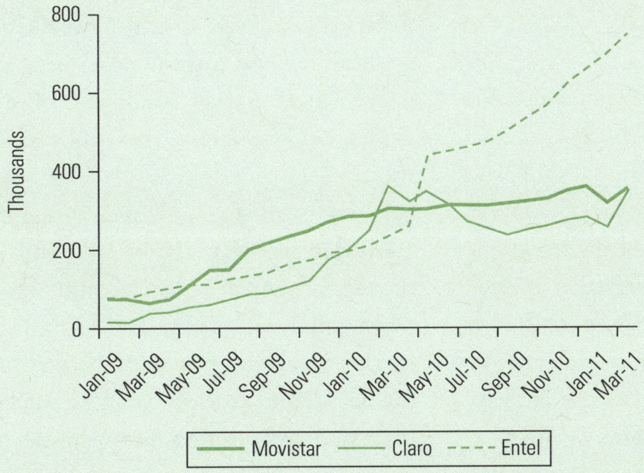

Source: Subtel.

a. Subsecretaria de Telecomunicaciones, Chile. 2010. Proyecto Bicentenario: "Red de Internet rural: Todo Chile comunicado." http://www.subtel.gob.cl/prontus_subtel/site/artic/20100819/asocfile/20100819103226/ppt_bicentenario_fdt_red_internet_rural.pdf; Entel, Todo Chile Comunicado. http://personas.entelpcs.cl/PortalPersonas/appmanager/entelpcs/personas?_nfpb=true&_pageLabel=P60015586312803495709 94.

spectrum caps in specific bands (see above) or sunset clauses in case the spectrum is not brought into timely use by a certain date. However, governments should be wary of imposing spectrum caps that are too stringent and might impede operators' ability to react to market demand. Broadband data traffic demand is expected to require increasing amounts of spectrum, especially in urban areas (Rysavy Research 2010). For this reason, it is advisable for governments to be flexible in using spectrum caps and monitor the market needs and competitive conditions as they evolve. If competition conditions are not in danger, regulators and policy-makers would be better off monitoring market conditions rather than establishing spectrum caps. Mobile broadband demand can grow very quickly as more and more applications are developed and handsets prices are reduced (see below). In this scenario, caps that are too stringent may result in underdevelopment of mobile broadband services, lower speeds, or limited quality of service.

Figure 7.3 Mobile data traffic by 2016, CISCO forecast

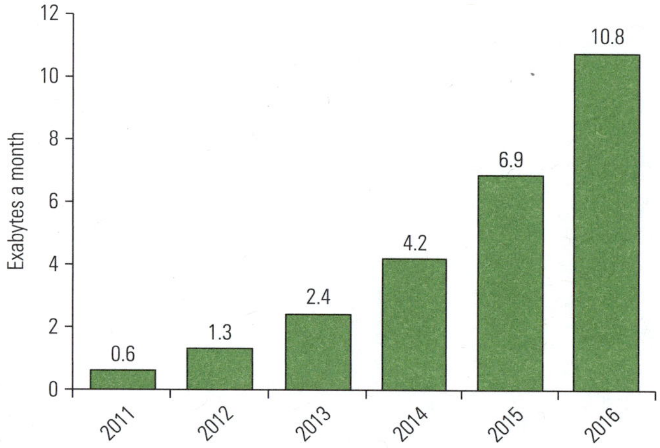

Source: CISCO 2012.
Note: The compound annual growth rate between 2011 and 2016 is projected to be 78 percent.

Foster the development of national broadband backbone networks. In contrast to voice mobile networks, mobile broadband networks require high bandwidth backbones to support the delivery of broadband to end users. To support rising volumes of mobile broadband traffic, the backbone networks of the mobile platform must be upgraded to fiber. Governments can support the development of backbone networks by enacting infrastructure sharing policies, allowing mobile operators to make rational build or lease decisions, streamlining procedures to obtain rights of ways (by issuing national rights of way, for example), and adopting other specific policies. In addition, governments can foster the development of backbone networks by coordinating with the private sector, providing seed capital for the development of backbone networks, and enabling public-private partnership (PPP) schemes. However, governments must be careful to avoid market distortions when intervening in the infrastructure market.

In addition, governments can also encourage the opening to broadband operators of fiber infrastructure deployed by other utilities, such as electricity, roads, or water. Many utilities have already deployed fiber networks for internal operational purposes, and their surplus capacity can be utilized for broadband development. Indeed, this surplus fiber capacity can serve to build or complement mobile broadband backbone networks (Arthur D. Little 2010).

Foster infrastructure and spectrum sharing. Policies that encourage infrastructure sharing allow operators to develop common networks, share costs, and hence lower investment requirements, all of which can result in lower prices for users.[6] In Kenya, instead of auctioning LTE-band spectrum to separate operators, the government is planning to implement a PPP model with a sole network with LTE-band spectrum available on an open-access basis. The possible risk is that by creating an effective monopoly, deployment may be slow and inefficient. On the other hand, by requiring companies to share a common infrastructure, the aim is to reduce duplicate investment and minimize competition distortion (Msimang 2011).

Demand-side policies

Demand-side policies aim at expanding adoption of broadband services by addressing the barriers to adoption and fostering the development of broadband-based services and applications and thereby promoting user demand. As with supply, local market conditions affect the effectiveness of demand-side policies, and policy-makers and regulators should take good note of those conditions. Two main barriers to entry are relatively common among developing countries, namely, the availability and affordability of broadband-enabled devices and service. In addition, the development of services and applications that address local market needs has proven to be a critical driver of demand for broadband services, because such services can improve their value proposition for businesses and consumers.

Ensure the availability and affordability of broadband-enabled devices. As mobile broadband has expanded globally, the reach of broadband-enabled devices, such as handsets and tablets, has increased, and their price has fallen. As penetration continues in developing countries, manufacturers are targeting these markets by providing low- and ultra-low-cost devices and designs tailored to these markets' needs. The global market for handsets has seen a continual reduction in prices even as performance increases. Mobile broadband handsets, or smartphones, have fallen in price from more than $300 in 2005 to less than $100 in 2011 for low-end models (IBM 2011; Kalavakunta 2007). Devices costing under $16 are forecast by 2015 (Scottsdale 2011).

However, barriers such as taxes, import restrictions, and duties may prevent consumers from benefiting from best global market prices (Katz et al. 2011). Direct sales taxes affect all legitimate handsets on sale within a country, and their level should be assessed carefully by policy-makers to avoid limiting broader access or spurring a profusion of "gray market" devices. Import restrictions and duties apply only to imported devices, but given that equipment manufacturing has become a global industry, virtually all devices are imported to some extent. The combination of sales taxes and import duties may increase prices to unaffordable levels for most of the population. For instance, in Bangladesh handsets are subject to a 12 percent import duty and an additional sales tax of 15 percent (Boakye et al 2010).

Subsidization of handsets by the mobile voice industry has made them affordable but has kept service prices high. As a result, a few countries, such as Finland, have made the practice illegal.[7] In the case of mobile broadband, though, high-end devices that make use of more efficient networks (such as LTE) may actually reduce unit prices for data. So, policy-makers should be prepared to show evidence of market distortion effects before imposing bans on subsidizing broadband-ready devices.

Finally, some countries have promoted domestic development of cheap handsets. For instance, India has fostered the development of cheap tablets coupled with a program of subsidies for the education sector, making tablets for education available for $35, less than 3 percent of that country's annual gross national income (GNI) per capita.[8] Not all countries have the manufacturing base, low labor costs, and large domestic market size of India, however, so policy-makers need to evaluate carefully the potential for success of these kinds of policies in their local markets. Without import protection, it is difficult to compete on cost and quality with the global market.

Enable increasing affordability of broadband services. Along with the cost of the handsets themselves, service costs may deter access to broadband. Mobile operators have generally been successful at reducing the total cost of ownership for mobile phones, in best practice cases to below $5 a month for a basket of services.[9] Prepaid offerings have been the most successful marketing strategy to increase the affordability of mobile services. In fact, prepaid service has been an important driver of mobile telephony in developing countries; for example, more than 80 percent of all users in Africa, Asia Pacific, and Latin America in the third quarter of 2011 bought prepaid service (Wireless Intelligence 2011).

A similar strategy is being applied to mobile broadband. Operators provide prepaid packages and other tailored services for mobile broadband services, such as offering a USB (universal serial bus) "dongle"[10] with a certain amount of data that can be used on laptops or PCs over cellular networks. For instance, in the Arab Republic of Egypt mobile operators are offering prepaid traffic-based mobile broadband access starting at $8, less than 4 percent of the monthly GNI per capita.[11] In Colombia operators offer prepaid mobile broadband for different prices based on duration and service access, ranging from $0.5 a day for chat or email access only to $25 a month for full broadband access, less than 6 percent of monthly GNI per capita at the highest offering.[12] Policy-makers and regulators should enable these practices and avoid distorting the market unnecessarily. Imposing a high level of taxes (particularly direct taxes) on mobile broadband service may reduce their affordability and deter adoption (Katz et al. 2011).

As the mobile voice market proved, competition among service providers is also a critical driver of price reductions and innovative offerings that increase affordability (Rossotto et al. 1999). Policy-makers and regulators should safeguard competitive conditions in the market and, when needed, increase competition (by reducing barriers to entry to the market, for example, or increasing the number of licenses).

Enable the development of broadband applications and content. Applications and content are drivers of broadband demand. Broadband in itself does not provide much value directly to business and consumers. It is the applications and content that can be accessed through broadband that consumers want. Mobile broadband has made this link even more evident. Adoption of mobile broadband services is closely followed by applications growth for this service (figure 7.4).

Mobile applications are easier to use than earlier web-based applications and allow additional features, such as geo-location of services, unique to mobile services. Coupled with social networks, applications are now the main demand drivers for mobile broadband. But most mobile broadband applications and services are developed in and for developed countries. For instance, the vast majority of downloads for the Android platform have occurred in the United States, followed by the Republic of Korea, Japan, and other developed countries (Empson 2011).

Figure 7.4 Mobile applications as a driver of mobile broadband demand

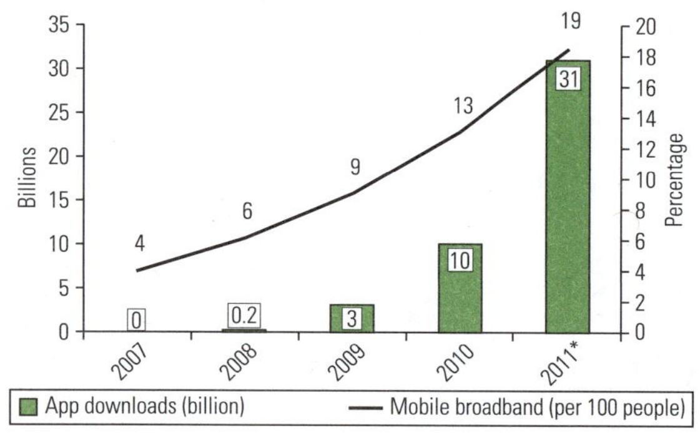

Source: Adapted from Apple, Google, and Wireless Intelligence.
Note: * Estimate.

To foster local demand for mobile broadband applications and content, policy-makers actively promote local capacity for development and customization. Policy-makers can develop policies to provide the right enabling environment for this industry and to actively foster its development through the creation of a mobile broadband innovation ecosystem. Co-creation platforms linking educational institutions and industry as well as technology hubs and crowdsourcing strategies are some of the tools for creating such an ecosystem. In addition, policy-makers can encourage government agencies to develop mGovernment applications (see chapter 6) and content for mobile broadband (through open data policies, for example), as well as acting as a consumer for sectoral applications (in education or health, for instance), in order to create a critical mass for the development of local applications and content.

Conclusions

Fostering mobile broadband diffusion in developing countries requires appropriate policy actions to enable and encourage both components of the mobile broadband ecosystem—supply and demand. Policy-makers should evaluate local conditions before applying specific policies, screening for bottlenecks or market failures on each of side of the ecosystem. The most common bottlenecks and market failures on the supply side are spectrum and backbone networks. On the demand side, limited availability of affordable broadband-enabled devices and services, as well as the lack of local applications and content, are the main bottlenecks and market failures. The policy recommendations described in this chapter provide guidance on how to address these common barriers.

This report has shown the potential of mobile applications to transform different sectors of the economy while benefiting the livelihoods and lifestyles of citizens and communities. Mobile broadband is an important element in that process, because it will offer the tools, from smartphones to services, that enable that transformation to take place: from access to apps.

Notes

1. http://www.itu.int/ITU-D/ict/statistics/.
2. Telegeography Inc., Globalcomms database, 2012.
3. Telecoms.com, 2010, "Sweden to Auction 800 MHz Spectrum in February" (December), http://www.telecoms.com/23770/sweden-to-auction-800mhz-spectrum-in-february-2011/; IT World, 2010, "Spectrum for Rural 4G Auctioned Off in Sweden" (March), http://www.itworld.com/mobile-amp-wireless/139121/spectrum-rural-4g-auctioned-sweden; Economist Intelligence Unit, 2011, "Germany/Sweden Telecoms: Fixing Mobile Broadband" (June), http://viewswire.eiu.com/index.asp?layout=ib3Article&pubtypeid=1162462501&article_id=1838266568&rf=0.
4. Mobile network optimization and self-organizing networks are expected to grow over 84 percent from 2010 to 2015 as LTE networks are deployed worldwide. See TotalTelecom (December 2011–January 2012), http://www.totaltele.com/.
5. For instance, the United Sates has limited the application of network neutrality principles to wireless operators. However, the government prohibits operators from blocking certain websites and applications. In France, network neutrality rules apply to all broadband operators (including wireless), although the regulator can still apply less stringent rules for traffic management for mobile operators based on objective reasons. In the Netherlands the Parliament passed a law forbidding mobile operators from blocking applications, particularly VoIP and text messaging. See http://www.iptelephonyusa.net/internet-protocol/2846-dutch-pass-law-to-ensure-open-internet-access.
6. Telecoms.com, 2009, "Tele2, Telenor to Build Swedish LTE Network" (April), http://www.telecoms.com/10423/tele2-telenor-to-build-swedish-lte-network/; Telegeography, 2011, "Telia and Telenor Share Danish Networks" (June), http://www.telegeography.com/products/commsupdate/articles/

2011/06/14/telia-and-telenor-share-danish-networks/; Unwired Insight, 2010, "LTE Leader TeliaSonera Launches 4G in Denmark" (December), http://www.unwiredinsight.com/teliasonera-4g.

7. "Market Analysis of Mobile Handset Subsidies," http://www.netlab.tkk.fi/tutkimus/lead/leaddocs/Daoud_Haemmaeinen_slides.pdf.
8. "India's Aakasha Tablet Soon to Be Free for Students" (February 2012), http://androidcommunity.com/indias-aakash-tablet-soon-to-be-free-for-students-20120208/.
9. Nokia and LIRNEasia conduct an annual survey of the total cost of ownership (TCO) of mobile, covering user prices for voice, SMS and data, a SIM card, taxes, and local handset costs in 50 countries. In the June 2011 study, Sri Lanka came out the cheapest, at $2.91 a month; 10 other countries had a TCO under $5 a month, excluding data. By contrast, in Morocco the same basket of services provided in Sri Lanka would cost $52.14; see: http://lirneasia.net/2011/06/nokia-annual-tco-total-cost-of-ownership-results-show-bangladesh-and-sri-lanka-as-cheapest/.
10. A "dongle" or data card is a piece of hardware that plugs into a PC, tablet, or other computing device to permit it to use mobile data services. Similar to Wi-Fi cards that proliferated in the early 2000s, the market for such devices is likely to disappear once the hardware is increasingly built into the device itself.
11. See Vodafone Egypt's offering: http://www.vodafone.com.eg/vodafoneportalWeb/en/P604978041288690285509.
12. See Movistar's prepaid offering: http://www.movistar.co/Personas/Internet_Movil/Planes/Internet_prepago/internet_para_telefonos/.

References

Arthur D. Little. 2010. "FTTH: Double Squeeze of Incumbents—Forced to Partner?" http://www.adl.com/uploads/tx_extthoughtleadership/ADL_Double_Squeeze.pdf.

Atkinson, R., and P. Weiser. 2006. "A 'Third Way' on Network Neutrality." Information Technology and Innovation Foundation (May 30). http://www.itif.org/files/netneutrality.pdf.

Bauer, J. 2002. "Spectrum Auctions: Pricing and Network Expansion in Wireless Telecommunications." http://arxiv.org/ftp/cs/papers/0109/0109108.pdf.

Boakye, K., et al. 2010. "Mobiles for Development." http://www.mobileactive.org/files/file_uploads/UNICEF%20Mobiles4Dev%20Report.pdf.

Brugger, R., and K. Oliver. 2010. "800 MHz Auctions and Implementation of the DD in Germany" (October). http://tech.ebu.ch/docs/events/ecs10/presentations/ebu_ecs10_workshop_brugger_kluth.pdf.

CISCO. 2012. "CISCO Visual Networking Index: Global Mobile Data, Traffic Forecast Update 2011–2016." http://www.cisco.com/en/US/solutions/collateral/ns341/ns525/ns537/ns705/ns827/white_paper_c11-520862.pdf.

Delian, A. 2001. "3G Mobile Licensing Policy: From GSM to IMT-2000, a Comparative Analysis." http://www.itu.int/osg/spu/ni/3G/casestudies/GSM-FINAL.pdf.

Digits. 2011. "The Facebook App Economy." Center for Digital Innovation Technology and Strategy, University of Maryland (September 19). http://www.rhsmith.umd.edu/digits/pdfs_docs/research/2011/AppEconomyImpact091911.pdf.

The Economist. 2011. "Apps on Tap." http://www.economist.com/node/21530920.

EC (European Commission). 2002. "Comparative Assessment of the Licensing Regimes for 3G Mobile Communications in the European Union and Their Impact on the Mobile Communications Sector," http://ec.europa.eu/information_society/topics/telecoms/radiospec/doc/pdf/mobiles/mckinsey_study/final_report.pdf.

Empson, R. 2011. "Android Global: South Korea Second Only to US in App Downloads." http://techcrunch.com/2011/11/16/android-global-south-korea-second-only-to-u-s-in-app-downloads/.

Friedrich, R., K. Sabbagh, B. El-Darwiche, and M. Singh. 2009. "Digital Highways: The Role of Government in 21st Century Infrastructure." Booz & Company. http://www.booz.com/media/uploads/Digital_Highways_Role_of_Government.pdf.

Gruber, H. 2006. "3G Mobile Telecommunications Licenses in Europe: A Critical Review" (November). http://papers.ssrn.com/sol3/papers.cfm?abstract_id=918003&download=yes.

Hazlett, T., and R. Munoz. 2008. "A Welfare Analysis of Spectrum Allocation Policies" (December). http://mason.gmu.edu/~thazlett/pubs/Hazlett.Munoz.RandJournalofEconomics.pdf.

———. 2010. "What Really Matters in Spectrum Allocation Design" (April). http://businessinnovation.berkeley.edu/Mobile_Impact/Hazlett-Munoz_Spectrum_Matters.pdf.

IBM. 2011. "Telco 2015: Five Telling Years, for Future Scenarios." http://www.ieee-iscc.org/2010/keynoteSlides/Franco%20Prampolini.pdf.

ITU (International Telecommunication Union). World Telecommunication/ICT Indicators Database. www.itu.int/ti.

Kalavakunta, R. 2007. "Low Cost Mobile Broadband Access." http://www.itu.int/ITU-D/afr/events/PPPF/PPPF-Africa2007/documents/presentations/Session2/Low_Cost_BWA_06_04_2007.pdf.

Katz, R., et al. 2011. "The Impact of Taxation on the Development of the Mobile Broadband Sector." GSM Association. http://www.gsma.com/documents/the-impact-of-taxation-on-the-development-of-the-mobile-broadband-sector/19669.

Katz, R., S. Vaterlaus, P. Zenhäusern, and S. Suter. 2010. "The Impact of Broadband on Jobs and the German Economy." *Intereconomics:*

Review of European Economic Policy 45 (1, January). http://www.polynomics.ch/dokumente/Polynomics_Broadband_Brochure_E.pdf.

Kelly, T., and C. Rossotto. 2011. *Broadband Strategies Handbook.* *info*Dev and World Bank. www.infodev.org/en/Document.1118.pdf.

Msimang, M. 2011. "Broadband in Kenya: Build It and They Will Come." *info*Dev. http://www.infodev.org/en/Publication.1108.html.

Patrick, X. 2001. "Licensing of Third Generation (3g) Mobile: Briefing Paper." http://www.itu.int/osg/spu/ni/3G/workshop/Briefing_paper.PDF.

Qiang, C., and C. Rossotto. 2009. "IC4D: Extending Reach and Increasing Impact." *Economic Impacts of Broadband*, ch. 3. Washington, DC: World Bank.

Qiang, C., and L. Xu. Forthcoming. "Telecommunications and Economic Performance: Macro and Micro Evidence." World Bank, Washington, DC.

Rossotto, C., et al. 1999. "Competition in Mobile Telecoms" (April). http://www-wds.worldbank.org/servlet/WDSContentServer/WDSP/IB/1999/09/14/000094946_99072807530926/Rendered/PDF/multi_page.pdf.

Rysavy Research. 2010. "Mobile Broadband Capacity Constraints and the Need for Optimization" (February). http://rysavy.com/Articles/2010_02_Rysavy_Mobile_Broadband_Capacity_Constraints.pdf.

Scottsdale, A. 2011. "Pent-Up Demand and Falling Price Drive Strong Entry-Level Market Growth." http://www.abiresearch.com/press/3573-Pentup+Demand+and+Falling+Prices+Drive+Strong+Entry-Level+Handset+Market+Growth.

Thomson, H., et al. 2011. "Economic Impacts of Mobile versus Fixed Broadband." *Telecommunications Policy.* http://www.sciencedirect.com/science/article/pii/S0308596111001339.

Wireless Intelligence. 2011. "Germany Rolls Out LTE to Rural Areas." https://www.wirelessintelligence.com/analysis/2011/06/germany-rolls-out-lte-to-rural-areas/.

Part II: Statistical Appendix

Key Trends in the Development of the Mobile Sector

Michael Minges

Access

Measuring mobile take-up

Mobile telephony has been one of the most quickly adopted technologies of all time. While 128 years passed before fixed telephone lines reached 1 billion users, mobile networks achieved this milestone in just over two decades (figure A.1). Even more astounding, mobile networks have roughly doubled in size every two years since 2002. By the end of 2011, there were 5.9 billion mobile cellular subscriptions worldwide.

This huge growth in mobile subscriptions has led to a significant increase in penetration. The traditional measure of mobile telephony penetration is the number of subscriptions per 100 people. By the end of 2011, more than 8 of every 10 people around the world had a mobile subscription, up from just over 1 in 10 in 2000, with particularly strong gains in middle-income countries (figure A.2). Over the span of a single decade, mobile telephones have changed from an elitist gadget that was mainly the preserve of high-income countries to a mass-market tool spanning the globe.

Some 90 economies—almost half of the member countries of the World Bank—had a mobile penetration exceeding 100 percent in 2011. Because there are more mobile subscriptions than inhabitants in these countries, these statistics do not reflect the number of people who actually have use of a mobile phone, because the same person may possess multiple SIM cards (for example, to avail themselves of lower on-network calling prices or to separate business from personal calls), and thus have multiple subscriptions. In the United Arab Emirates in 2010, for example, 28 percent of subscriptions were duplicates, mainly for these reasons, as well as for roaming and gaining better coverage in different parts of the country (UAE 2011).

Another factor skewing the figures in some countries is the number of mobile cellular subscriptions taken by people residing in bordering nations. Subscriptions can also be inactive, with the length of time that must pass before the subscription elapses varying by operator. At the same time, an increasing number of devices are connected to mobile networks that do not use voice services or do not interface with humans. These include laptop computers, as well as equipment such as automated teller machines. In Spain, these types of subscriptions accounted for 10 percent of the mobile market in 2010.

Although methods other than counting subscriptions may be more precise for measuring access to mobile phones, subscription data are most widely available. For example, another useful measure could be the number of persons with access to a mobile phone. But gathering that data requires the use of surveys, which are conducted only infrequently or, in many countries, not conducted at all.

Another measure is the number of households where at least one household member has a mobile phone. This metric is useful because it is precise: it cannot exceed 100. If a mobile phone exists in a household, then all members

Figure A.1 Worldwide fixed and mobile telephone subscriptions

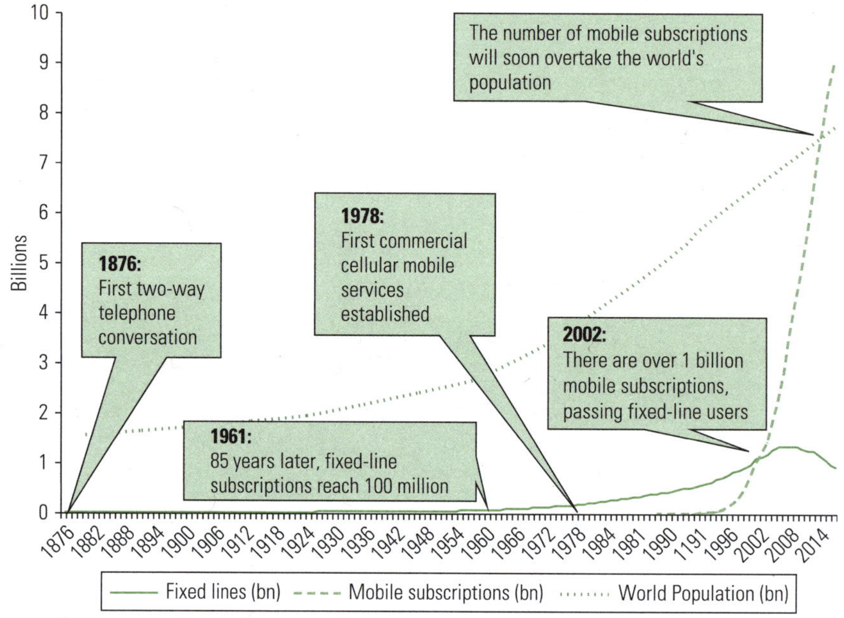

Source: Adapted from ITU, World Bank estimates.
Note: Log scale.

Figure A.2 Mobile cellular subscriptions per 100 people, by income group

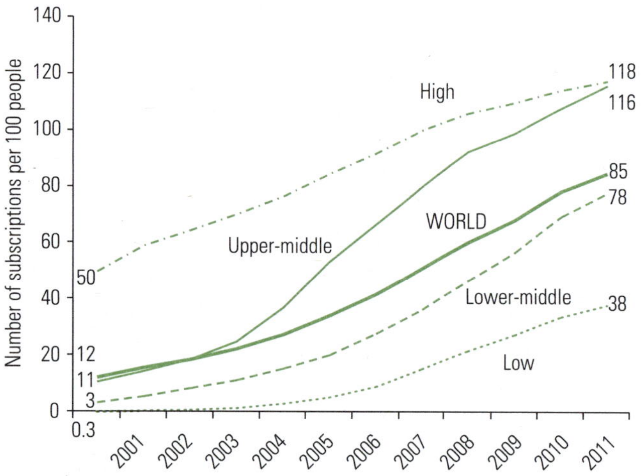

Source: Adapted from ITU, and author's own estimates.

could theoretically use it, thereby extending access. Household availability has thus been the traditional indicator for measuring universal service. This indicator is collected by a growing number of countries through ongoing household surveys, as well as special health surveys. The United Nations recommended in 2008 that the question "Household having mobile cellular telephone(s)" be included in the questionnaires used for the 2010 round of censuses.[1] Based on the surveys carried out by a significant number of countries, almost three of four households were estimated to have mobile phone service in 2010.

Another factor to consider is household size. Individual use surveys tend to exaggerate subscription penetration rates in developed economies, while household surveys suggest that the level of access to mobile is higher in many developing countries than subscription penetration figures would suggest. Access is particularly high in countries with large households. Take Senegal, where the subscription penetration was 57 per 100 people in 2009, but household penetration was estimated to be 30 points higher at 87 (figure A.3a). This larger household size can dramatically extend access to mobile phones, considering that on average nine persons are in each Senegalese household. Several low-income nations have higher mobile phone home penetration than some developed economies. For example, Senegal, along with some other low- and middle-income economies, has a higher proportion of homes with mobile phones than either Canada or the United States (figure A.3b).

Figure A.3 **Mobile household penetration, Senegal and other selected countries, 2009**

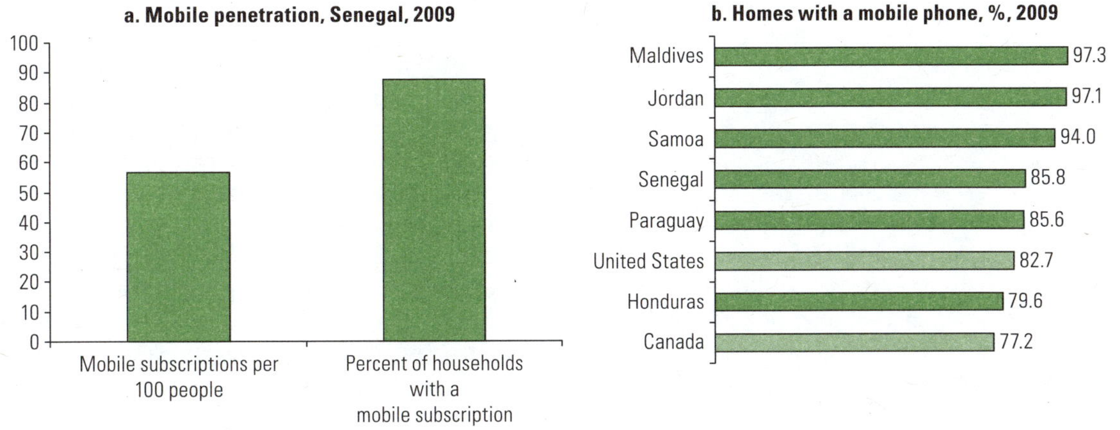

Source: Adapted from Autorité de Régulation des Télécommunications et des Posts (Senegal) and national household and health surveys.

Reaching the base of the pyramid

At the turn of the new millennium, most analysts would have considered a world with 6 billion cellular phones impossible. At the time, there were some 700 million mobile subscriptions, 70 percent of which were located in high-income economies. This link between mobile penetration and national income gave rise to a belief that there was a price below which mobile service would be unprofitable, thereby making it commercially unviable and unaffordable for many in lower-income countries. After all, fixed telephones had been in existence for more than a century, yet penetration rates were still less than 1 in 100 in many developing nations.

The mobile industry has defied that theory. Every year, it expands its user base, reaching more and more low-income users. This has been made possible by cheaper equipment, falling handset prices, prepaid subscriptions, flexible regulation, competition for marginal users as markets become saturated at the top, and rising incomes.[2] A recent study carried out in three provinces in China found that 95 percent of rural households had a mobile telephone (box A.1). Nonetheless, a significant proportion of the world's population has no mobile connection. Of the some 5.9 billion mobile subscriptions in the world, 3.4 billion were in low- and middle-income economies (figure A.4). Given some 4.8 billion residing in those countries, that leaves a gap of 1.4 billion without a mobile subscription. The number of people living on less than $1.25 a day (purchasing power parity) in low- and middle-income economies, estimated to be 1.1 billion, might be considered outside the target market.[3] That leaves an addressable unserved population of just 300 million people worldwide at the start of 2012.

Mobile equipment manufacturer Nokia has calculated a total cost of ownership (TCO) measure that factors in the cost of the handset, service charges, and taxes (Nokia 2009). The TCO needs to be adjusted by income, given that levels of income vary between countries. Even if users can afford service, they still need signal coverage. Figure A.5 illustrates affordability and coverage for selected developing countries. The relationship lends itself to four scenarios bounded by affordability of 10 percent (that is, where mobile services are either less or more than 10 percent of income) and coverage (where mobile covers either less or more than 9 percent of the population). These scenarios are reflected in the four quadrants in figure A.5:

1. high affordability and high coverage (upper left quadrant)
2. high affordability and low coverage (lower left quadrant)
3. low affordability but high coverage (upper right quadrant)
4. low affordability and low coverage (lower right quadrant)

Countries where mobile services cost less than 10 percent of income and cover at least 90 percent of the population

Box A.1 Mobile use in rural China

An ongoing World Bank project has been investigating attitudes, use, and impact of information and communication technologies in rural China. Funded by the Bill and Melinda Gates Foundation, one of the activities was a survey in rural areas of three provinces (Jilin, Guizhou, and Shandong). Some 58 percent of the population in these provinces is rural; the combined rural population is 88 million, which would make the three provinces the 13th largest country in the world (about the size of Vietnam). The survey, carried out in October 2011, found very high use of mobiles, with 95 percent of rural households reporting having one. Individual ownership was lower at 85 percent, but over half of individuals without their own mobile reported they did not have one because they could use someone else's or they had no need. Around half of mobile phone owners reported sending text messages, and some 13 percent use the internet from their cell phone. One interesting finding was the relatively large amount spent on mobile services. Average monthly mobile phone service expenditure was 13 percent of income, with users willing to devote up to 18 percent of their income to mobile services.

Box Figure A.1.1 Mobile usage in rural areas of three Chinese provinces, 2011

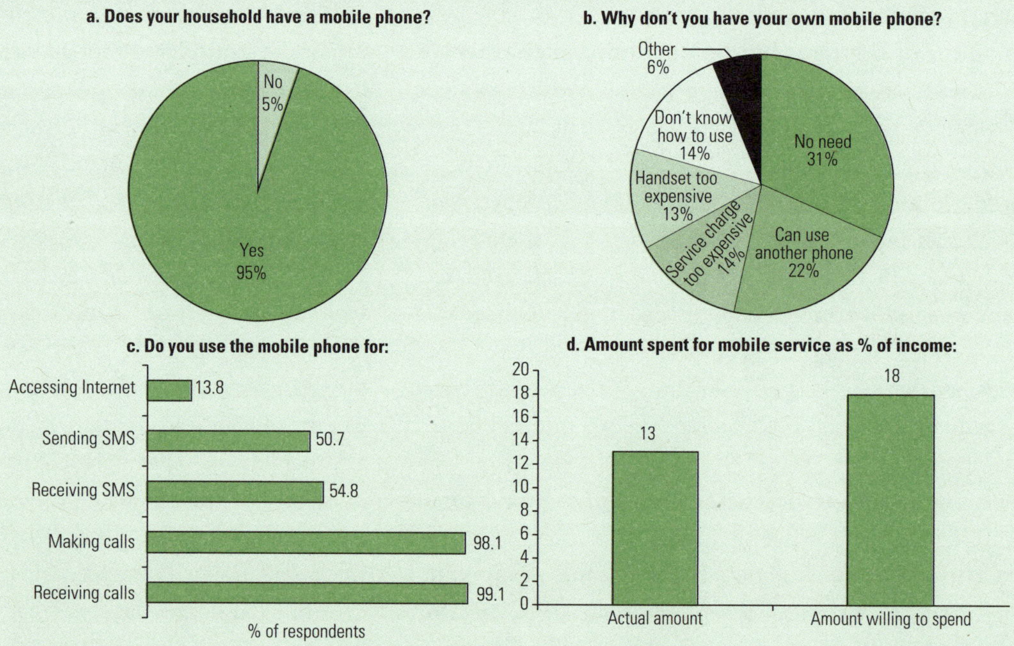

Source: World Bank.

tend to have high levels of access (measured by the availability of mobile phones in households).

Service charges alone do not explain the problem. Consider Angola, which has relatively low tariffs but also low coverage. Mainly because of lack of competition, Angola has not been successful in expanding mobile coverage compared with peer countries. While two operators have worked in the country for more than a decade, they used different technologies, which drove up equipment costs and made it difficult for subscribers to switch from one operator to the other. A mobile technology (GSM) common to both operators has been available only since November 2010. Further, the market remains a duopoly.

Figure A.4 Population, mobile subscriptions, and poverty headcount in low- and middle-income economies

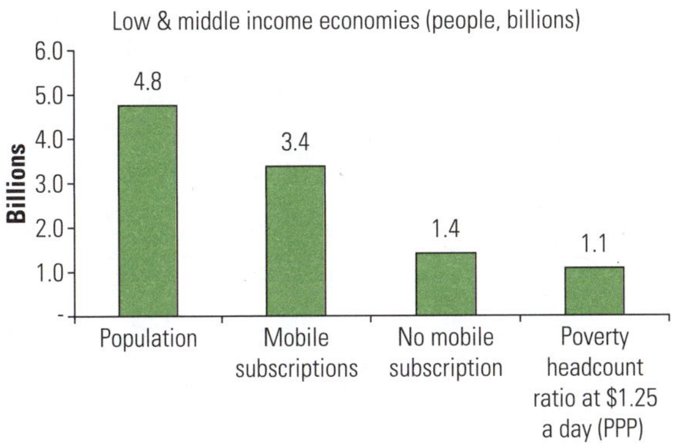

Sources: ITU and World Bank data and World Bank estimates.
Note: PPP = purchasing power parity.

Figure A.5 Affordability and coverage in developing economies

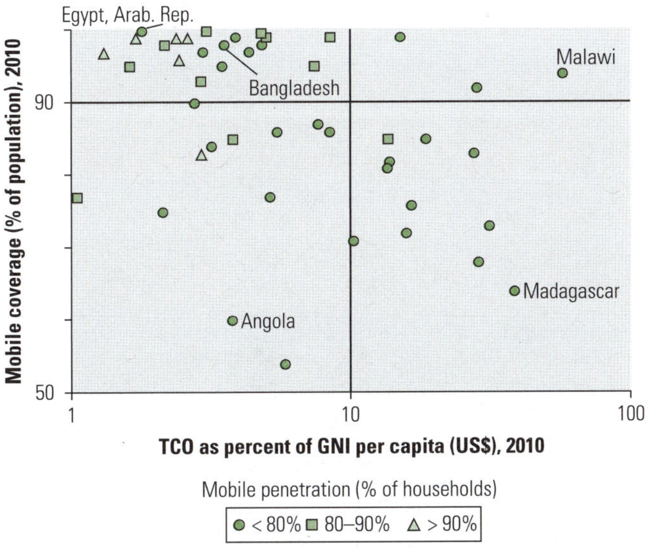

Sources: Adapted from Nokia (2009), ITU and World Bank estimates.
Note: Horizontal scale is logarithmic. TCO is "total cost of ownership," reflecting the average costs, by country, of handset purchase, service charges, and taxes. GNI is gross national income per capita. The ratio of TCO to GNI per capita is therefore an approximate measure of affordability per capita.

At the same time, other countries have relatively high coverage along with relatively high prices. Consider Malawi, where mobile networks are estimated to cover more than 90 percent of the population but where the Nokia annual TCO amounts to more than half of per capita income. Densely populated and relatively small, Malawi has been relatively easy to cover. Attempts to introduce additional competition have not been completely successful, however, and the market remains dominated by two operators. The least desirable position is to have high tariffs and low coverage. In Madagascar, the Nokia TCO amounts to over one-third of income, and only around three-fifths of the population is covered. Although there are three operators, competition has been affected by high interconnection charges.

In contrast, some countries have a high degree of affordability and coverage but relatively low take-up. The Arab Republic of Egypt has a high penetration of fixed telephone lines that provide an alternative to mobile. In Bangladesh mobile calls cost about one U.S. cent a minute, and, according to Nokia, its mobile tariffs are among the lowest in the world.[4] Coverage is high at 99 percent of the population. Despite these extremely low prices and very high coverage, household penetration stood at around 64 percent in 2010. According to mobile operator Grameenphone, its attempts to expand access are difficult because of the high SIM tax, which has remained "the biggest barrier to the growth of mobile telephone industry in Bangladesh" (Grameenphone 2011). The tax of Tk 800 ($11.60) on new SIM cards has a huge negative impact on low-end subscribers. If the SIM tax were eliminated, an estimated 90 percent of Bangladeshi households could afford mobile service. The GSM Association has called on the Bangladesh government to end the SIM tax, citing it as the "single largest obstacle to the acquisition of new subscribers." (GSMA 2009)

Operators are looking at innovative ways to widen access, including lowering recharge values, conducting more consumer research among bottom-of-the-pyramid populations, and developing low or alternative energy base stations. Another possibility is through virtual telephony using emerging cloud networks. Users would not need to buy a handset and would instead be allocated a number that they can use on a borrowed phone. Their contacts and voice mail would be stored on the cloud, where there would also be a gateway to mobile money services. Virtual telephony also lowers the cost of acquiring new users; for example, a trial network in Madagascar claims it costs operators just $0.20 to establish cloud-based virtual telephony services, compared with $14–$21 to deliver a SIM card.[5]

The barriers to increasing access to mobile communications for every household in the world are more of a regulatory and policy nature rather than technical. Introducing and strengthening competition and eliminating special "mobile" taxes could significantly narrow the range of those not served by mobile communications. The remaining few households without access could then be captured through universal access programs. It is also important to ensure that those at the bottom of the pyramid also enjoy access to value-added services, which requires capacity building to understand how these services can benefit their lives and how to use them.

Mobile broadband

Using the ITU/OECD definition of broadband—networks with a minimum download speed of 256 kilobits per second (kbit/s)—the first mobile broadband networks were launched in late 2000 in Japan (W-CDMA) and in 2001 in the Republic of Korea (EV-DO). According to industry sources, there were 939 million mobile broadband subscriptions worldwide in June 2011 (figure A.6a). This number implies that just over 15 percent of the global subscription base can theoretically use mobile network services at high speeds.

A number of these subscriptions are not *active* users of mobile broadband (that is, they do not use the internet at mobile broadband speeds, even though they are equipped to do so). Users could have a theoretical ability to use mobile broadband by having coverage and a mobile-broadband-enabled device, but they may not necessarily be using high-speed services, perhaps because of high prices. They could also be subscribing to mobile broadband and using a high-speed mobile service (such as video telephony), but not necessarily accessing the internet. Alternatively, they could be using mobile broadband to access the internet over handsets, as well as through laptops or tablets.

This definitional challenge presents analytical difficulties with interpreting mobile broadband statistics. The issue is whether to count and include theoretical access, active access to any high-speed service, active access using internet browsers, or active access via data cards (figure A.6b). Intergovernmental agencies have called for more clarity on mobile broadband statistics (OECD 2010). However, most countries report their mobile broadband statistics in insufficient detail,

Figure A.6 Mobile broadband

a. Estimated mobile broadband subscriptions, June 2011

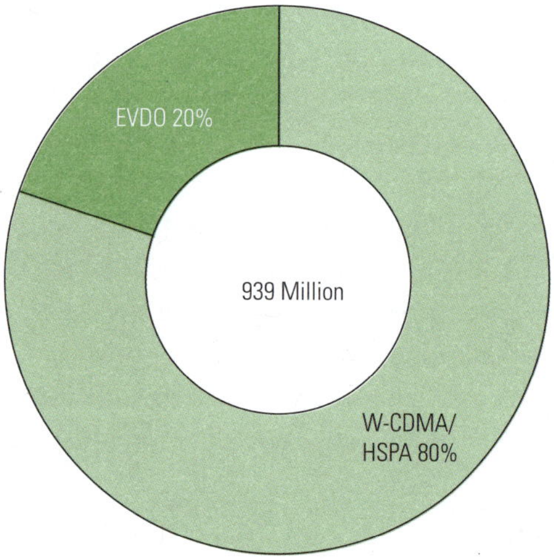

b. Definitional "layers" of mobile broadband

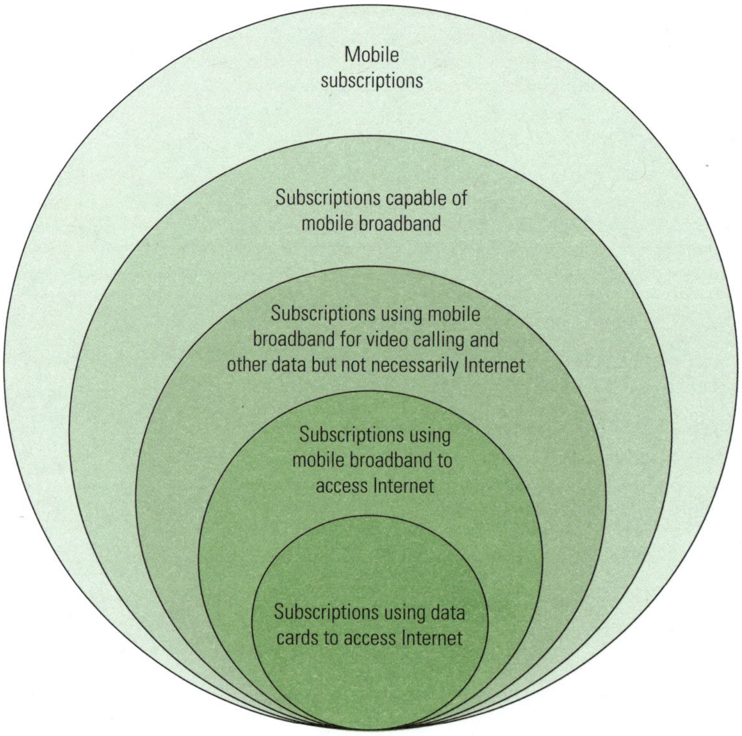

Source: Adapted from CDMA Development Group and Global Mobile Suppliers Association (figure A.6a).
Note: Not including LTE (estimated at 2 million subscriptions) or WiMAX (estimated at 20 million).

so data comparability remains limited. Given these definitional issues, some countries have gone with the lowest common denominator, counting only internet access through data cards as mobile broadband (denoted by the innermost circle in figure A.6b).

Despite confusion over statistical definitions, mobile broadband is already concretely impacting a number of developing countries, allowing them to leapfrog a lack of fixed broadband infrastructure. Based on the more certain yardstick of data cards (arguably the most direct comparison with wired broadband subscriptions), then mobile broadband far surpasses fixed broadband in nations such as the Philippines and South Africa (figure A.7). And if the wider definition for mobile broadband of plain internet access is applied, then the combination of wireless networks such as GPRS, EDGE, CDMA2000 1x, mobile broadband, and WiMAX greatly exceeds wired connections in most developing countries. An estimated 750 million people around the world accessed the internet from their mobiles in 2010, up from some 180 million in 2005. Developing countries in Asia account for over half of this total, with some two out of five mobile internet users in China alone.

Figure A.7 Broadband subscriptions in the Philippines and South Africa

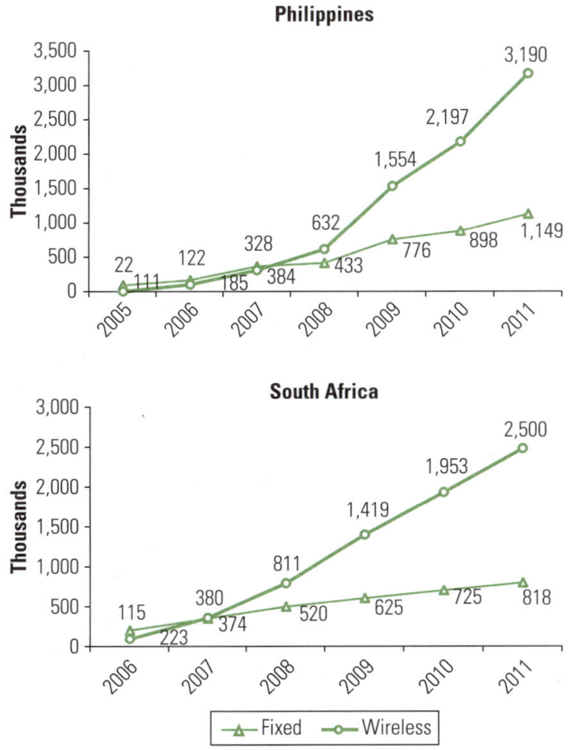

Sources: Adapted from Globe Telecom, PLDT, MTN, Telkom, and Vodacom.
Note: "Wireless" refers to data cards only and not to access directly from handsets. Data are for major operators only. Figures for South Africa have been adjusted to calendar years.

In addition to the statistical challenge of measuring active mobile broadband users, there are often significant shortfalls between the theoretical and actual speeds of data throughput. Manufacturers and operators cite ever-increasing bandwidth, but the average speeds fall far short. According to Akamai's analysis of 96 mobile networks across 58 economies carried out in the third quarter of 2011, peak speeds were around 8.9 megabits per second (Mbit/s), but average speeds were 1.8 Mbit/s (Akamai 2012). In contrast, Akamai reported average download speeds of 4.7 Mbit/s for fixed broadband networks. Further, usage over mobile broadband networks is generally "capped"; if users exceed a preset amount of data transfer, then they no longer have access to data services or their speed may be reduced or they will have to pay overage charges. Mobile data usage varies tremendously around the world. In the third quarter of 2011 it averaged 536 megabytes (MBs) per month across networks in 58 countries with a low of 22 MB per month and a high of 4,906 MB per month (table A.1).

While high-speed wireless holds promise for reducing the broadband divide, countries need to allocate spectrum and license operators to provide services. At the end of 2011, 46 World Bank members—almost all developing countries—had not commercially deployed mobile broadband services. And in a number of developing countries, a high-speed wireless service may technically exist, but it is often available only as a fixed wireless option.

Devices

According to Gartner, global sales of personal computers (PCs) numbered 353 million in 2011.[6] Assuming a PC is replaced on average every five years,[7] an estimated total of 1.6 billion PCs were in use around the world at the end of 2011. In comparison, some 1.8 billion mobile handsets were sold in 2011 alone (figure A.8a).[8] In other words, more mobile phones were sold in 2011 than the entire base of installed PCs. Sales of smartphones rose 59 percent in 2011 to more than 470 million units, about one of every four mobile handsets.

Another entry into the device world came in April 2010. The Apple iPad, which straddles the boundary

Table A.1 Mobile data speeds and volumes, Q3 2011

Economy	Network	Average speed (kbit/s)	Peak kbit/s	Average data usage (MB/month)
Australia	AU-3	1,553	7,878	222
Austria	AT-1	2,903	10,722	142
Belgium	BE-2	1,938	5,277	22
Bulgaria	BG-1	1,715	7,499	127
Canada	CA-2	1,171	2,923	608
Chile	CL-3	1,560	11,207	133
China	CN-1	1,475	3,927	247
Colombia	CO-1	1,003	6,541	156
Czech Republic	CZ-1	1,709	8,630	87
Egypt, Arab Rep.	EG-1	575	3,344	155
El Salvador	SV-3	926	4,782	353
Estonia	EE-1	1,401	7,487	264
France	FR-2	2,382	8,542	1,714
Germany	DE-1	967	3,720	93
Greece	GR-2	1,199	4,179	132
Guam	GU-1	957	4,663	101
Guatemala	GT-1	1,441	7,379	411
Hong Kong SAR, China	HK-2	1,925	10,842	583
Hungary	HU-1	1,863	8,481	130
India	IN-1	1,597	9,443	274
Indonesia	ID-1	475	7,172	4,906
Ireland	IE-1	2,880	14,055	725
Israel	IL-1	1,435	6,419	69
Italy	IT-4	1,413	8,693	219
Kuwait	KW-1	1,444	6,979	252
Lithuania	LT-2	1,973	11,945	414
Malaysia	MY-3	1,024	7,598	361
Mexico	MX-1	1,233	6,938	94
Moldova	MD-1	1,791	7,183	142
Morocco	MA-1	1,256	10,925	322
Netherlands	NL-1	1,763	4,871	36
New Caledonia	NC-1	1,070	4,757	854
New Zealand	NZ-2	1,880	9,988	768
Nicaragua	NI-1	1,551	7,886	754
Nigeria	NG-1	254	5,024	514
Norway	NO-2	2,071	6,752	58
Pakistan	PK-1	691	4,682	332
Paraguay	PY-1	643	5,850	163
Poland	PL-2	1,511	7,593	78
Portugal	PT-1	880	4,277	200
Puerto Rico	PR-1	2,639	10,975	2,703
Qatar	QA-1	1,620	10,074	281
Romania	RO-1	884	4,250	91
Russian Federation	RU-3	995	3,990	117
Saudi Arabia	SA-1	1,672	8,713	357
Singapore	SG-4	1,585	9,490	289
Slovakia	SK-1	327	2,077	38
Slovenia	SI-1	2,189	8,687	54
South Africa	ZA-1	438	1,386	168
Spain	ES-2	1,089	8,648	149

(continued next page)

Table A.1 *continued*

Economy	Network	Average speed (kbit/s)	Peak kbit/s	Average data usage (MB/month)
Sri Lanka	LK-1	894	7,373	327
Thailand	TH-1	149	1,412	135
Turkey	TR-1	1,771	7,975	203
Ukraine	UA-1	2,227	7,500	128
United Kingdom	UK-3	4,009	19,334	81
United States	US-2	1,072	4,411	47
Uruguay	UY-2	542	4,712	63
Venezuela, RB	VE-1	911	6,146	178
AVERAGE		**1,818**	**8,960**	**536**

Source: Akamai 2012.

Figure A.8 **Global sales of mobile and computing devices**

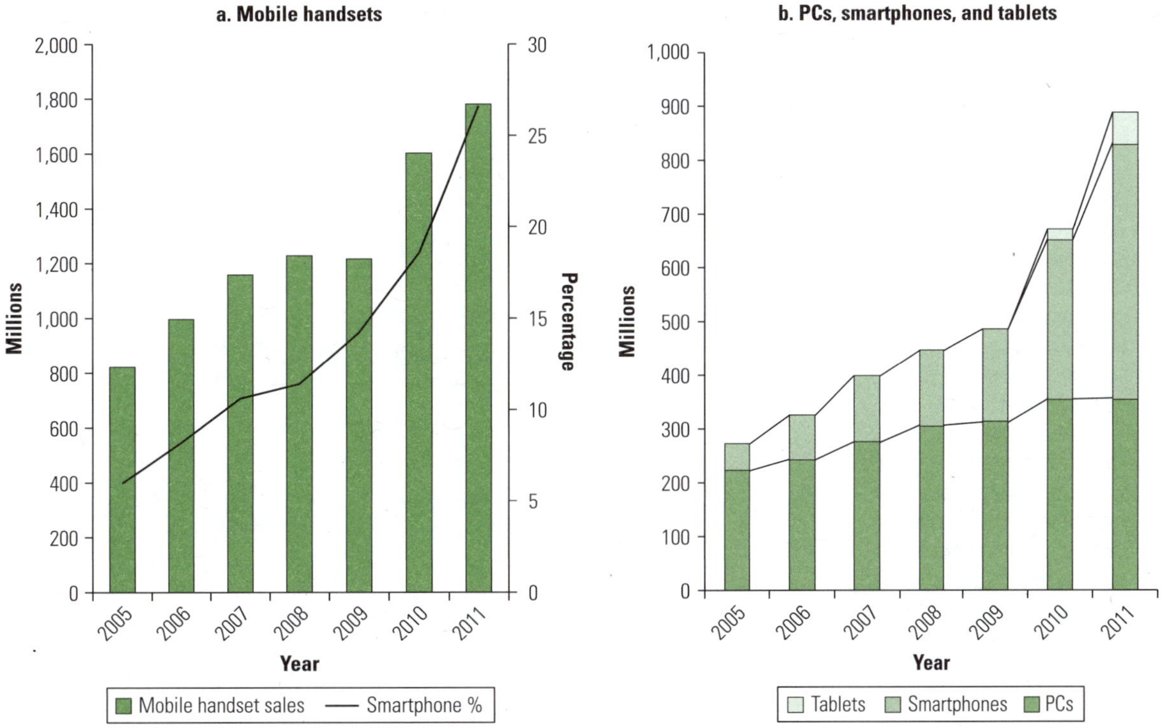

Source: Adapted from Gartner Inc.
Note: In these figures, PC includes desk-based and mobile PCs, including mini-notebooks, but not tablets.

between smartphones and laptop computers, created a new category of "tablet" computers. Just over 14 million iPads were sold in 2011. The launch of the iPad has helped attract more competitors into the tablet arena, and sales of all brands are expected to be close to 300 million by 2015. Combined global sales of smartphones and tablet computers exceeded those of PCs in 2011 (figure A.8b). The outlook for internet connectivity is clearly through a more portable and convenient device than a personal computer, with smartphones enjoying stellar growth in popularity (figure A.9).

Most mobile internet subscribers in developing countries are using low-end mobile handsets with minimal features, which limits their functionality, particularly for the development of advanced information and communication technology for development applications. For

Figure A.9 Smartphone penetration as a share of population, 2011

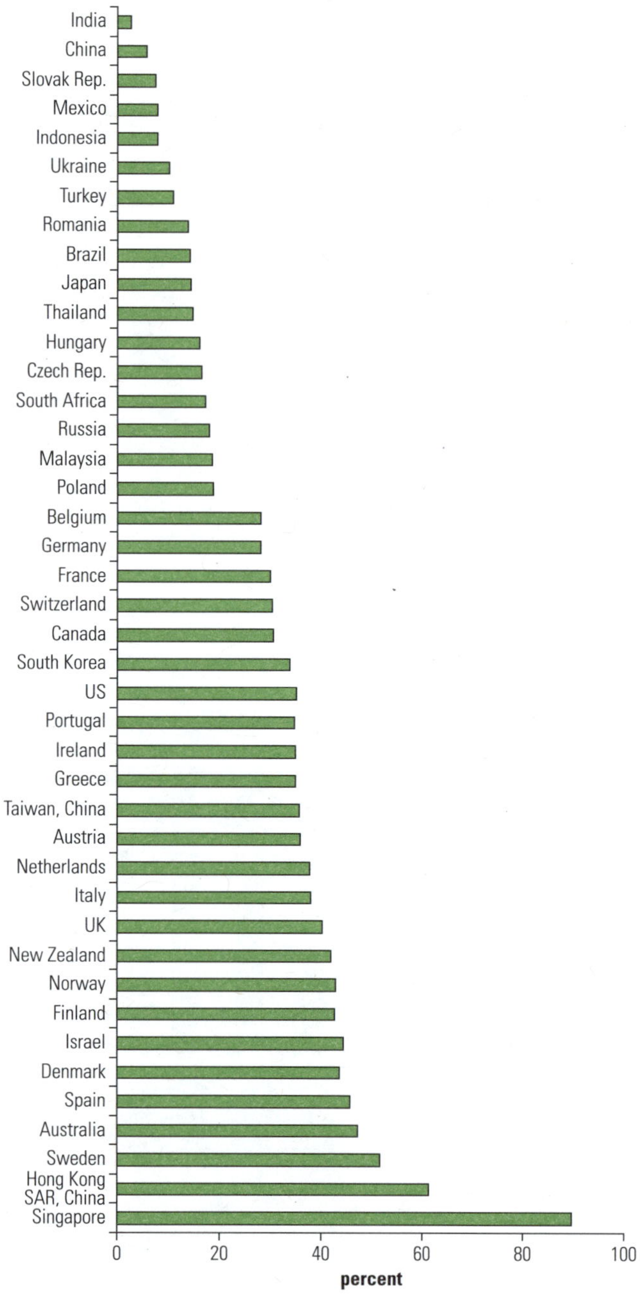

Source: Tomi Ahonen Consulting Analysis, December 2011. http://communities-dominate.blogs.com/brands/2011/12/smartphone-penetration-rates-by-country-we-have-good-data-finally.html.

smartphones and tablets to spread more widely and be adopted more rapidly in developing economies, their price needs to fall. Google is interested in developing a mass-market smartphone for emerging nations. It has been working with Indian handset manufacturers to develop an Android smartphone to be sold for $150, with the price eventually dropping to under $100. Although those prices will widen the potential target market considerably, such smartphones will still prove expensive for many Indians, who "can buy less advanced phones for $40 that have cameras and basic data services" (Sharma 2010).

Mobile industry

Mobile economy

The mobile industry is a significant player in many national economies. Mobile telecommunication operators generated an estimated $848 billion in revenue in 2011 (figure A.10).[9] That is around 1.2 percent of total global annual gross domestic product (GDP) and 56 percent of overall telecommunication revenues. The direct economic impact of the mobile industry varies across regions. While revenue remains at a consistent ratio of around 1 percent of GDP in most regions, in some developing regions, its direct impact is far higher. For instance, in Kenya, financial transactions via the M-PESA platform are estimated to equate to up to 20 percent of national GDP (World Bank 2010).

Mobile communications also has an indirect impact beyond its direct impact on the economy. The consultancy and accountancy firm Deloitte has developed a framework to illustrate the wider impact of the mobile communications services sector on the mobile ecosystem (Deloitte 2008).

Figure A.10 Global telecommunication services market

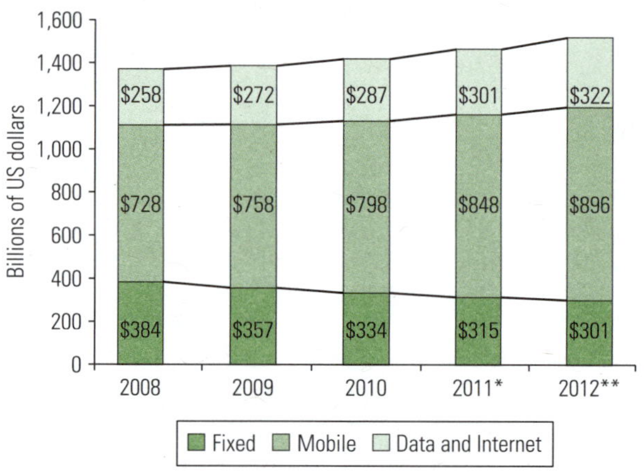

Source: Adapted from IDATE.
*estimate.
**forecast.

This ecosystem includes equipment suppliers, support services, resellers, and retail shops, as well as significant contributions to government in the form of taxes (figure A.11). Deloitte considers that mobile communications has three indirect economic impacts:

1. An impact on other industries related to mobile services, including network and handset suppliers, airtime resellers, and the like
2. An impact on end users from improved productivity, such as reductions in travel costs, improved job opportunities, and greater market efficiency
3. An impact on society related to such benefits as social cohesion, the extension of communications to low-income users, stimulation of local content, and disaster relief assistance

In addition, there are multiplier effects throughout the broader economy, as the initial spending related to mobile communications ripples through other sectors of the economy. Based on various economic studies, Deloitte estimates that this multiplier effect ranges from 1.1 to 1.7.

The model has been used to calculate the economic impact of mobile communications in a number of countries. One study of six countries in 2007 concluded that the direct economic impact of mobile communications ranged from 3.7 to 6.2 percent of national GDP (Deloitte 2008).

The employment impact of mobile communications is also significant. In addition to the direct employment of mobile operators, the Deloitte model includes related industries (such as equipment suppliers and airtime resellers), as well as spillover employment generated by government taxes and employment created from the consumption expenditures of personnel in mobile-related industries.

Strategic investors

The mobile services industry is one of the most globalized in the world. Practically every developing country has experienced foreign investment in its mobile cellular market, as have many developed nations. Opening up markets to

Figure A.11 Mobile value chain

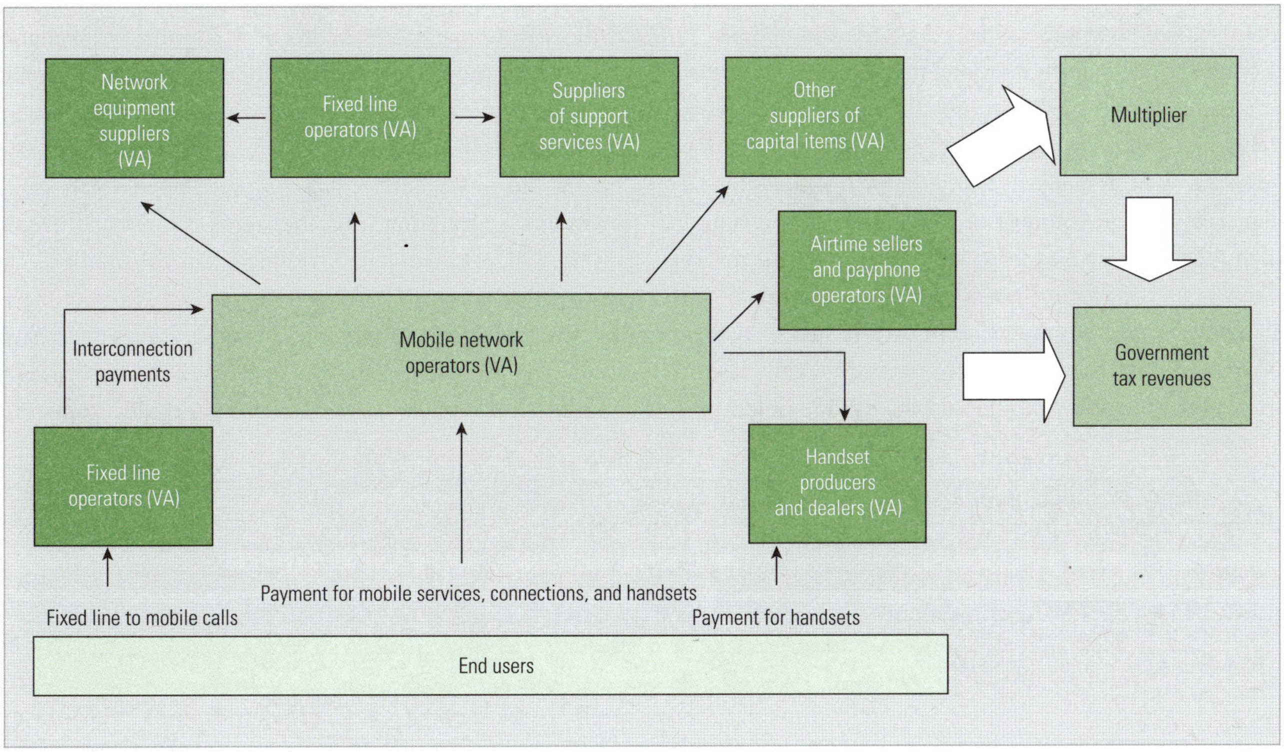

Source: Deloitte.
Note: Value added (VA) is specific to a national economy and does not show international value added.

privatization and foreign investment has been a major factor driving the growth of the mobile industry in emerging markets. According to the World Bank's Private Participation in Infrastructure database, between 1990 and 2010, some 329 projects in the mobile telecommunication sector in developing regions attracted $441 billion in private sector investment—much of it foreign and most of it from strategic mobile multinational groups (table A.2). No matter the size of a country, its political system, where it is located, or its income, private and foreign companies are willing to invest in mobile communications.

The relationship between foreign investors and host countries has changed considerably in recent years. Publicly held corporations now have to abide by a greater range of regulations and scrutiny relating to management, accounting, reporting, and governance than in the past. Deviating from these rules can have severe repercussions with investors, governments, and the public. Telenor, the Norwegian strategic investor, was rocked by reports of poor labor practices in firms that supply its mobile operations in Bangladesh. It immediately implemented reforms to remedy the situation.[10] At the same time, multinationals are responding to pressing social issues such as the environment and poverty by instituting recycling, corporate social responsibility, and similar programs and policies.

Today, mobile communications markets in many developing countries have achieved notable scale. Indeed, some developing country subsidiaries now enjoy larger subscriber bases than their foreign investors' home markets. One is Vodafone, where the number of mobile subscriptions in its Indian subsidiary is seven times larger than its home market of the United Kingdom. Growth in overseas markets means that investors are responding more to the needs of these overseas markets than previously and leveraging lessons learned abroad to apply throughout their group holdings.

Many investors are developing a growing geographic focus and specialization in certain regions. Although only a few strategic investors are engaged around the world, most focus on a specific region or geography. South Africa's MTN, for example, has investments throughout Sub-Saharan Africa and the Middle East. It has grown from operations in five countries in 2000 and just 2 million subscribers to operations in 21 countries and 142 million subscriptions in 2010. Digicel, which focuses on islands in the Caribbean and Pacific, has investments in 32 countries. Some multinationals channel their investments through regional holdings; for example, Vivendi works through Maroc Telecom, France Telecom through Senegal's Sonatel, and Vodafone through South Africa's Vodacom. This trend toward regional specialization makes investors better informed about their markets and enhances roaming, potential economies of scale, and platform-sharing.

A Mobile analytical tool

This edition of the World Bank Group's *Information and Communications for Development* report features a number of mobile indicators in both the chapter text and the statistical appendix. The wide variety of indicators used can make it difficult to gauge and benchmark country performance. Combining several significant indicators into a smaller number of composite indicators and tracking changes in them over time can provide a useful analytical tool for evaluating the outcomes of different investments and policy measures. These composite indicators can also be used to diagnose the strengths and weaknesses of the mobile sector in a particular country and thereby can serve as a useful tool for future policy development. The publication of an analytical tool is consistent with the previous edition of the report, which introduced a series of ICT performance measures,

Table A.2 Private participation in mobile networks, 1990–2010

Region	Number of projects	Investment commitments in physical assets (millions of current US$)
East Asia and Pacific	45	54,194
Europe and Central Asia	75	87,445
Latin America and the Caribbean	52	153,944
Middle East and North Africa	21	23,538
South Asia	31	69,286
Sub-Saharan Africa	105	52,305
TOTAL	**329**	**440,7132**

Sources: World Bank and PPIAF, PPI Project Database. (http://ppi.worldbank.org).

based on country groupings (World Bank 2009). Although the indicators used in this analytical tool are focused only on the mobile sector, the full range of ICT indicators used in the performance measures can be found in the World Bank's *Little Data Book on ICT*, the 2012 edition of which is being published in conjunction with this report.

There have been several methodological approaches and compilations of composite mobile indicators. The International Telecommunication Union (ITU) compiled a one-time "Mobile/Internet Index" in 2002 (ITU 2002). A framework for a composite mobile indicator with a focus on the internet has been proposed (Minges 2005). The ITU Digital Opportunity Index contained several mobile variables and allowed disaggregation into a mobile-only subcategory (ITU 2006) and was updated in 2007 (ITU and UNCTAD 2007). A mobile broadband composite indicator has recently been compiled for Latin American nations (A. T. Kearney 2012). None of these composite mobile indicators is particularly appropriate for this report because they have either been one-off, are not confined to mobile, or are limited to a particular region. Therefore a specially constructed mobile analytical tool, based on a series of composite indicators and building on the foundation of this earlier work, can help to fill the void.

In the context of the development orientation of the report, the Mobile Analytical Tool measures, on a country-by-country basis, the affordability and coverage of mobile networks (**universality**), the degree to which operators provide voice and advanced network services (**supply**), and the ownership and usage of mobile phones (**demand**). Each composite indicator is constructed from two separate indicators that measure these three components with equal weight given to each (figure A.12). The three composites could also be combined into a single measure if researchers found this useful, but that is not the intention here. The indicators are reproduced in the statistical appendix of this report, providing transparency and allowing users to recreate the analysis.[11]

The methodology is similar to that used for the United Nations Development Programme's Human Development Index (HDI). Each indicator has an equal weight. Indicators are converted to standardized values based on a logical 100 percent "goalpost." This is a straightforward conversion except for affordability, which is subtracted from 1 to reflect best performance. Although the affordability value may never reach 1 (where mobile service would be free), in 2010 there were twenty-six economies where the price of a mobile basket was less than 1 percent of per capita income (ITU 2011).

Table A.3 uses the data for Morocco to provide an example of the construction of the Mobile Analytical Tool.

The analytical tool has been applied to a representative range of 100 economies with data availability for the years 2005 and 2010 (table A.4 at the end of the appendix). The results provide some interesting insights into the development of mobile networks over that critical time period.

Figure A.12 Mobile analytical tool: indicators and categories

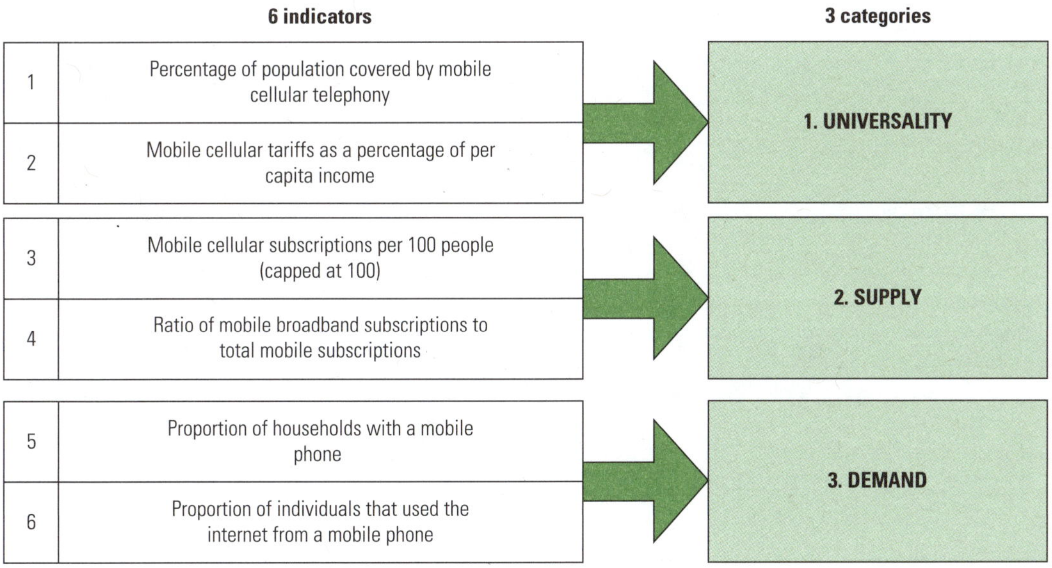

Table A.3 Worked example of the mobile analytical tool, Morocco

Indicator	Indicator value		Component scores		Category
	2005	2010	2005	2010	
Percentage of population covered by mobile cellular telephony	98	98			
Mobile cellular tariffs as a percentage of per capita income	20.1	14.3	0.88	0.93	**Universality**
Inverted (100-tariff/GNI)	79.9	85.7			
Mobile cellular subscriptions per 100 people (capped at 100)	41	100	0.21	0.52	**Supply**
Ratio of mobile broadband subscriptions to total mobile subscriptions	0	4			
Proportion of households with a mobile telephone	59	84	0.29	0.44	**Demand**
Proportion of individuals that used the mobile internet	0.04	3.4			

Source: Based on Agence Nationale de Réglementation des Télécommunications (ANRT) and Maroc Telecom.

Table A.4 Mobile analytical tool components for 100 selected economies, 2005 and 2010

Country	Universality			Supply			Demand		
	2005	2010	Change (%)	2005	2010	Change (%)	2005	2010	Change (%)
Albania	0.86	0.95	10	0.24	0.50	108	0.15	0.49	227
Algeria	0.90	0.96	7	0.21	0.46	119	0.25	0.48	92
Argentina	0.96	0.98	2	0.27	0.52	93	0.35	0.48	37
Armenia	0.95	0.95	0	0.11	0.51	364	0.16	0.49	206
Australia	0.98	0.99	1	0.47	0.70	49	0.44	0.51	16
Austria	0.98	0.99	1	0.55	0.57	4	0.46	0.56	22
Azerbaijan	0.92	0.99	8	0.13	0.42	223	0.26	0.45	73
Bahrain	0.99	0.99	0	0.50	0.51	2	0.50	0.54	8
Bangladesh	0.84	0.97	15	0.03	0.21	600	0.06	0.34	467
Belarus	0.94	0.98	4	0.21	0.51	143	0.15	0.39	160
Belgium	0.99	0.99	0	0.42	0.54	29	0.46	0.49	7
Bolivia	0.73	0.73	0	0.13	0.35	169	0.15	0.40	167
Bosnia and Herzegovina	0.93	0.97	4	0.20	0.43	115	0.27	0.43	59
Brazil	0.90	0.97	8	0.23	0.54	135	0.30	0.43	43
Bulgaria	0.95	0.97	2	0.40	0.52	30	0.36	0.45	25
Cambodia	0.71	0.89	25	0.04	0.34	750	0.11	0.32	191
Cameroon	0.69	0.82	19	0.06	0.20	233	0.14	0.22	57
Canada	0.97	0.99	2	0.26	0.48	85	0.33	0.48	45
Chile	0.97	0.98	1	0.33	0.53	61	0.36	0.48	33
China	0.91	0.98	8	0.15	0.35	133	0.28	0.58	107
Colombia	0.89	0.90	1	0.25	0.50	100	0.28	0.47	68
Costa Rica	0.87	0.87	0	0.13	0.37	185	0.25	0.40	60
Croatia	0.98	0.99	1	0.40	0.52	30	0.40	0.54	35
Czech Republic	0.98	0.99	1	0.50	0.54	8	0.43	0.50	16
Denmark	0.99	0.99	0	0.51	0.73	43	0.50	0.59	18
Ecuador	0.89	0.90	1	0.24	0.52	117	0.19	0.41	116
Egypt, Arab Rep.	0.94	0.98	4	0.09	0.46	411	0.14	0.39	179
Estonia	0.98	0.99	1	0.50	0.59	18	0.42	0.47	12
Finland	0.99	0.99	0	0.51	0.60	18	0.52	0.62	19
France	0.99	0.99	0	0.40	0.64	60	0.38	0.52	37
Georgia	0.93	0.96	3	0.17	0.56	229	0.15	0.43	187
Germany	0.98	0.99	1	0.50	0.59	18	0.39	0.51	31
Ghana	0.61	0.85	39	0.06	0.37	517	0.10	0.37	270

(continued next page)

Table A.4 *continued*

Country	Universality			Supply			Demand		
	2005	2010	Change (%)	2005	2010	Change (%)	2005	2010	Change (%)
Greece	0.99	0.99	0	0.50	0.54	8	0.38	0.47	24
Hong Kong SAR, China	1.00	0.99	−1	0.54	0.72	33	0.46	0.60	30
Hungary	0.98	0.98	0	0.46	0.53	15	0.40	0.48	20
India	0.72	0.86	19	0.03	0.31	933	0.07	0.27	286
Indonesia	0.91	0.96	5	0.10	0.46	360	0.15	0.39	160
Ireland	0.99	0.99	0	0.50	0.65	30	0.45	0.52	16
Israel	0.98	0.99	1	0.51	0.56	10	0.49	0.55	12
Italy	0.99	0.99	0	0.57	0.59	4	0.41	0.55	34
Jamaica	0.96	0.98	2	0.50	0.51	2	0.47	0.50	6
Japan	0.99	0.99	0	0.51	0.94	84	0.72	0.78	8
Jordan	0.95	0.98	3	0.29	0.51	76	0.27	0.55	104
Kazakhstan	0.94	0.97	3	0.16	0.50	213	0.14	0.43	207
Kenya	0.59	0.85	44	0.08	0.36	350	0.11	0.36	227
Korea, Rep.	0.99	0.99	0	0.56	0.96	71	0.58	0.69	19
Kyrgyzstan	0.59	0.90	53	0.05	0.45	800	0.05	0.45	800
Latvia	0.97	0.99	2	0.43	0.65	51	0.42	0.53	26
Lebanon	0.94	0.97	3	0.13	0.37	185	0.25	0.42	68
Lithuania	0.98	0.99	1	0.50	0.58	16	0.37	0.51	38
Macedonia, FYR	0.94	0.96	2	0.31	0.57	84	0.33	0.44	33
Malaysia	0.97	0.99	2	0.38	0.59	55	0.29	0.53	83
Mali	0.19	0.58	205	0.03	0.26	767	0.08	0.11	38
Mauritius	0.99	0.99	0	0.27	0.54	100	0.32	0.51	59
Mexico	0.92	0.95	3	0.22	0.43	95	0.21	0.35	67
Moldova	0.89	0.95	7	0.15	0.47	213	0.16	0.36	125
Morocco	0.88	0.93	6	0.21	0.52	148	0.29	0.44	52
Namibia	0.91	0.96	5	0.10	0.51	410	0.20	0.31	55
Nepal	0.48	0.85	77	0.01	0.16	1,500	0.02	0.30	1,400
Netherlands	0.99	1.00	1	0.50	0.59	18	0.50	0.55	10
New Zealand	0.98	0.98	0	0.46	0.63	37	0.37	0.54	46
Nigeria	0.63	0.79	25	0.07	0.29	314	0.20	0.31	55
Norway	1.00	1.00	0	0.50	0.58	16	0.50	0.56	12
Pakistan	0.71	0.94	32	0.06	0.31	417	0.17	0.25	47
Paraguay	0.84	0.95	13	0.13	0.49	277	0.25	0.44	76
Peru	0.75	0.88	17	0.10	0.49	390	0.11	0.39	255
Philippines	0.89	0.97	9	0.20	0.51	155	0.24	0.43	79
Poland	0.98	0.99	1	0.38	0.66	74	0.33	0.46	39
Portugal	0.99	0.99	0	0.53	0.58	9	0.42	0.47	12
Qatar	0.97	0.99	2	0.45	0.58	29	0.55	0.66	20
Romania	0.96	0.98	2	0.31	0.54	74	0.25	0.41	64
Russian Federation	0.96	0.98	2	0.44	0.52	18	0.16	0.49	206
Rwanda	0.44	0.83	89	0.01	0.19	1,800	0.03	0.20	567
Saudi Arabia	0.95	0.95	0	0.31	0.53	71	0.48	0.53	10
Senegal	0.72	0.85	18	0.07	0.32	357	0.15	0.43	187
Serbia	0.95	0.96	1	0.35	0.54	54	0.36	0.43	19
Singapore	1.00	1.00	0	0.52	0.59	13	0.55	0.61	11
Slovak Republic	0.98	0.98	0	0.42	0.63	50	0.45	0.53	18
Slovenia	0.99	0.99	0	0.45	0.61	36	0.44	0.54	23
South Africa	0.96	0.97	1	0.33	0.53	61	0.31	0.47	52

(continued next page)

Table A.4 *continued*

Country	Universality 2005	Universality 2010	Universality Change (%)	Supply 2005	Supply 2010	Supply Change (%)	Demand 2005	Demand 2010	Demand Change (%)
Spain	0.99	0.98	–1	0.51	0.59	16	0.42	0.54	29
Sri Lanka	0.87	0.97	11	0.09	0.45	400	0.10	0.32	220
Sweden	0.99	0.99	0	0.52	0.62	19	0.50	0.59	18
Switzerland	0.99	0.99	0	0.48	0.62	29	0.50	0.56	12
Tajikistan	0.20	0.94	370	0.02	0.35	1,650	0.06	0.41	583
Tanzania	0.49	0.77	57	0.05	0.26	420	0.04	0.23	475
Thailand	0.93	0.97	4	0.23	0.51	122	0.36	0.52	44
Turkey	0.97	0.97	0	0.32	0.44	38	0.36	0.49	36
Uganda	0.69	0.85	23	0.03	0.19	533	0.08	0.27	238
Ukraine	0.92	0.98	7	0.32	0.52	63	0.22	0.43	95
United Arab Emirates	0.99	0.99	0	0.50	0.71	42	0.50	0.56	12
United Kingdom	0.99	0.99	0	0.53	0.61	15	0.50	0.57	14
United States	0.99	0.99	0	0.36	0.61	69	0.29	0.56	93
Uruguay	0.95	0.99	4	0.17	0.55	224	0.23	0.37	61
Uzbekistan	0.79	0.95	20	0.01	0.40	3900	0.25	0.44	76
Venezuela, RB	0.90	0.93	3	0.24	0.51	113	0.18	0.26	44
Vietnam	0.89	0.96	8	0.05	0.53	960	0.15	0.29	93
Zambia	0.56	0.78	39	0.04	0.20	400	0.08	0.30	275
Zimbabwe	0.82	0.63	–23	0.03	0.35	1067	0.05	0.27	440

Source: Authors' analysis.

The mean score for all components added together increased by 30 percent between 2005 and 2010, from 0.49 to 0.63, attesting to the rapid growth and improvement in mobile networks over that period (figure A.13). The highest increase was among low-income countries, where significant gains in coverage were coupled with falling prices from intensified competition. Regionally, the highest growth was in South Asia, followed by Sub-Saharan Africa. The highest absolute increase was in Tajikistan, where the score rose by 0.47 points to 0.57. Mobile competition intensified in Tajikistan between 2005 and 2010, with four GSM (Global System for Mobile communications) and several CDMA (Code Division Multiple Access) operators and a number of panregional mobile groups entering the market, including TeliaSonera and Vimpelcom. Investment soared, leading to plummeting prices, an expansion of coverage, and skyrocketing access. Several other Central Asian countries also had among the highest growth in their scores between 2005 and 2010.

Looking at each of the three components individually, some 80 countries have achieved a high degree of universality (a subindex value of 0.9 or higher). Most developed and middle-income nations had already achieved near universality by 2005, and gains since then have been marginal. Although many developing nations had large increases in universality between 2005 and 2010, many still remain below the 0.9 threshold. Universal access to mobile networks remains constrained in these countries because of relatively high tariffs, incomplete mobile coverage, or both. In Mali, for instance, mobile service covers less than half the population, and the price of a monthly basket of mobile services is one-quarter of per capita income. In Rwanda, pricing is a barrier: mobile networks cover more than 90 percent of the Rwandan population, but a monthly mobile basket is 30 percent of per capita income. In India the bottleneck is coverage: a mobile basket is just 3 percent of per capita income but only three-quarters of the population is covered.

The supply component showed the greatest increase between 2005 and 2010, with the mean value nearly doubling from 0.28 to 0.50. In developed countries the increase was chiefly attributable to the deployment of mobile broadband networks, whereas gains in developing countries came from the provision of basic voice services. Around half of the countries are "stuck" at a supply

Figure A.13 Mobile analytical tool scores, 2005 and 2010, by income and region group

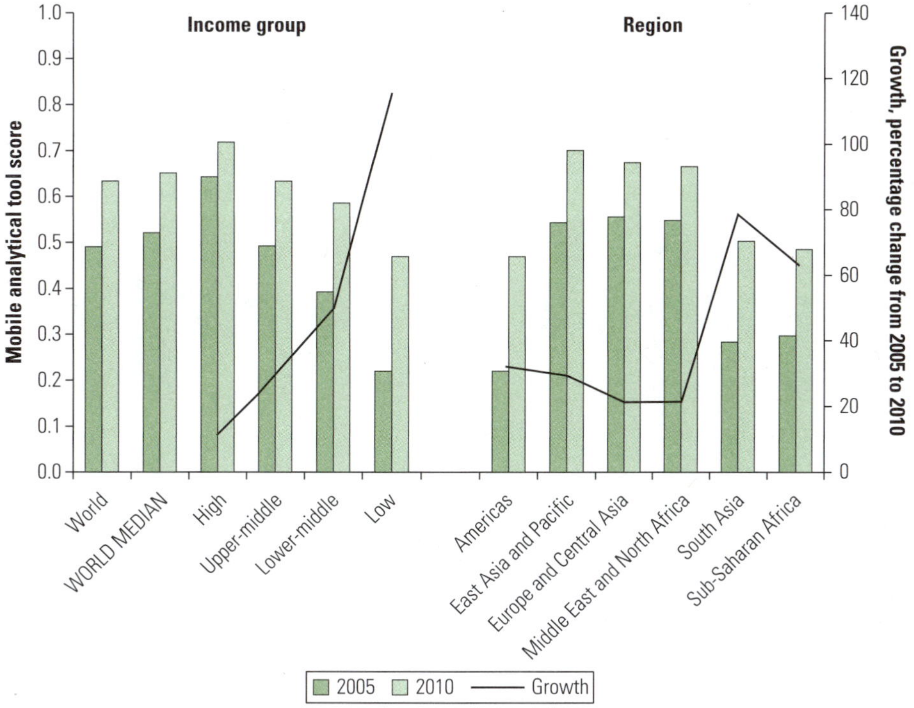

Source: Author analysis.
Note: Scores shown are the mean of the three components. Group averages are the mean of the group.

component value of 0.5; there are more SIM cards than people, but the share of mobile broadband is low.

The score achieved on the Mobile Analytical Tool is closely related to gross national income (GNI) per capita (figure A.14). None of the high-income economies had a score under 0.65, and several upper-middle-income countries exceeded that value although with much lower incomes. These include the Russian Federation and South Africa along with Argentina, Jamaica, Jordan, Lithuania, Macedonia, Malaysia, Mauritius, and Thailand. Japan and the Republic of Korea stand out as outliers—their score is significantly above where it should be considering their income. Most countries at very low per capita income averages (less than $1,000 a year) fall below the regression line, suggesting that a certain level of economic development is necessary for balanced mobile growth. Regional clusters are also noticeable: lower-middle-income economies in Sub-Saharan Africa tend to be performing poorly whereas the opposite is true in Central Asia. A number of Latin American upper-middle-income economies are also not doing as well as expected.

Figure A.14 Mobile analytical tool and GNI per capita, 2010

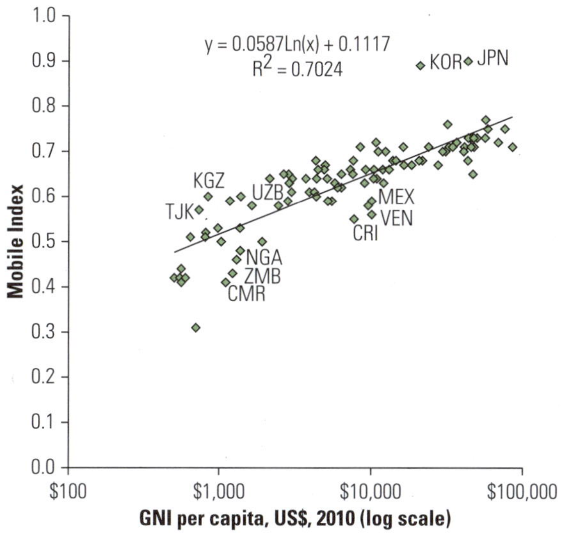

Source: Authors' analysis.
Note: Scores shown are the mean of the three components. Each point represents one economy with outliers highlighted: CMR = Cameroon; CRI = Costa Rica; JPN = Japan; KOR = Republic of Korea; KGZ = Kyrgyzstan; MEX = Mexico; NGA = Nigeria; TJK = Tajikistan; UZB = Uzbekistan; VEN = Venezuela, RB; ZMB = Zambia.

Figure A.15 Mobile analytical tool: China and Sri Lanka compared

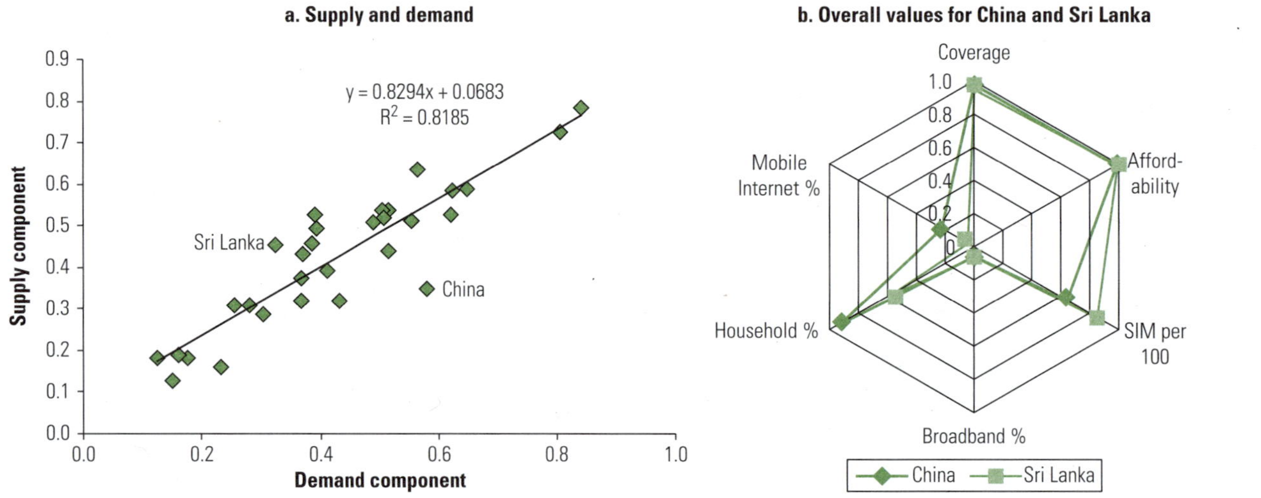

Source: Authors' analysis.

As might be expected, there is a close relationship between the supply and demand categories (figure A.15a). Outliers illustrate mismatches between supply and demand. For example in China the demand component (0.58) is higher than the supply component (0.35), suggesting further room for growth (figure A.15b). Over 90 percent of Chinese households have a mobile phone, the second-highest level among the developing countries used in the Mobile Analytical Tool (Jordan has the highest home mobile penetration among this group). China also has a relatively high level of internet access through mobile phones. According to a recent 21-country survey, some 37 percent of Chinese mobile phone owners use their handset to access the internet, a higher ratio than in France, Germany, or Spain.[12] On the supply side, China's SIM card penetration is only 64 per 100 people, relatively low because there are few incentives for multiple SIM card ownership thanks to inexpensive cross network pricing. China is relatively new to mobile broadband with networks having launched only in 2009. Subscriptions to high-speed mobile networks have grown rapidly, and by the end of 2010 China had the third-largest number of mobile broadband users in the world (after Japan and the United States). Nevertheless, mobile broadband still accounted for only 5 percent of total mobile subscriptions, with the result that most Chinese mobile internet users were accessing the web over narrowband mobile connections.

In contrast, Sri Lanka scores higher on the supply component (0.45) than on the demand one (0.32). On the supply side, there is a high degree of competition in the Sri Lankan mobile market with SIM card penetration at 85 per 100 people. Further, Sri Lanka was the first South Asian nation to launch mobile broadband networks. On the demand side, however, the penetration of mobile phones in Sri Lankan homes is only 60 percent and just 5 of every 100 people use a mobile phone to access the internet (see figure A.15b). The mismatch suggests that efforts here need to be devoted to boosting demand.

Figure A.16 illustrates the relationship between the three composite indicators and underlying indicators of the Mobile Analytical Tool. The values for high-income economies are contrasted with the world and low- and middle-income averages. As noted, high degrees of universality have been achieved with high affordability and second-generation (2G) coverage. There have also been large gains in supply of 2G networks and household demand between 2005 and 2010. However, levels of mobile broadband networks and internet usage are low, and these will be the growth areas in the future. Care is needed to ensure that an advanced mobile digital divide does not develop as a result of restricted mobile broadband coverage (such as poor coverage in rural areas) and limited mobile broadband affordability.

Figure A.16 Mobile analytical tool components summarized

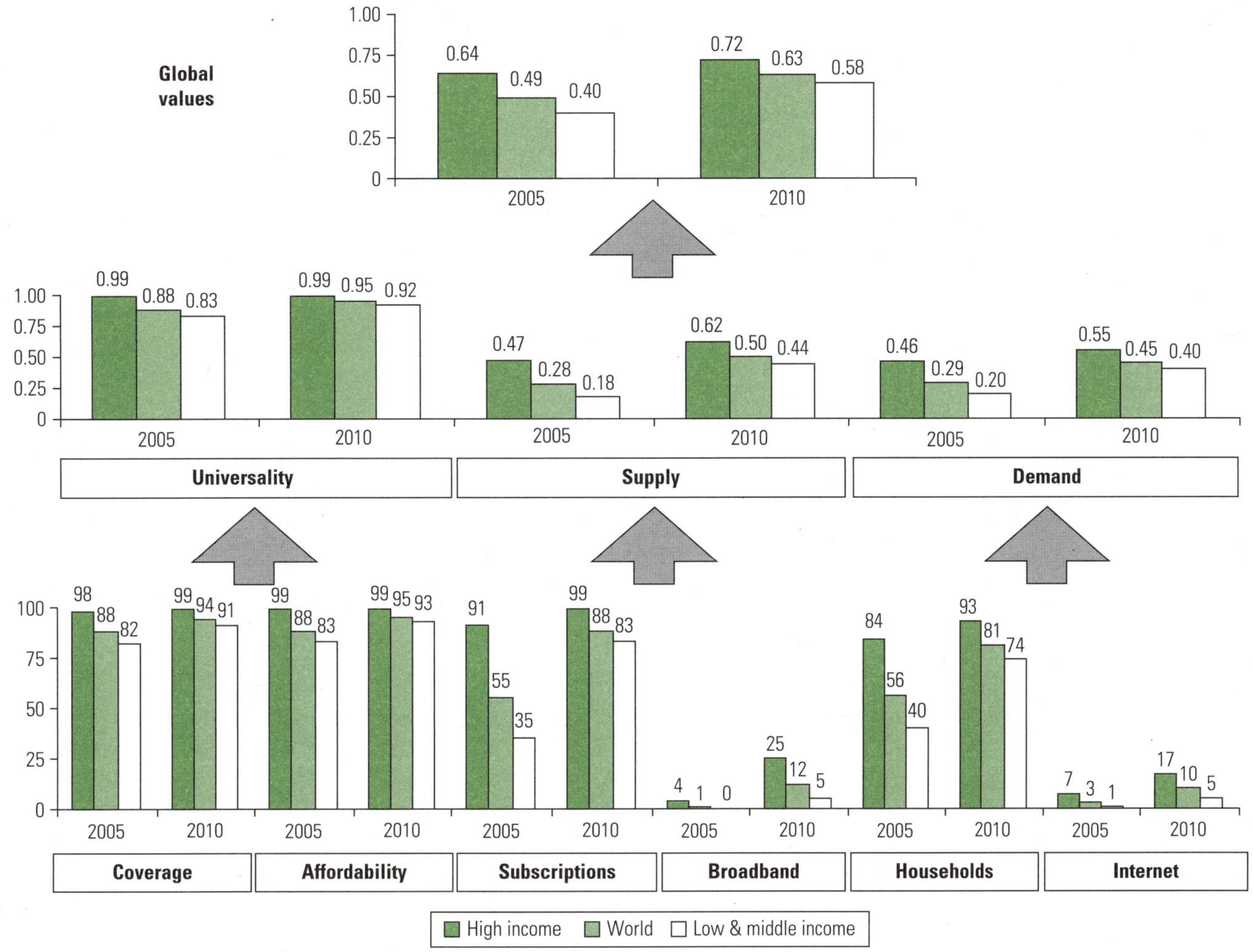

Source: Authors' analysis.
Note: The scores shown at the top are based on the mean of the three components of the Mobile Analytical Tool.

The Mobile Analytical Tool provides different insights into the availability and demand for mobile communications. It overcomes the limitations of using a single indicator to gauge mobile performance. For example, the number of mobile subscriptions is often used as a comparator of development, but it can be misleading because of underlying variations in multiple SIM card ownership, which in turn reflects interoperator pricing strategies. As the case of China and Sri Lanka illustrated, Sri Lanka has a higher SIM card penetration but much lower rates of actual mobile ownership and internet browsing from cell phones. The Mobile Analytical Tool provides a holistic and integrated perspective of country mobile network development compared with using single indicators to measure performance.

Notes

1. "The importance of availability of information communication technology (ICT) devices is increasing significantly in contemporary society. These devices provide a set of services that are changing the structure and pattern of major social and economic phenomena. The housing census provides an outstanding opportunity to assess the availability of these devices to the household" (United Nations 2008, 215).
2. A major factor has been the development of low-cost models: "... the spread of mobile phones in developing countries has been

accompanied by the rise of homegrown mobile operators in China, India, Africa and the Middle East that rival or exceed the industry's Western incumbents in size. These operators have developed new business models and industry structures that enable them to make a profit serving low-spending customers that Western firms would not bother with" (Standage 2009).

3. This is derived from the 22 percent of the population in low- and middle-income economies who lived on less than $1.25 a day in 2008 (at 2005 international prices); see http://data.worldbank.org/indicator/SI.POV.DDAY.
4. http://lirneasia.net/2011/06/nokia-annual-tco-total-cost-of-ownership-results-show-bangladesh-and-sri-lanka-as-cheapest/.
5. "Movirtu Rolls Out a Cloud Phone Aimed at Low-Income Users: First Market Is Madagascar, Others Will Follow." *Balancing Act*, June 24, 2011. http://www.balancingact-africa.com/news/en/issue-no-560/top-story/movirtu-rolls-out-a/en.
6. http://www.gartner.com/it/page.jsp?id=1893523.
7. http://www.c-i-a.com/methodology.htm#computeruse.
8. http://www.gartner.com/it/page.jsp?id=1924314.
9. http://blog.idate.fr/?p=133.
10. "Improving HSSE Standards: The Case of Bangladesh." http://www.telenor.com/en/corporate-responsibility/initiatives-worldwide/improving-hsse-standards-bangladesh.
11. Note that the Mobile Analytical Tool was calculated prior to final data updates and the results for some countries would differ if there were later data revisions.
12. http://www.pewglobal.org/2011/12/20/global-digital-communication-texting-social-networking-popular-worldwide/.

References

Akamai. 2012. "The State of the Internet: 3rd Quarter, 2011 Report." http://www.akamai.com/html/about/press/releases/2012/press_013112.html.

A. T. Kearney. 2012. "Latin American Mobile Observatory 2011." GSMA. http://www.gsma.com/documents/download-full-report-english-pdf-5-3-mb/21905.

Deloitte. 2008. "Economic Impact of Mobile Communications in Serbia, Ukraine, Malaysia, Thailand, Bangladesh and Pakistan." http://www.telenor.rs/media/TelenorSrbija/fondacija/economic_impact_of_mobile_communications.pdf.

Grameenphone. 2011. *Annual Report 2010*. http://investor-relations.grameenphone.com/Annual-Reports.html.

GSMA. 2009. "GSMA Urges Bangladesh Government to Eliminate SIM Card Tax ~ GSM World." Press release (July 22). http://www.gsmworld.com/newsroom/press-releases/2009/3493.htm.

ITU (International Telecommunication Union). 2002. "Hong Kong (China) and Denmark Top ITU Mobile/Internet Index." Press release (September 17). http://www.itu.int/newsroom/press_releases/2002/20.html.

———. 2006. "Digital Opportunity Index (DOI)." http://www.itu.int/ITU-D/ict/doi/index.html.

———. 2011. "Measuring the Information Society." http://www.itu.int/ITU-D/ict/publications/idi/index.html.

ITU and UNCTAD (United Nations Conference for Trade and Development). 2007. "World Information Society Report: Beyond WSIS." http://www.itu.int/osg/spu/publications/worldinformationsociety/2007/.

Minges, M. 2005. "Is the Internet Mobile? Measurements from the Asia-Pacific Region." *Telecommunications Policy* 29 (2–3): 113–125. doi:10.1016/j.telpol.2004.11.002.

Nokia. 2009. "Tailoring Mobile Costs to the Pockets of a Billion New Customers." *Expanding Horizons*. Q2 2009 edition. http://www.nokiasiemensnetworks.com/sites/default/files/document/Expanding_Horizons_2_2009_0.pdf.

OECD (Organisation for Economic Co-operation and Development). 2010. *Wireless Broadband Indicator Methodology*. Paris (March 18). http://www.oecd.org/LongAbstract/0,3425,en_2649_34225_44930927_119666_1_1_1,00.html.

Sharma, Amol. 2010. "Google Bets on Cheap Smartphones for India." *WSJ.com*, October 12. http://online.wsj.com/article/SB10001424052748703794104575545963108615120.html.

Standage, T. 2009. "Mobile Marvels." *The Economist*, September 26.

UAE Telecommunications Regulatory Authority. 2011. *ICT in the UAE: Household Survey, 2010.*

United Nations. 2008. "Principles and Recommendations for Population and Housing Censuses, Revision 2." http://unstats.un.org/unsd/demographic/sources/census/census3.htm.

Vodacom. 2011. "Integrated Report for the Year Ended 31 March 2011."

World Bank. 2009. "ICT Performance Measures: Methodology and Findings." In *Information and Communication for Development 2009. Extending Reach and Increasing Impact*. Washington, DC. www.worldbank.org/ic4d.

———. 2010. "Kenya Economic Update: Kenya at the Tipping Point." No. 3, Washington (December). http://siteresources.worldbank.org/KENYAEXTN/Resources/KEU-Dec_2010_with_cover_e-version.pdf.

———. 2011. "Little Data Book on Information and Communication Technology." http://siteresources.worldbank.org/INFORMATIONANDCOMMUNICATIONANDTECHNOLOGIES/Resources/ICT_Little_Data2011.pdf.

Data Notes

Kaoru Kimura and Michael Minges

The World Bank's *Mobile At-a-Glance Country Tables* present in one place the most recent country-specific mobile cellular data from many sources. The data offer a snapshot of the economic and social context and the structure and performance of the mobile cellular sector in some 152 economies.

Tables

Economies are presented alphabetically. Data are shown for 152 economies with populations of more than 1 million for which timely and reliable information exists. The table *Key Mobile Indicators for Other Economies* presents data for 64 additional economies—those with sparse data, smaller economies with populations of between 30,000 and 1 million, and others that are not members of the World Bank Group.

The data in the tables are categorized into three sections:

- *Economic and social context* provides a snapshot of the economy's macroeconomic and social environment. Several indicators have been included that relate to the different sectors discussed in the report.
- *Sector structure* provides an overview of the competitive market status in the mobile cellular sector.
- *Sector performance* provides statistical data on the mobile cellular sector with indicators for access, usage, and affordability.

Aggregate measures for income groups and regions

The aggregate measures for income groups include 216 economies (those economies listed in the At-a-Glance Country Tables plus those in the Other Economies table) wherever data are available.

The aggregate measures for regions include only low- and middle-income economies (note that these measures include developing economies with populations of less than 1 million, including those listed in the Other Economies table). The country composition of regions is based on the World Bank's analytical regions and may differ from common geographic usage.

Charts

The Mobile Cellular Subscriptions chart shows the number of mobile subscribers (per 100 people) from 2005 to 2011. Country and region information is presented when available.

The mobile basket chart shows the mobile prepaid tariff basket (% of GNI per capita) in the country from 2005 to 2010. Country and region information is presented when available.

Data consistency and reliability

Considerable effort has been made to standardize the data collected. Full comparability of data among countries cannot be ensured, however, and care must be taken in interpreting the indicators.

Many factors affect availability, comparability, and reliability: statistical systems in some developing countries are weak; statistical methods, coverage, practices, and definitions differ widely among countries; and cross-country and intertemporal comparisons involve complex technical and conceptual problems that cannot be unequivocally resolved. Data coverage may not be complete because of special circumstances or because economies are experiencing problems (such as those stemming from conflicts) that affect the collection and reporting of data. For these reasons, although data are drawn from the sources thought to be most authoritative, they should be construed only as indicating trends and characterizing major differences among economies rather than offering precise quantitative measures of those differences.

Administrative subscription-based data generally refer to the end of the calendar year. If end-of-year data are not available, the most recent data for that year are used. Survey-based data refer to the year the survey was carried out. In some cases estimates have been made when there is sufficient historical data.

The cut-off date for data inclusion was March 31, 2012.

Data sources

Data are drawn from ictDATA.org, International Monetary Fund (IMF), International Telecommunication Union (ITU), United Nations; United Nations Educational, Scientific and Cultural Organization (UNESCO), Institute for Statistics (UIS), Wireless Intelligence, World Health Organization (WHO), and the World Bank.

Classification of economies

For operational and analytical purposes, the World Bank's main criterion for classifying economies is GNI (gross national income) per capita. Every economy is classified as low income, middle income (these are subdivided into lower middle and upper middle), or high income. Note that classification by incomes does not necessarily reflect development status. Because GNI per capita changes over time, the country composition of income groups may change, but one consistent classification, based on GNI per capita in 2010, is used throughout this publication.

Low-income economies are those with a GNI per capita of $1,005 or less in 2010. Middle-income economies are those with a GNI per capita of more than $1,005 but less than $12,276. Lower-middle-income and upper-middle-income economies are separated at a GNI per capita of $3,975. High-income economies are those with a GNI per capita of $12,276 or more.

For more information on these classifications, see the Classification of Economies by Income and Region table below and the World Bank's country classification website: http://data.worldbank.org/about/country-classifications.

Symbols

The following symbols are used throughout the At-a-Glance tables:

— This symbol means that data are not available or that aggregates cannot be calculated because of missing data in the year shown.

0 or 0.0 means zero or less than half the unit shown.

$ refers to U.S. dollars, unless otherwise stated.

Classification of Economies by Region and Income, FY 2012

East Asia and the Pacific
American Samoa (UMC)
Cambodia (LIC)
China (UMC)
Fiji (LMC)
Indonesia (LMC)
Kiribati (LMC)
Korea, Dem. Rep. (LIC)
Lao PDR (LMC)
Malaysia (UMC)
Marshall Islands (LMC)
Micronesia, Fed. Sts. (LMC)
Mongolia (LMC)
Myanmar (LIC)
Palau (UMC)
Papua New Guinea (LMC)
Philippines (LMC)
Samoa (LMC)
Solomon Islands (LMC)
Thailand (UMC)
Timor-Leste (LMC)
Tonga (LMC)
Tuvalu (LMC)
Vanuatu (LMC)
Vietnam (LMC)

Europe and Central Asia
Albania (UMC)
Armenia (LMC)
Azerbaijan (UMC)
Belarus (UMC)
Bosnia and Herzegovina (UMC)
Bulgaria (UMC)
Georgia (LMC)
Kazakhstan (UMC)
Kosovo (LMC)
Kyrgyz Republic (LIC)
Latvia (UMC)
Lithuania (UMC)
Macedonia, FYR (UMC)
Moldova (LMC)
Montenegro (UMC)
Romania (UMC)
Russian Federation (UMC)
Serbia (UMC)
Tajikistan (LIC)
Turkey (UMC)
Turkmenistan (LMC)
Ukraine (LMC)
Uzbekistan (LMC)

Latin America and the Caribbean
Antigua and Barbuda (UMC)
Argentina (UMC)
Belize (LMC)
Bolivia (LMC)
Brazil (UMC)
Chile (UMC)
Colombia (UMC)
Costa Rica (UMC)
Cuba (UMC)
Dominica (UMC)
Dominican Republic (UMC)
Ecuador (UMC)
El Salvador (LMC)
Grenada (UMC)
Guatemala (LMC)
Guyana (LMC)
Haiti (LIC)
Honduras (LMC)
Jamaica (UMC)
Mexico (UMC)
Nicaragua (LMC)
Panama (UMC)
Paraguay (LMC)
Peru (UMC)
St. Kitts and Nevis (UMC)
St. Lucia (UMC)
St. Vincent and the Grenadines (UMC)
Suriname (UMC)
Uruguay (UMC)
Venezuela, RB (UMC)

Middle East and North Africa
Algeria (UMC)
Djibouti (LMC)
Egypt, Arab Rep. (LMC)
Iran, Islamic Rep. (UMC)
Iraq (LMC)
Jordan (UMC)
Lebanon (UMC)
Libya (UMC)
Morocco (LMC)
Syrian Arab Republic (LMC)
Tunisia (UMC)
West Bank and Gaza (LMC)
Yemen, Rep. (LMC)

South Asia
Afghanistan (LIC)
Bangladesh (LIC)
Bhutan (LMC)
India (LMC)
Maldives (UMC)
Nepal (LIC)
Pakistan (LMC)
Sri Lanka (LMC)

Sub-Saharan Africa
Angola (LMC)
Benin (LIC)
Botswana (UMC)
Burkina Faso (LIC)
Burundi (LIC)
Cameroon (LMC)
Cape Verde (LMC)
Central African Republic (LIC)
Chad (LIC)
Comoros (LIC)
Congo, Dem. Rep. (LIC)
Congo, Rep. (LMC)
Côte d'Ivoire (LMC)
Eritrea (LIC)
Ethiopia (LIC)
Gabon (UMC)
Gambia, The (LIC)
Ghana (LMC)
Guinea (LIC)
Guinea-Bissau (LIC)
Kenya (LIC)
Lesotho (LMC)
Liberia (LIC)
Madagascar (LIC)
Malawi (LIC)
Mali (LIC)
Mauritania (LMC)
Mauritius (UMC)
Mayotte (UMC)
Mozambique (LIC)
Namibia (UMC)
Niger (LIC)
Nigeria (LMC)
Rwanda (LIC)
São Tomé and Principe (LMC)
Senegal (LMC)
Seychelles (UMC)
Sierra Leone (LIC)
Somalia (LIC)
South Africa (UMC)
South Sudan (LMC)
Sudan (LMC)
Swaziland (LMC)
Tanzania (LIC)
Togo (LIC)
Uganda (LIC)
Zambia (LMC)
Zimbabwe (LIC)

High income OECD
Australia
Austria
Belgium
Canada
Czech Republic
Denmark
Estonia
Finland
France
Germany
Greece
Hungary
Iceland
Ireland
Israel
Italy
Japan

(continued next page)

Classification of Economies by Region and Income, FY 2012 *continued*

Korea, Rep.	Bahrain	Macao SAR, China
Luxembourg	Barbados	Malta
Netherlands	Bermuda	Monaco
New Zealand	Brunei Darussalam	New Caledonia
Norway	Cayman Islands	Northern Mariana Islands
Poland	Channel Islands	Oman
Portugal	Croatia	Puerto Rico
Slovak Republic	Curaçao	Qatar
Slovenia	Cyprus	San Marino
Spain	Equatorial Guinea	Saudi Arabia
Sweden	Faeroe Islands	Singapore
Switzerland	French Polynesia	Sint Maarten (Dutch part)
United Kingdom	Gibraltar	St. Martin (French part)
United States	Greenland	Taiwan, China
	Guam	Trinidad and Tobago
Other high income	Hong Kong SAR, China	Turks and Caicos Islands
Andorra	Isle of Man	United Arab Emirates
Aruba	Kuwait	Virgin Islands (U.S.)
Bahamas, The	Liechtenstein	

Source: World Bank.
Note: This table classifies all World Bank member economies and all other economies with populations of more than 30,000. Economies are divided among income groups according to 2010 GNI per capita, calculated using the World Bank Atlas method. The groups are: low-income countries (LIC), $1,005 or less; lower-middle-income countries (LMC), $1,006–$3,975; upper-middle-income countries (UMC), $3,976–$12,275; and high-income countries, $12,276 or more.

Definitions and data sources

This section provides definitions and sources of the indicators used in the *World Bank's Mobile At-a-Glance Country Tables.*

Economic and social context

Population (total, million) is based on the de facto definition of population, which counts all residents regardless of legal status or citizenship—except for refugees not permanently settled in the country of asylum, who are generally considered part of the population of their country of origin. The values shown are mid-year estimates. (World Bank)

GNI per capita, World Bank Atlas method (current US$) is gross national income converted to U.S. dollars using the World Bank Atlas method, divided by the mid-year population. GNI is the sum of value added by all resident producers plus any product taxes (less subsidies) not included in the valuation of output plus net receipts of primary income (compensation of employees and property income) from abroad. (World Bank)

Rural population (% of total) refers to people living in rural areas as defined by national statistical offices. It is calculated as the difference between total population and urban population. (United Nations)

Expected years of schooling (years) are the number of years a child of school entrance age is expected to spend at school, or university, including years spent on repetition. It is the sum of the age-specific enrollment ratios for primary, secondary, postsecondary nontertiary, and tertiary education. (UNESCO Institute for Statistics)

Physicians density (per 1,000 people) refers to the number of physicians (including generalists and specialist medical practitioners) (WHO).

Depositors with commercial banks (per 1,000 adults) are the reported number of deposit account holders at commercial banks and other resident banks functioning as commercial banks that are resident nonfinancial corporations (public and private). For many countries data cover the total number of deposit accounts because information on account holders is lacking. The major types of deposits are checking accounts, savings accounts, and time deposits. (IMF)

Sector structure

Number of mobile operators refers to licensed mobile cellular service providers that have their own network infrastructure as opposed to other mobile service providers who lease it (for example, Mobile Virtual Network Operators). The data refer to nationwide operators. (ictDATA.org)

Herfindahl–Hirschman Index (HHI) (scale = 0–10,000) refers to the level of market concentration. It is calculated on the basis of the market shares of each company operating in the industry. The market share for each company is squared; these are then added up to get the HHI. A monopoly market would have an HHI of 10,000; a duopoly with each operator having half the market would have an HHI of 5,000; and a market with four operators each having the same market share would have an HHI of 2,500. The HHI is computed for the mobile market based on the number of subscribers. (ictDATA.org)

Sector performance

Access

Mobile cellular subscriptions (per 100 people) are subscriptions to a public mobile telephone service using cellular technology, which provide access to the public switched telephone network. Postpaid and prepaid subscriptions are included. Note that data is not strictly comparable because of differences in the period in which a subscriber is considered active and whether nonhuman subscriptions (such as data cards for laptop access or automatic teller machines) are included. For these reasons and others, mobile subscriptions do not reflect actual mobile phone ownership since there can be multiple subscriptions. (ITU, ictDATA.org)

Mobile cellular subscriptions (% prepaid) refer to the total number of mobile cellular telephone subscriptions that use prepaid refills. These are subscriptions where, instead of paying an ongoing monthly fee, users purchase blocks of usage time. Only active subscriptions should be included (those used at least once in the last three months for making or receiving a call or carrying out a nonvoice activity such as sending or reading an SMS or accessing the internet). The number of prepaid subscriptions is divided by total mobile cellular telephone subscriptions. (Wireless Intelligence)

Population covered by a mobile-cellular network (%) is the percentage of people within range of a mobile cellular signal regardless of whether they are subscribers. (ITU)

Mobile broadband subscriptions (per 100 people) are the sum of the number of subscriptions using the following technologies: CDMA2000 1xEV-DO, W-CDMA, TD-SCDMA, LTE, and mobile WiMAX. (Wireless Intelligence)

Mobile broadband (% of total mobile subscriptions) is the number of mobile broadband subscribers (defined above) divided by the total mobile cellular subscriptions in a country. (Wireless Intelligence)

Usage

Households with a mobile telephone (%) refers to the percentage of households reporting ownership of a mobile cellular telephone. (ictDATA.org)

Mobile voice usage (minutes per user per month) measures the minutes of use per mobile user per month. (Wireless Intelligence)

Population using mobile internet (%) refers to the share of people using a mobile phone to access the internet (regardless of speed or technology). The data are derived from both mobile and internet user surveys and therefore the figure is shown as a percentage of the total population. In cases where survey data are not available, subscription data (that is, mobile internet subscribers) have been used. (ictDATA.org)

Short message service (SMS) users (% of mobile users) refers to the percentage of mobile users who send SMS text messages. (ictDATA.org)

Affordability

Mobile basket (US$ a month) is based on the Organisation for Economic Co-operation and Development's updated basket for low users (retrofitted also for 2005), which includes the cost of monthly mobile use for 30 outgoing calls a month spread over the same mobile network, other mobile networks, and mobile-to-fixed-line calls and during peak, off-peak, and weekend times as well as 100 text messages a month. (ictDATA.org for 2005 data, ITU for 2010 data). For more information on the definition, see ITU, 2011, *Measuring the Information Society*. Annex Table 2.1, p 144–45, http://www .itu.int/net/pressoffice/backgrounders/general/pdf/5.pdf.

Mobile tariff basket (% of GNI per capita) refers to the mobile cellular prepaid monthly tariff basket divided by GNI per capita. (ictDATA.org, ITU, and World Bank)

Albania

	Albania 2005	Albania 2010	Upper-middle-income group 2010	Europe & Central Asia Region 2010
Economic and social context				
Population (total, million)	3	3	2,452	405
GNI per capita, World Bank Atlas method (current US$)	2,580	3,960	5,886	7,272
Rural population (% of total)	55	52	43	36
Expected years of schooling (years)	*11*	—	*13*	*13*
Physicians density (per 1,000 people)	*1.2*	*1.2*	1.7	3.2
Depositors with commercial banks (per 1,000 adults)	—	—	—	894
Sector structure				
Number of mobile operators	—	4		
Herfindahl-Hirschman Index (HHI) (scale = 0–10,000)	—	3,661		
Sector performance				
Access				
Mobile cellular subscriptions (per 100 people)	49	138	92[a]	125[a]
Mobile cellular subscriptions (% prepaid)	97	91[a]	81[a]	82[a]
Population covered by a mobile-cellular network (%)	91	98	*99*	96
Mobile broadband subscriptions (per 100 people)	—	5.6[a]	14.3[a]	22.6[a]
Mobile broadband (% of total mobile subscriptions)	—	3.5[a]	15.4[a]	18.0[a]
Usage				
Households with a mobile telephone (%)	31	*94*	84	82
Mobile voice usage (minutes per user per month)	64	*103*	325[a]	288[a]
Population using mobile Internet (%)	—	3.4	22.9[a]	8.5
Short Message Service (SMS) users (% of mobile users)	—	66.0	74.4[a]	69.8[a]
Affordability				
Mobile basket (% of GNI per capita)	22.6	7.8	2.9	3.1

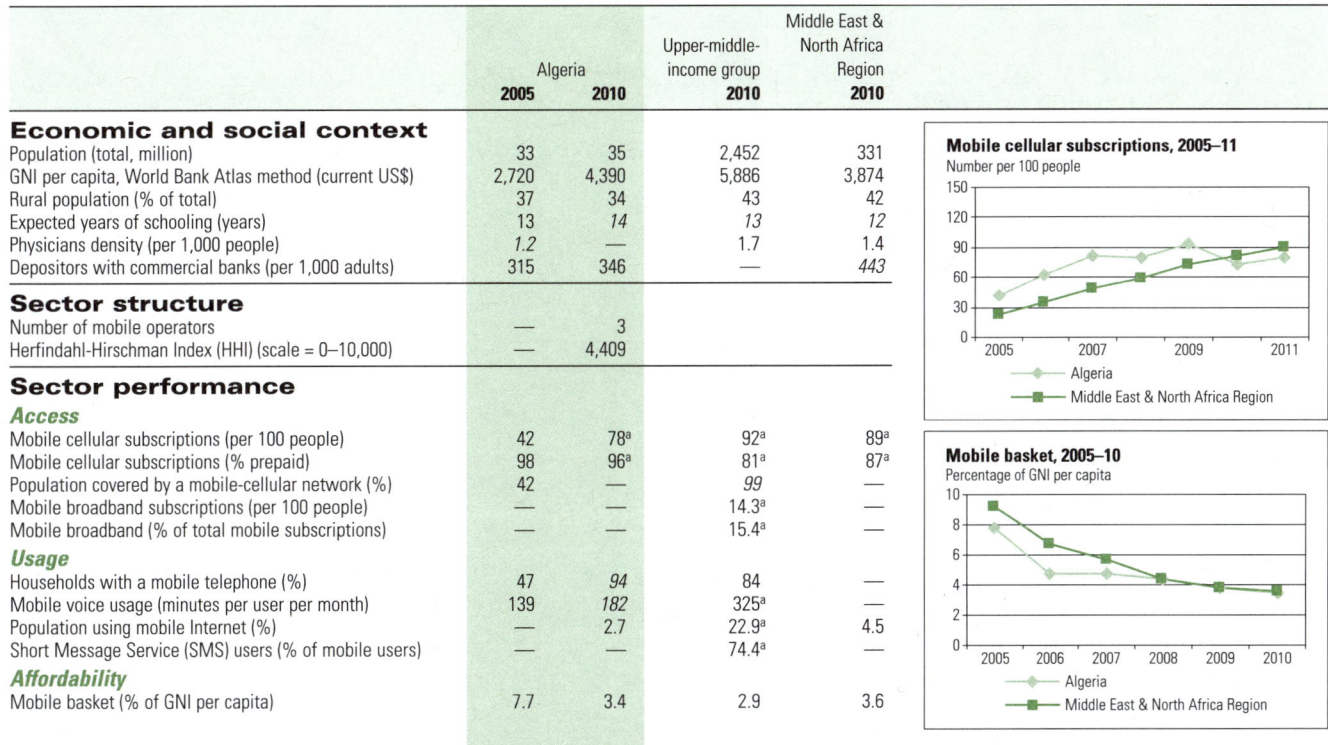

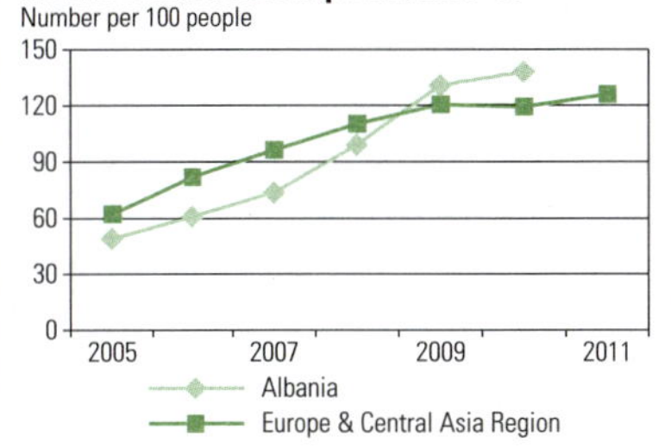

Mobile cellular subscriptions, 2005–11

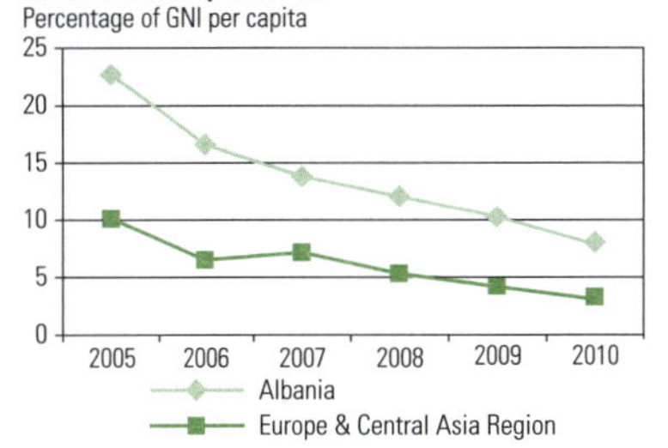

Mobile basket, 2005–10

Algeria

	Algeria 2005	Algeria 2010	Upper-middle-income group 2010	Middle East & North Africa Region 2010
Economic and social context				
Population (total, million)	33	35	2,452	331
GNI per capita, World Bank Atlas method (current US$)	2,720	4,390	5,886	3,874
Rural population (% of total)	37	34	43	42
Expected years of schooling (years)	13	*14*	*13*	*12*
Physicians density (per 1,000 people)	*1.2*	—	1.7	1.4
Depositors with commercial banks (per 1,000 adults)	315	346	—	*443*
Sector structure				
Number of mobile operators	—	3		
Herfindahl-Hirschman Index (HHI) (scale = 0–10,000)	—	4,409		
Sector performance				
Access				
Mobile cellular subscriptions (per 100 people)	42	78[a]	92[a]	89[a]
Mobile cellular subscriptions (% prepaid)	98	96[a]	81[a]	87[a]
Population covered by a mobile-cellular network (%)	42	—	*99*	—
Mobile broadband subscriptions (per 100 people)	—	—	14.3[a]	—
Mobile broadband (% of total mobile subscriptions)	—	—	15.4[a]	—
Usage				
Households with a mobile telephone (%)	47	*94*	84	—
Mobile voice usage (minutes per user per month)	139	*182*	325[a]	—
Population using mobile Internet (%)	—	2.7	22.9[a]	4.5
Short Message Service (SMS) users (% of mobile users)	—	—	74.4[a]	—
Affordability				
Mobile basket (% of GNI per capita)	7.7	3.4	2.9	3.6

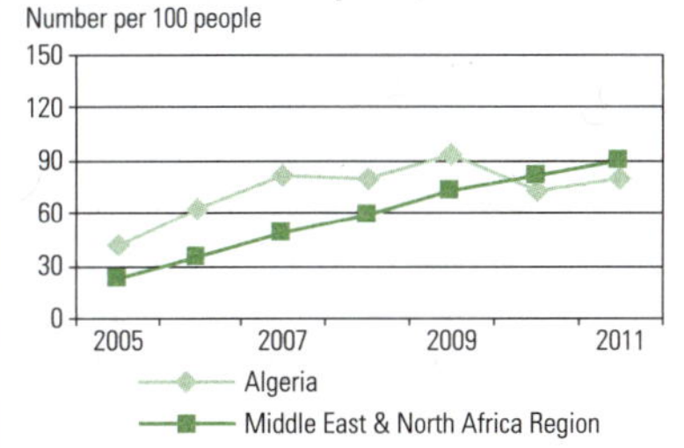

Mobile cellular subscriptions, 2005–11

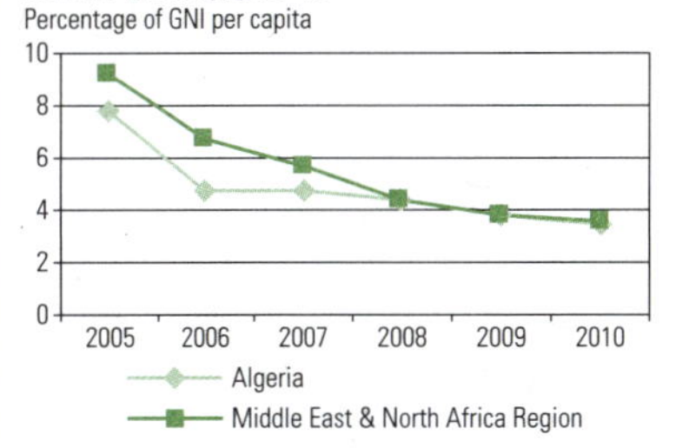

Mobile basket, 2005–10

Sources: Economic and social context: IMF, UIS, UN, WHO and World Bank; Sector structure: ictDATA.org; Sector performance: ictDATA.org, ITU; Wireless Intelligence, and World Bank.
Notes: Use of italics in the column entries indicates years or periods other than those specified. — Not available. GNI = gross national income.
a. Data are for 2011.

Angola

	Angola 2005	Angola 2010	Lower-middle-income group 2010	Sub-Saharan Africa Region 2010
Economic and social context				
Population (total, million)	16	19	2,519	853
GNI per capita, World Bank Atlas method (current US$)	1,330	3,960	1,623	1,188
Rural population (% of total)	46	42	61	63
Expected years of schooling (years)	*9*	—	*10*	*9*
Physicians density (per 1,000 people)	*0.1*	—	0.8	0.2
Depositors with commercial banks (per 1,000 adults)	35	97	—	*167*
Sector structure				
Number of mobile operators	—	2		
Herfindahl-Hirschman Index (HHI) (scale = 0–10,000)	—	5,638		
Sector performance				
Access				
Mobile cellular subscriptions (per 100 people)	10	58[a]	78[a]	57[a]
Mobile cellular subscriptions (% prepaid)	99	99[a]	96[a]	96[a]
Population covered by a mobile-cellular network (%)	*40*	—	*86*	72
Mobile broadband subscriptions (per 100 people)	*0.1*	10.1[a]	7.3[a]	5.6[a]
Mobile broadband (% of total mobile subscriptions)	*0.3*	16.5[a]	9.0[a]	10.1[a]
Usage				
Households with a mobile telephone (%)	*26*	52	77	52
Mobile voice usage (minutes per user per month)	*141*	*108*	276[a]	—
Population using mobile Internet (%)	—	—	2.9	—
Short Message Service (SMS) users (% of mobile users)	—	—	61.9[a]	—
Affordability				
Mobile basket (% of GNI per capita)	*10.0*	5.8	7.2	19.5

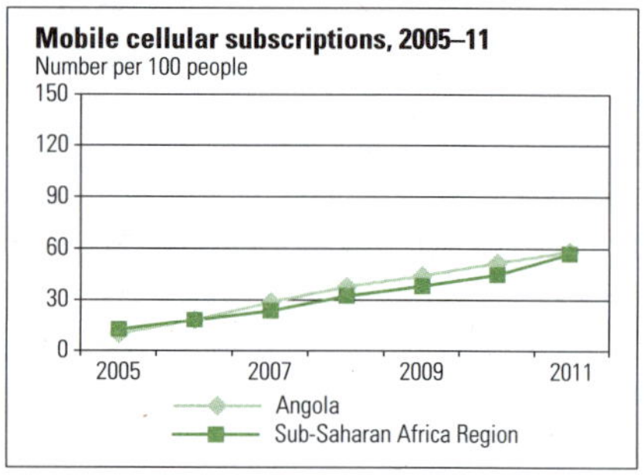

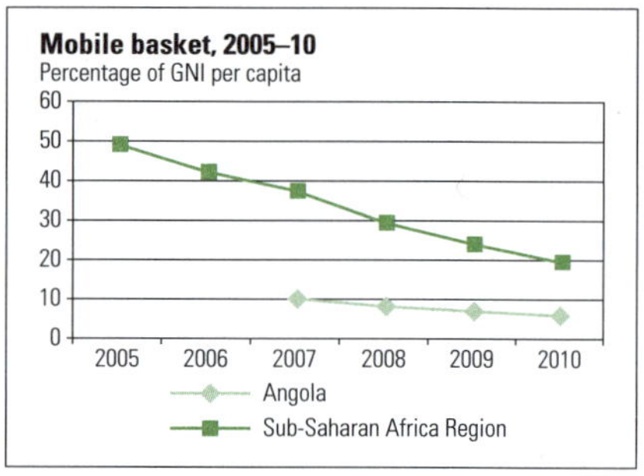

Argentina

	Argentina 2005	Argentina 2010	Upper-middle-income group 2010	Latin America & the Caribbean Region 2010
Economic and social context				
Population (total, million)	39	40	2,452	583
GNI per capita, World Bank Atlas method (current US$)	4,460	8,620	5,886	7,741
Rural population (% of total)	9	8	43	21
Expected years of schooling (years)	15	*16*	*13*	*14*
Physicians density (per 1,000 people)	3.2	—	1.7	1.8
Depositors with commercial banks (per 1,000 adults)	524	702	—	—
Sector structure				
Number of mobile operators	—	3		
Herfindahl-Hirschman Index (HHI) (scale = 0–10,000)	—	3,351		
Sector performance				
Access				
Mobile cellular subscriptions (per 100 people)	57	141[a]	92[a]	109[a]
Mobile cellular subscriptions (% prepaid)	70	71[a]	81[a]	81[a]
Population covered by a mobile-cellular network (%)	94	—	*99*	98
Mobile broadband subscriptions (per 100 people)	—	18.8[a]	14.3[a]	16.1[a]
Mobile broadband (% of total mobile subscriptions)	—	13.7[a]	15.4[a]	15.2[a]
Usage				
Households with a mobile telephone (%)	67	86	84	84
Mobile voice usage (minutes per user per month)	*139*	100[a]	325[a]	141[a]
Population using mobile Internet (%)	—	10.6	22.9[a]	4.4
Short Message Service (SMS) users (% of mobile users)	—	97.0[a]	74.4[a]	—
Affordability				
Mobile basket (% of GNI per capita)	3.6	4.3	2.9	3.7

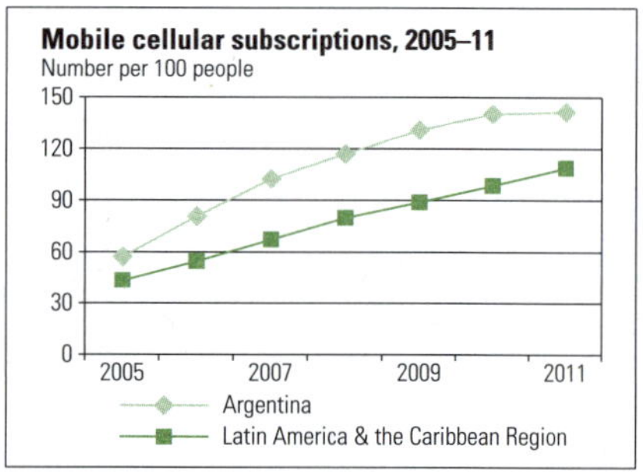

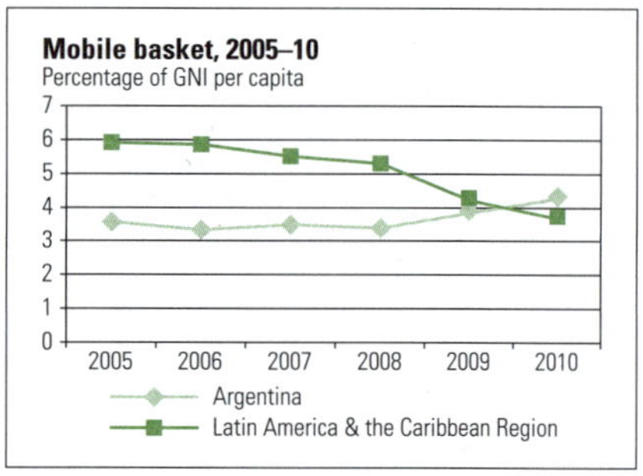

Sources: Economic and social context: IMF, UIS, UN, WHO and World Bank; Sector structure: ictDATA.org; Sector performance: ictDATA.org, ITU; Wireless Intelligence, and World Bank.
Notes: Use of italics in the column entries indicates years or periods other than those specified. — Not available. GNI = gross national income.
a. Data are for 2011.

Armenia

	Armenia 2005	Armenia 2010	Lower-middle-income group 2010	Europe & Central Asia Region 2010
Economic and social context				
Population (total, million)	3	3	2,519	405
GNI per capita, World Bank Atlas method (current US$)	1,470	3,200	1,623	7,272
Rural population (% of total)	36	36	61	36
Expected years of schooling (years)	11	*12*	*10*	*13*
Physicians density (per 1,000 people)	*3.7*	*3.8*	0.8	3.2
Depositors with commercial banks (per 1,000 adults)	357	589	—	894
Sector structure				
Number of mobile operators	—	3		
Herfindahl-Hirschman Index (HHI) (scale = 0–10,000)	—	4,993		
Sector performance				
Access				
Mobile cellular subscriptions (per 100 people)	10	122[a]	78[a]	125[a]
Mobile cellular subscriptions (% prepaid)	87	86[a]	96[a]	82[a]
Population covered by a mobile-cellular network (%)	85	99	*86*	96
Mobile broadband subscriptions (per 100 people)	—	9.8[a]	7.3[a]	22.6[a]
Mobile broadband (% of total mobile subscriptions)	—	9.3[a]	9.0[a]	18.0[a]
Usage				
Households with a mobile telephone (%)	33	91	77	82
Mobile voice usage (minutes per user per month)	*121*	344[a]	276[a]	288[a]
Population using mobile Internet (%)	—	7.4	2.9	8.5
Short Message Service (SMS) users (% of mobile users)	—	31.0	61.9[a]	69.8[a]
Affordability				
Mobile basket (% of GNI per capita)	17.8	3.3	7.2	3.1

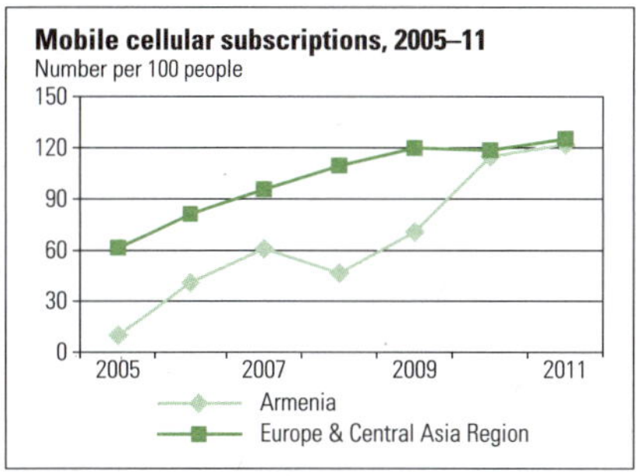

Mobile cellular subscriptions, 2005–11

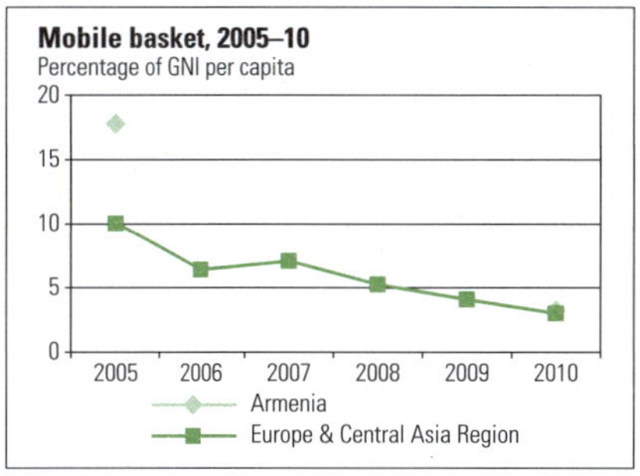

Mobile basket, 2005–10

Australia

	Australia 2005	Australia 2010	High-income group 2010
Economic and social context			
Population (total, million)	20	22	1,127
GNI per capita, World Bank Atlas method (current US$)	30,440	46,200	38,746
Rural population (% of total)	12	11	22
Expected years of schooling (years)	20	*20*	*16*
Physicians density (per 1,000 people)	*1.0*	*3.0*	2.8
Depositors with commercial banks (per 1,000 adults)	—	—	—
Sector structure			
Number of mobile operators	—	3	
Herfindahl-Hirschman Index (HHI) (scale = 0–10,000)	—	3,433	
Sector performance			
Access			
Mobile cellular subscriptions (per 100 people)	90	130[a]	118[a]
Mobile cellular subscriptions (% prepaid)	49	47[a]	36[a]
Population covered by a mobile-cellular network (%)	98	99	100
Mobile broadband subscriptions (per 100 people)	4.2	97.7[a]	69.6[a]
Mobile broadband (% of total mobile subscriptions)	4.5	74.4[a]	57.6[a]
Usage			
Households with a mobile telephone (%)	83	88	93
Mobile voice usage (minutes per user per month)	109	131[a]	339
Population using mobile Internet (%)	4.9	13.9	24.3
Short Message Service (SMS) users (% of mobile users)	—	86.0[a]	78.2[a]
Affordability			
Mobile basket (% of GNI per capita)	1.3	0.7	1.0

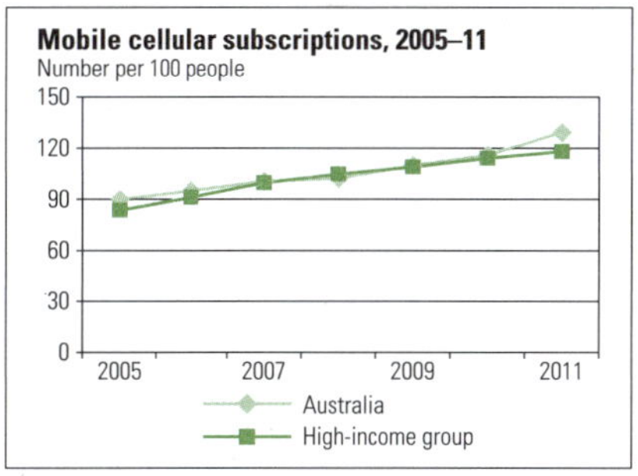

Mobile cellular subscriptions, 2005–11

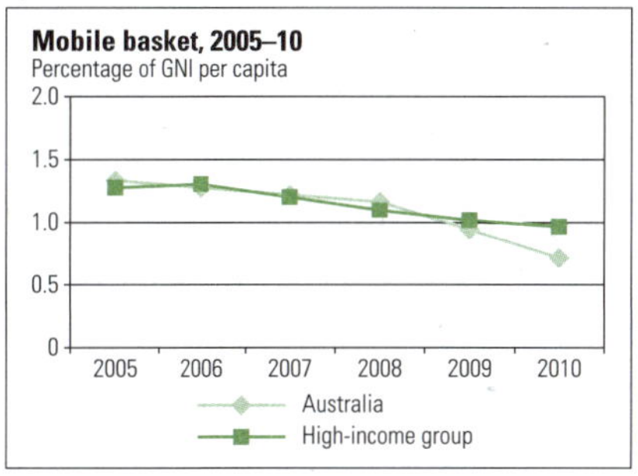

Mobile basket, 2005–10

Sources: Economic and social context: IMF, UIS, UN, WHO and World Bank; Sector structure: ictDATA.org; Sector performance: ictDATA.org, ITU; Wireless Intelligence, and World Bank.
Notes: Use of italics in the column entries indicates years or periods other than those specified. — Not available. GNI = gross national income.
a. Data are for 2011.

Austria

	Austria 2005	Austria 2010	High-income group 2010
Economic and social context			
Population (total, million)	8	8	1,127
GNI per capita, World Bank Atlas method (current US$)	37,210	47,030	38,746
Rural population (% of total)	34	32	22
Expected years of schooling (years)	15	*15*	*16*
Physicians density (per 1,000 people)	*3.7*	4.9	2.8
Depositors with commercial banks (per 1,000 adults)	1,420	1,376	—
Sector structure			
Number of mobile operators	—	4	
Herfindahl-Hirschman Index (HHI) (scale = 0–10,000)	—	3,339	
Sector performance			
Access			
Mobile cellular subscriptions (per 100 people)	105	157[a]	118[a]
Mobile cellular subscriptions (% prepaid)	36	26[a]	36[a]
Population covered by a mobile-cellular network (%)	99	*99*	100
Mobile broadband subscriptions (per 100 people)	10.7	83.3[a]	69.6[a]
Mobile broadband (% of total mobile subscriptions)	10.4	54.9[a]	57.6[a]
Usage			
Households with a mobile telephone (%)	88	91	93
Mobile voice usage (minutes per user per month)	—	181[a]	339
Population using mobile Internet (%)	3.6	20.3	24.3
Short Message Service (SMS) users (% of mobile users)	—	—	78.2[a]
Affordability			
Mobile basket (% of GNI per capita)	1.7	0.4	1.0

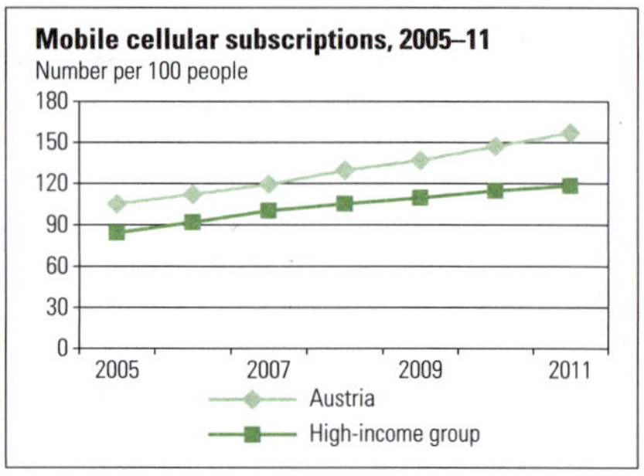

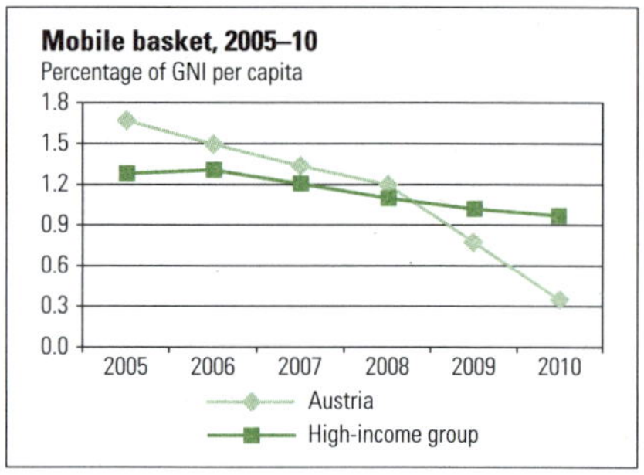

Azerbaijan

	Azerbaijan 2005	Azerbaijan 2010	Upper-middle-income group 2010	Europe & Central Asia Region 2010
Economic and social context				
Population (total, million)	8	9	2,452	405
GNI per capita, World Bank Atlas method (current US$)	1,270	5,330	5,886	7,272
Rural population (% of total)	49	48	43	36
Expected years of schooling (years)	12	*12*	*13*	*13*
Physicians density (per 1,000 people)	*3.6*	*3.8*	1.7	3.2
Depositors with commercial banks (per 1,000 adults)	18	41	—	894
Sector structure				
Number of mobile operators	—	3		
Herfindahl-Hirschman Index (HHI) (scale = 0–10,000)	—	3,780		
Sector performance				
Access				
Mobile cellular subscriptions (per 100 people)	27	84[a]	92[a]	125[a]
Mobile cellular subscriptions (% prepaid)	96	94[a]	81[a]	82[a]
Population covered by a mobile-cellular network (%)	99	100	*99*	96
Mobile broadband subscriptions (per 100 people)	—	4.5[a]	14.3[a]	22.6[a]
Mobile broadband (% of total mobile subscriptions)	—	4.6[a]	15.4[a]	18.0[a]
Usage				
Households with a mobile telephone (%)	50	80	84	82
Mobile voice usage (minutes per user per month)	66	*114*	325[a]	288[a]
Population using mobile Internet (%)	—	1.3	22.9[a]	8.5
Short Message Service (SMS) users (% of mobile users)	—	26.0	74.4[a]	69.8[a]
Affordability				
Mobile basket (% of GNI per capita)	18.5	1.6	2.9	3.1

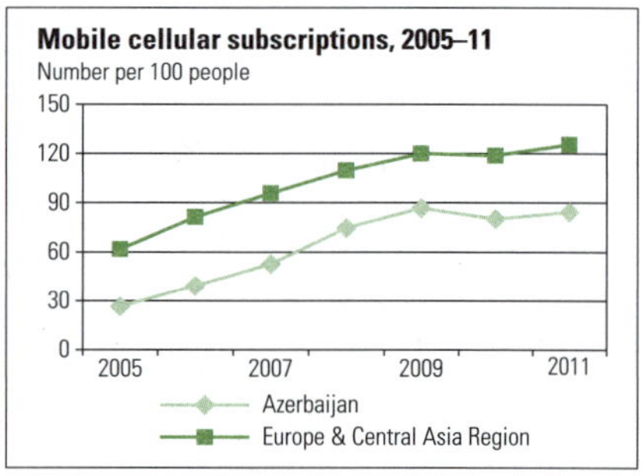

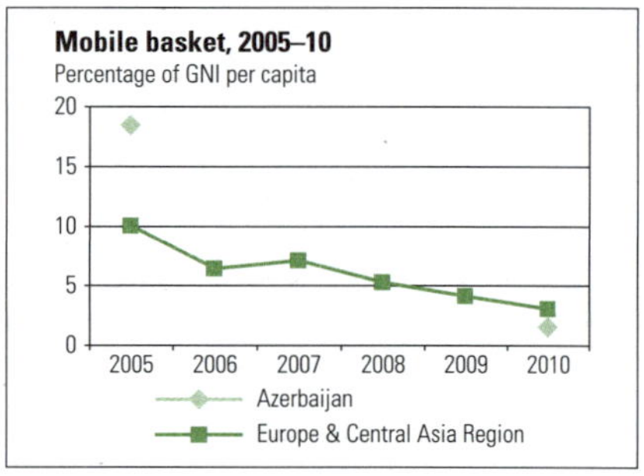

Sources: Economic and social context: IMF, UIS, UN, WHO and World Bank; Sector structure: ictDATA.org; Sector performance: ictDATA.org, ITU; Wireless Intelligence, and World Bank.
Notes: Use of italics in the column entries indicates years or periods other than those specified. — Not available. GNI = gross national income.
a. Data are for 2011.

Bahrain

	Bahrain 2005	Bahrain 2010	High-income group 2010
Economic and social context			
Population (total, million)	0.72	1	1,127
GNI per capita, World Bank Atlas method (current US$)	17,400	*18,730*	38,746
Rural population (% of total)	12	11	22
Expected years of schooling (years)	—	—	*16*
Physicians density (per 1,000 people)	2.7	*1.4*	2.8
Depositors with commercial banks (per 1,000 adults)	—	—	—
Sector structure			
Number of mobile operators	—	3	
Herfindahl-Hirschman Index (HHI) (scale = 0–10,000)	—	3,354	
Sector performance			
Access			
Mobile cellular subscriptions (per 100 people)	106	128[a]	118[a]
Mobile cellular subscriptions (% prepaid)	83	79[a]	36[a]
Population covered by a mobile-cellular network (%)	100	100	100
Mobile broadband subscriptions (per 100 people)	0.6	41.8[a]	69.6[a]
Mobile broadband (% of total mobile subscriptions)	0.6	29.8[a]	57.6[a]
Usage			
Households with a mobile telephone (%)	95	99	93
Mobile voice usage (minutes per user per month)	—	—	339
Population using mobile Internet (%)	5.5	8.7	24.3
Short Message Service (SMS) users (% of mobile users)	—	80.0[a]	78.2[a]
Affordability			
Mobile basket (% of GNI per capita)	1.0	*0.9*	1.0

Mobile cellular subscriptions, 2005–11
Number per 100 people

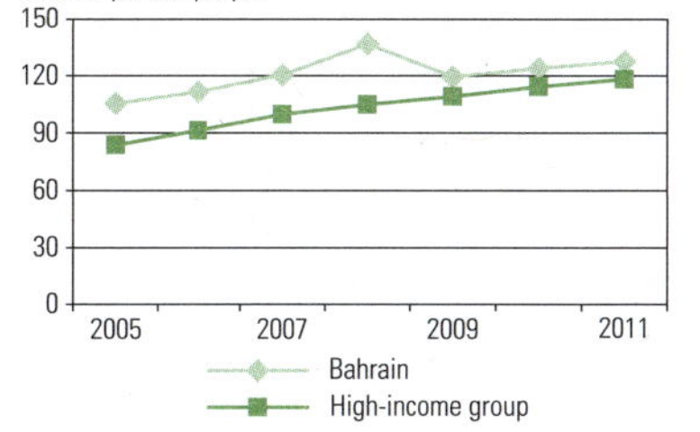

Mobile basket, 2005–10
Percentage of GNI per capita

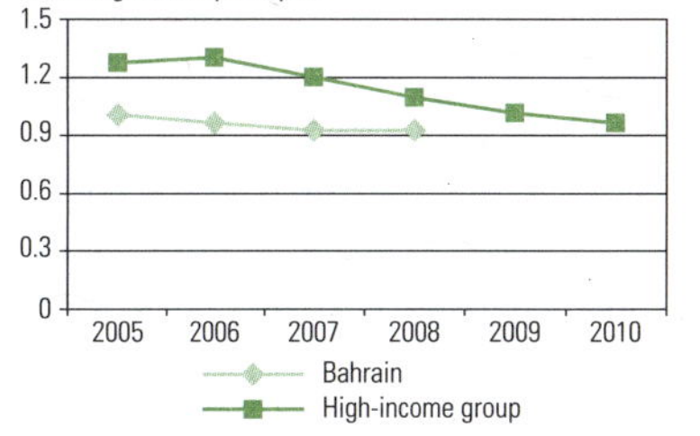

Bangladesh

	Bangladesh 2005	Bangladesh 2010	Low-income group 2010	South Asia Region 2010
Economic and social context				
Population (total, million)	141	149	796	1,633
GNI per capita, World Bank Atlas method (current US$)	480	700	530	1,176
Rural population (% of total)	74	72	72	70
Expected years of schooling (years)	8	—	*9*	*10*
Physicians density (per 1,000 people)	0.3	—	0.2	0.6
Depositors with commercial banks (per 1,000 adults)	321	418	—	249
Sector structure				
Number of mobile operators	—	6		
Herfindahl-Hirschman Index (HHI) (scale = 0–10,000)	—	3,067		
Sector performance				
Access				
Mobile cellular subscriptions (per 100 people)	6	57[a]	43[a]	67[a]
Mobile cellular subscriptions (% prepaid)	95	98[a]	98[a]	96[a]
Population covered by a mobile-cellular network (%)	80	—	—	*84*
Mobile broadband subscriptions (per 100 people)	—	—	—	3.3[a]
Mobile broadband (% of total mobile subscriptions)	—	—	—	4.6[a]
Usage				
Households with a mobile telephone (%)	11	64	43	54
Mobile voice usage (minutes per user per month)	235	210[a]	—	305[a]
Population using mobile Internet (%)	—	—	—	3.3[a]
Short Message Service (SMS) users (% of mobile users)	—	*30.0*	—	47.0[a]
Affordability				
Mobile basket (% of GNI per capita)	16.3	3.4	28.8	3.2

Mobile cellular subscriptions, 2005–11
Number per 100 people

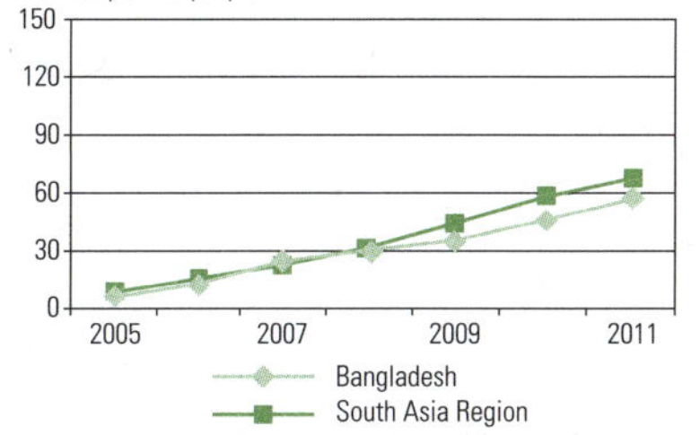

Mobile basket, 2005–10
Percentage of GNI per capita

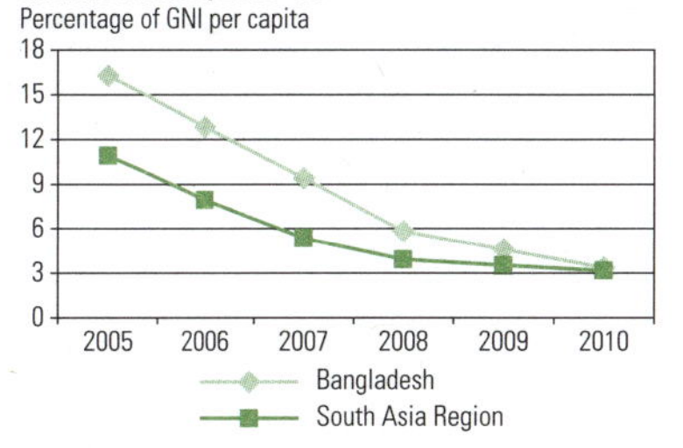

Sources: Economic and social context: IMF, UIS, UN, WHO and World Bank; Sector structure: ictDATA.org; Sector performance: ictDATA.org, ITU; Wireless Intelligence, and World Bank.
Notes: Use of italics in the column entries indicates years or periods other than those specified. — Not available. GNI = gross national income.
a. Data are for 2011.

Belarus

	Belarus 2005	Belarus 2010	Upper-middle-income group 2010	Europe & Central Asia Region 2010
Economic and social context				
Population (total, million)	10	9	2,452	405
GNI per capita, World Bank Atlas method (current US$)	2,780	5,950	5,886	7,272
Rural population (% of total)	28	26	43	36
Expected years of schooling (years)	14	—	*13*	*13*
Physicians density (per 1,000 people)	4.7	*5.2*	1.7	3.2
Depositors with commercial banks (per 1,000 adults)	—	—	—	894
Sector structure				
Number of mobile operators	—	3		
Herfindahl-Hirschman Index (HHI) (scale = 0–10,000)	—	3,889		
Sector performance				
Access				
Mobile cellular subscriptions (per 100 people)	42	119[a]	92[a]	125[a]
Mobile cellular subscriptions (% prepaid)	59	64[a]	81[a]	82[a]
Population covered by a mobile-cellular network (%)	90	100	*99*	96
Mobile broadband subscriptions (per 100 people)	0.0	20.8[a]	14.3[a]	22.6[a]
Mobile broadband (% of total mobile subscriptions)	0.01	17.1[a]	15.4[a]	18.0[a]
Usage				
Households with a mobile telephone (%)	30	76	84	82
Mobile voice usage (minutes per user per month)	450	347[a]	325[a]	288[a]
Population using mobile Internet (%)	—	—	22.9[a]	8.5
Short Message Service (SMS) users (% of mobile users)	—	*42.0*	74.4[a]	69.8[a]
Affordability				
Mobile basket (% of GNI per capita)	2.3	1.6	2.9	3.1

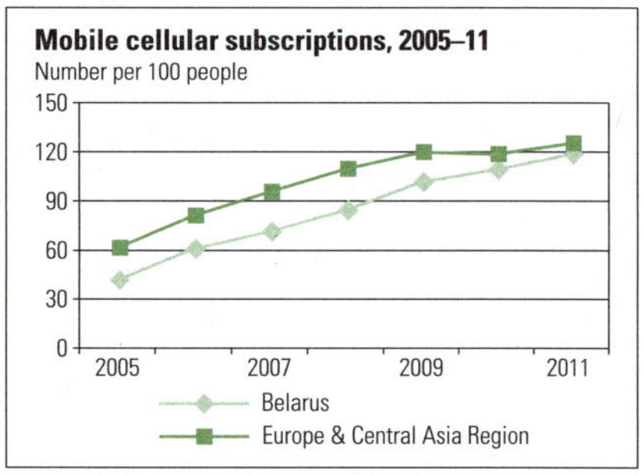

Mobile cellular subscriptions, 2005–11

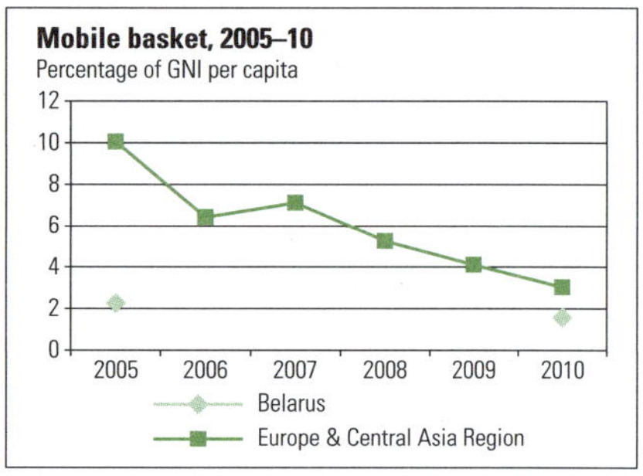

Mobile basket, 2005–10

Belgium

	Belgium 2005	Belgium 2010	High-income group 2010
Economic and social context			
Population (total, million)	10	11	1,127
GNI per capita, World Bank Atlas method (current US$)	36,600	45,840	38,746
Rural population (% of total)	3	3	22
Expected years of schooling (years)	16	*16*	*16*
Physicians density (per 1,000 people)	*4.2*	*3.0*	2.8
Depositors with commercial banks (per 1,000 adults)	—	—	—
Sector structure			
Number of mobile operators	—	3	
Herfindahl-Hirschman Index (HHI) (scale = 0–10,000)	—	3,457	
Sector performance			
Access			
Mobile cellular subscriptions (per 100 people)	92	122[a]	118[a]
Mobile cellular subscriptions (% prepaid)	62	49[a]	36[a]
Population covered by a mobile-cellular network (%)	99	100	100
Mobile broadband subscriptions (per 100 people)	0.3	33.8[a]	69.6[a]
Mobile broadband (% of total mobile subscriptions)	0.3	27.4[a]	57.6[a]
Usage			
Households with a mobile telephone (%)	*85*	91	93
Mobile voice usage (minutes per user per month)	153	147[a]	339
Population using mobile Internet (%)	—	7.7	24.3
Short Message Service (SMS) users (% of mobile users)	—	—	78.2[a]
Affordability			
Mobile basket (% of GNI per capita)	1.6	1.1	1.0

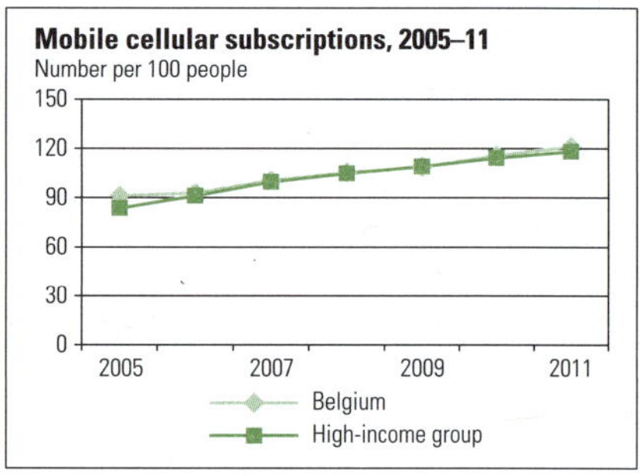

Mobile cellular subscriptions, 2005–11

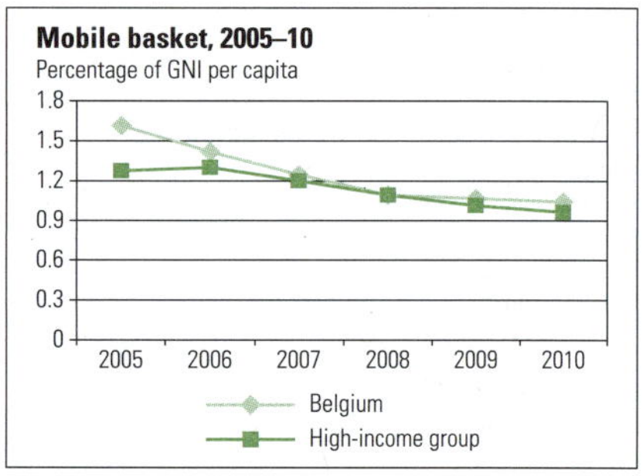

Mobile basket, 2005–10

Sources: Economic and social context: IMF, UIS, UN, WHO and World Bank; Sector structure: ictDATA.org; Sector performance: ictDATA.org, ITU; Wireless Intelligence, and World Bank.
Notes: Use of italics in the column entries indicates years or periods other than those specified. — Not available. GNI = gross national income.
a. Data are for 2011.

Benin

	Benin 2005	Benin 2010	Low-income group 2010	Sub-Saharan Africa Region 2010
Economic and social context				
Population (total, million)	8	9	796	853
GNI per capita, World Bank Atlas method (current US$)	570	780	530	1,188
Rural population (% of total)	60	58	72	63
Expected years of schooling (years)	9	—	*9*	*9*
Physicians density (per 1,000 people)	*0.04*	*0.1*	0.2	0.2
Depositors with commercial banks (per 1,000 adults)	—	—	—	*167*
Sector structure				
Number of mobile operators	—	5		
Herfindahl-Hirschman Index (HHI) (scale = 0–10,000)	—	2,536		
Sector performance				
Access				
Mobile cellular subscriptions (per 100 people)	8	80	43[a]	57[a]
Mobile cellular subscriptions (% prepaid)	98	99[a]	98[a]	96[a]
Population covered by a mobile-cellular network (%)	43	90	—	72
Mobile broadband subscriptions (per 100 people)	—	—	—	5.6[a]
Mobile broadband (% of total mobile subscriptions)	—	—	—	10.1[a]
Usage				
Households with a mobile telephone (%)	*24*	—	43	52
Mobile voice usage (minutes per user per month)	—	33	—	—
Population using mobile Internet (%)	—	—	—	—
Short Message Service (SMS) users (% of mobile users)	—	—	—	—
Affordability				
Mobile basket (% of GNI per capita)	47.1	20.0	28.8	19.5

Mobile cellular subscriptions, 2005–11
Number per 100 people

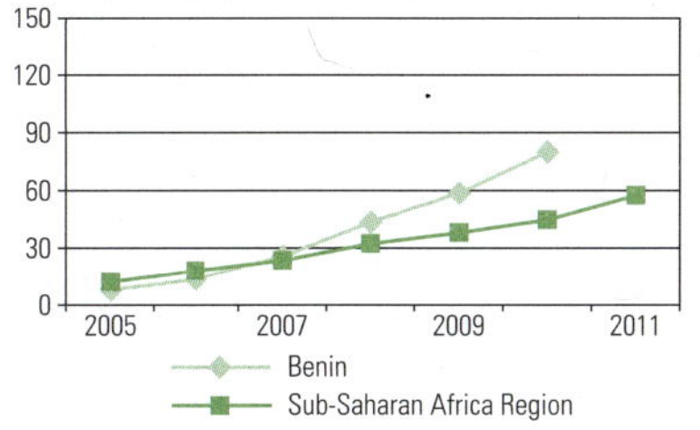

Mobile basket, 2005–10
Percentage of GNI per capita

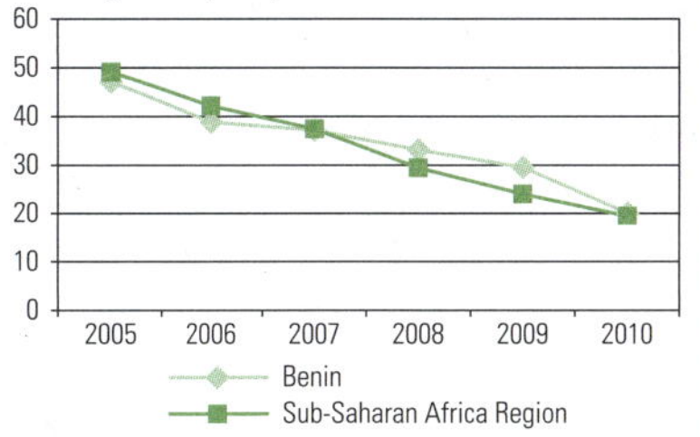

Bolivia

	Bolivia 2005	Bolivia 2010	Lower-middle-income group 2010	Latin America & the Caribbean Region 2010
Economic and social context				
Population (total, million)	9	10	2,519	583
GNI per capita, World Bank Atlas method (current US$)	1,030	1,810	1,623	7,741
Rural population (% of total)	36	34	61	21
Expected years of schooling (years)	*14*	—	*10*	*14*
Physicians density (per 1,000 people)	—	—	0.8	1.8
Depositors with commercial banks (per 1,000 adults)	—	—	—	—
Sector structure				
Number of mobile operators	—	3		
Herfindahl-Hirschman Index (HHI) (scale = 0–10,000)	—	3,450		
Sector performance				
Access				
Mobile cellular subscriptions (per 100 people)	26	72	78[a]	109[a]
Mobile cellular subscriptions (% prepaid)	90	91[a]	96[a]	81[a]
Population covered by a mobile-cellular network (%)	46	—	*86*	98
Mobile broadband subscriptions (per 100 people)	—	3.1[a]	7.3[a]	16.1[a]
Mobile broadband (% of total mobile subscriptions)	—	4.0[a]	9.0[a]	15.2[a]
Usage				
Households with a mobile telephone (%)	39	74	77	84
Mobile voice usage (minutes per user per month)	—	—	276[a]	141[a]
Population using mobile Internet (%)	—	—	2.9	4.4
Short Message Service (SMS) users (% of mobile users)	—	—	61.9[a]	—
Affordability				
Mobile basket (% of GNI per capita)	14.3	7.5	7.2	3.7

Mobile cellular subscriptions, 2005–11
Number per 100 people

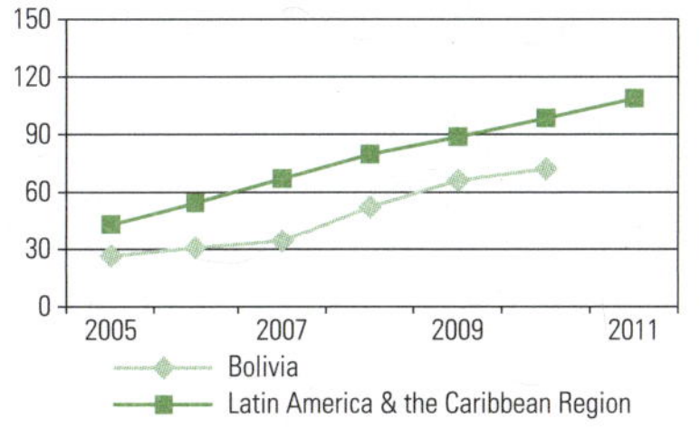

Mobile basket, 2005–10
Percentage of GNI per capita

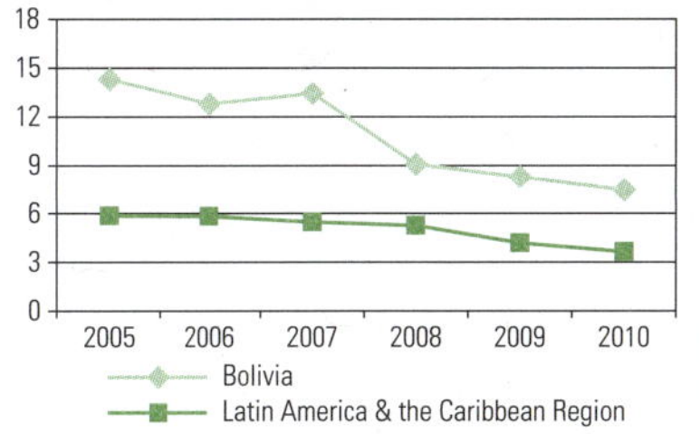

Sources: Economic and social context: IMF, UIS, UN, WHO and World Bank; Sector structure: ictDATA.org; Sector performance: ictDATA.org, ITU; Wireless Intelligence, and World Bank.
Notes: Use of italics in the column entries indicates years or periods other than those specified. — Not available. GNI = gross national income.
a. Data are for 2011.

Bosnia and Herzegovina

	Bosnia and Herzegovina 2005	Bosnia and Herzegovina 2010	Upper-middle-income group 2010	Europe & Central Asia Region 2010
Economic and social context				
Population (total, million)	4	4	2,452	405
GNI per capita, World Bank Atlas method (current US$)	3,000	4,770	5,886	7,272
Rural population (% of total)	54	51	43	36
Expected years of schooling (years)	*13*	*14*	*13*	*13*
Physicians density (per 1,000 people)	1.4	*1.6*	1.7	3.2
Depositors with commercial banks (per 1,000 adults)	573	914	—	894
Sector structure				
Number of mobile operators	—	3		
Herfindahl-Hirschman Index (HHI) (scale = 0–10,000)	—	4,013		
Sector performance				
Access				
Mobile cellular subscriptions (per 100 people)	42	87	92[a]	125[a]
Mobile cellular subscriptions (% prepaid)	86	86[a]	81[a]	82[a]
Population covered by a mobile-cellular network (%)	97	100	*99*	96
Mobile broadband subscriptions (per 100 people)	—	20.1[a]	14.3[a]	22.6[a]
Mobile broadband (% of total mobile subscriptions)	—	24.1[a]	15.4[a]	18.0[a]
Usage				
Households with a mobile telephone (%)	53	82	84	82
Mobile voice usage (minutes per user per month)	—	—	325[a]	288[a]
Population using mobile Internet (%)	—	4.5	22.9[a]	8.5
Short Message Service (SMS) users (% of mobile users)	—	64.0[a]	74.4[a]	69.8[a]
Affordability				
Mobile basket (% of GNI per capita)	7.1	3.9	2.9	3.1

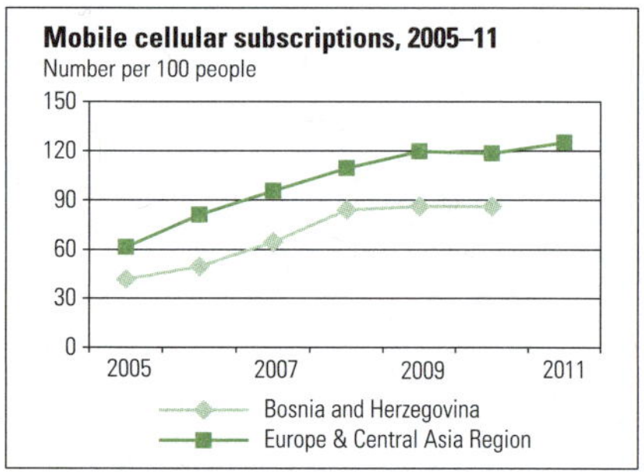

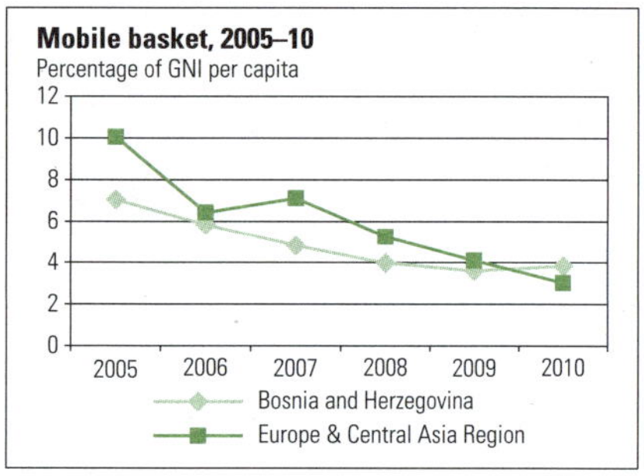

Botswana

	Botswana 2005	Botswana 2010	Upper-middle-income group 2010	Sub-Saharan Africa Region 2010
Economic and social context				
Population (total, million)	2	2	2,452	853
GNI per capita, World Bank Atlas method (current US$)	5,070	6,740	5,886	1,188
Rural population (% of total)	43	39	43	63
Expected years of schooling (years)	12	—	*13*	*9*
Physicians density (per 1,000 people)	*0.3*	—	1.7	0.2
Depositors with commercial banks (per 1,000 adults)	345	*496*	—	*167*
Sector structure				
Number of mobile operators	—	3		
Herfindahl-Hirschman Index (HHI) (scale = 0–10,000)	—	4,079		
Sector performance				
Access				
Mobile cellular subscriptions (per 100 people)	30	144[a]	92[a]	57[a]
Mobile cellular subscriptions (% prepaid)	98	98[a]	81[a]	96[a]
Population covered by a mobile-cellular network (%)	99	*99*	*99*	72
Mobile broadband subscriptions (per 100 people)	—	10.0[a]	14.3[a]	5.6[a]
Mobile broadband (% of total mobile subscriptions)	—	6.8[a]	15.4[a]	10.1[a]
Usage				
Households with a mobile telephone (%)	—	62	84	52
Mobile voice usage (minutes per user per month)	—	—	325[a]	—
Population using mobile Internet (%)	—	—	22.9[a]	—
Short Message Service (SMS) users (% of mobile users)	—	—	74.4[a]	—
Affordability				
Mobile basket (% of GNI per capita)	*3.4*	2.4	2.9	19.5

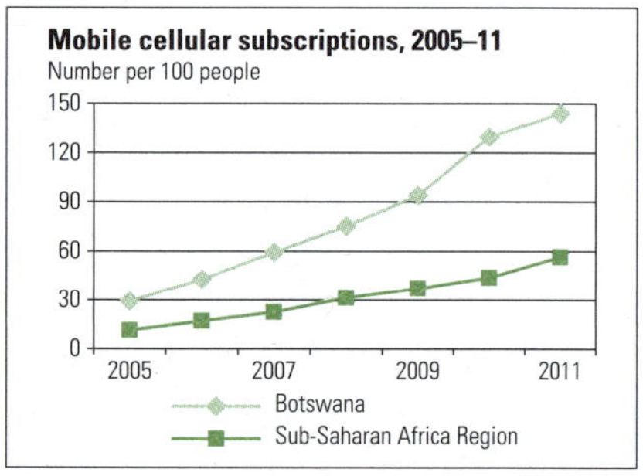

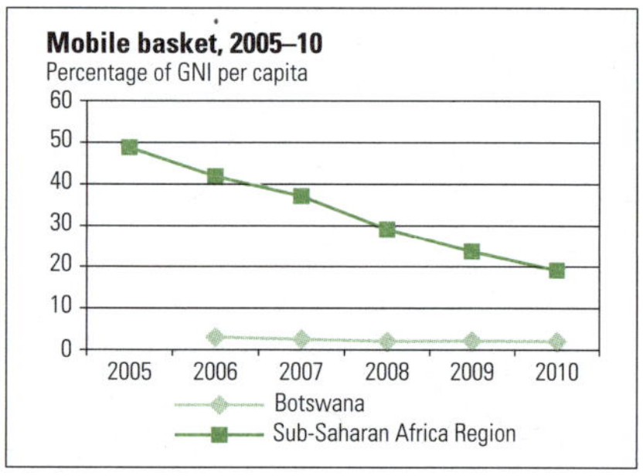

Sources: Economic and social context: IMF, UIS, UN, WHO and World Bank; Sector structure: ictDATA.org; Sector performance: ictDATA.org, ITU; Wireless Intelligence, and World Bank.
Notes: Use of italics in the column entries indicates years or periods other than those specified. — Not available. GNI = gross national income.
a. Data are for 2011.

Brazil

	Brazil 2005	Brazil 2010	Upper-middle-income group 2010	Latin America & the Caribbean Region 2010
Economic and social context				
Population (total, million)	186	195	2,452	583
GNI per capita, World Bank Atlas method (current US$)	3,960	9,390	5,886	7,741
Rural population (% of total)	16	14	43	21
Expected years of schooling (years)	14	*14*	*13*	*14*
Physicians density (per 1,000 people)	*1.7*	*1.8*	1.7	1.8
Depositors with commercial banks (per 1,000 adults)	—	—	—	—
Sector structure				
Number of mobile operators	—	4		
Herfindahl-Hirschman Index (HHI) (scale = 0–10,000)	—	2,537		
Sector performance				
Access				
Mobile cellular subscriptions (per 100 people)	46	123[a]	92[a]	109[a]
Mobile cellular subscriptions (% prepaid)	81	80[a]	81[a]	81[a]
Population covered by a mobile-cellular network (%)	88	100	*99*	98
Mobile broadband subscriptions (per 100 people)	0.2	20.9[a]	14.3[a]	16.1[a]
Mobile broadband (% of total mobile subscriptions)	0.3	16.7[a]	15.4[a]	15.2[a]
Usage				
Households with a mobile telephone (%)	59	92	84	84
Mobile voice usage (minutes per user per month)	88	118[a]	325[a]	141[a]
Population using mobile Internet (%)	1.5	2.7	22.9[a]	4.4
Short Message Service (SMS) users (% of mobile users)	—	49.0	74.4[a]	—
Affordability				
Mobile basket (% of GNI per capita)	11.7	7.3	2.9	3.7

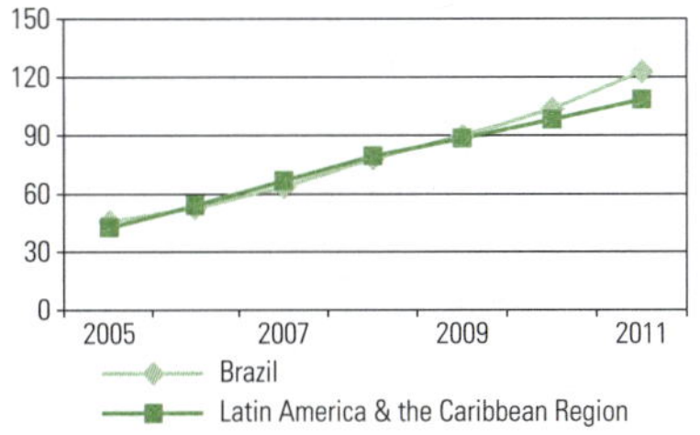

Mobile cellular subscriptions, 2005–11
Number per 100 people

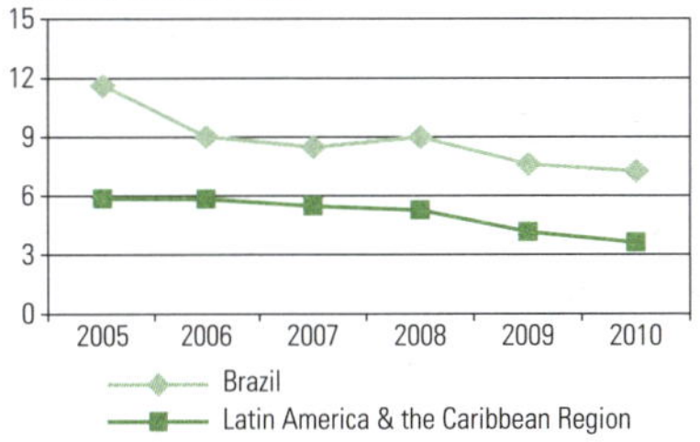

Mobile basket, 2005–10
Percentage of GNI per capita

Bulgaria

	Bulgaria 2005	Bulgaria 2010	Upper-middle-income group 2010	Europe & Central Asia Region 2010
Economic and social context				
Population (total, million)	8	8	2,452	405
GNI per capita, World Bank Atlas method (current US$)	3,640	6,280	5,886	7,272
Rural population (% of total)	30	28	43	36
Expected years of schooling (years)	13	*14*	*13*	*13*
Physicians density (per 1,000 people)	*3.7*	*3.7*	1.7	3.2
Depositors with commercial banks (per 1,000 adults)	1,466	1,958	—	894
Sector structure				
Number of mobile operators	—	3		
Herfindahl-Hirschman Index (HHI) (scale = 0–10,000)	—	3,866		
Sector performance				
Access				
Mobile cellular subscriptions (per 100 people)	81	151[a]	92[a]	125[a]
Mobile cellular subscriptions (% prepaid)	67	39[a]	81[a]	82[a]
Population covered by a mobile-cellular network (%)	100	100	*99*	96
Mobile broadband subscriptions (per 100 people)	0.3	40.8[a]	14.3[a]	22.6[a]
Mobile broadband (% of total mobile subscriptions)	0.4	25.8[a]	15.4[a]	18.0[a]
Usage				
Households with a mobile telephone (%)	*64*	80	84	82
Mobile voice usage (minutes per user per month)	—	118[a]	325[a]	288[a]
Population using mobile Internet (%)	—	—	22.9[a]	8.5
Short Message Service (SMS) users (% of mobile users)	—	—	74.4[a]	69.8[a]
Affordability				
Mobile basket (% of GNI per capita)	10.1	5.9	2.9	3.1

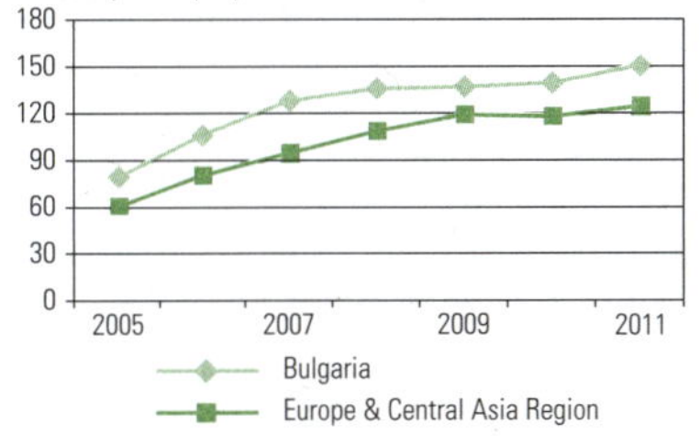

Mobile cellular subscriptions, 2005–11
Number per 100 people

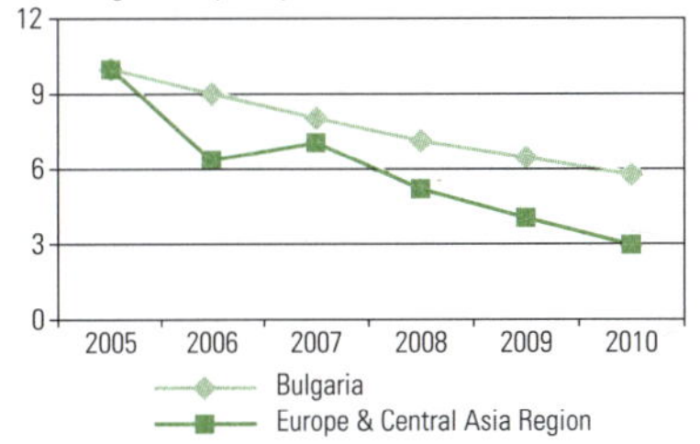

Mobile basket, 2005–10
Percentage of GNI per capita

Sources: Economic and social context: IMF, UIS, UN, WHO and World Bank; Sector structure: ictDATA.org; Sector performance: ictDATA.org, ITU; Wireless Intelligence, and World Bank.
Notes: Use of italics in the column entries indicates years or periods other than those specified. — Not available. GNI = gross national income.
a. Data are for 2011.

Burkina Faso

	Burkina Faso 2005	Burkina Faso 2010	Low-income group 2010	Sub-Saharan Africa Region 2010
Economic and social context				
Population (total, million)	14	16	796	853
GNI per capita, World Bank Atlas method (current US$)	390	550	530	1,188
Rural population (% of total)	82	80	72	63
Expected years of schooling (years)	5	*6*	*9*	*9*
Physicians density (per 1,000 people)	*0.1*	*0.1*	0.2	0.2
Depositors with commercial banks (per 1,000 adults)	—	—	—	*167*
Sector structure				
Number of mobile operators	—	3		
Herfindahl-Hirschman Index (HHI) (scale = 0–10,000)	—	4,047		
Sector performance				
Access				
Mobile cellular subscriptions (per 100 people)	4	43[a]	43[a]	57[a]
Mobile cellular subscriptions (% prepaid)	99	99[a]	98[a]	96[a]
Population covered by a mobile-cellular network (%)	26	—	—	72
Mobile broadband subscriptions (per 100 people)	—	—	—	5.6[a]
Mobile broadband (% of total mobile subscriptions)	—	—	—	10.1[a]
Usage				
Households with a mobile telephone (%)	18	—	43	52
Mobile voice usage (minutes per user per month)	—	—	—	—
Population using mobile Internet (%)	—	—	—	—
Short Message Service (SMS) users (% of mobile users)	—	—	—	—
Affordability				
Mobile basket (% of GNI per capita)	72.8	46.3	28.8	19.5

Mobile cellular subscriptions, 2005–11
Number per 100 people

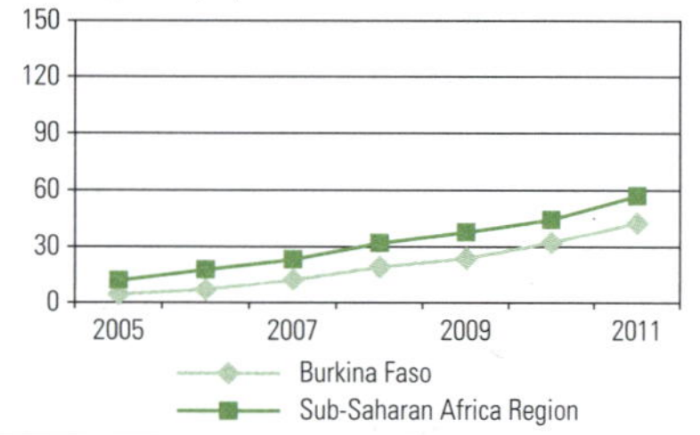

Mobile basket, 2005–10
Percentage of GNI per capita

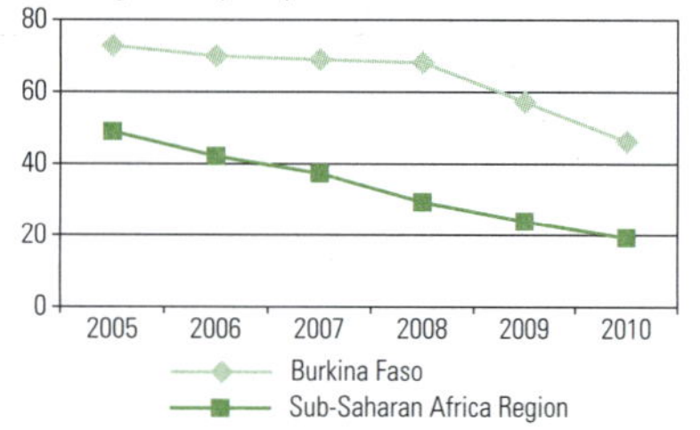

Burundi

	Burundi 2005	Burundi 2010	Low-income group 2010	Sub-Saharan Africa Region 2010
Economic and social context				
Population (total, million)	7	8	796	853
GNI per capita, World Bank Atlas method (current US$)	100	170	530	1,188
Rural population (% of total)	91	89	72	63
Expected years of schooling (years)	6	*10*	*9*	*9*
Physicians density (per 1,000 people)	*0.03*	—	0.2	0.2
Depositors with commercial banks (per 1,000 adults)	—	—	—	*167*
Sector structure				
Number of mobile operators	—	5		
Herfindahl-Hirschman Index (HHI) (scale = 0–10,000)	—	—		
Sector performance				
Access				
Mobile cellular subscriptions (per 100 people)	2	25[a]	43[a]	57[a]
Mobile cellular subscriptions (% prepaid)	99	100[a]	98[a]	96[a]
Population covered by a mobile-cellular network (%)	*82*	*83*	—	72
Mobile broadband subscriptions (per 100 people)	—	0.1[a]	—	5.6[a]
Mobile broadband (% of total mobile subscriptions)	—	0.4[a]	—	10.1[a]
Usage				
Households with a mobile telephone (%)	—	32	43	52
Mobile voice usage (minutes per user per month)	—	—	—	—
Population using mobile Internet (%)	—	0.6	—	—
Short Message Service (SMS) users (% of mobile users)	—	25.0	—	—
Affordability				
Mobile basket (% of GNI per capita)	—	—	28.8	19.5

Mobile cellular subscriptions, 2005–11
Number per 100 people

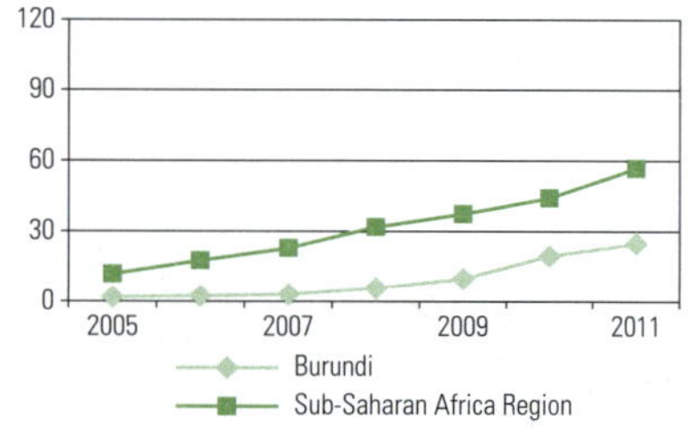

Mobile basket, 2005–10
Percentage of GNI per capita

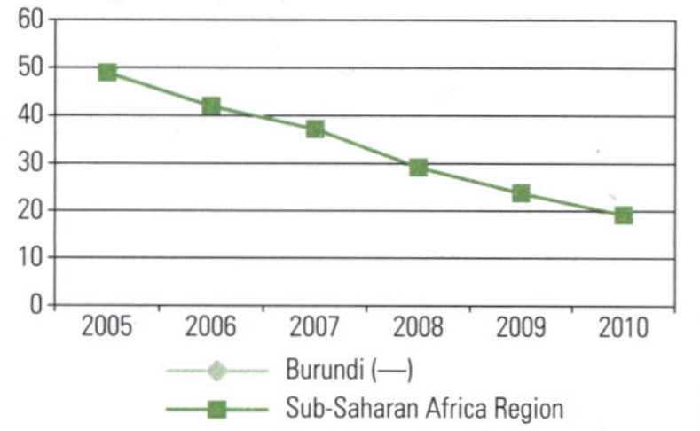

Sources: Economic and social context: IMF, UIS, UN, WHO and World Bank; Sector structure: ictDATA.org; Sector performance: ictDATA.org, ITU; Wireless Intelligence, and World Bank.
Notes: Use of italics in the column entries indicates years or periods other than those specified. — Not available. GNI = gross national income.
a. Data are for 2011.

Cambodia

	Cambodia 2005	Cambodia 2010	Low-income group 2010	East Asia & Pacific Region 2010
Economic and social context				
Population (total, million)	13	14	796	1,962
GNI per capita, World Bank Atlas method (current US$)	460	750	530	3,696
Rural population (% of total)	80	77	72	54
Expected years of schooling (years)	*10*	—	*9*	*12*
Physicians density (per 1,000 people)	—	0.2	0.2	1.2
Depositors with commercial banks (per 1,000 adults)	—	108	—	—
Sector structure				
Number of mobile operators	—	7		
Herfindahl-Hirschman Index (HHI) (scale = 0–10,000)	—	2,354		
Sector performance				
Access				
Mobile cellular subscriptions (per 100 people)	8	53	43[a]	83[a]
Mobile cellular subscriptions (% prepaid)	94	98[a]	98[a]	85[a]
Population covered by a mobile-cellular network (%)	75	*99*	—	*99*
Mobile broadband subscriptions (per 100 people)	*0.04*	9.1[a]	—	11.6[a]
Mobile broadband (% of total mobile subscriptions)	*0.37*	8.4[a]	—	14.4[a]
Usage				
Households with a mobile telephone (%)	20	62	43	83
Mobile voice usage (minutes per user per month)	—	—	—	367[a]
Population using mobile Internet (%)	—	2.8	—	22.4[a]
Short Message Service (SMS) users (% of mobile users)	—	*26.1*	—	84.0[a]
Affordability				
Mobile basket (% of GNI per capita)	17.9	10.7	28.8	5.7

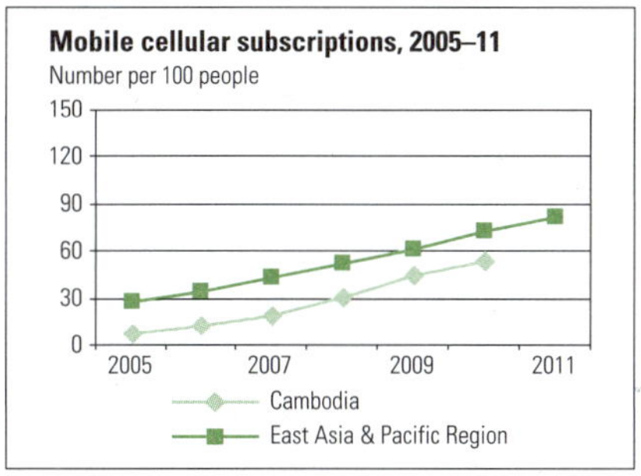

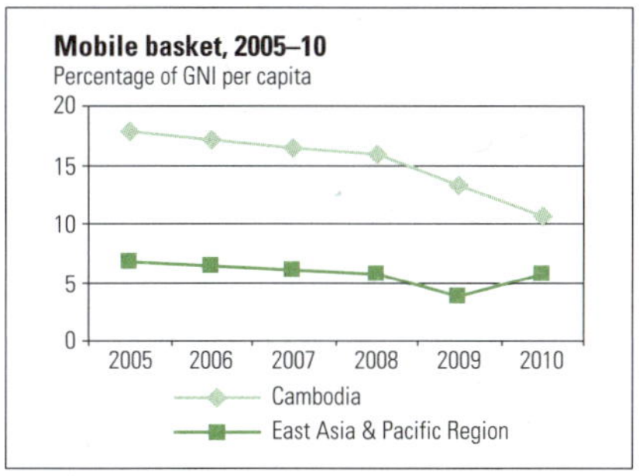

Cameroon

	Cameroon 2005	Cameroon 2010	Lower-middle-income group 2010	Sub-Saharan Africa Region 2010
Economic and social context				
Population (total, million)	18	20	2,519	853
GNI per capita, World Bank Atlas method (current US$)	930	1,200	1,623	1,188
Rural population (% of total)	46	42	61	63
Expected years of schooling (years)	9	*10*	*10*	*9*
Physicians density (per 1,000 people)	*0.2*	—	0.8	0.2
Depositors with commercial banks (per 1,000 adults)	36	*72*	—	*167*
Sector structure				
Number of mobile operators	—	2		
Herfindahl-Hirschman Index (HHI) (scale = 0–10,000)	—	4,949		
Sector performance				
Access				
Mobile cellular subscriptions (per 100 people)	13	53[a]	78[a]	57[a]
Mobile cellular subscriptions (% prepaid)	98	100[a]	96[a]	96[a]
Population covered by a mobile-cellular network (%)	54	—	*86*	72
Mobile broadband subscriptions (per 100 people)	—	—	7.3[a]	5.6[a]
Mobile broadband (% of total mobile subscriptions)	—	—	9.0[a]	10.1[a]
Usage				
Households with a mobile telephone (%)	27	43	77	52
Mobile voice usage (minutes per user per month)	—	42[a]	276[a]	—
Population using mobile Internet (%)	—	1.3	2.9	—
Short Message Service (SMS) users (% of mobile users)	—	—	61.9[a]	—
Affordability				
Mobile basket (% of GNI per capita)	40.0	20.1	7.2	19.5

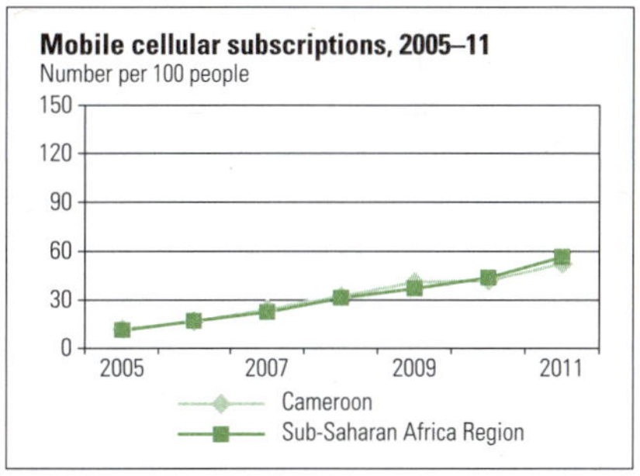

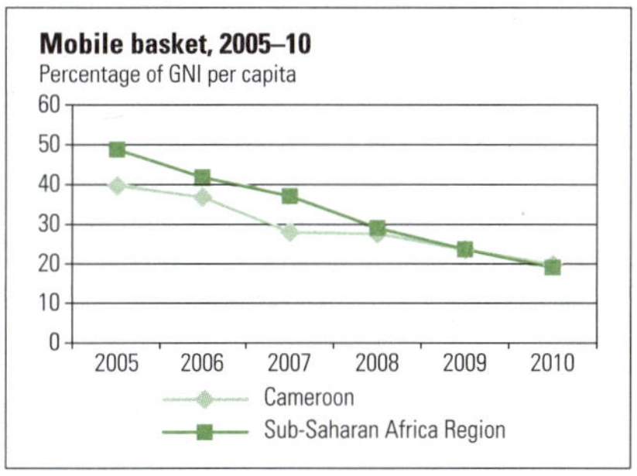

Sources: Economic and social context: IMF, UIS, UN, WHO and World Bank; Sector structure: ictDATA.org; Sector performance: ictDATA.org, ITU; Wireless Intelligence, and World Bank.
Notes: Use of italics in the column entries indicates years or periods other than those specified. — Not available. GNI = gross national income.
a. Data are for 2011.

Canada

	Canada 2005	Canada 2010	High-income group 2010
Economic and social context			
Population (total, million)	32	34	1,127
GNI per capita, World Bank Atlas method (current US$)	33,110	43,250	38,746
Rural population (% of total)	20	19	22
Expected years of schooling (years)	—	—	*16*
Physicians density (per 1,000 people)	*1.9*	*2.0*	2.8
Depositors with commercial banks (per 1,000 adults)	—	—	—
Sector structure			
Number of mobile operators	—	3	
Herfindahl-Hirschman Index (HHI) (scale = 0–10,000)	—	3,019	
Sector performance			
Access			
Mobile cellular subscriptions (per 100 people)	53	74[a]	118[a]
Mobile cellular subscriptions (% prepaid)	22	21[a]	36[a]
Population covered by a mobile-cellular network (%)	97	*99*	100
Mobile broadband subscriptions (per 100 people)	0.1	32.6[a]	69.6[a]
Mobile broadband (% of total mobile subscriptions)	0.2	42.4[a]	57.6[a]
Usage			
Households with a mobile telephone (%)	64	77	93
Mobile voice usage (minutes per user per month)	326	376[a]	339
Population using mobile Internet (%)	2.8	21.5[a]	24.3
Short Message Service (SMS) users (% of mobile users)	—	67.4[a]	78.2[a]
Affordability			
Mobile basket (% of GNI per capita)	0.5	1.0	1.0

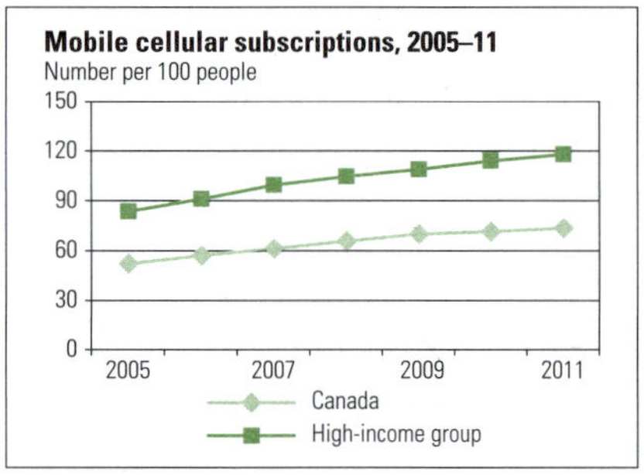

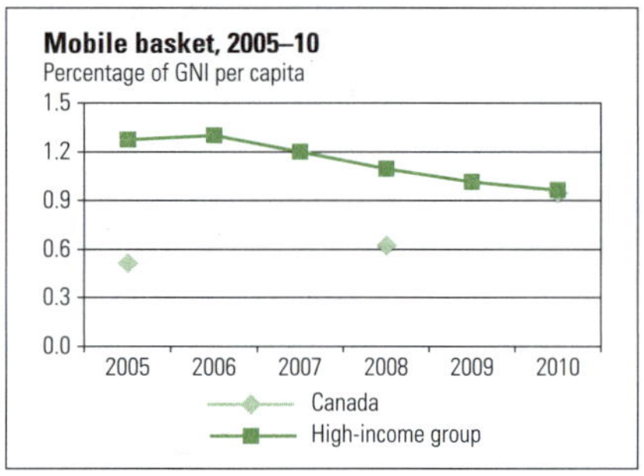

Central African Republic

	Central African Republic 2005	Central African Republic 2010	Low-income group 2010	Sub-Saharan Africa Region 2010
Economic and social context				
Population (total, million)	4	4	796	853
GNI per capita, World Bank Atlas method (current US$)	340	470	530	1,188
Rural population (% of total)	62	61	72	63
Expected years of schooling (years)	—	*7*	*9*	*9*
Physicians density (per 1,000 people)	*0.1*	—	0.2	0.2
Depositors with commercial banks (per 1,000 adults)	3	*3*	—	*167*
Sector structure				
Number of mobile operators	—	4		
Herfindahl-Hirschman Index (HHI) (scale = 0–10,000)	—	—		
Sector performance				
Access				
Mobile cellular subscriptions (per 100 people)	2	17	43[a]	57[a]
Mobile cellular subscriptions (% prepaid)	100	100[a]	98[a]	96[a]
Population covered by a mobile-cellular network (%)	19	55	—	72
Mobile broadband subscriptions (per 100 people)	—	—	—	5.6[a]
Mobile broadband (% of total mobile subscriptions)	—	—	—	10.1[a]
Usage				
Households with a mobile telephone (%)	—	16	43	52
Mobile voice usage (minutes per user per month)	—	—	—	—
Population using mobile Internet (%)	—	—	—	—
Short Message Service (SMS) users (% of mobile users)	—	—	—	—
Affordability				
Mobile basket (% of GNI per capita)	—	*34.5*	28.8	19.5

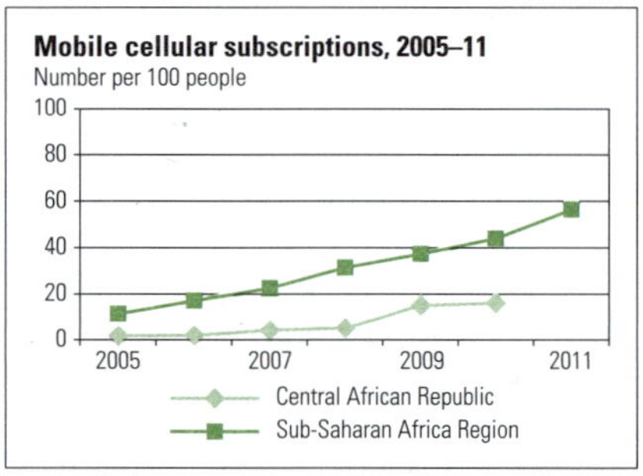

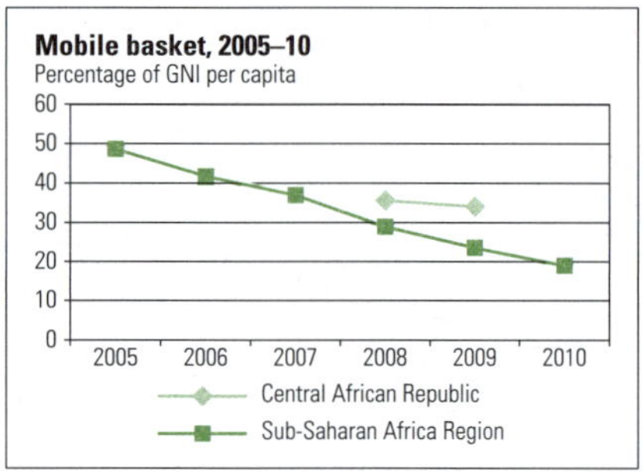

Sources: Economic and social context: IMF, UIS, UN, WHO and World Bank; Sector structure: ictDATA.org; Sector performance: ictDATA.org, ITU; Wireless Intelligence, and World Bank.
Notes: Use of italics in the column entries indicates years or periods other than those specified. — Not available. GNI = gross national income.
a. Data are for 2011.

Chad

	Chad 2005	Chad 2010	Low-income group 2010	Sub-Saharan Africa Region 2010
Economic and social context				
Population (total, million)	10	11	796	853
GNI per capita, World Bank Atlas method (current US$)	430	620	530	1,188
Rural population (% of total)	75	72	72	63
Expected years of schooling (years)	6	*7*	*9*	*9*
Physicians density (per 1,000 people)	*0.04*	—	0.2	0.2
Depositors with commercial banks (per 1,000 adults)	7	*24*	—	*167*
Sector structure				
Number of mobile operators	—	2		
Herfindahl-Hirschman Index (HHI) (scale = 0–10,000)	—	5,095		
Sector performance				
Access				
Mobile cellular subscriptions (per 100 people)	2	34[a]	43[a]	57[a]
Mobile cellular subscriptions (% prepaid)	100	100[a]	98[a]	96[a]
Population covered by a mobile-cellular network (%)	24	—	—	72
Mobile broadband subscriptions (per 100 people)	—	—	—	5.6[a]
Mobile broadband (% of total mobile subscriptions)	—	—	—	10.1[a]
Usage				
Households with a mobile telephone (%)	—	32	43	52
Mobile voice usage (minutes per user per month)	—	—	—	—
Population using mobile Internet (%)	—	—	—	—
Short Message Service (SMS) users (% of mobile users)	—	—	—	—
Affordability				
Mobile basket (% of GNI per capita)	*57.7*	29.8	28.8	19.5

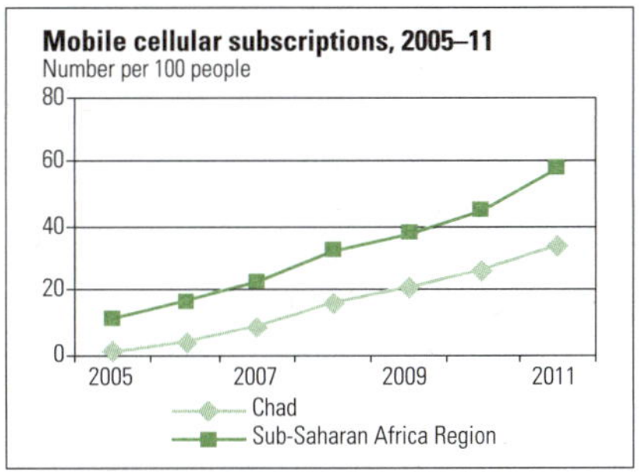

Mobile cellular subscriptions, 2005–11

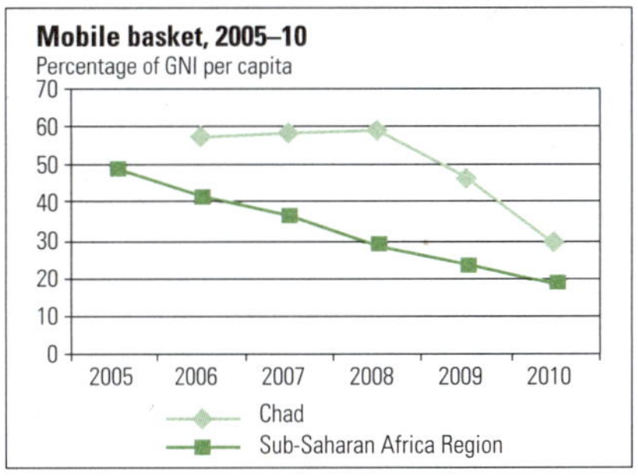

Mobile basket, 2005–10

Chile

	Chile 2005	Chile 2010	Upper-middle-income group 2010	Latin America & the Caribbean Region 2010
Economic and social context				
Population (total, million)	16	17	2,452	583
GNI per capita, World Bank Atlas method (current US$)	5,920	10,120	5,886	7,741
Rural population (% of total)	12	11	43	21
Expected years of schooling (years)	14	*15*	*13*	*14*
Physicians density (per 1,000 people)	*1.3*	*1.0*	1.7	1.8
Depositors with commercial banks (per 1,000 adults)	1,425	2,134	—	—
Sector structure				
Number of mobile operators	—	4		
Herfindahl-Hirschman Index (HHI) (scale = 0–10,000)	—	3,509		
Sector performance				
Access				
Mobile cellular subscriptions (per 100 people)	65	124[a]	92[a]	109[a]
Mobile cellular subscriptions (% prepaid)	83	72[a]	81[a]	81[a]
Population covered by a mobile-cellular network (%)	100	100	*99*	98
Mobile broadband subscriptions (per 100 people)	*0.0*	13.1[a]	14.3[a]	16.1[a]
Mobile broadband (% of total mobile subscriptions)	*0.0*	9.2[a]	15.4[a]	15.2[a]
Usage				
Households with a mobile telephone (%)	*61*	91	84	84
Mobile voice usage (minutes per user per month)	131	169[a]	325[a]	141[a]
Population using mobile Internet (%)	—	5.3	22.9[a]	4.4
Short Message Service (SMS) users (% of mobile users)	—	—	74.4[a]	—
Affordability				
Mobile basket (% of GNI per capita)	5.5	2.8	2.9	3.7

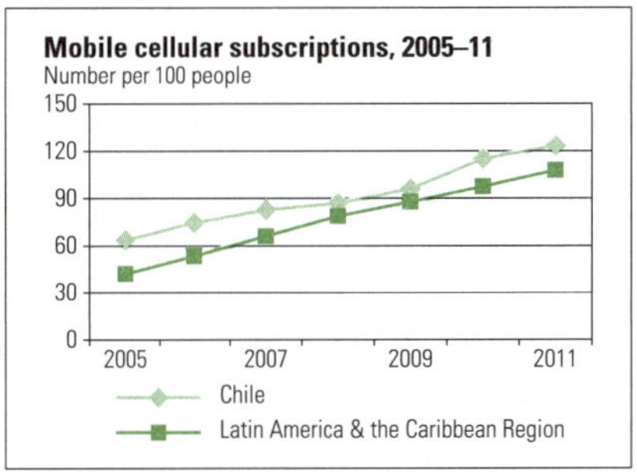

Mobile cellular subscriptions, 2005–11

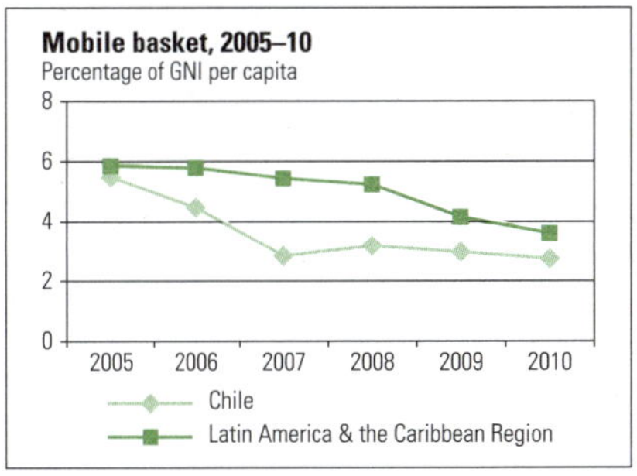

Mobile basket, 2005–10

Sources: Economic and social context: IMF, UIS, UN, WHO and World Bank; Sector structure: ictDATA.org; Sector performance: ictDATA.org, ITU; Wireless Intelligence, and World Bank.
Notes: Use of italics in the column entries indicates years or periods other than those specified. — Not available. GNI = gross national income.
a. Data are for 2011.

China

	China 2005	China 2010	Upper-middle-income group 2010	East Asia & Pacific Region 2010
Economic and social context				
Population (total, million)	1,304	1,338	2,452	1,962
GNI per capita, World Bank Atlas method (current US$)	1,740	4,270	5,886	3,696
Rural population (% of total)	60	55	43	54
Expected years of schooling (years)	*11*	*12*	*13*	*12*
Physicians density (per 1,000 people)	1.5	*1.4*	1.7	1.2
Depositors with commercial banks (per 1,000 adults)	—	—	—	—
Sector structure				
Number of mobile operators	—	3		
Herfindahl-Hirschman Index (HHI) (scale = 0–10,000)	—	5,323		
Sector performance				
Access				
Mobile cellular subscriptions (per 100 people)	30	73[a]	92[a]	83[a]
Mobile cellular subscriptions (% prepaid)	71	81[a]	81[a]	85[a]
Population covered by a mobile-cellular network (%)	*95*	*99*	*99*	*99*
Mobile broadband subscriptions (per 100 people)	—	9.5[a]	14.3[a]	11.6[a]
Mobile broadband (% of total mobile subscriptions)	—	13.1[a]	15.4[a]	14.4[a]
Usage				
Households with a mobile telephone (%)	55	93	84	83
Mobile voice usage (minutes per user per month)	299	450[a]	325[a]	367[a]
Population using mobile Internet (%)	0.5	26.5[a]	22.9[a]	22.4[a]
Short Message Service (SMS) users (% of mobile users)	—	80.0[a]	74.4[a]	84.0[a]
Affordability				
Mobile basket (% of GNI per capita)	3.4	1.7	2.9	5.7

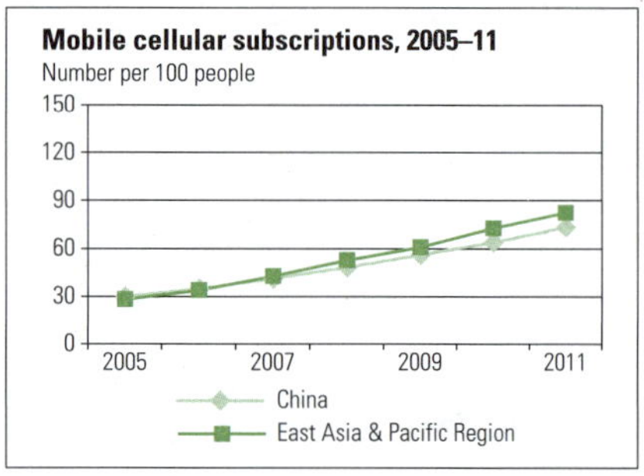

Mobile cellular subscriptions, 2005–11

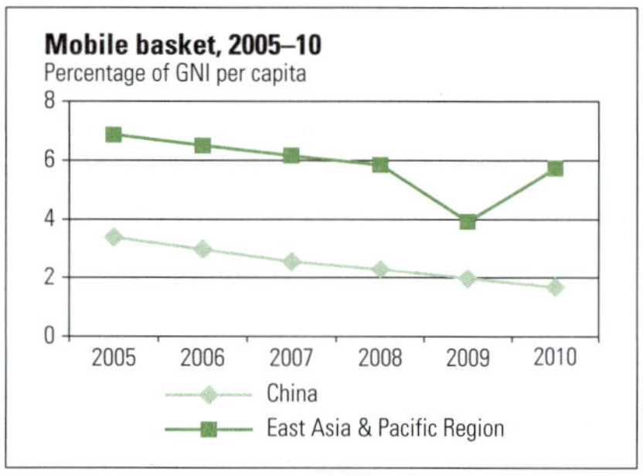

Mobile basket, 2005–10

Colombia

	Colombia 2005	Colombia 2010	Upper-middle-income group 2010	Latin America & the Caribbean Region 2010
Economic and social context				
Population (total, million)	43	46	2,452	583
GNI per capita, World Bank Atlas method (current US$)	2,940	5,510	5,886	7,741
Rural population (% of total)	26	25	43	21
Expected years of schooling (years)	13	*14*	*13*	*14*
Physicians density (per 1,000 people)	1.4	0.1	1.7	1.8
Depositors with commercial banks (per 1,000 adults)	—	—	—	—
Sector structure				
Number of mobile operators	—	3		
Herfindahl-Hirschman Index (HHI) (scale = 0–10,000)	—	4,973		
Sector performance				
Access				
Mobile cellular subscriptions (per 100 people)	51	102[a]	92[a]	109[a]
Mobile cellular subscriptions (% prepaid)	83	81[a]	81[a]	81[a]
Population covered by a mobile-cellular network (%)	82	—	*99*	98
Mobile broadband subscriptions (per 100 people)	—	9.0[a]	14.3[a]	16.1[a]
Mobile broadband (% of total mobile subscriptions)	—	9.4[a]	15.4[a]	15.2[a]
Usage				
Households with a mobile telephone (%)	56	91	84	84
Mobile voice usage (minutes per user per month)	116	191[a]	325[a]	141[a]
Population using mobile Internet (%)	—	—	22.9[a]	4.4
Short Message Service (SMS) users (% of mobile users)	—	—	74.4[a]	—
Affordability				
Mobile basket (% of GNI per capita)	6.4	3.7	2.9	3.7

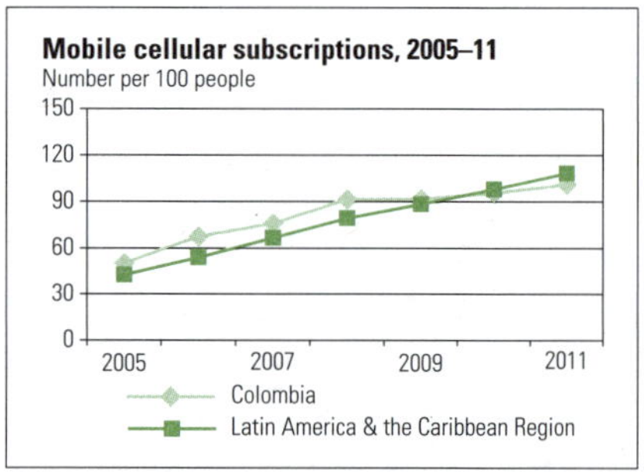

Mobile cellular subscriptions, 2005–11

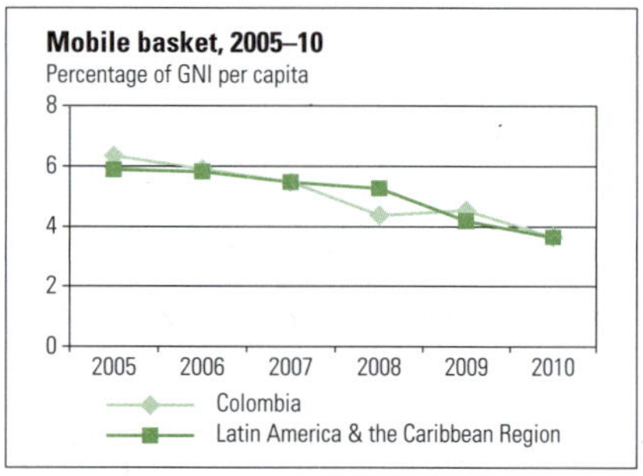

Mobile basket, 2005–10

Sources: Economic and social context: IMF, UIS, UN, WHO and World Bank; Sector structure: ictDATA.org; Sector performance: ictDATA.org, ITU; Wireless Intelligence, and World Bank.
Notes: Use of italics in the column entries indicates years or periods other than those specified. — Not available. GNI = gross national income.
a. Data are for 2011.

Congo, Dem. Rep.

	Congo, Dem. Rep. 2005	Congo, Dem. Rep. 2010	Low-income group 2010	Sub-Saharan Africa Region 2010
Economic and social context				
Population (total, million)	57	66	796	853
GNI per capita, World Bank Atlas method (current US$)	120	180	530	1,188
Rural population (% of total)	68	65	72	63
Expected years of schooling (years)	*8*	*8*	*9*	*9*
Physicians density (per 1,000 people)	*0.1*	—	0.2	0.2
Depositors with commercial banks (per 1,000 adults)	—	—	—	*167*
Sector structure				
Number of mobile operators	—	5		
Herfindahl-Hirschman Index (HHI) (scale = 0–10,000)	—	3,242		
Sector performance				
Access				
Mobile cellular subscriptions (per 100 people)	5	14	43[a]	57[a]
Mobile cellular subscriptions (% prepaid)	99	99[a]	98[a]	96[a]
Population covered by a mobile-cellular network (%)	50	50	—	72
Mobile broadband subscriptions (per 100 people)	—	—	—	5.6[a]
Mobile broadband (% of total mobile subscriptions)	—	—	—	10.1[a]
Usage				
Households with a mobile telephone (%)	*21*	—	43	52
Mobile voice usage (minutes per user per month)	—	—	—	—
Population using mobile Internet (%)	—	—	—	—
Short Message Service (SMS) users (% of mobile users)	—	—	—	—
Affordability				
Mobile basket (% of GNI per capita)	—	—	28.8	19.5

Mobile cellular subscriptions, 2005–11
Number per 100 prople

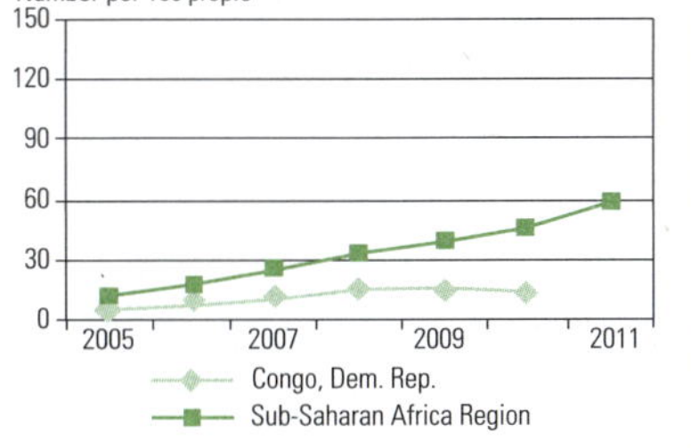

Mobile basket, 2005–10
Percentage of GNI per capita

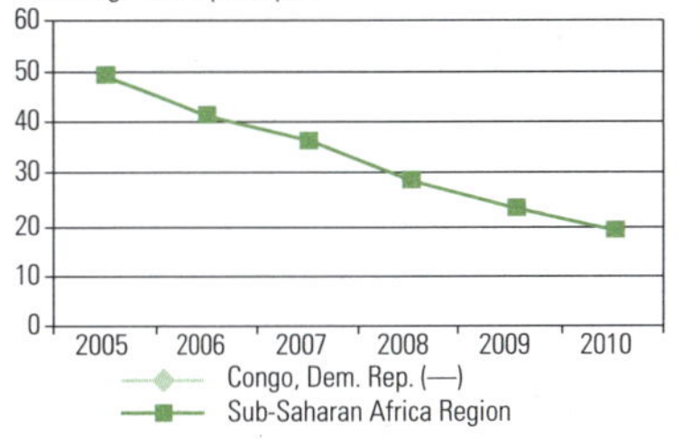

Congo, Rep.

	Congo, Rep. 2005	Congo, Rep. 2010	Lower-middle-income group 2010	Sub-Saharan Africa Region 2010
Economic and social context				
Population (total, million)	4	4	2,519	853
GNI per capita, World Bank Atlas method (current US$)	980	2,240	1,623	1,188
Rural population (% of total)	40	38	61	63
Expected years of schooling (years)	*10*	—	*10*	*9*
Physicians density (per 1,000 people)	*0.2*	—	0.8	0.2
Depositors with commercial banks (per 1,000 adults)	5	*20*	—	*167*
Sector structure				
Number of mobile operators	—	3		
Herfindahl-Hirschman Index (HHI) (scale = 0–10,000)	—	4,078		
Sector performance				
Access				
Mobile cellular subscriptions (per 100 people)	16	94[a]	78[a]	57[a]
Mobile cellular subscriptions (% prepaid)	99	99[a]	96[a]	96[a]
Population covered by a mobile-cellular network (%)	39	—	*86*	72
Mobile broadband subscriptions (per 100 people)	—	0.6[a]	7.3[a]	5.6[a]
Mobile broadband (% of total mobile subscriptions)	—	0.6[a]	9.0[a]	10.1[a]
Usage				
Households with a mobile telephone (%)	34	*77*	77	52
Mobile voice usage (minutes per user per month)	—	50[a]	276[a]	—
Population using mobile Internet (%)	—	—	2.9	—
Short Message Service (SMS) users (% of mobile users)	—	—	61.9[a]	—
Affordability				
Mobile basket (% of GNI per capita)	—	—	7.2	19.5

Mobile cellular subscriptions, 2005–11
Number per 100 people

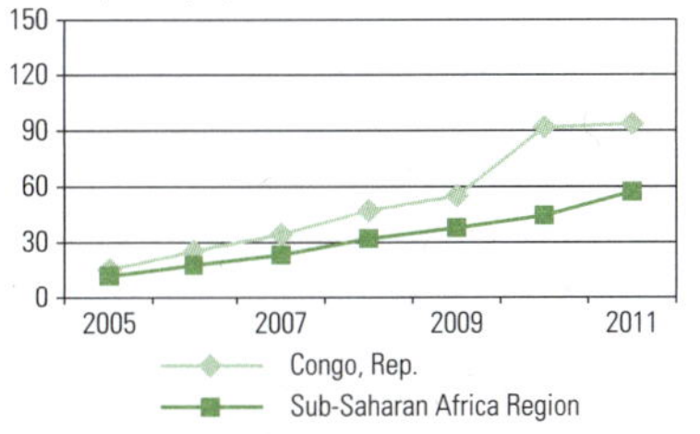

Mobile basket, 2005–10
Percentage of GNI per capita

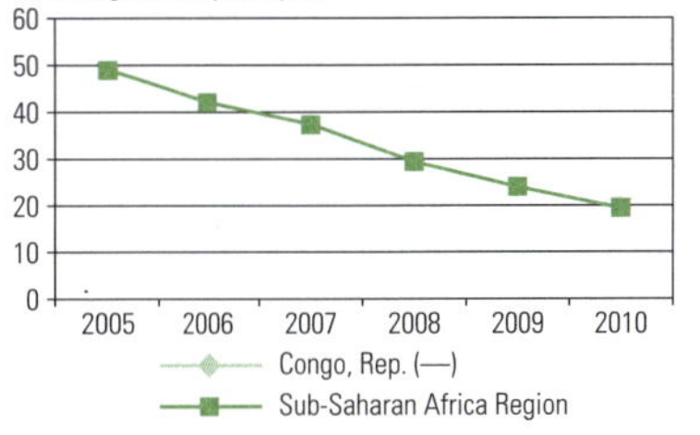

Sources: Economic and social context: IMF, UIS, UN, WHO and World Bank; Sector structure: ictDATA.org; Sector performance: ictDATA.org, ITU; Wireless Intelligence, and World Bank.
Notes: Use of italics in the column entries indicates years or periods other than those specified. — Not available. GNI = gross national income.
a. Data are for 2011.

Costa Rica

	Costa Rica 2005	Costa Rica 2010	Upper-middle-income group 2010	Latin America & the Caribbean Region 2010
Economic and social context				
Population (total, million)	4	5	2,452	583
GNI per capita, World Bank Atlas method (current US$)	4,680	6,810	5,886	7,741
Rural population (% of total)	38	36	43	21
Expected years of schooling (years)	12	—	*13*	*14*
Physicians density (per 1,000 people)	—	—	1.7	1.8
Depositors with commercial banks (per 1,000 adults)	—	—	—	—
Sector structure				
Number of mobile operators	—	1		
Herfindahl-Hirschman Index (HHI) (scale = 0–10,000)	—	10,000		
Sector performance				
Access				
Mobile cellular subscriptions (per 100 people)	26	85[a]	92[a]	109[a]
Mobile cellular subscriptions (% prepaid)	—	52[a]	81[a]	81[a]
Population covered by a mobile-cellular network (%)	86	—	*99*	98
Mobile broadband subscriptions (per 100 people)	—	16.5[a]	14.3[a]	16.1[a]
Mobile broadband (% of total mobile subscriptions)	—	19.3[a]	15.4[a]	15.2[a]
Usage				
Households with a mobile telephone (%)	50	74	84	84
Mobile voice usage (minutes per user per month)	*305*	—	325[a]	141[a]
Population using mobile Internet (%)	—	6.4	22.9[a]	4.4
Short Message Service (SMS) users (% of mobile users)	—	—	74.4[a]	—
Affordability				
Mobile basket (% of GNI per capita)	1.6	0.6	2.9	3.7

Mobile cellular subscriptions, 2005–11
Number per 100 people

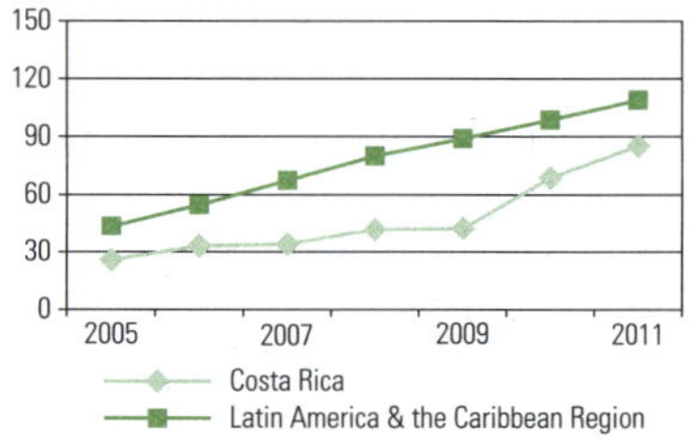

Mobile basket, 2005–10
Percentage of GNI per capita

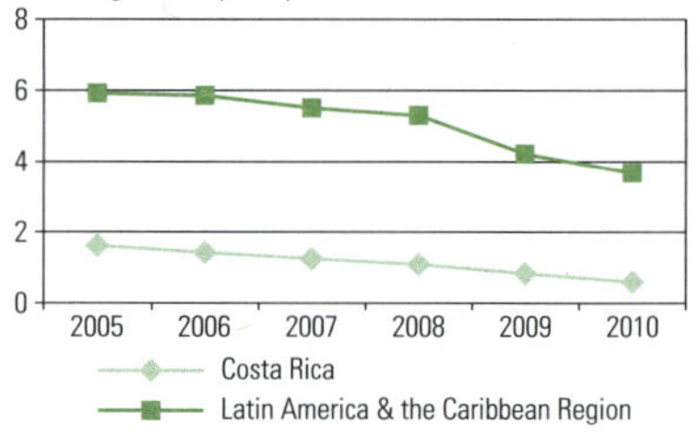

Côte d'Ivoire

	Côte d'Ivoire 2005	Côte d'Ivoire 2010	Lower-middle-income group 2010	Sub-Saharan Africa Region 2010
Economic and social context				
Population (total, million)	18	20	2,519	853
GNI per capita, World Bank Atlas method (current US$)	870	1,160	1,623	1,188
Rural population (% of total)	53	50	61	63
Expected years of schooling (years)	—	—	*10*	*9*
Physicians density (per 1,000 people)	*0.1*	*0.1*	0.8	0.2
Depositors with commercial banks (per 1,000 adults)	—	—	—	*167*
Sector structure				
Number of mobile operators	—	5		
Herfindahl-Hirschman Index (HHI) (scale = 0–10,000)	—	2,849		
Sector performance				
Access				
Mobile cellular subscriptions (per 100 people)	13	85[a]	78[a]	57[a]
Mobile cellular subscriptions (% prepaid)	98	99[a]	96[a]	96[a]
Population covered by a mobile-cellular network (%)	55	*92*	*86*	72
Mobile broadband subscriptions (per 100 people)	—	—	7.3[a]	5.6[a]
Mobile broadband (% of total mobile subscriptions)	—	—	9.0[a]	10.1[a]
Usage				
Households with a mobile telephone (%)	23	—	77	52
Mobile voice usage (minutes per user per month)	—	—	276[a]	—
Population using mobile Internet (%)	—	—	2.9	—
Short Message Service (SMS) users (% of mobile users)	—	—	61.9[a]	—
Affordability				
Mobile basket (% of GNI per capita)	62.1	14.1	7.2	19.5

Mobile cellular subscriptions, 2005–11
Number per 100 people

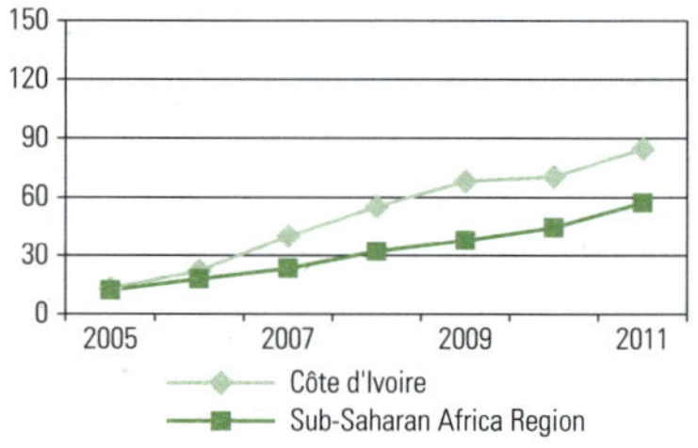

Mobile basket, 2005–10
Percentage of GNI per capita

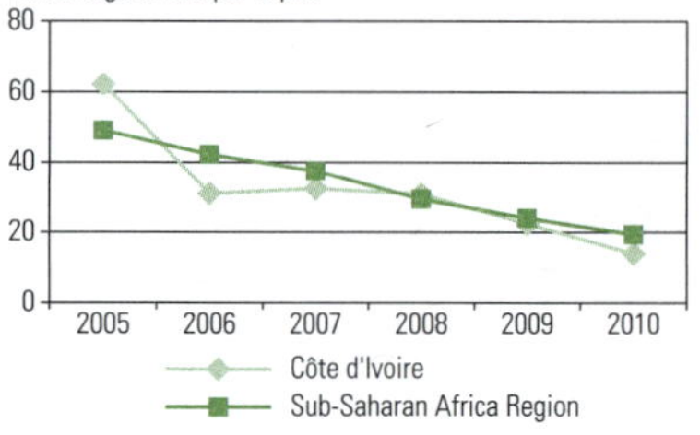

Sources: Economic and social context: IMF, UIS, UN, WHO and World Bank; Sector structure: ictDATA.org; Sector performance: ictDATA.org, ITU; Wireless Intelligence, and World Bank.
Notes: Use of italics in the column entries indicates years or periods other than those specified. — Not available. GNI = gross national income.
a. Data are for 2011.

Croatia

	Croatia 2005	Croatia 2010	High-income group 2010
Economic and social context			
Population (total, million)	4	4	1,127
GNI per capita, World Bank Atlas method (current US$)	9,730	13,890	38,746
Rural population (% of total)	44	42	22
Expected years of schooling (years)	14	*14*	*16*
Physicians density (per 1,000 people)	2.5	*2.6*	2.8
Depositors with commercial banks (per 1,000 adults)	—	—	—
Sector structure			
Number of mobile operators	—	3	
Herfindahl-Hirschman Index (HHI) (scale = 0–10,000)	—	4,046	
Sector performance			
Access			
Mobile cellular subscriptions (per 100 people)	82	117[a]	118[a]
Mobile cellular subscriptions (% prepaid)	81	61[a]	36[a]
Population covered by a mobile-cellular network (%)	100	*100*	100
Mobile broadband subscriptions (per 100 people)	2.0	33.9[a]	69.6[a]
Mobile broadband (% of total mobile subscriptions)	2.4	29.0[a]	57.6[a]
Usage			
Households with a mobile telephone (%)	79	95	93
Mobile voice usage (minutes per user per month)	—	114[a]	339
Population using mobile Internet (%)	0.7	13.6	24.3
Short Message Service (SMS) users (% of mobile users)	—	—	78.2[a]
Affordability			
Mobile basket (% of GNI per capita)	2.8	1.5	1.0

Mobile cellular subscriptions, 2005–11
Number per 100 people

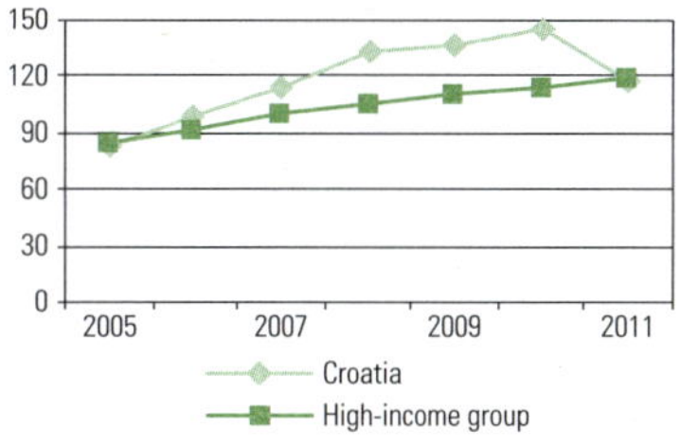

Mobile basket, 2005–10
Percentage of GNI per capita

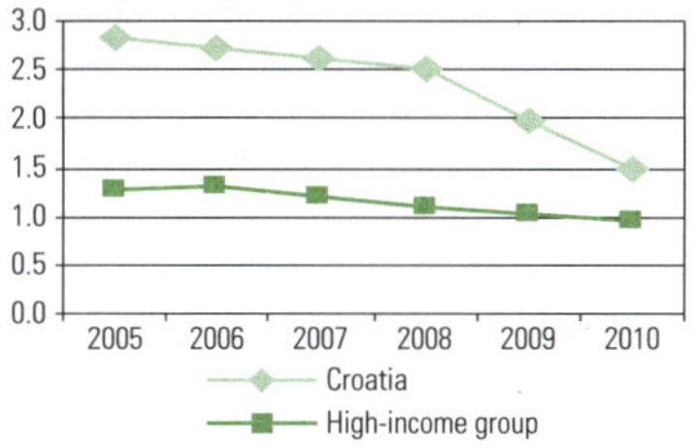

Cuba

	Cuba 2005	Cuba 2010	Upper-middle-income group 2010	Latin America & the Caribbean Region 2010
Economic and social context				
Population (total, million)	11	11	2,452	583
GNI per capita, World Bank Atlas method (current US$)	3,960	*5,460*	5,886	7,741
Rural population (% of total)	24	24	43	21
Expected years of schooling (years)	15	*18*	*13*	*14*
Physicians density (per 1,000 people)	*6.4*	6.7	1.7	1.8
Depositors with commercial banks (per 1,000 adults)	—	—	—	—
Sector structure				
Number of mobile operators	—	1		
Herfindahl-Hirschman Index (HHI) (scale = 0–10,000)	—	10,000		
Sector performance				
Access				
Mobile cellular subscriptions (per 100 people)	1	9	92[a]	109[a]
Mobile cellular subscriptions (% prepaid)	87	90[a]	81[a]	81[a]
Population covered by a mobile-cellular network (%)	71	78	*99*	98
Mobile broadband subscriptions (per 100 people)	—	—	14.3[a]	16.1[a]
Mobile broadband (% of total mobile subscriptions)	—	—	15.4[a]	15.2[a]
Usage				
Households with a mobile telephone (%)	*1*	—	84	84
Mobile voice usage (minutes per user per month)	—	—	325[a]	141[a]
Population using mobile Internet (%)	—	—	22.9[a]	4.4
Short Message Service (SMS) users (% of mobile users)	—	—	74.4[a]	—
Affordability				
Mobile basket (% of GNI per capita)	—	—	2.9	3.7

Mobile cellular subscriptions, 2005–11
Number per 100 people

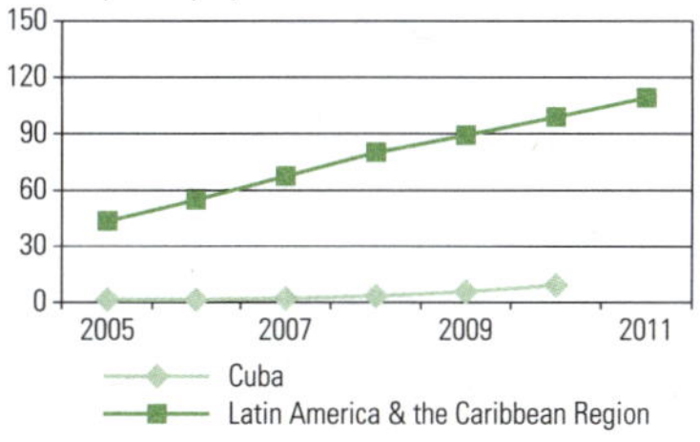

Mobile basket, 2005–10
Percentage of GNI per capita

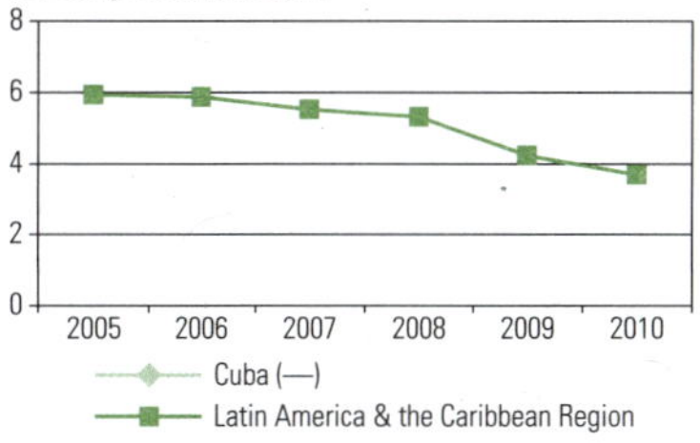

Sources: Economic and social context: IMF, UIS, UN, WHO and World Bank; Sector structure: ictDATA.org; Sector performance: ictDATA.org, ITU; Wireless Intelligence, and World Bank.
Notes: Use of italics in the column entries indicates years or periods other than those specified. — Not available. GNI = gross national income.
a. Data are for 2011.

Cyprus

	Cyprus 2005	Cyprus 2010	High-income group 2010
Economic and social context			
Population (total, million)	1	1	1,127
GNI per capita, World Bank Atlas method (current US$)	21,490	29,430	38,746
Rural population (% of total)	31	30	22
Expected years of schooling (years)	14	*15*	*16*
Physicians density (per 1,000 people)	*2.3*	*2.6*	2.8
Depositors with commercial banks (per 1,000 adults)	—	—	—
Sector structure			
Number of mobile operators	—	2	
Herfindahl-Hirschman Index (HHI) (scale = 0–10,000)	—	6,429	
Sector performance			
Access			
Mobile cellular subscriptions (per 100 people)	76	84[a]	118[a]
Mobile cellular subscriptions (% prepaid)	56	59[a]	36[a]
Population covered by a mobile-cellular network (%)	100	100	100
Mobile broadband subscriptions (per 100 people)	0.6	58.6[a]	69.6[a]
Mobile broadband (% of total mobile subscriptions)	0.6	42.0[a]	57.6[a]
Usage			
Households with a mobile telephone (%)	85	—	93
Mobile voice usage (minutes per user per month)	—	—	339
Population using mobile Internet (%)	—	3.6	24.3
Short Message Service (SMS) users (% of mobile users)	—	—	78.2[a]
Affordability			
Mobile basket (% of GNI per capita)	1.2	0.3	1.0

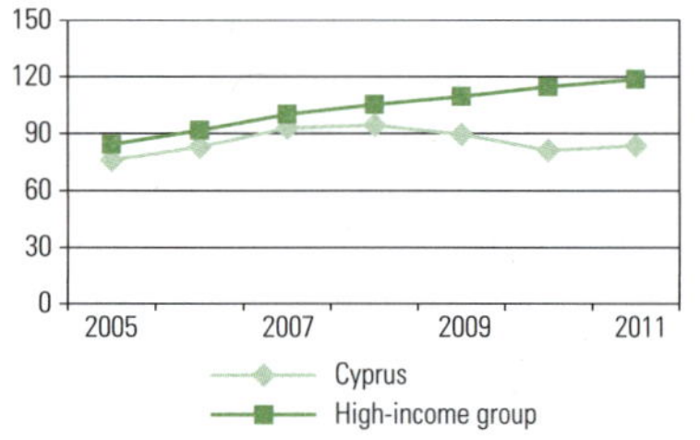

Mobile cellular subscriptions, 2005–11
Number per 100 people

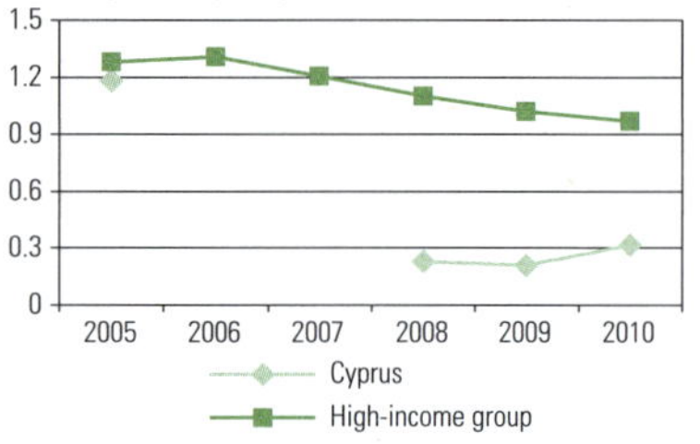

Mobile basket, 2005–10
Percentage of GNI per capita

Czech Republic

	Czech Republic 2005	Czech Republic 2010	High-income group 2010
Economic and social context			
Population (total, million)	10	11	1,127
GNI per capita, World Bank Atlas method (current US$)	11,330	17,890	38,746
Rural population (% of total)	27	27	22
Expected years of schooling (years)	15	*16*	*16*
Physicians density (per 1,000 people)	*3.6*	*3.7*	2.8
Depositors with commercial banks (per 1,000 adults)	—	—	—
Sector structure			
Number of mobile operators	—	3	
Herfindahl-Hirschman Index (HHI) (scale = 0–10,000)	—	3,489	
Sector performance			
Access			
Mobile cellular subscriptions (per 100 people)	115	129[a]	118[a]
Mobile cellular subscriptions (% prepaid)	66	43[a]	36[a]
Population covered by a mobile-cellular network (%)	100	100	100
Mobile broadband subscriptions (per 100 people)	0.7	34.7[a]	69.6[a]
Mobile broadband (% of total mobile subscriptions)	0.7	26.6[a]	57.6[a]
Usage			
Households with a mobile telephone (%)	81	95	93
Mobile voice usage (minutes per user per month)	194	141[a]	339
Population using mobile Internet (%)	—	4.8	24.3
Short Message Service (SMS) users (% of mobile users)	—	—	78.2[a]
Affordability			
Mobile basket (% of GNI per capita)	3.3	1.9	1.0

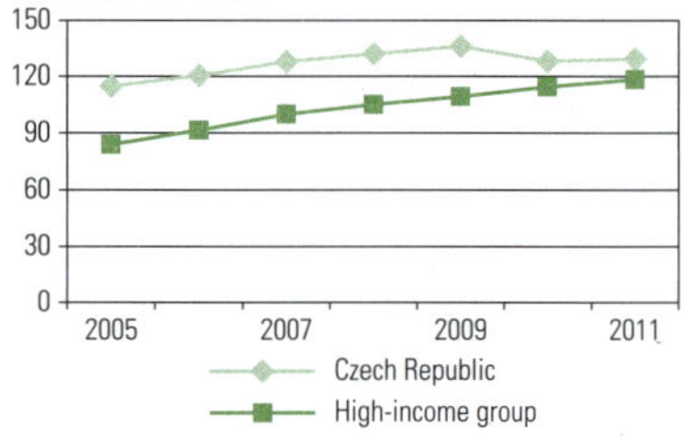

Mobile cellular subscriptions, 2005–11
Number per 100 people

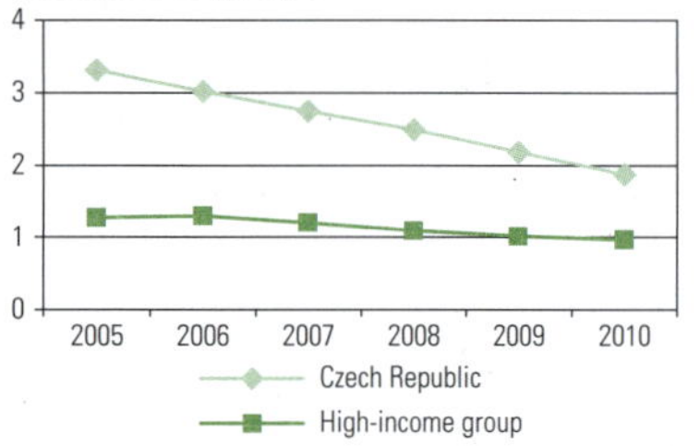

Mobile basket, 2005–10
Percentage of GNI per capita

Sources: Economic and social context: IMF, UIS, UN, WHO and World Bank; Sector structure: ictDATA.org; Sector performance: ictDATA.org, ITU; Wireless Intelligence, and World Bank.
Notes: Use of italics in the column entries indicates years or periods other than those specified. — Not available. GNI = gross national income.
a. Data are for 2011.

Denmark

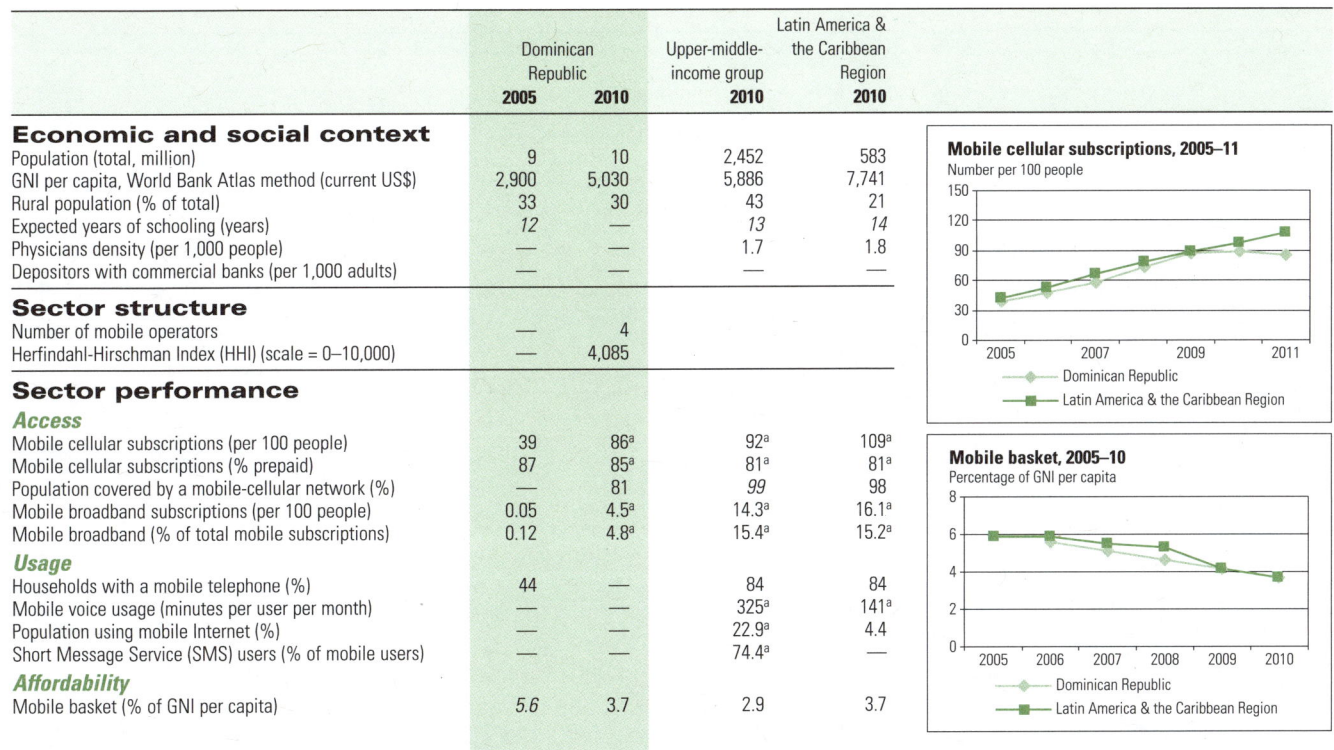

	Denmark 2005	Denmark 2010	High-income group 2010
Economic and social context			
Population (total, million)	5	6	1,127
GNI per capita, World Bank Atlas method (current US$)	48,590	59,400	38,746
Rural population (% of total)	14	13	22
Expected years of schooling (years)	17	*17*	*16*
Physicians density (per 1,000 people)	*3.2*	*3.4*	2.8
Depositors with commercial banks (per 1,000 adults)	—	—	—
Sector structure			
Number of mobile operators	—	4	
Herfindahl-Hirschman Index (HHI) (scale = 0–10,000)	—	2,401	
Sector performance			
Access			
Mobile cellular subscriptions (per 100 people)	101	141[a]	118[a]
Mobile cellular subscriptions (% prepaid)	23	15[a]	36[a]
Population covered by a mobile-cellular network (%)	—	—	100
Mobile broadband subscriptions (per 100 people)	2.3	84.4[a]	69.6[a]
Mobile broadband (% of total mobile subscriptions)	2.1	55.5[a]	57.6[a]
Usage			
Households with a mobile telephone (%)	91	97	93
Mobile voice usage (minutes per user per month)	159	173[a]	339
Population using mobile Internet (%)	8.6	21.6	24.3
Short Message Service (SMS) users (% of mobile users)	—	—	78.2[a]
Affordability			
Mobile basket (% of GNI per capita)	0.9	0.2	1.0

Mobile cellular subscriptions, 2005–11
Number per 100 people

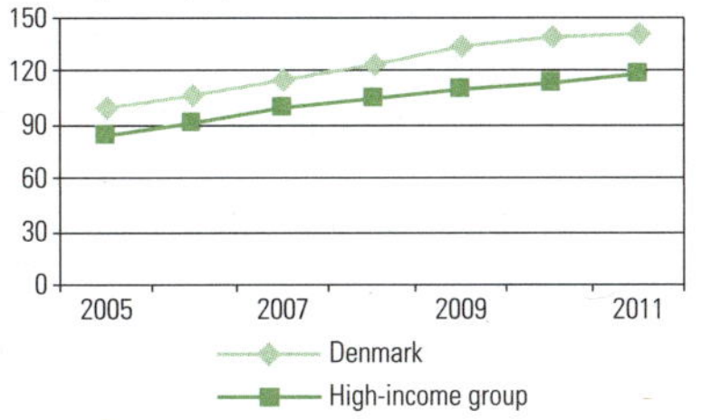

Mobile basket, 2005–10
Percentage of GNI per capita

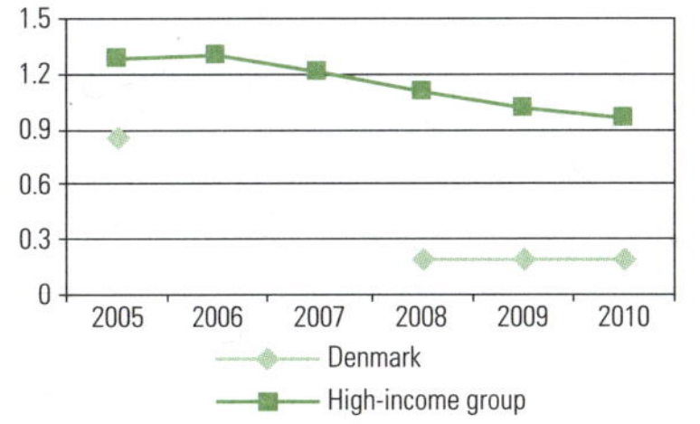

Dominican Republic

	Dominican Republic 2005	Dominican Republic 2010	Upper-middle-income group 2010	Latin America & the Caribbean Region 2010
Economic and social context				
Population (total, million)	9	10	2,452	583
GNI per capita, World Bank Atlas method (current US$)	2,900	5,030	5,886	7,741
Rural population (% of total)	33	30	43	21
Expected years of schooling (years)	*12*	—	*13*	*14*
Physicians density (per 1,000 people)	—	—	1.7	1.8
Depositors with commercial banks (per 1,000 adults)	—	—	—	—
Sector structure				
Number of mobile operators	—	4		
Herfindahl-Hirschman Index (HHI) (scale = 0–10,000)	—	4,085		
Sector performance				
Access				
Mobile cellular subscriptions (per 100 people)	39	86[a]	92[a]	109[a]
Mobile cellular subscriptions (% prepaid)	87	85[a]	81[a]	81[a]
Population covered by a mobile-cellular network (%)	—	81	*99*	98
Mobile broadband subscriptions (per 100 people)	0.05	4.5[a]	14.3[a]	16.1[a]
Mobile broadband (% of total mobile subscriptions)	0.12	4.8[a]	15.4[a]	15.2[a]
Usage				
Households with a mobile telephone (%)	44	—	84	84
Mobile voice usage (minutes per user per month)	—	—	325[a]	141[a]
Population using mobile Internet (%)	—	—	22.9[a]	4.4
Short Message Service (SMS) users (% of mobile users)	—	—	74.4[a]	—
Affordability				
Mobile basket (% of GNI per capita)	*5.6*	3.7	2.9	3.7

Mobile cellular subscriptions, 2005–11
Number per 100 people

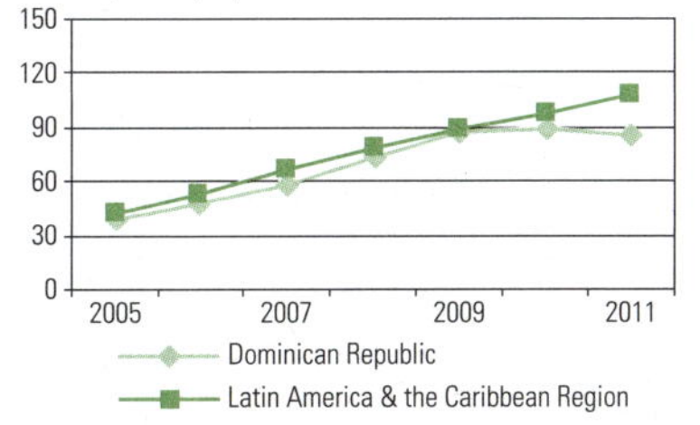

Mobile basket, 2005–10
Percentage of GNI per capita

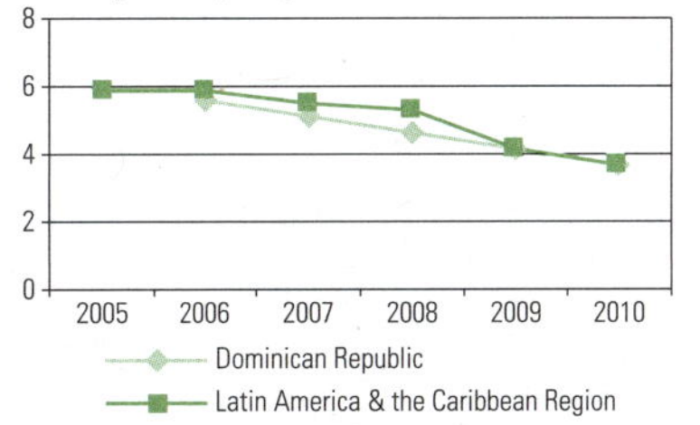

Sources: Economic and social context: IMF, UIS, UN, WHO and World Bank; Sector structure: ictDATA.org; Sector performance: ictDATA.org, ITU; Wireless Intelligence, and World Bank.
Notes: Use of italics in the column entries indicates years or periods other than those specified. — Not available. GNI = gross national income.
a. Data are for 2011.

Ecuador

	Ecuador 2005	Ecuador 2010	Upper-middle-income group 2010	Latin America & the Caribbean Region 2010
Economic and social context				
Population (total, million)	13	14	2,452	583
GNI per capita, World Bank Atlas method (current US$)	2,620	3,850	5,886	7,741
Rural population (% of total)	36	33	43	21
Expected years of schooling (years)	*13*	*14*	*13*	*14*
Physicians density (per 1,000 people)	—	*1.7*	1.7	1.8
Depositors with commercial banks (per 1,000 adults)	—	—	—	—
Sector structure				
Number of mobile operators	—	3		
Herfindahl-Hirschman Index (HHI) (scale = 0–10,000)	—	5,625		
Sector performance				
Access				
Mobile cellular subscriptions (per 100 people)	47	107[a]	92[a]	109[a]
Mobile cellular subscriptions (% prepaid)	87	84[a]	81[a]	81[a]
Population covered by a mobile-cellular network (%)	80	93	*99*	98
Mobile broadband subscriptions (per 100 people)	0.03	9.0[a]	14.3[a]	16.1[a]
Mobile broadband (% of total mobile subscriptions)	0.07	8.3[a]	15.4[a]	15.2[a]
Usage				
Households with a mobile telephone (%)	*64*	75	84	84
Mobile voice usage (minutes per user per month)	46	145[a]	325[a]	141[a]
Population using mobile Internet (%)	—	2.4	22.9[a]	4.4
Short Message Service (SMS) users (% of mobile users)	—	—	74.4[a]	—
Affordability				
Mobile basket (% of GNI per capita)	10.8	4.3	2.9	3.7

Mobile cellular subscriptions, 2005–11
Number per 100 people

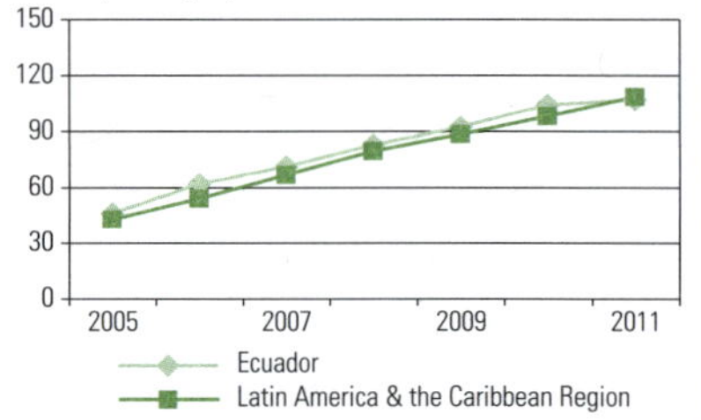

Mobile basket, 2005–10
Percentage of GNI per capita

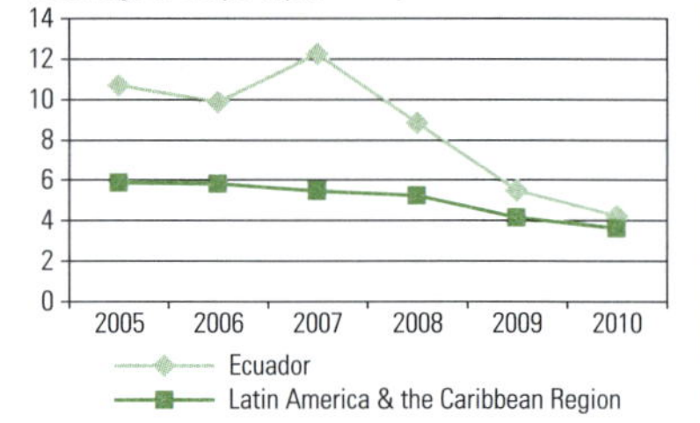

Egypt, Arab Rep.

	Egypt, Arab Rep. 2005	Egypt, Arab Rep. 2010	Lower-middle-income group 2010	Middle East & North Africa Region 2010
Economic and social context				
Population (total, million)	74	81	2,519	331
GNI per capita, World Bank Atlas method (current US$)	1,250	2,420	1,623	3,874
Rural population (% of total)	57	57	61	42
Expected years of schooling (years)	*11*	—	*10*	*12*
Physicians density (per 1,000 people)	2.4	*2.8*	0.8	1.4
Depositors with commercial banks (per 1,000 adults)	—	—	—	*443*
Sector structure				
Number of mobile operators	—	3		
Herfindahl-Hirschman Index (HHI) (scale = 0–10,000)	—	4,003		
Sector performance				
Access				
Mobile cellular subscriptions (per 100 people)	17	97[a]	78[a]	89[a]
Mobile cellular subscriptions (% prepaid)	88	96[a]	96[a]	87[a]
Population covered by a mobile-cellular network (%)	92	100	*86*	—
Mobile broadband subscriptions (per 100 people)	—	11.6[a]	7.3[a]	—
Mobile broadband (% of total mobile subscriptions)	—	11.4[a]	9.0[a]	—
Usage				
Households with a mobile telephone (%)	25	79	77	—
Mobile voice usage (minutes per user per month)	128	178[a]	276[a]	—
Population using mobile Internet (%)	—	6.4	2.9	4.5
Short Message Service (SMS) users (% of mobile users)	—	72.0[a]	61.9[a]	—
Affordability				
Mobile basket (% of GNI per capita)	10.7	3.5	7.2	3.6

Mobile cellular subscriptions, 2005–11
Number per 100 people

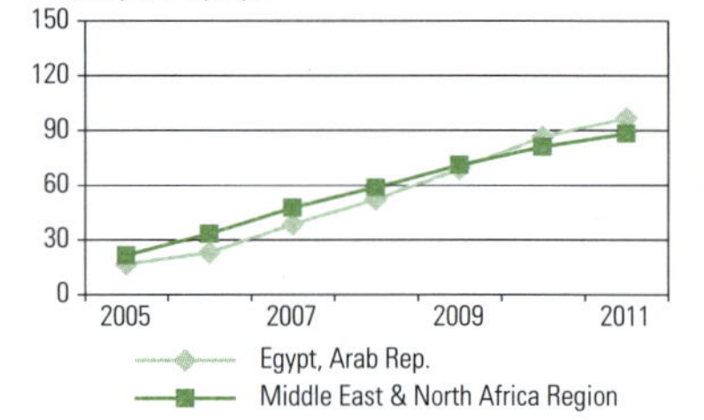

Mobile basket, 2005–10
Percentage of GNI per capita

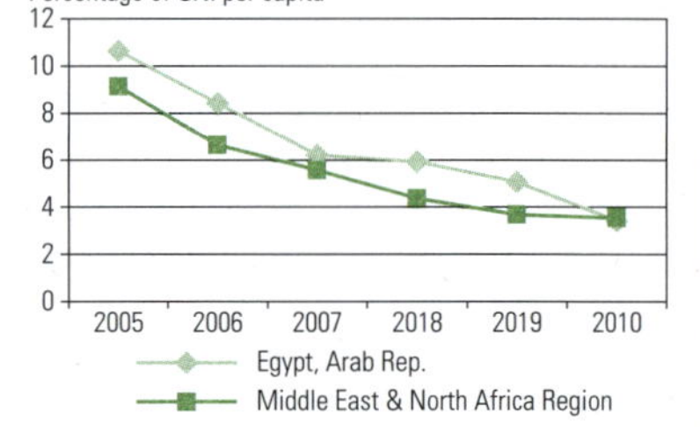

Sources: Economic and social context: IMF, UIS, UN, WHO and World Bank; Sector structure: ictDATA.org; Sector performance: ictDATA.org, ITU; Wireless Intelligence, and World Bank.
Notes: Use of italics in the column entries indicates years or periods other than those specified. — Not available. GNI = gross national income.
a. Data are for 2011.

El Salvador

	El Salvador 2005	El Salvador 2010	Lower-middle-income group 2010	Latin America & the Caribbean Region 2010
Economic and social context				
Population (total, million)	6	6	2,519	583
GNI per capita, World Bank Atlas method (current US$)	2,820	3,380	1,623	7,741
Rural population (% of total)	40	39	61	21
Expected years of schooling (years)	12	12	*10*	*14*
Physicians density (per 1,000 people)	*1.5*	*1.6*	0.8	1.8
Depositors with commercial banks (per 1,000 adults)	—	—	—	—
Sector structure				
Number of mobile operators	—	5		
Herfindahl-Hirschman Index (HHI) (scale = 0–10,000)	—	—		
Sector performance				
Access				
Mobile cellular subscriptions (per 100 people)	40	126	78[a]	109[a]
Mobile cellular subscriptions (% prepaid)	83	89[a]	96[a]	81[a]
Population covered by a mobile-cellular network (%)	95	—	*86*	98
Mobile broadband subscriptions (per 100 people)	—	7.0[a]	7.3[a]	16.1[a]
Mobile broadband (% of total mobile subscriptions)	—	5.0[a]	9.0[a]	15.2[a]
Usage				
Households with a mobile telephone (%)	35	87	77	84
Mobile voice usage (minutes per user per month)	—	—	276[a]	141[a]
Population using mobile Internet (%)	—	—	2.9	4.4
Short Message Service (SMS) users (% of mobile users)	—	—	61.9[a]	—
Affordability				
Mobile basket (% of GNI per capita)	*5.1*	3.4	7.2	3.7

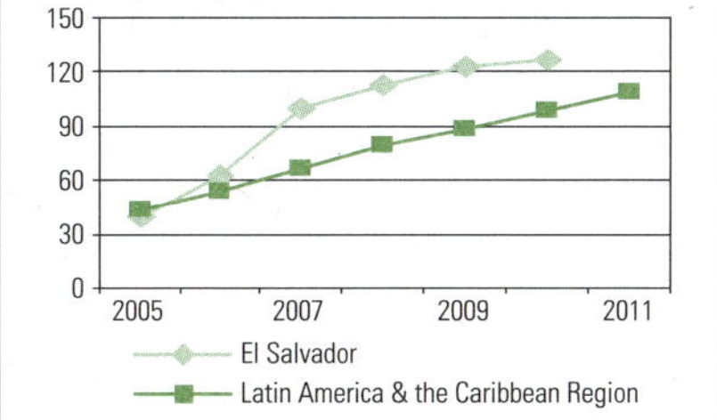
Mobile cellular subscriptions, 2005–11
Number per 100 people

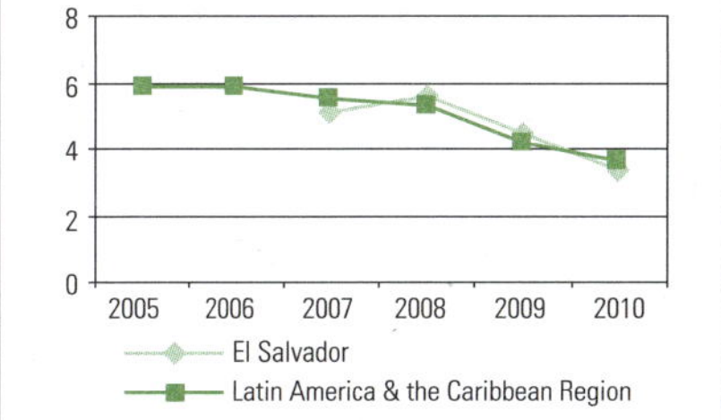
Mobile basket, 2005–10
Percentage of GNI per capita

Eritrea

	Eritrea 2005	Eritrea 2010	Low-income group 2010	Sub-Saharan Africa Region 2010
Economic and social context				
Population (total, million)	4	5	796	853
GNI per capita, World Bank Atlas method (current US$)	250	340	530	1,188
Rural population (% of total)	81	78	72	63
Expected years of schooling (years)	6	5	*9*	*9*
Physicians density (per 1,000 people)	*0.1*	—	0.2	0.2
Depositors with commercial banks (per 1,000 adults)	—	—	—	*167*
Sector structure				
Number of mobile operators	—	1		
Herfindahl-Hirschman Index (HHI) (scale = 0–10,000)	—	10,000		
Sector performance				
Access				
Mobile cellular subscriptions (per 100 people)	1	4	43[a]	57[a]
Mobile cellular subscriptions (% prepaid)	99	99[a]	98[a]	96[a]
Population covered by a mobile-cellular network (%)	*1.3*	90	—	72
Mobile broadband subscriptions (per 100 people)	—	—	—	5.6[a]
Mobile broadband (% of total mobile subscriptions)	—	—	—	10.1[a]
Usage				
Households with a mobile telephone (%)	—	—	43	52
Mobile voice usage (minutes per user per month)	—	—	—	—
Population using mobile Internet (%)	—	—	—	—
Short Message Service (SMS) users (% of mobile users)	—	—	—	—
Affordability				
Mobile basket (% of GNI per capita)	—	—	28.8	19.5

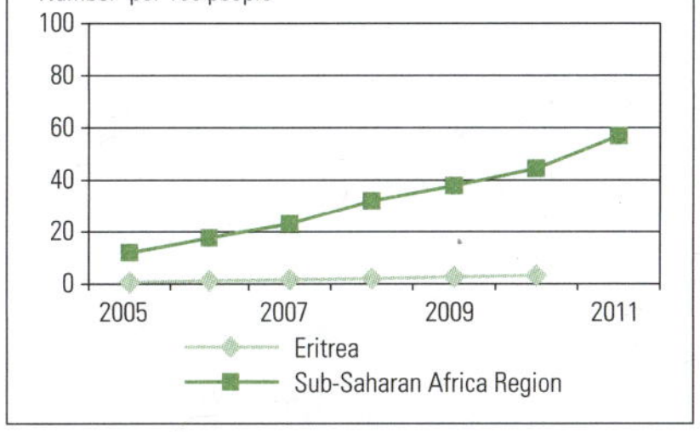
Mobile cellular subscriptions, 2005–11
Number per 100 people

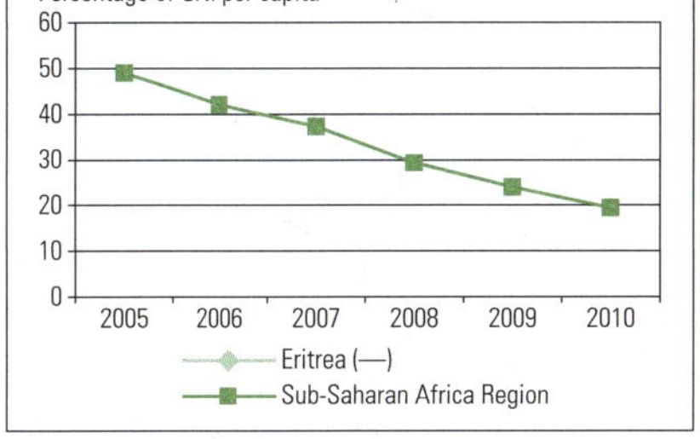
Mobile basket, 2005–10
Percentage of GNI per capita

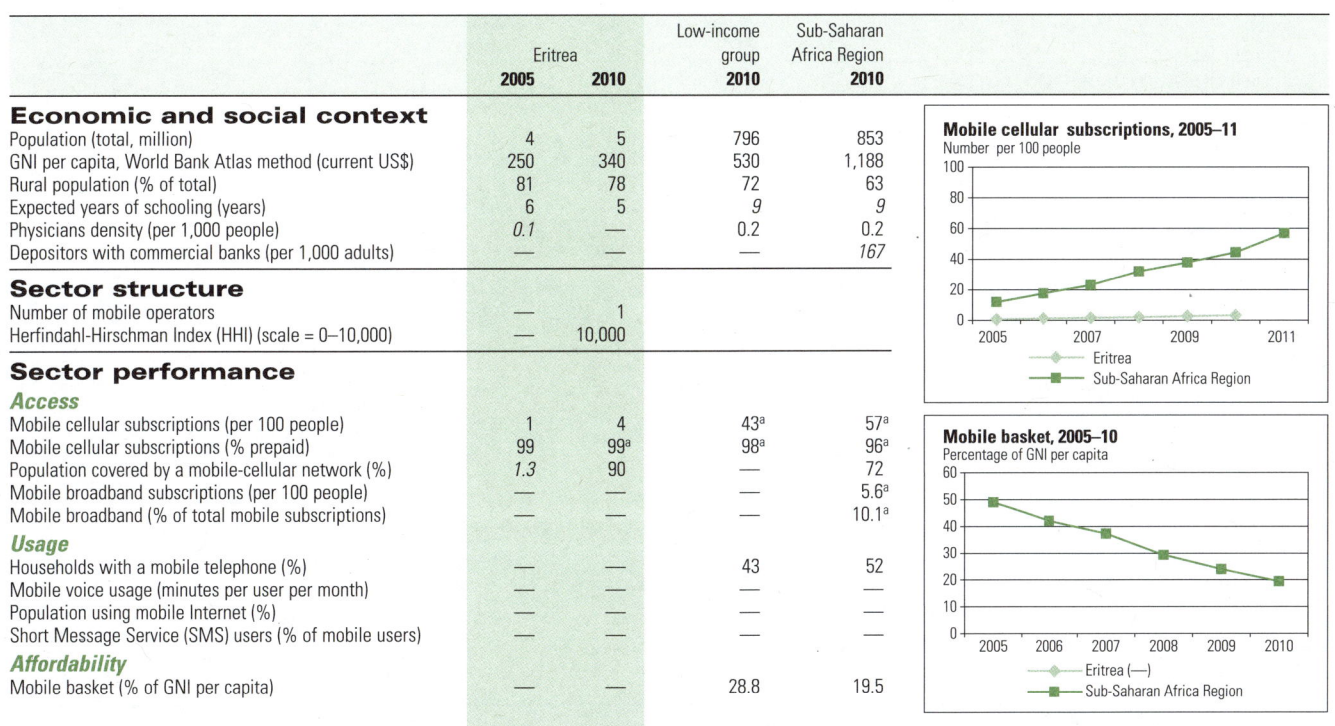

Sources: Economic and social context: IMF, UIS, UN, WHO and World Bank; Sector structure: ictDATA.org; Sector performance: ictDATA.org, ITU; Wireless Intelligence, and World Bank.
Notes: Use of italics in the column entries indicates years or periods other than those specified. — Not available. GNI = gross national income.
a. Data are for 2011.

Estonia

	Estonia 2005	Estonia 2010	High-income group 2010
Economic and social context			
Population (total, million)	1	1	1,127
GNI per capita, World Bank Atlas method (current US$)	9,760	14,460	38,746
Rural population (% of total)	31	31	22
Expected years of schooling (years)	16	16	16
Physicians density (per 1,000 people)	*3.3*	*3.3*	2.8
Depositors with commercial banks (per 1,000 adults)	—	1,993	—
Sector structure			
Number of mobile operators	—	3	
Herfindahl-Hirschman Index (HHI) (scale = 0–10,000)	—	3,674	
Sector performance			
Access			
Mobile cellular subscriptions (per 100 people)	107	135[a]	118[a]
Mobile cellular subscriptions (% prepaid)	56	55[a]	36[a]
Population covered by a mobile-cellular network (%)	99	100	100
Mobile broadband subscriptions (per 100 people)	0.05	12.8[a]	69.6[a]
Mobile broadband (% of total mobile subscriptions)	0.05	9.5[a]	57.6[a]
Usage			
Households with a mobile telephone (%)	81	91	93
Mobile voice usage (minutes per user per month)	—	—	339
Population using mobile Internet (%)	3.7	3.0	24.3
Short Message Service (SMS) users (% of mobile users)	—	—	78.2[a]
Affordability			
Mobile basket (% of GNI per capita)	2.9	1.9	1.0

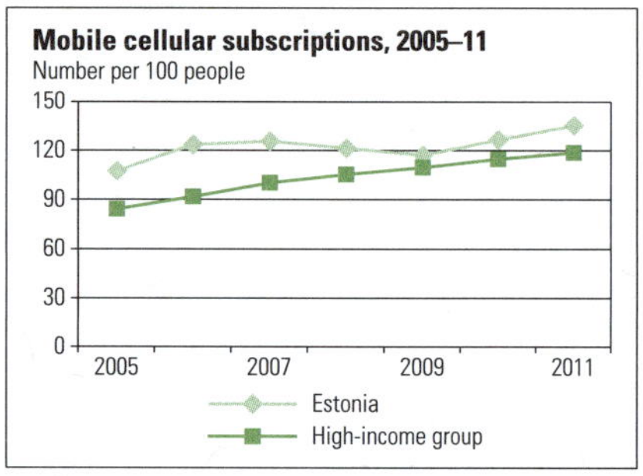

Mobile cellular subscriptions, 2005–11

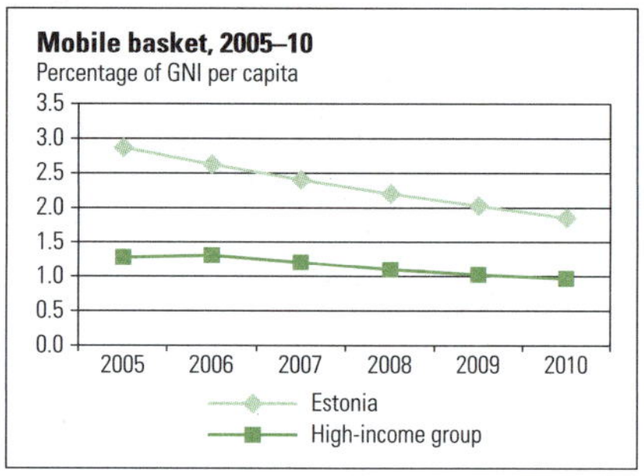

Mobile basket, 2005–10

Ethiopia

	Ethiopia 2005	Ethiopia 2010	Low-income group 2010	Sub-Saharan Africa Region 2010
Economic and social context				
Population (total, million)	74	83	796	853
GNI per capita, World Bank Atlas method (current US$)	160	390	530	1,188
Rural population (% of total)	84	82	72	63
Expected years of schooling (years)	7	*9*	*9*	*9*
Physicians density (per 1,000 people)	*0.02*	—	0.2	0.2
Depositors with commercial banks (per 1,000 adults)	*66*	107	—	*167*
Sector structure				
Number of mobile operators	—	1		
Herfindahl-Hirschman Index (HHI) (scale = 0–10,000)	—	10,000		
Sector performance				
Access				
Mobile cellular subscriptions (per 100 people)	1	12[a]	43[a]	57[a]
Mobile cellular subscriptions (% prepaid)	92	99[a]	98[a]	96[a]
Population covered by a mobile-cellular network (%)	10	—	—	72
Mobile broadband subscriptions (per 100 people)	—	1.1[a]	—	5.6[a]
Mobile broadband (% of total mobile subscriptions)	—	6.7[a]	—	10.1[a]
Usage				
Households with a mobile telephone (%)	2	25	43	52
Mobile voice usage (minutes per user per month)	—	—	—	—
Population using mobile Internet (%)	—	—	—	—
Short Message Service (SMS) users (% of mobile users)	—	—	—	—
Affordability				
Mobile basket (% of GNI per capita)	*40.2*	12.6	28.8	19.5

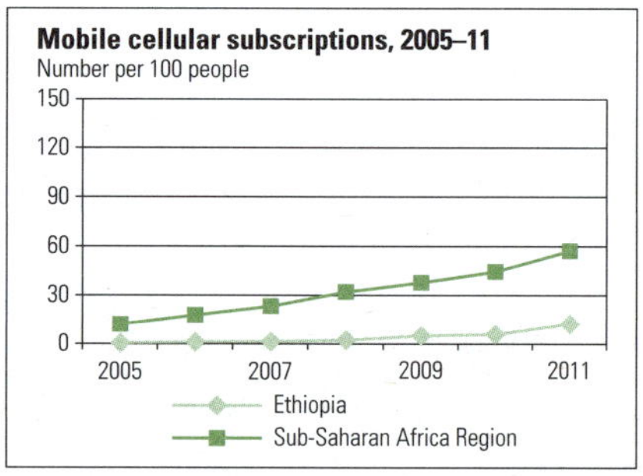

Mobile cellular subscriptions, 2005–11

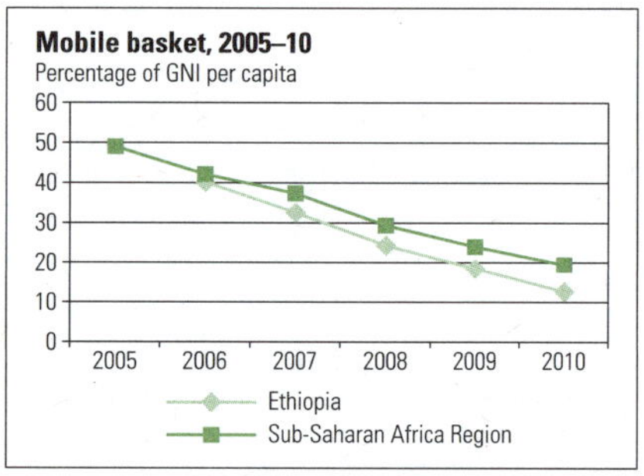

Mobile basket, 2005–10

Sources: Economic and social context: IMF, UIS, UN, WHO and World Bank; Sector structure: ictDATA.org; Sector performance: ictDATA.org, ITU; Wireless Intelligence, and World Bank.
Notes: Use of italics in the column entries indicates years or periods other than those specified. — Not available. GNI = gross national income.
a. Data are for 2011.

Finland

	Finland 2005	Finland 2010	High-income group 2010
Economic and social context			
Population (total, million)	5	5	1,127
GNI per capita, World Bank Atlas method (current US$)	38,550	47,570	38,746
Rural population (% of total)	38	36	22
Expected years of schooling (years)	17	*17*	*16*
Physicians density (per 1,000 people)	*3.3*	*2.9*	2.8
Depositors with commercial banks (per 1,000 adults)	—	—	—
Sector structure			
Number of mobile operators	—	3	
Herfindahl-Hirschman Index (HHI) (scale = 0–10,000)	—	3,350	
Sector performance			
Access			
Mobile cellular subscriptions (per 100 people)	100	163[a]	118[a]
Mobile cellular subscriptions (% prepaid)	6	9[a]	36[a]
Population covered by a mobile-cellular network (%)	99	*100*	100
Mobile broadband subscriptions (per 100 people)	1.1	96.4[a]	69.6[a]
Mobile broadband (% of total mobile subscriptions)	1.0	55.2[a]	57.6[a]
Usage			
Households with a mobile telephone (%)	96	99	93
Mobile voice usage (minutes per user per month)	236	206[a]	339
Population using mobile Internet (%)	9.5	24.2	24.3
Short Message Service (SMS) users (% of mobile users)	—	—	78.2[a]
Affordability			
Mobile basket (% of GNI per capita)	0.5	0.3	1.0

Mobile cellular subscriptions, 2005–11
Number per 100 people

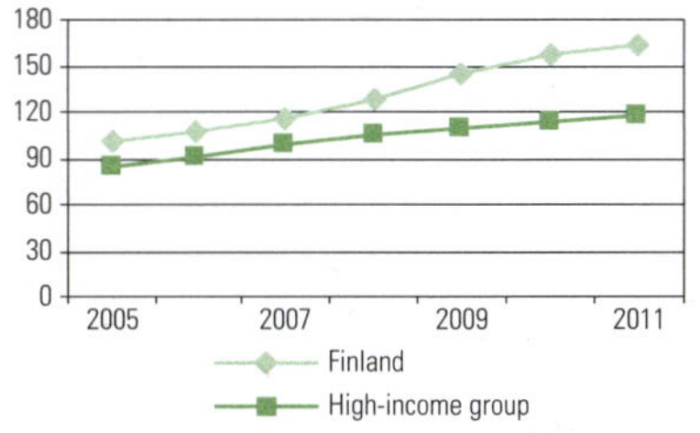

Mobile basket, 2005–10
Percentage of GNI per capita

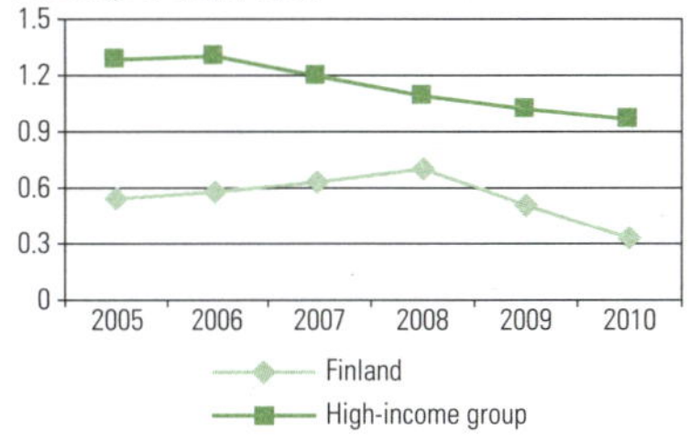

France

	France 2005	France 2010	High-income group 2010
Economic and social context			
Population (total, million)	63	65	1,127
GNI per capita, World Bank Atlas method (current US$)	34,890	42,370	38,746
Rural population (% of total)	23	22	22
Expected years of schooling (years)	16	*16*	*16*
Physicians density (per 1,000 people)	*3.4*	3.4	2.8
Depositors with commercial banks (per 1,000 adults)	—	—	—
Sector structure			
Number of mobile operators	—	3	
Herfindahl-Hirschman Index (HHI) (scale = 0–10,000)	—	3,223	
Sector performance			
Access			
Mobile cellular subscriptions (per 100 people)	76	98[a]	118[a]
Mobile cellular subscriptions (% prepaid)	36	30[a]	36[a]
Population covered by a mobile-cellular network (%)	99	99	100
Mobile broadband subscriptions (per 100 people)	2.1	41.3[a]	69.6[a]
Mobile broadband (% of total mobile subscriptions)	2.9	42.3[a]	57.6[a]
Usage			
Households with a mobile telephone (%)	*81*	88	93
Mobile voice usage (minutes per user per month)	233	218[a]	339
Population using mobile Internet (%)	4.4	18.9[a]	24.3
Short Message Service (SMS) users (% of mobile users)	—	81.8	78.2[a]
Affordability			
Mobile basket (% of GNI per capita)	1.7	1.4	1.0

Mobile cellular subscriptions, 2005–11
Number per 100 people

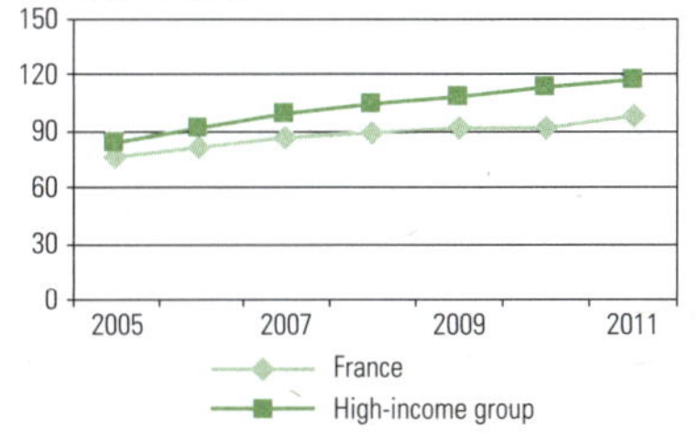

Mobile basket, 2005–10
Percentage of GNI per capita

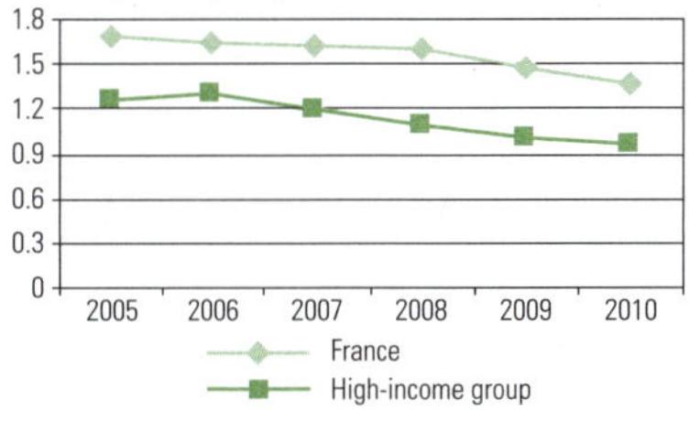

Sources: Economic and social context: IMF, UIS, UN, WHO and World Bank; Sector structure: ictDATA.org; Sector performance: ictDATA.org, ITU; Wireless Intelligence, and World Bank.
Notes: Use of italics in the column entries indicates years or periods other than those specified. — Not available. GNI = gross national income.
a. Data are for 2011.

Gabon

	Gabon 2005	Gabon 2010	Upper-middle-income group 2010	Sub-Saharan Africa Region 2010
Economic and social context				
Population (total, million)	1	2	2,452	853
GNI per capita, World Bank Atlas method (current US$)	5,110	7,650	5,886	1,188
Rural population (% of total)	16	14	43	63
Expected years of schooling (years)	—	—	*13*	*9*
Physicians density (per 1,000 people)	*0.3*	—	1.7	0.2
Depositors with commercial banks (per 1,000 adults)	43	*95*	—	*167*
Sector structure				
Number of mobile operators	—	4		
Herfindahl-Hirschman Index (HHI) (scale = 0–10,000)	—	4,584		
Sector performance				
Access				
Mobile cellular subscriptions (per 100 people)	54	165[a]	92[a]	57[a]
Mobile cellular subscriptions (% prepaid)	99	99[a]	81[a]	96[a]
Population covered by a mobile-cellular network (%)	78	—	*99*	72
Mobile broadband subscriptions (per 100 people)	—	—	14.3[a]	5.6[a]
Mobile broadband (% of total mobile subscriptions)	—	—	15.4[a]	10.1[a]
Usage				
Households with a mobile telephone (%)	—	—	84	52
Mobile voice usage (minutes per user per month)	—	—	325[a]	—
Population using mobile Internet (%)	—	—	22.9[a]	—
Short Message Service (SMS) users (% of mobile users)	—	—	74.4[a]	—
Affordability				
Mobile basket (% of GNI per capita)	—	—	2.9	19.5

Mobile cellular subscriptions, 2005–11
Number per 100 people

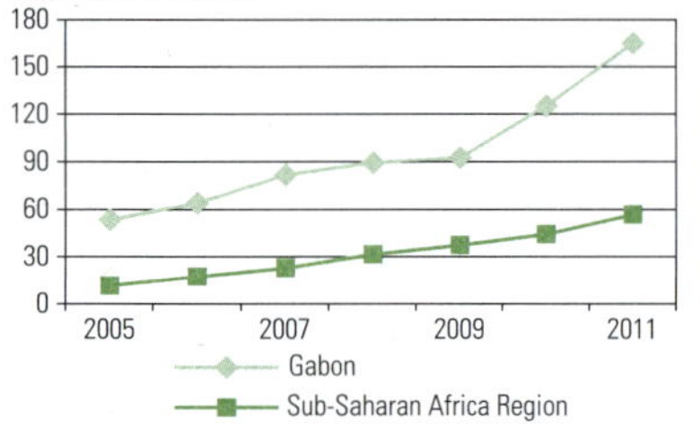

Mobile basket, 2005–10
Percentage of GNI per capita

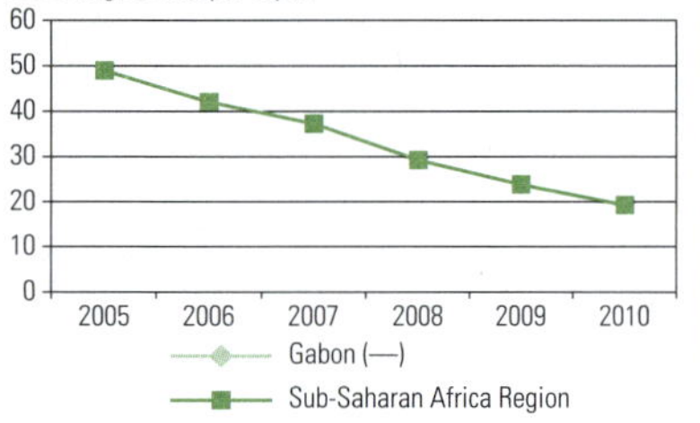

Gambia, The

	Gambia, The 2005	Gambia, The 2010	Low-income group 2010	Sub-Saharan Africa Region 2010
Economic and social context				
Population (total, million)	2	2	796	853
GNI per capita, World Bank Atlas method (current US$)	270	450	530	1,188
Rural population (% of total)	46	42	72	63
Expected years of schooling (years)	—	*9*	*9*	*9*
Physicians density (per 1,000 people)	*0.1*	*0.0*	0.2	0.2
Depositors with commercial banks (per 1,000 adults)	—	—	—	*167*
Sector structure				
Number of mobile operators	—	4		
Herfindahl-Hirschman Index (HHI) (scale = 0–10,000)	—	—		
Sector performance				
Access				
Mobile cellular subscriptions (per 100 people)	16	86	43[a]	57[a]
Mobile cellular subscriptions (% prepaid)	99	100[a]	98[a]	96[a]
Population covered by a mobile-cellular network (%)	70	—	—	72
Mobile broadband subscriptions (per 100 people)	—	1.5[a]	—	5.6[a]
Mobile broadband (% of total mobile subscriptions)	—	1.3[a]	—	10.1[a]
Usage				
Households with a mobile telephone (%)	—	—	43	52
Mobile voice usage (minutes per user per month)	*51*	—	—	—
Population using mobile Internet (%)	—	—	—	—
Short Message Service (SMS) users (% of mobile users)	—	—	—	—
Affordability				
Mobile basket (% of GNI per capita)	—	*17.1*	28.8	19.5

Mobile cellular subscriptions, 2005–11
Number per 100 people

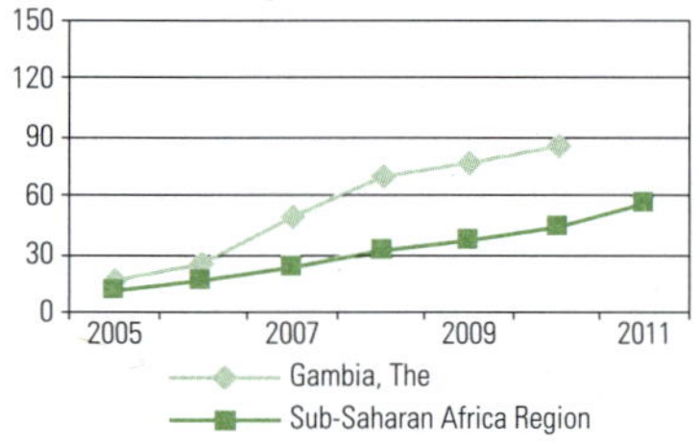

Mobile basket, 2005–10
Percentage of GNI per capita

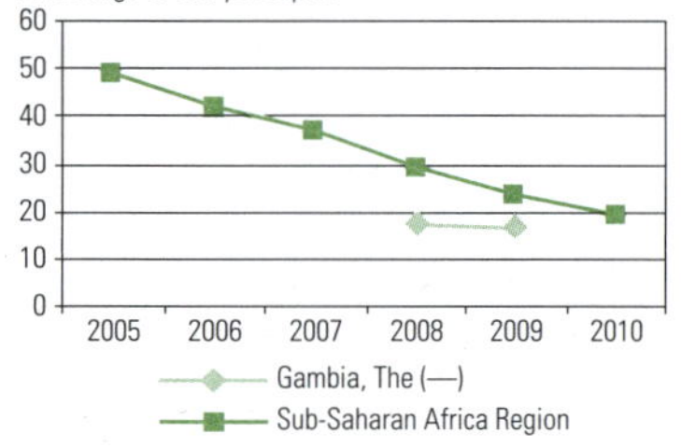

Sources: Economic and social context: IMF, UIS, UN, WHO and World Bank; Sector structure: ictDATA.org; Sector performance: ictDATA.org, ITU; Wireless Intelligence, and World Bank.
Notes: Use of italics in the column entries indicates years or periods other than those specified. — Not available. GNI = gross national income.
a. Data are for 2011.

Georgia

	Georgia 2005	Georgia 2010	Lower-middle-income group 2010	Europe & Central Asia Region 2010
Economic and social context				
Population (total, million)	4	4	2,519	405
GNI per capita, World Bank Atlas method (current US$)	1,360	2,690	1,623	7,272
Rural population (% of total)	48	47	61	36
Expected years of schooling (years)	13	*13*	*10*	*13*
Physicians density (per 1,000 people)	*4.7*	*4.8*	0.8	3.2
Depositors with commercial banks (per 1,000 adults)	363	697	—	894
Sector structure				
Number of mobile operators	—	5		
Herfindahl-Hirschman Index (HHI) (scale = 0–10,000)	—	3,465		
Sector performance				
Access				
Mobile cellular subscriptions (per 100 people)	27	111[a]	78[a]	125[a]
Mobile cellular subscriptions (% prepaid)	88	89[a]	96[a]	82[a]
Population covered by a mobile-cellular network (%)	95	99	*86*	96
Mobile broadband subscriptions (per 100 people)	*0.4*	17.2[a]	7.3[a]	22.6[a]
Mobile broadband (% of total mobile subscriptions)	*1.0*	15.0[a]	9.0[a]	18.0[a]
Usage				
Households with a mobile telephone (%)	30	80	77	82
Mobile voice usage (minutes per user per month)	91	148[a]	276[a]	288[a]
Population using mobile Internet (%)	—	—	2.9	8.5
Short Message Service (SMS) users (% of mobile users)	—	38.0	61.9[a]	69.8[a]
Affordability				
Mobile basket (% of GNI per capita)	10.1	5.2	7.2	3.1

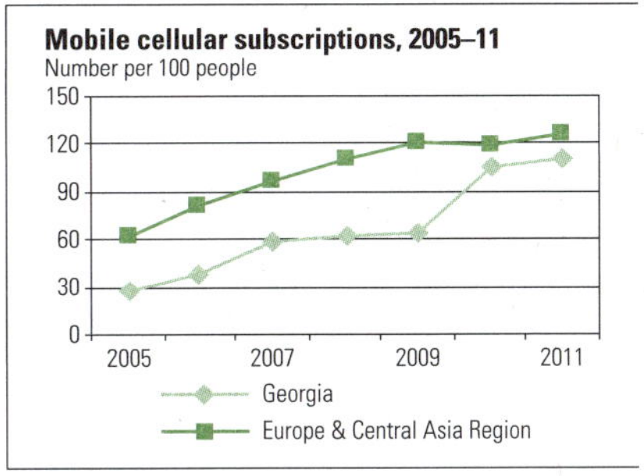

Mobile cellular subscriptions, 2005–11

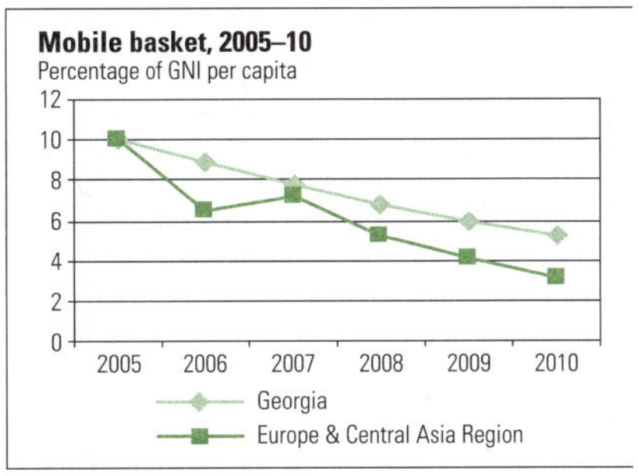

Mobile basket, 2005–10

Germany

	Germany 2005	Germany 2010	High-income group 2010
Economic and social context			
Population (total, million)	82	82	1,127
GNI per capita, World Bank Atlas method (current US$)	34,780	43,070	38,746
Rural population (% of total)	27	26	22
Expected years of schooling (years)	—	—	*16*
Physicians density (per 1,000 people)	*3.4*	*3.6*	2.8
Depositors with commercial banks (per 1,000 adults)	—	—	—
Sector structure			
Number of mobile operators	—	4	
Herfindahl-Hirschman Index (HHI) (scale = 0–10,000)	—	2,749	
Sector performance			
Access			
Mobile cellular subscriptions (per 100 people)	96	140[a]	118[a]
Mobile cellular subscriptions (% prepaid)	51	56[a]	36[a]
Population covered by a mobile-cellular network (%)	99	99	100
Mobile broadband subscriptions (per 100 people)	2.4	51.1[a]	69.6[a]
Mobile broadband (% of total mobile subscriptions)	2.5	36.5[a]	57.6[a]
Usage			
Households with a mobile telephone (%)	84	89	93
Mobile voice usage (minutes per user per month)	90	116[a]	339
Population using mobile Internet (%)	1.8	16.4[a]	24.3
Short Message Service (SMS) users (% of mobile users)	76.2	79.8	78.2[a]
Affordability			
Mobile basket (% of GNI per capita)	1.8	0.4	1.0

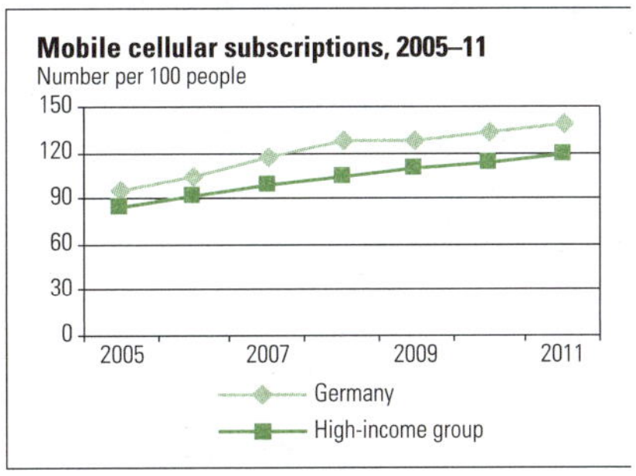

Mobile cellular subscriptions, 2005–11

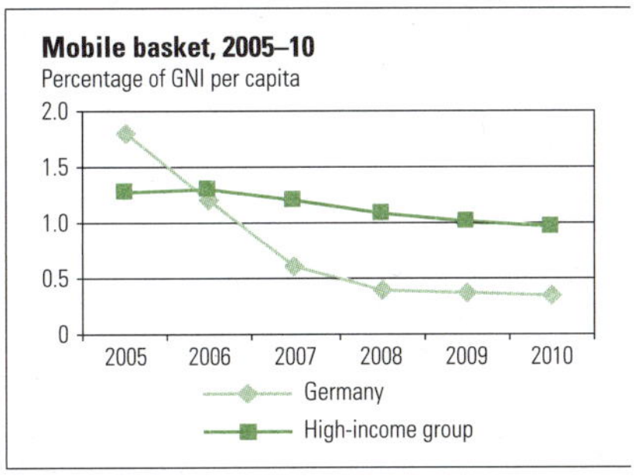

Mobile basket, 2005–10

Sources: Economic and social context: IMF, UIS, UN, WHO and World Bank; Sector structure: ictDATA.org; Sector performance: ictDATA.org, ITU; Wireless Intelligence, and World Bank.
Notes: Use of italics in the column entries indicates years or periods other than those specified. — Not available. GNI = gross national income.
a. Data are for 2011.

Ghana

	Ghana 2005	Ghana 2010	Lower-middle-income group 2010	Sub-Saharan Africa Region 2010
Economic and social context				
Population (total, million)	22	24	2,519	853
GNI per capita, World Bank Atlas method (current US$)	460	1,250	1,623	1,188
Rural population (% of total)	52	49	61	63
Expected years of schooling (years)	9	*10*	*10*	*9*
Physicians density (per 1,000 people)	*0.2*	*0.1*	0.8	0.2
Depositors with commercial banks (per 1,000 adults)	173	*324*	—	*167*
Sector structure				
Number of mobile operators	—	5		
Herfindahl-Hirschman Index (HHI) (scale = 0–10,000)	—	3,374		
Sector performance				
Access				
Mobile cellular subscriptions (per 100 people)	13	85[a]	78[a]	57[a]
Mobile cellular subscriptions (% prepaid)	99	102[a]	96[a]	96[a]
Population covered by a mobile-cellular network (%)	59	77	*86*	72
Mobile broadband subscriptions (per 100 people)	—	1.7[a]	7.3[a]	5.6[a]
Mobile broadband (% of total mobile subscriptions)	—	2.0[a]	9.0[a]	10.1[a]
Usage				
Households with a mobile telephone (%)	20	72	77	52
Mobile voice usage (minutes per user per month)	*104*	114	276[a]	—
Population using mobile Internet (%)	—	—	2.9	—
Short Message Service (SMS) users (% of mobile users)	—	*20.0*	61.9[a]	—
Affordability				
Mobile basket (% of GNI per capita)	42.2	7.1	7.2	19.5

Mobile cellular subscriptions, 2005–11
Number per 100 people

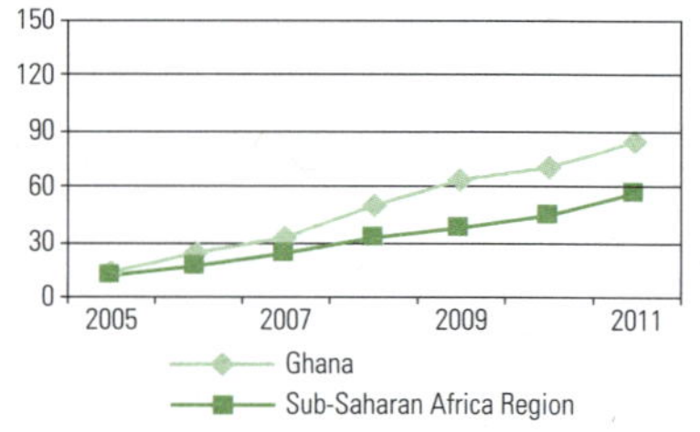

Mobile basket, 2005–10
Percentage of GNI per capita

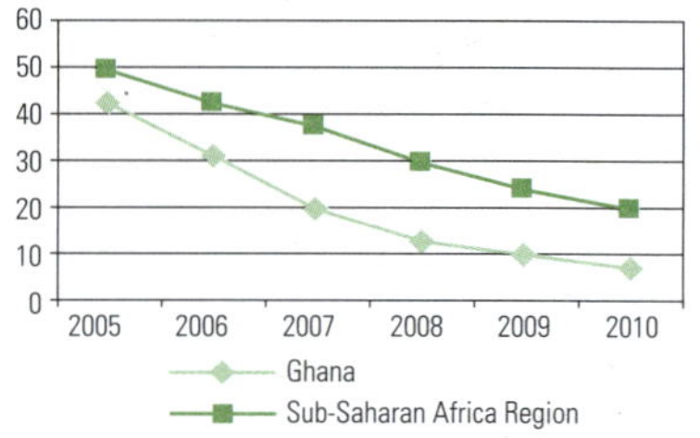

Greece

	Greece 2005	Greece 2010	High-income group 2010
Economic and social context			
Population (total, million)	11	11	1,127
GNI per capita, World Bank Atlas method (current US$)	21,400	26,950	38,746
Rural population (% of total)	40	39	22
Expected years of schooling (years)	17	—	*16*
Physicians density (per 1,000 people)	5.0	*6.2*	2.8
Depositors with commercial banks (per 1,000 adults)	—	—	—
Sector structure			
Number of mobile operators	—	3	
Herfindahl-Hirschman Index (HHI) (scale = 0–10,000)	—	3,747	
Sector performance			
Access			
Mobile cellular subscriptions (per 100 people)	92	141[a]	118[a]
Mobile cellular subscriptions (% prepaid)	67	69[a]	36[a]
Population covered by a mobile-cellular network (%)	100	100	100
Mobile broadband subscriptions (per 100 people)	2.1	60.0[a]	69.6[a]
Mobile broadband (% of total mobile subscriptions)	1.8	43.7[a]	57.6[a]
Usage			
Households with a mobile telephone (%)	76	93	93
Mobile voice usage (minutes per user per month)	138	245[a]	339
Population using mobile Internet (%)	0.9	1.7	24.3
Short Message Service (SMS) users (% of mobile users)	—	—	78.2[a]
Affordability			
Mobile basket (% of GNI per capita)	1.9	1.7	1.0

Mobile cellular subscriptions, 2005–11
Number per 100 people

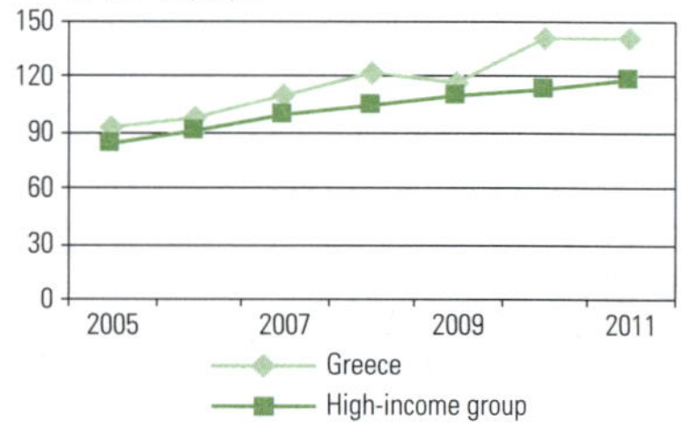

Mobile basket, 2005–10
Percentage of GNI per capita

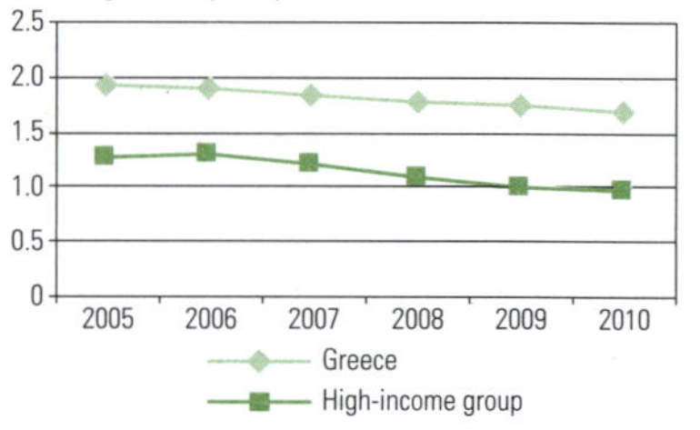

Sources: Economic and social context: IMF, UIS, UN, WHO and World Bank; Sector structure: ictDATA.org; Sector performance: ictDATA.org, ITU; Wireless Intelligence, and World Bank.
Notes: Use of italics in the column entries indicates years or periods other than those specified. — Not available. GNI = gross national income.
a. Data are for 2011.

Guatemala

	Guatemala 2005	Guatemala 2010	Lower-middle-income group 2010	Latin America & the Caribbean Region 2010
Economic and social context				
Population (total, million)	13	14	2,519	583
GNI per capita, World Bank Atlas method (current US$)	2,070	2,740	1,623	7,741
Rural population (% of total)	53	51	61	21
Expected years of schooling (years)	10	—	*10*	*14*
Physicians density (per 1,000 people)	—	—	0.8	1.8
Depositors with commercial banks (per 1,000 adults)	—	—	—	—
Sector structure				
Number of mobile operators	—	3		
Herfindahl-Hirschman Index (HHI) (scale = 0–10,000)	—	3,481		
Sector performance				
Access				
Mobile cellular subscriptions (per 100 people)	35	126[a]	78[a]	109[a]
Mobile cellular subscriptions (% prepaid)	94	95[a]	96[a]	81[a]
Population covered by a mobile-cellular network (%)	*76*	—	*86*	98
Mobile broadband subscriptions (per 100 people)	0.1	6.2[a]	7.3[a]	16.1[a]
Mobile broadband (% of total mobile subscriptions)	0.3	6.4[a]	9.0[a]	15.2[a]
Usage				
Households with a mobile telephone (%)	*55*	—	77	84
Mobile voice usage (minutes per user per month)	—	—	276[a]	141[a]
Population using mobile Internet (%)	—	—	2.9	4.4
Short Message Service (SMS) users (% of mobile users)	—	—	61.9[a]	—
Affordability				
Mobile basket (% of GNI per capita)	4.5	3.4	7.2	3.7

Mobile cellular subscriptions, 2005–11

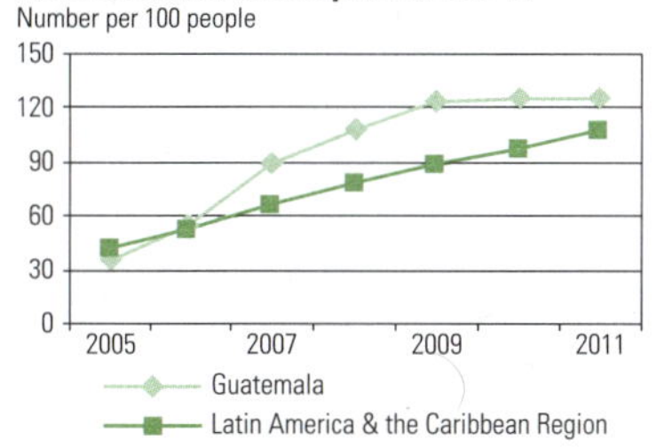

Mobile basket, 2005–10

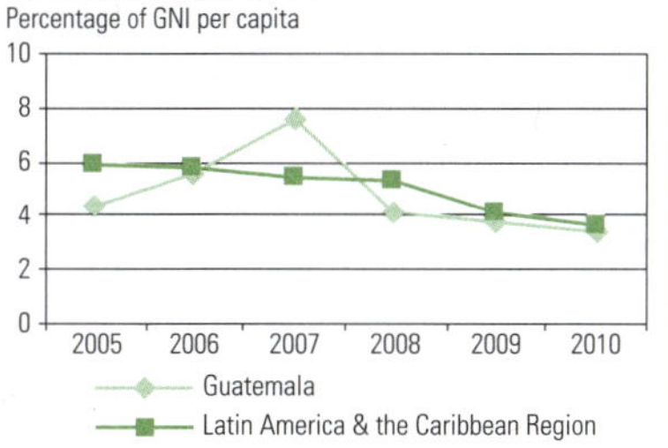

Guinea

	Guinea 2005	Guinea 2010	Low-income group 2010	Sub-Saharan Africa Region 2010
Economic and social context				
Population (total, million)	9	10	796	853
GNI per capita, World Bank Atlas method (current US$)	360	400	530	1,188
Rural population (% of total)	67	65	72	63
Expected years of schooling (years)	7	*9*	*9*	*9*
Physicians density (per 1,000 people)	0.1	—	0.2	0.2
Depositors with commercial banks (per 1,000 adults)	—	—	—	*167*
Sector structure				
Number of mobile operators	—	5		
Herfindahl-Hirschman Index (HHI) (scale = 0–10,000)	—	2,699		
Sector performance				
Access				
Mobile cellular subscriptions (per 100 people)	2	46[a]	43[a]	57[a]
Mobile cellular subscriptions (% prepaid)	98	99[a]	98[a]	96[a]
Population covered by a mobile-cellular network (%)	*80*	*80*	—	72
Mobile broadband subscriptions (per 100 people)	—	—	—	5.6[a]
Mobile broadband (% of total mobile subscriptions)	—	—	—	10.1[a]
Usage				
Households with a mobile telephone (%)	—	—	43	52
Mobile voice usage (minutes per user per month)	—	—	—	—
Population using mobile Internet (%)	—	—	—	—
Short Message Service (SMS) users (% of mobile users)	—	*17.0*	—	—
Affordability				
Mobile basket (% of GNI per capita)	*25.4*	12.3	28.8	19.5

Mobile cellular subscriptions, 2005–11

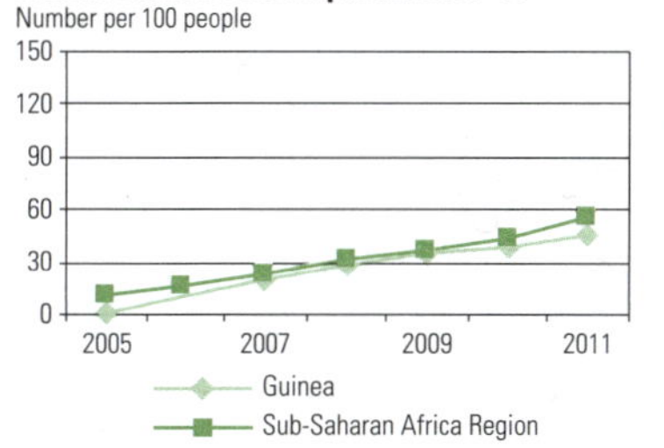

Mobile basket, 2005–10

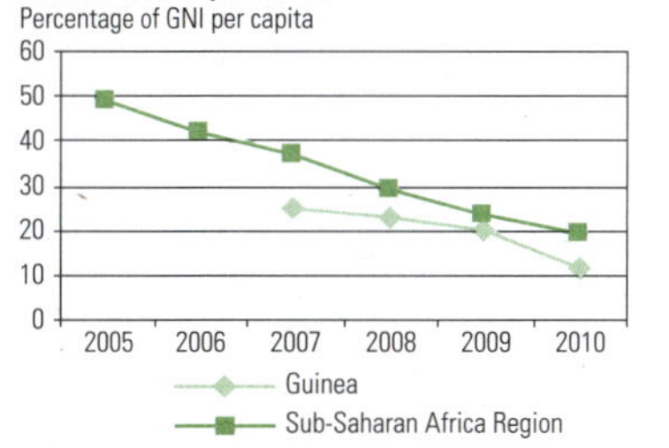

Sources: Economic and social context: IMF, UIS, UN, WHO and World Bank; Sector structure: ictDATA.org; Sector performance: ictDATA.org, ITU; Wireless Intelligence, and World Bank.
Notes: Use of italics in the column entries indicates years or periods other than those specified. — Not available. GNI = gross national income.
a. Data are for 2011.

Guinea-Bissau

	Guinea-Bissau 2005	Guinea-Bissau 2010	Low-income group 2010	Sub-Saharan Africa Region 2010
Economic and social context				
Population (total, million)	1	2	796	853
GNI per capita, World Bank Atlas method (current US$)	410	590	530	1,188
Rural population (% of total)	70	70	72	63
Expected years of schooling (years)	9	—	*9*	*9*
Physicians density (per 1,000 people)	*0.12*	*0.05*	0.2	0.2
Depositors with commercial banks (per 1,000 adults)	—	—	—	*167*
Sector structure				
Number of mobile operators	—	3		
Herfindahl-Hirschman Index (HHI) (scale = 0–10,000)	—	6,250		
Sector performance				
Access				
Mobile cellular subscriptions (per 100 people)	7	56[a]	43[a]	57[a]
Mobile cellular subscriptions (% prepaid)	98	99[a]	98[a]	96[a]
Population covered by a mobile-cellular network (%)	*65*	—	—	72
Mobile broadband subscriptions (per 100 people)	—	—	—	5.6[a]
Mobile broadband (% of total mobile subscriptions)	—	—	—	10.1[a]
Usage				
Households with a mobile telephone (%)	—	—	43	52
Mobile voice usage (minutes per user per month)	—	23[a]	—	—
Population using mobile Internet (%)	—	—	—	—
Short Message Service (SMS) users (% of mobile users)	—	—	—	—
Affordability				
Mobile basket (% of GNI per capita)	—	—	28.8	19.5

Mobile cellular subscriptions, 2005–11
Number per 100 people

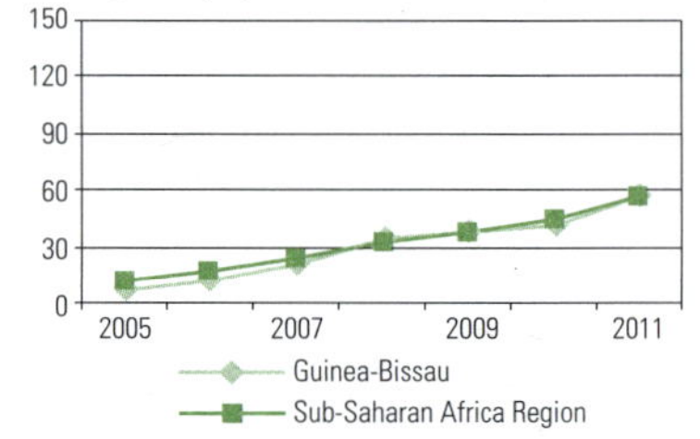

Mobile basket, 2005–10
Percentage of GNI per capita

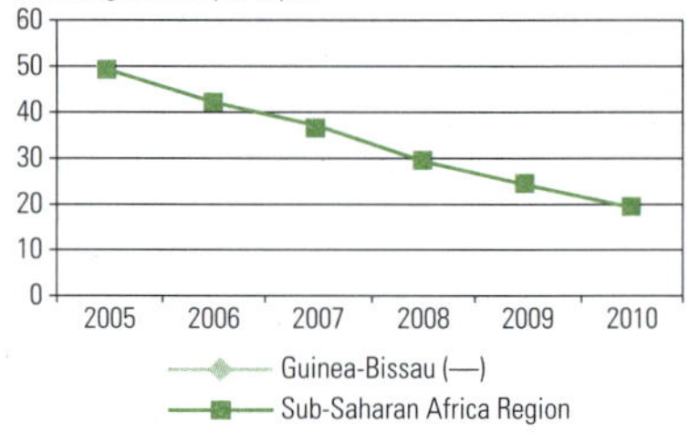

Haiti

	Haiti 2005	Haiti 2010	Low-income group 2010	Latin America & the Caribbean Region 2010
Economic and social context				
Population (total, million)	9	10	796	583
GNI per capita, World Bank Atlas method (current US$)	400	*670*	530	7,741
Rural population (% of total)	57	50	72	21
Expected years of schooling (years)	—	—	*9*	*14*
Physicians density (per 1,000 people)	—	—	0.2	1.8
Depositors with commercial banks (per 1,000 adults)	233	*339*	—	—
Sector structure				
Number of mobile operators	—	—		
Herfindahl-Hirschman Index (HHI) (scale = 0–10,000)	—	—		
Sector performance				
Access				
Mobile cellular subscriptions (per 100 people)	5	55[a]	43[a]	109[a]
Mobile cellular subscriptions (% prepaid)	95	96[a]	98[a]	81[a]
Population covered by a mobile-cellular network (%)	—	—	—	98
Mobile broadband subscriptions (per 100 people)	*0.3*	3.0[a]	—	16.1[a]
Mobile broadband (% of total mobile subscriptions)	*1.3*	5.4[a]	—	15.2[a]
Usage				
Households with a mobile telephone (%)	*17*	—	43	84
Mobile voice usage (minutes per user per month)	—	—	—	141[a]
Population using mobile Internet (%)	—	—	—	4.4
Short Message Service (SMS) users (% of mobile users)	—	*73.0*	—	—
Affordability				
Mobile basket (% of GNI per capita)	—	—	28.8	3.7

Mobile cellular subscriptions, 2005–11
Number per 100 people

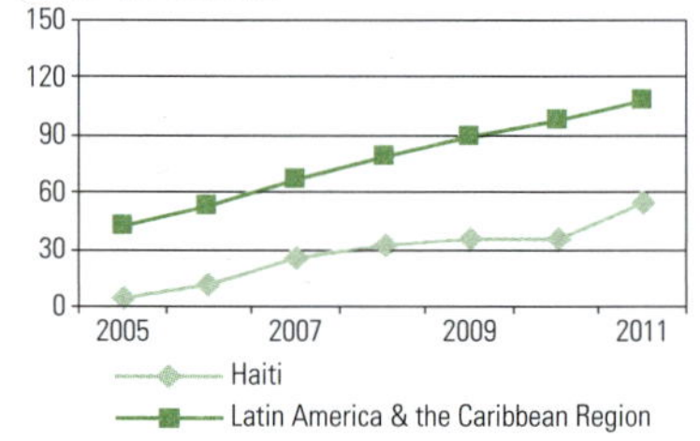

Mobile basket, 2005–10
Percentage of GNI per capita

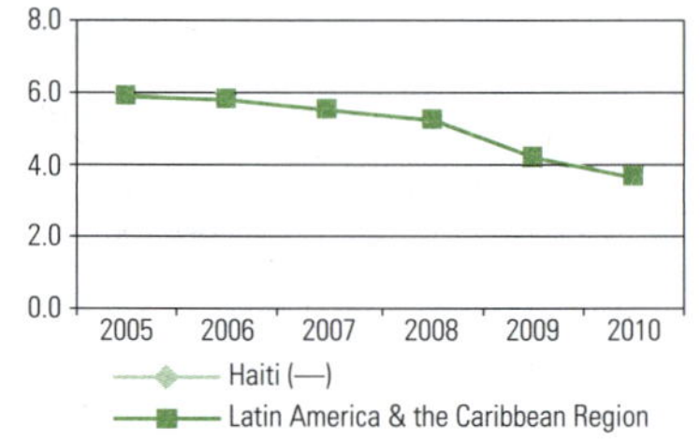

Sources: Economic and social context: IMF, UIS, UN, WHO and World Bank; Sector structure: ictDATA.org; Sector performance: ictDATA.org, ITU; Wireless Intelligence, and World Bank.
Notes: Use of italics in the column entries indicates years or periods other than those specified. — Not available. GNI = gross national income.
a. Data are for 2011.

Honduras

	Honduras 2005	Honduras 2010	Lower-middle-income group 2010	Latin America & the Caribbean Region 2010
Economic and social context				
Population (total, million)	7	8	2,519	583
GNI per capita, World Bank Atlas method (current US$)	1,400	1,870	1,623	7,741
Rural population (% of total)	54	51	61	21
Expected years of schooling (years)	—	*11*	*10*	*14*
Physicians density (per 1,000 people)	—	—	0.8	1.8
Depositors with commercial banks (per 1,000 adults)	—	—	—	—
Sector structure				
Number of mobile operators	—	4		
Herfindahl-Hirschman Index (HHI) (scale = 0–10,000)	—	4,822		
Sector performance				
Access				
Mobile cellular subscriptions (per 100 people)	19	103[a]	78[a]	109[a]
Mobile cellular subscriptions (% prepaid)	92	93[a]	96[a]	81[a]
Population covered by a mobile-cellular network (%)	*57*	—	*86*	98
Mobile broadband subscriptions (per 100 people)	—	6.2[a]	7.3[a]	16.1[a]
Mobile broadband (% of total mobile subscriptions)	—	6.5[a]	9.0[a]	15.2[a]
Usage				
Households with a mobile telephone (%)	*41*	81	77	84
Mobile voice usage (minutes per user per month)	—	—	276[a]	141[a]
Population using mobile Internet (%)	—	—	2.9	4.4
Short Message Service (SMS) users (% of mobile users)	—	—	61.9[a]	—
Affordability				
Mobile basket (% of GNI per capita)	*11.4*	5.7	7.2	3.7

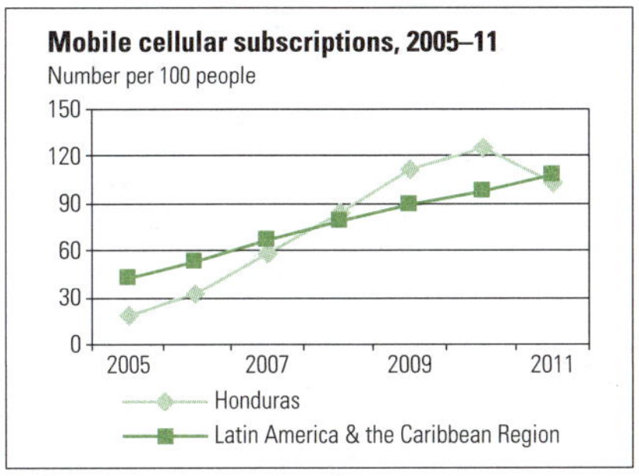

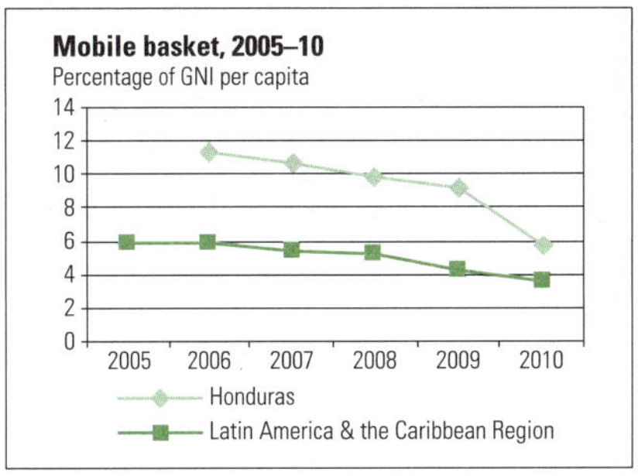

Hong Kong SAR, China

	Hong Kong SAR, China 2005	Hong Kong SAR, China 2010	High-income group 2010
Economic and social context			
Population (total, million)	7	7	1,127
GNI per capita, World Bank Atlas method (current US$)	28,150	32,780	38,746
Rural population (% of total)	0	0	22
Expected years of schooling (years)	14	*16*	*16*
Physicians density (per 1,000 people)	—	—	2.8
Depositors with commercial banks (per 1,000 adults)	—	—	—
Sector structure			
Number of mobile operators	—	5	
Herfindahl-Hirschman Index (HHI) (scale = 0–10,000)	—	—	
Sector performance			
Access			
Mobile cellular subscriptions (per 100 people)	114	183[a]	118[a]
Mobile cellular subscriptions (% prepaid)	39	49[a]	36[a]
Population covered by a mobile-cellular network (%)	100	100	100
Mobile broadband subscriptions (per 100 people)	9.3	89.5[a]	69.6[a]
Mobile broadband (% of total mobile subscriptions)	8.2	53.2[a]	57.6[a]
Usage			
Households with a mobile telephone (%)	88	98	93
Mobile voice usage (minutes per user per month)	—	—	339
Population using mobile Internet (%)	3.6	21.2	24.3
Short Message Service (SMS) users (% of mobile users)	—	—	78.2[a]
Affordability			
Mobile basket (% of GNI per capita)	0.2	0.1	1.0

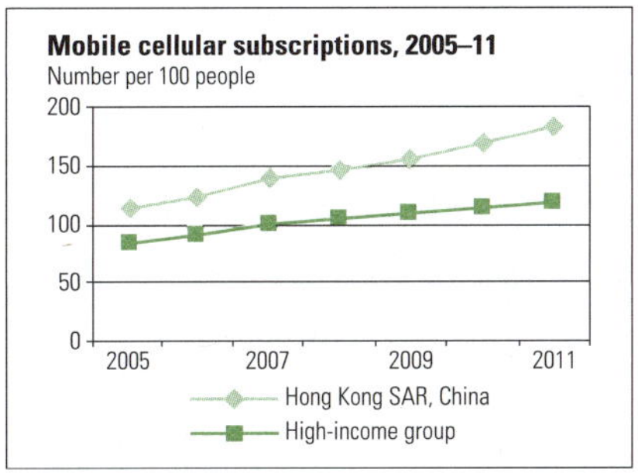

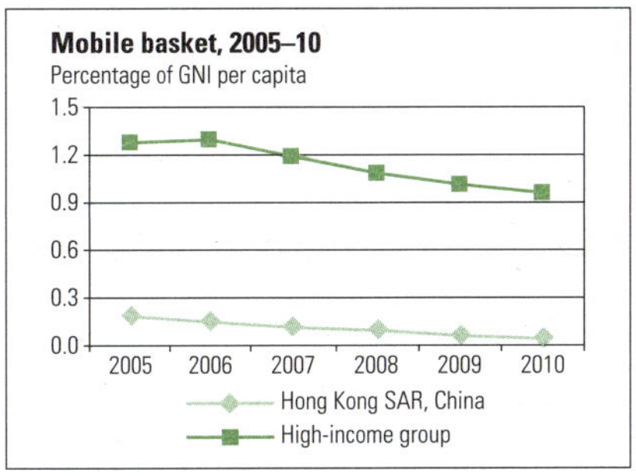

Sources: Economic and social context: IMF, UIS, UN, WHO and World Bank; Sector structure: ictDATA.org; Sector performance: ictDATA.org, ITU; Wireless Intelligence, and World Bank.
Notes: Use of italics in the column entries indicates years or periods other than those specified. — Not available. GNI = gross national income.
a. Data are for 2011.

Hungary

	Hungary 2005	Hungary 2010	High-income group 2010
Economic and social context			
Population (total, million)	10	10	1,127
GNI per capita, World Bank Atlas method (current US$)	10,220	12,860	38,746
Rural population (% of total)	34	32	22
Expected years of schooling (years)	15	*15*	*16*
Physicians density (per 1,000 people)	*3.0*	*3.0*	2.8
Depositors with commercial banks (per 1,000 adults)	840	1,072	—
Sector structure			
Number of mobile operators	—	3	
Herfindahl-Hirschman Index (HHI) (scale = 0–10,000)	—	3,555	
Sector performance			
Access			
Mobile cellular subscriptions (per 100 people)	92	111[a]	118[a]
Mobile cellular subscriptions (% prepaid)	68	52[a]	36[a]
Population covered by a mobile-cellular network (%)	99	99	100
Mobile broadband subscriptions (per 100 people)	0.4	39.3[a]	69.6[a]
Mobile broadband (% of total mobile subscriptions)	0.5	36.1[a]	57.6[a]
Usage			
Households with a mobile telephone (%)	80	87	93
Mobile voice usage (minutes per user per month)	144	174[a]	339
Population using mobile Internet (%)	0.1	9.0	24.3
Short Message Service (SMS) users (% of mobile users)	—	—	78.2[a]
Affordability			
Mobile basket (% of GNI per capita)	3.4	2.4	1.0

Mobile cellular subscriptions, 2005–11
Number per 100 people

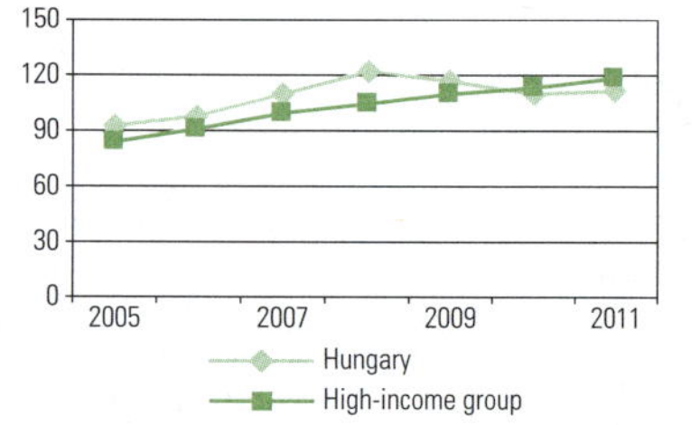

Mobile basket, 2005–10
Percentage of GNI per capita

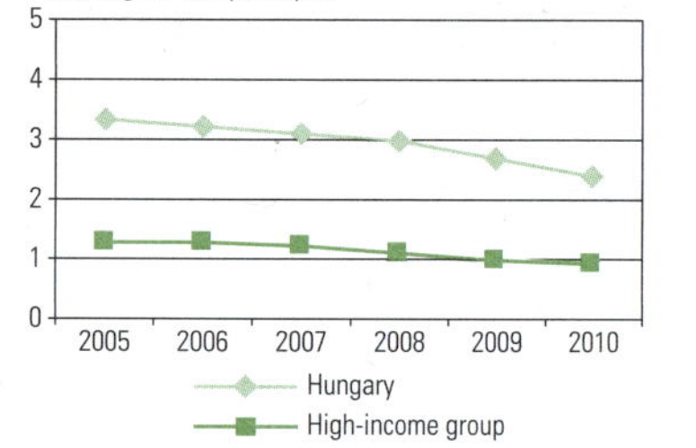

India

	India 2005	India 2010	Lower-middle-income group 2010	South Asia Region 2010
Economic and social context				
Population (total, million)	1,140	1,225	2,519	1,633
GNI per capita, World Bank Atlas method (current US$)	720	1,270	1,623	1,176
Rural population (% of total)	71	70	61	70
Expected years of schooling (years)	10	—	*10*	*10*
Physicians density (per 1,000 people)	0.6	*0.6*	0.8	0.6
Depositors with commercial banks (per 1,000 adults)	637	*747*	—	249
Sector structure				
Number of mobile operators	—	8		
Herfindahl-Hirschman Index (HHI) (scale = 0–10,000)	—	1,393		
Sector performance				
Access				
Mobile cellular subscriptions (per 100 people)	8	70[a]	78[a]	67[a]
Mobile cellular subscriptions (% prepaid)	80	96[a]	96[a]	96[a]
Population covered by a mobile-cellular network (%)	31	*83*	*86*	*84*
Mobile broadband subscriptions (per 100 people)	—	3.3[a]	7.3[a]	3.3[a]
Mobile broadband (% of total mobile subscriptions)	—	4.6[a]	9.0[a]	4.6[a]
Usage				
Households with a mobile telephone (%)	13	53	77	54
Mobile voice usage (minutes per user per month)	*425*	330[a]	276[a]	305[a]
Population using mobile Internet (%)	0.1	3.3[a]	2.9	3.3[a]
Short Message Service (SMS) users (% of mobile users)	—	49.0[a]	61.9[a]	47.0[a]
Affordability				
Mobile basket (% of GNI per capita)	11.0	3.2	7.2	3.2

Mobile cellular subscriptions, 2005–11
Number per 100 people

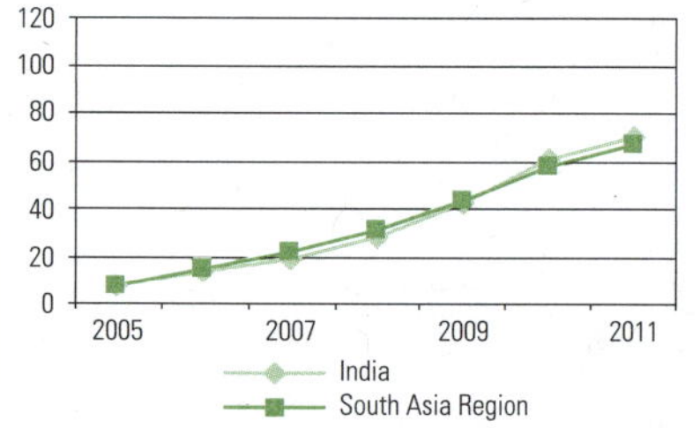

Mobile basket, 2005–10
Percentage of GNI per capita

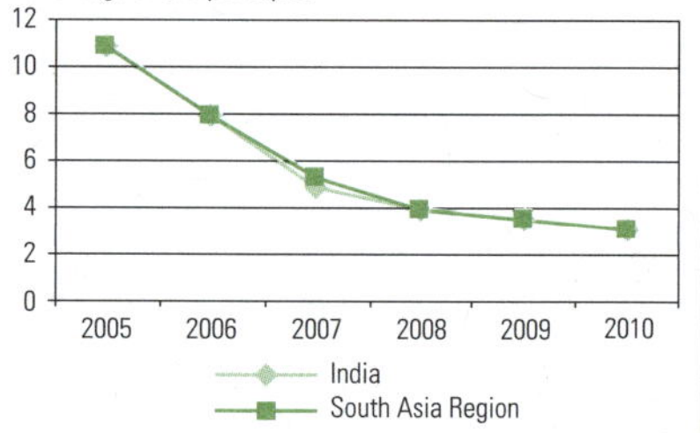

Sources: Economic and social context: IMF, UIS, UN, WHO and World Bank; Sector structure: ictDATA.org; Sector performance: ictDATA.org, ITU; Wireless Intelligence, and World Bank.
Notes: Use of italics in the column entries indicates years or periods other than those specified. — Not available. GNI = gross national income.
a. Data are for 2011.

Indonesia

	Indonesia 2005	Indonesia 2010	Lower-middle-income group 2010	East Asia & Pacific Region 2010
Economic and social context				
Population (total, million)	227	240	2,519	1,962
GNI per capita, World Bank Atlas method (current US$)	1,220	2,500	1,623	3,696
Rural population (% of total)	52	46	61	54
Expected years of schooling (years)	12	*13*	*10*	*12*
Physicians density (per 1,000 people)	*0.1*	—	0.8	1.2
Depositors with commercial banks (per 1,000 adults)	—	—	—	—
Sector structure				
Number of mobile operators	—	5		
Herfindahl-Hirschman Index (HHI) (scale = 0–10,000)	—	3,229		
Sector performance				
Access				
Mobile cellular subscriptions (per 100 people)	21	103[a]	78[a]	83[a]
Mobile cellular subscriptions (% prepaid)	95	98[a]	96[a]	85[a]
Population covered by a mobile-cellular network (%)	90	—	*86*	*99*
Mobile broadband subscriptions (per 100 people)	*0.05*	16.1[a]	7.3[a]	11.6[a]
Mobile broadband (% of total mobile subscriptions)	*0.17*	15.7[a]	9.0[a]	14.4[a]
Usage				
Households with a mobile telephone (%)	20	72	77	83
Mobile voice usage (minutes per user per month)	*50*	191[a]	276[a]	367[a]
Population using mobile Internet (%)	0.04	8.0[a]	2.9	22.4[a]
Short Message Service (SMS) users (% of mobile users)	—	96.0[a]	61.9[a]	84.0[a]
Affordability				
Mobile basket (% of GNI per capita)	8.2	3.7	7.2	5.7

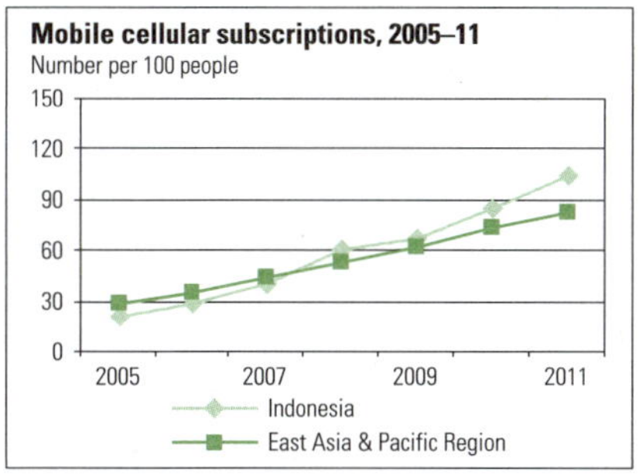

Mobile cellular subscriptions, 2005–11
Number per 100 people

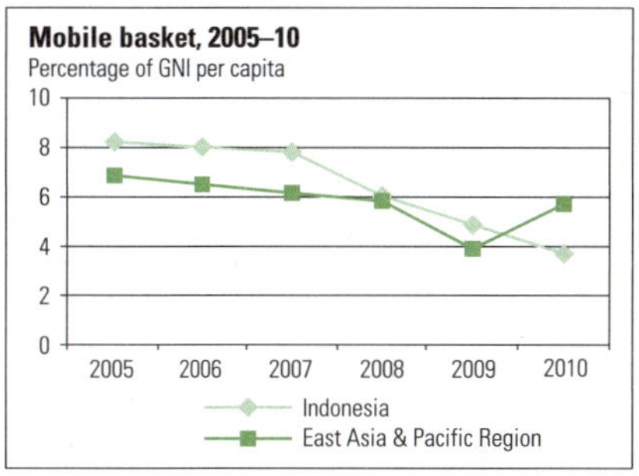

Mobile basket, 2005–10
Percentage of GNI per capita

Iran, Islamic Rep.

	Iran, Islamic Rep. 2005	Iran, Islamic Rep. 2010	Upper-middle-income group 2010	Middle East & North Africa Region 2010
Economic and social context				
Population (total, million)	70	74	2,452	331
GNI per capita, World Bank Atlas method (current US$)	2,550	*4,520*	5,886	3,874
Rural population (% of total)	33	31	43	42
Expected years of schooling (years)	12	*13*	*13*	*12*
Physicians density (per 1,000 people)	0.9	—	1.7	1.4
Depositors with commercial banks (per 1,000 adults)	—	—	—	*443*
Sector structure				
Number of mobile operators	—	3		
Herfindahl-Hirschman Index (HHI) (scale = 0–10,000)	—	—		
Sector performance				
Access				
Mobile cellular subscriptions (per 100 people)	12	103[a]	92[a]	89[a]
Mobile cellular subscriptions (% prepaid)	3	64[a]	81[a]	87[a]
Population covered by a mobile-cellular network (%)	75	—	*99*	—
Mobile broadband subscriptions (per 100 people)	—	—	14.3[a]	—
Mobile broadband (% of total mobile subscriptions)	—	—	15.4[a]	—
Usage				
Households with a mobile telephone (%)	—	*71*	84	—
Mobile voice usage (minutes per user per month)	—	—	325[a]	—
Population using mobile Internet (%)	—	3.0	22.9[a]	4.5
Short Message Service (SMS) users (% of mobile users)	—	*56.0*	74.4[a]	—
Affordability				
Mobile basket (% of GNI per capita)	—	*0.9*	2.9	3.6

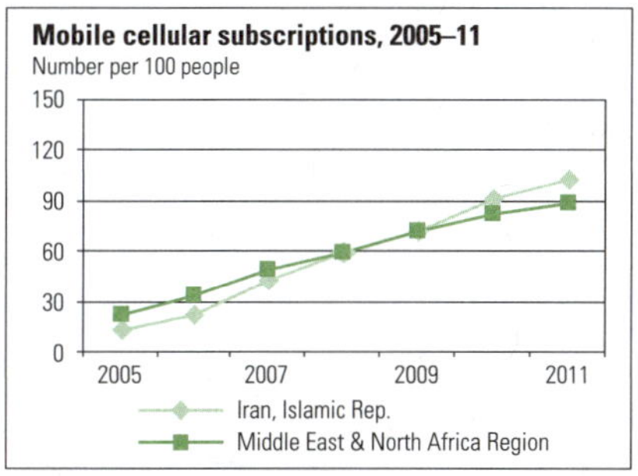

Mobile cellular subscriptions, 2005–11
Number per 100 people

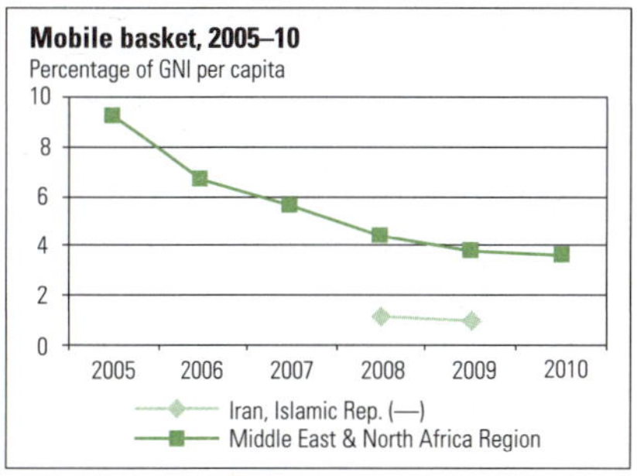

Mobile basket, 2005–10
Percentage of GNI per capita

Sources: Economic and social context: IMF, UIS, UN, WHO and World Bank; Sector structure: ictDATA.org; Sector performance: ictDATA.org, ITU; Wireless Intelligence, and World Bank.
Notes: Use of italics in the column entries indicates years or periods other than those specified. — Not available. GNI = gross national income.
a. Data are for 2011.

Iraq

	Iraq 2005	Iraq 2010	Lower-middle-income group 2010	Middle East & North Africa Region 2010
Economic and social context				
Population (total, million)	28	32	2,519	331
GNI per capita, World Bank Atlas method (current US$)	*1,150*	2,340	1,623	3,874
Rural population (% of total)	33	34	61	42
Expected years of schooling (years)	*10*	—	*10*	*12*
Physicians density (per 1,000 people)	0.7	*0.7*	0.8	1.4
Depositors with commercial banks (per 1,000 adults)	—	—	—	*443*
Sector structure				
Number of mobile operators	—	3		
Herfindahl-Hirschman Index (HHI) (scale = 0–10,000)	—	4,518		
Sector performance				
Access				
Mobile cellular subscriptions (per 100 people)	6	71[a]	78[a]	89[a]
Mobile cellular subscriptions (% prepaid)	97	100[a]	96[a]	87[a]
Population covered by a mobile-cellular network (%)	68	—	*86*	—
Mobile broadband subscriptions (per 100 people)	—	—	7.3[a]	—
Mobile broadband (% of total mobile subscriptions)	—	—	9.0[a]	—
Usage				
Households with a mobile telephone (%)	—	*94*	77	—
Mobile voice usage (minutes per user per month)	—	—	276[a]	—
Population using mobile Internet (%)	—	5.0	2.9	4.5
Short Message Service (SMS) users (% of mobile users)	—	62.0	61.9[a]	—
Affordability				
Mobile basket (% of GNI per capita)	—	—	7.2	3.6

Mobile cellular subscriptions, 2005–11
Number per 100 people

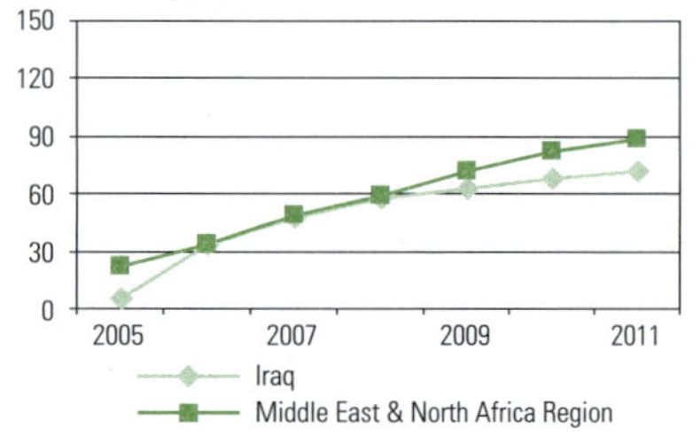

Mobile basket, 2005–10
Percentage of GNI per capita

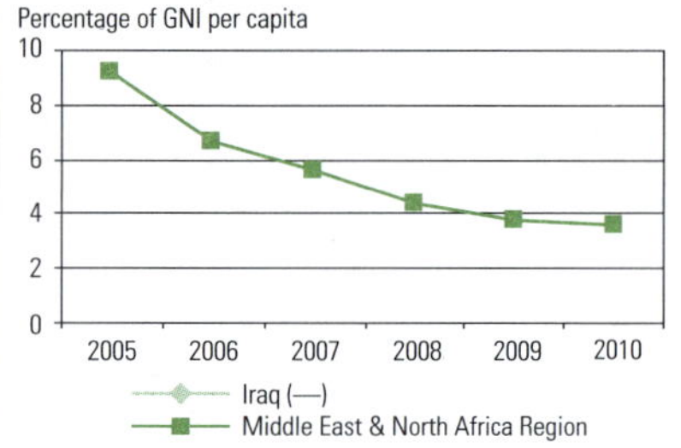

Ireland

	Ireland 2005	Ireland 2010	High-income group 2010
Economic and social context			
Population (total, million)	4	4	1,127
GNI per capita, World Bank Atlas method (current US$)	42,380	41,820	38,746
Rural population (% of total)	40	38	22
Expected years of schooling (years)	18	*18*	*16*
Physicians density (per 1,000 people)	*2.9*	3.2	2.8
Depositors with commercial banks (per 1,000 adults)	—	—	—
Sector structure			
Number of mobile operators	—	4	
Herfindahl-Hirschman Index (HHI) (scale = 0–10,000)	—	3,357	
Sector performance			
Access			
Mobile cellular subscriptions (per 100 people)	100	122[a]	118[a]
Mobile cellular subscriptions (% prepaid)	76	62[a]	36[a]
Population covered by a mobile-cellular network (%)	99	99	100
Mobile broadband subscriptions (per 100 people)	4.4	53.6[a]	69.6[a]
Mobile broadband (% of total mobile subscriptions)	4.3	45.2[a]	57.6[a]
Usage			
Households with a mobile telephone (%)	89	92	93
Mobile voice usage (minutes per user per month)	217	*245*	339
Population using mobile Internet (%)	0.7	17.7[a]	24.3
Short Message Service (SMS) users (% of mobile users)	—	*71.0*	78.2[a]
Affordability			
Mobile basket (% of GNI per capita)	1.1	1.1	1.0

Mobile cellular subscriptions, 2005–11
Number per 100 people

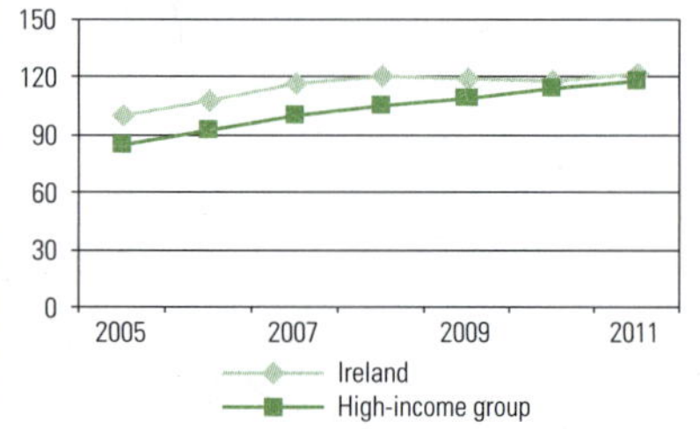

Mobile basket, 2005–10
Percentage of GNI per capita

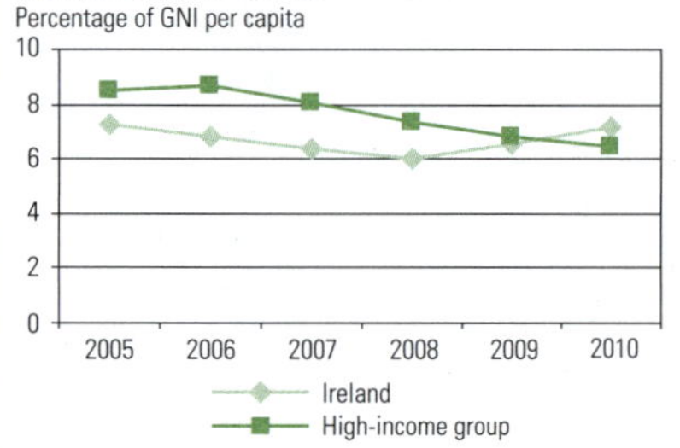

Sources: Economic and social context: IMF, UIS, UN, WHO and World Bank; Sector structure: ictDATA.org; Sector performance: ictDATA.org, ITU; Wireless Intelligence, and World Bank.
Notes: Use of italics in the column entries indicates years or periods other than those specified. — Not available. GNI = gross national income.
a. Data are for 2011.

Israel

	Israel 2005	Israel 2010	High-income group 2010
Economic and social context			
Population (total, million)	7	8	1,127
GNI per capita, World Bank Atlas method (current US$)	20,250	27,180	38,746
Rural population (% of total)	8	8	22
Expected years of schooling (years)	15	*16*	*16*
Physicians density (per 1,000 people)	*3.7*	3.7	2.8
Depositors with commercial banks (per 1,000 adults)	—	—	—
Sector structure			
Number of mobile operators	—	—	
Herfindahl-Hirschman Index (HHI) (scale = 0–10,000)	—	—	
Sector performance			
Access			
Mobile cellular subscriptions (per 100 people)	112	128[a]	118[a]
Mobile cellular subscriptions (% prepaid)	22	22[a]	36[a]
Population covered by a mobile-cellular network (%)	99	*100*	100
Mobile broadband subscriptions (per 100 people)	5.5	65.8[a]	69.6[a]
Mobile broadband (% of total mobile subscriptions)	4.9	51.9[a]	57.6[a]
Usage			
Households with a mobile telephone (%)	86	92	93
Mobile voice usage (minutes per user per month)	298	380[a]	339
Population using mobile Internet (%)	—	17.1	24.3
Short Message Service (SMS) users (% of mobile users)	—	73.0[a]	78.2[a]
Affordability			
Mobile basket (% of GNI per capita)	1.1	1.5	1.0

Mobile cellular subscriptions, 2005–11
Number per 100 people

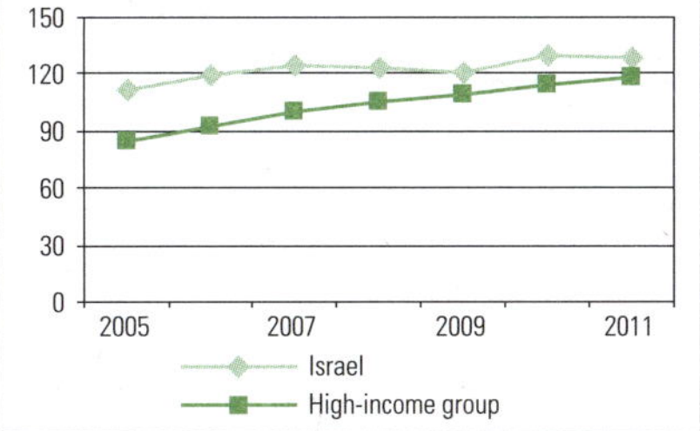

Mobile basket, 2005–10
Percentage of GNI per capita

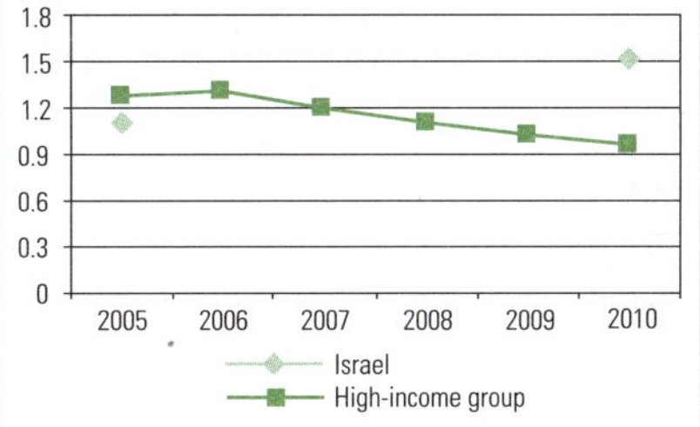

Italy

	Italy 2005	Italy 2010	High-income group 2010
Economic and social context			
Population (total, million)	59	60	1,127
GNI per capita, World Bank Atlas method (current US$)	30,880	35,700	38,746
Rural population (% of total)	32	32	22
Expected years of schooling (years)	16	*16*	*16*
Physicians density (per 1,000 people)	*3.7*	*3.5*	2.8
Depositors with commercial banks (per 1,000 adults)	749	1,307	—
Sector structure			
Number of mobile operators	—	4	
Herfindahl-Hirschman Index (HHI) (scale = 0–10,000)	—	3,011	
Sector performance			
Access			
Mobile cellular subscriptions (per 100 people)	122	153[a]	118[a]
Mobile cellular subscriptions (% prepaid)	90	82[a]	36[a]
Population covered by a mobile-cellular network (%)	100	100	100
Mobile broadband subscriptions (per 100 people)	17.6	69.5[a]	69.6[a]
Mobile broadband (% of total mobile subscriptions)	14.4	45.5[a]	57.6[a]
Usage			
Households with a mobile telephone (%)	88	93	93
Mobile voice usage (minutes per user per month)	117	161[a]	339
Population using mobile Internet (%)	1.7	15.9	24.3
Short Message Service (SMS) users (% of mobile users)	—	77.7	78.2[a]
Affordability			
Mobile basket (% of GNI per capita)	1.1	1.0	1.0

Mobile cellular subscriptions, 2005–11
Number per 100 people

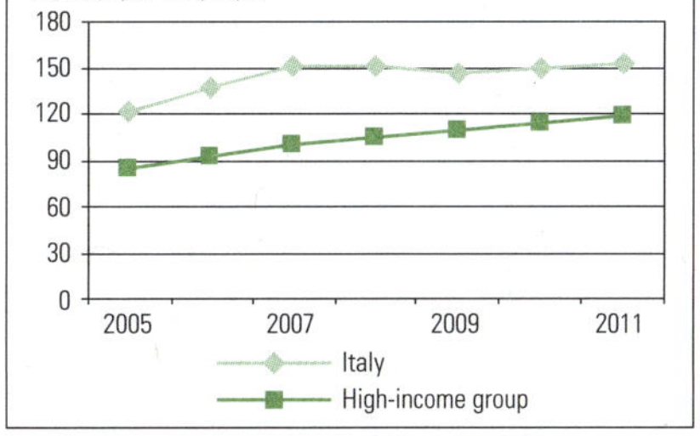

Mobile basket, 2005–10
Percentage of GNI per capita

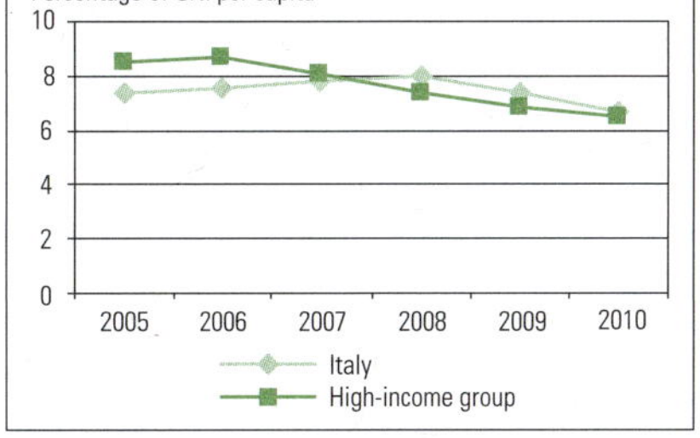

Sources: Economic and social context: IMF, UIS, UN, WHO and World Bank; Sector structure: ictDATA.org; Sector performance: ictDATA.org, ITU; Wireless Intelligence, and World Bank.
Notes: Use of italics in the column entries indicates years or periods other than those specified. — Not available. GNI = gross national income.
a. Data are for 2011.

Jamaica

	Jamaica 2005	Jamaica 2010	Upper-middle-income group 2010	Latin America & the Caribbean Region 2010
Economic and social context				
Population (total, million)	3	3	2,452	583
GNI per capita, World Bank Atlas method (current US$)	3,910	4,800	5,886	7,741
Rural population (% of total)	47	46	43	21
Expected years of schooling (years)	*12*	*14*	*13*	*14*
Physicians density (per 1,000 people)	*0.9*	—	1.7	1.8
Depositors with commercial banks (per 1,000 adults)	—	—	—	—
Sector structure				
Number of mobile operators	—	3		
Herfindahl-Hirschman Index (HHI) (scale = 0–10,000)	—	5,042		
Sector performance				
Access				
Mobile cellular subscriptions (per 100 people)	75	110	92[a]	109[a]
Mobile cellular subscriptions (% prepaid)	96	97[a]	81[a]	81[a]
Population covered by a mobile-cellular network (%)	95	—	*99*	98
Mobile broadband subscriptions (per 100 people)	*0.06*	13.8[a]	14.3[a]	16.1[a]
Mobile broadband (% of total mobile subscriptions)	*0.07*	10.2[a]	15.4[a]	15.2[a]
Usage				
Households with a mobile telephone (%)	*69*	92	84	84
Mobile voice usage (minutes per user per month)	—	—	25[a]	141[a]
Population using mobile Internet (%)	—	8.5	22.9[a]	4.4
Short Message Service (SMS) users (% of mobile users)	—	—	74.4[a]	—
Affordability				
Mobile basket (% of GNI per capita)	4.1	3.0	2.9	3.7

Mobile cellular subscriptions, 2005–11
Number per 100 people

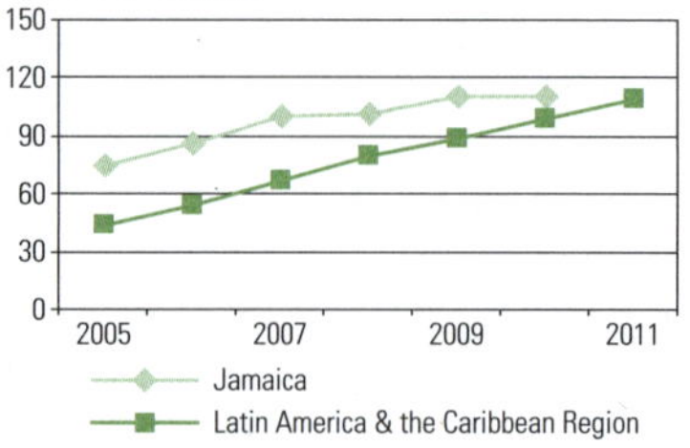

Mobile basket, 2005–10
Percentage of GNI per capita

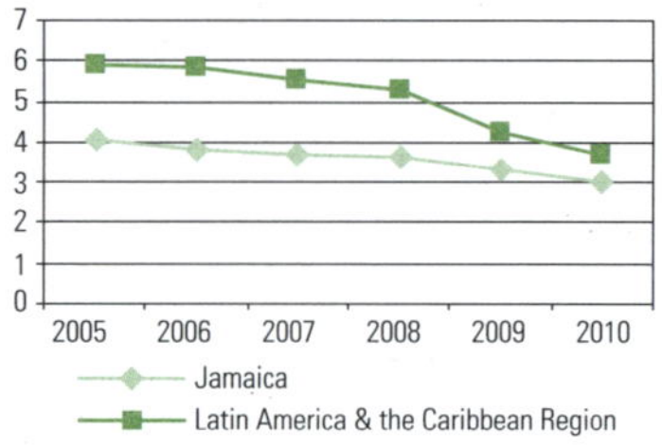

Japan

	Japan 2005	Japan 2010	High-income group 2010
Economic and social context			
Population (total, million)	128	127	1,127
GNI per capita, World Bank Atlas method (current US$)	38,950	41,850	38,746
Rural population (% of total)	34	33	22
Expected years of schooling (years)	15	*15*	*16*
Physicians density (per 1,000 people)	*2.1*	*2.1*	2.8
Depositors with commercial banks (per 1,000 adults)	7,827	7,169	—
Sector structure			
Number of mobile operators	—	4	
Herfindahl-Hirschman Index (HHI) (scale = 0–10,000)	—	3,601	
Sector performance			
Access			
Mobile cellular subscriptions (per 100 people)	76	95[a]	118[a]
Mobile cellular subscriptions (% prepaid)	3	1[a]	36[a]
Population covered by a mobile-cellular network (%)	99	100	100
Mobile broadband subscriptions (per 100 people)	22.8	98.1[a]	69.6[a]
Mobile broadband (% of total mobile subscriptions)	31.0	95.3[a]	57.6[a]
Usage			
Households with a mobile telephone (%)	90	93	93
Mobile voice usage (minutes per user per month)	149	134[a]	339
Population using mobile Internet (%)	54.2	61.8	24.3
Short Message Service (SMS) users (% of mobile users)	—	81.0[a]	78.2[a]
Affordability			
Mobile basket (% of GNI per capita)	1.3	1.6	1.0

Mobile cellular subscriptions, 2005–11
Number per 100 people

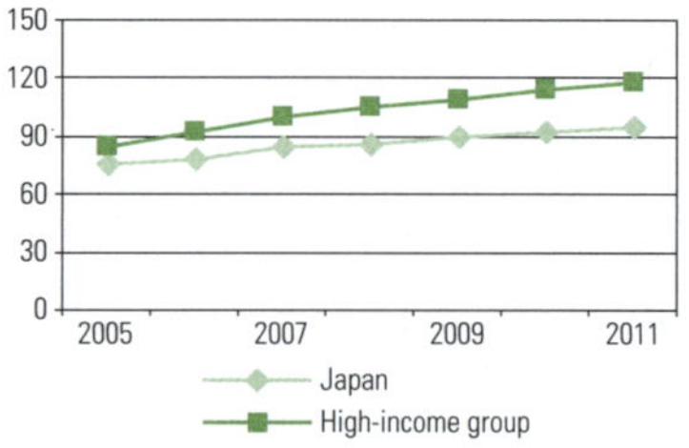

Mobile basket, 2005–10
Percentage of GNI per capita

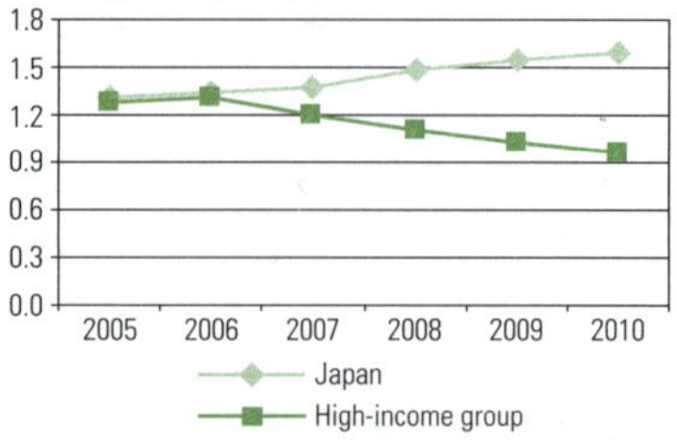

Sources: Economic and social context: IMF, UIS, UN, WHO and World Bank; Sector structure: ictDATA.org; Sector performance: ictDATA.org, ITU; Wireless Intelligence, and World Bank.
Notes: Use of italics in the column entries indicates years or periods other than those specified. — Not available. GNI = gross national income.
a. Data are for 2011.

Jordan

	Jordan 2005	Jordan 2010	Upper-middle-income group 2010	Middle East & North Africa Region 2010
Economic and social context				
Population (total, million)	5	6	2,452	331
GNI per capita, World Bank Atlas method (current US$)	2,490	4,340	5,886	3,874
Rural population (% of total)	22	22	43	42
Expected years of schooling (years)	13	*13*	*13*	*12*
Physicians density (per 1,000 people)	2.4	*2.5*	1.7	1.4
Depositors with commercial banks (per 1,000 adults)	—	—	—	*443*
Sector structure				
Number of mobile operators	—	3		
Herfindahl-Hirschman Index (HHI) (scale = 0–10,000)	—	3,402		
Sector performance				
Access				
Mobile cellular subscriptions (per 100 people)	58	121[a]	92[a]	89[a]
Mobile cellular subscriptions (% prepaid)	87	91[a]	81[a]	87[a]
Population covered by a mobile-cellular network (%)	99	99	*99*	—
Mobile broadband subscriptions (per 100 people)	—	26.0[a]	14.3[a]	—
Mobile broadband (% of total mobile subscriptions)	—	20.8[a]	15.4[a]	—
Usage				
Households with a mobile telephone (%)	51	98	84	—
Mobile voice usage (minutes per user per month)	—	—	325[a]	—
Population using mobile Internet (%)	—	13.2	22.9[a]	4.5
Short Message Service (SMS) users (% of mobile users)	—	63.0[a]	74.4[a]	—
Affordability				
Mobile basket (% of GNI per capita)	7.6	2.9	2.9	3.6

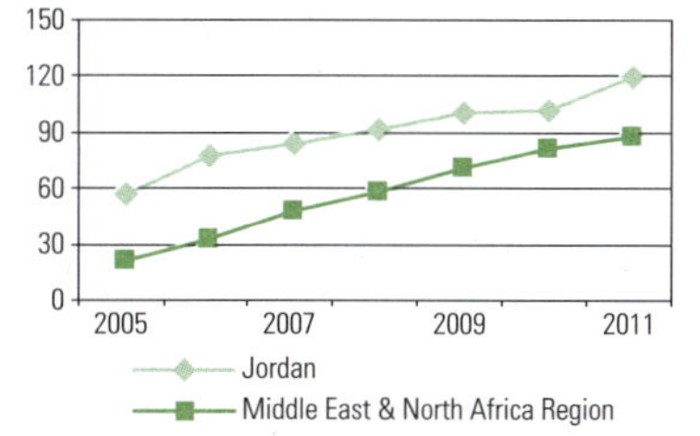

Mobile cellular subscriptions, 2005–11
Number per 100 people

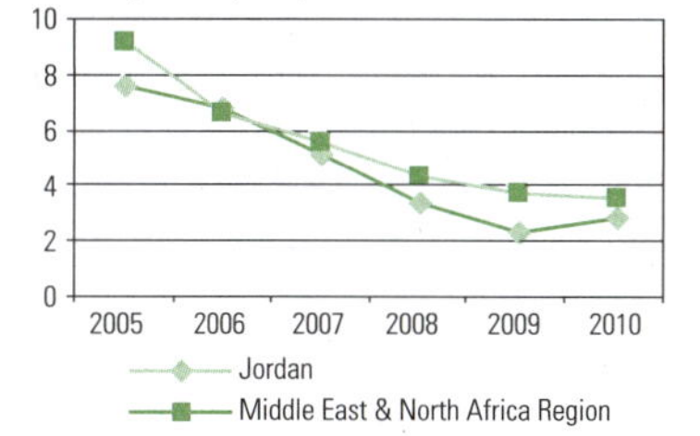

Mobile basket, 2005–10
Percentage of GNI per capita

Kazakhstan

	Kazakhstan 2005	Kazakhstan 2010	Upper-middle-income group 2010	Europe & Central Asia Region 2010
Economic and social context				
Population (total, million)	15	16	2,452	405
GNI per capita, World Bank Atlas method (current US$)	2,930	7,580	5,886	7,272
Rural population (% of total)	43	42	43	36
Expected years of schooling (years)	15	15	*13*	*13*
Physicians density (per 1,000 people)	3.7	*4.1*	1.7	3.2
Depositors with commercial banks (per 1,000 adults)	831	874	—	894
Sector structure				
Number of mobile operators	—	4		
Herfindahl-Hirschman Index (HHI) (scale = 0–10,000)	—	4,236		
Sector performance				
Access				
Mobile cellular subscriptions (per 100 people)	36	137[a]	92[a]	125[a]
Mobile cellular subscriptions (% prepaid)	80	91[a]	81[a]	82[a]
Population covered by a mobile-cellular network (%)	*51*	95	*99*	96
Mobile broadband subscriptions (per 100 people)	—	3.4[a]	14.3[a]	22.6[a]
Mobile broadband (% of total mobile subscriptions)	—	2.7[a]	15.4[a]	18.0[a]
Usage				
Households with a mobile telephone (%)	27	81	84	82
Mobile voice usage (minutes per user per month)	59	157[a]	325[a]	288[a]
Population using mobile Internet (%)	—	4.8	22.9[a]	8.5
Short Message Service (SMS) users (% of mobile users)	—	*50.0*	74.4[a]	69.8[a]
Affordability				
Mobile basket (% of GNI per capita)	8.4	2.3	2.9	3.1

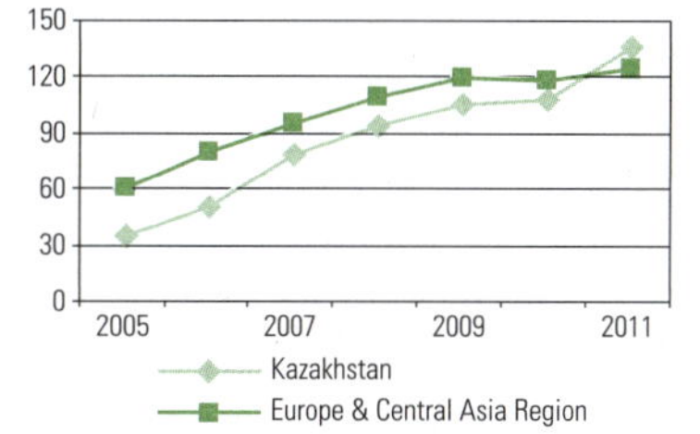

Mobile cellular subscriptions, 2005–11
Number per 100 people

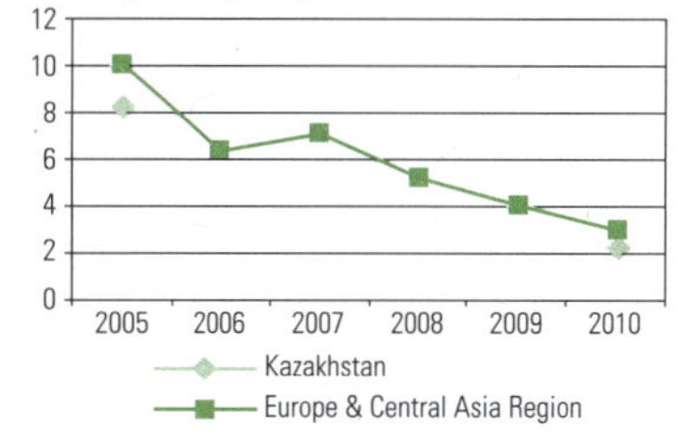

Mobile basket, 2005–10
Percentage of GNI per capita

Sources: Economic and social context: IMF, UIS, UN, WHO and World Bank; Sector structure: ictDATA.org; Sector performance: ictDATA.org, ITU; Wireless Intelligence, and World Bank.
Notes: Use of italics in the column entries indicates years or periods other than those specified. — Not available. GNI = gross national income.
a. Data are for 2011.

Kenya

	Kenya 2005	Kenya 2010	Low-income group 2010	Sub-Saharan Africa Region 2010
Economic and social context				
Population (total, million)	36	41	796	853
GNI per capita, World Bank Atlas method (current US$)	520	810	530	1,188
Rural population (% of total)	79	78	72	63
Expected years of schooling (years)	10	*11*	*9*	*9*
Physicians density (per 1,000 people)	*0.1*	—	0.2	0.2
Depositors with commercial banks (per 1,000 adults)	115	*370*	—	*167*
Sector structure				
Number of mobile operators	—	4		
Herfindahl-Hirschman Index (HHI) (scale = 0–10,000)	—	5,229		
Sector performance				
Access				
Mobile cellular subscriptions (per 100 people)	15	67[a]	43[a]	57[a]
Mobile cellular subscriptions (% prepaid)	98	99[a]	98[a]	96[a]
Population covered by a mobile-cellular network (%)	62	89	—	72
Mobile broadband subscriptions (per 100 people)	—	6.9[a]	—	5.6[a]
Mobile broadband (% of total mobile subscriptions)	—	10.3[a]	—	10.1[a]
Usage				
Households with a mobile telephone (%)	21	65	43	52
Mobile voice usage (minutes per user per month)	—	82	—	—
Population using mobile Internet (%)	0.03	11.3[a]	—	—
Short Message Service (SMS) users (% of mobile users)	—	89.0[a]	—	—
Affordability				
Mobile basket (% of GNI per capita)	48.6	16.0	28.8	19.5

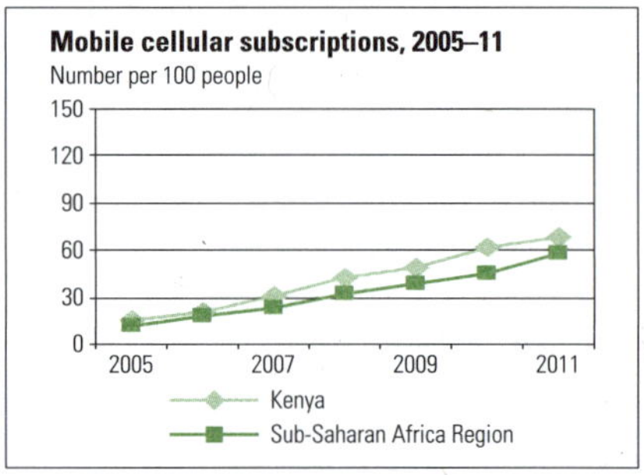

Mobile cellular subscriptions, 2005–11

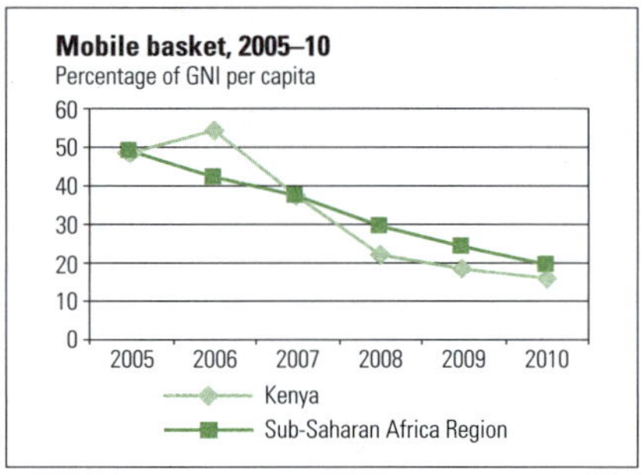

Mobile basket, 2005–10

Korea, Rep.

	Korea, Rep. 2005	Korea, Rep. 2010	High-income group 2010
Economic and social context			
Population (total, million)	48	49	1,127
GNI per capita, World Bank Atlas method (current US$)	16,900	19,890	38,746
Rural population (% of total)	19	18	22
Expected years of schooling (years)	16	*17*	*16*
Physicians density (per 1,000 people)	*1.7*	2.0	2.8
Depositors with commercial banks (per 1,000 adults)	3,997	4,522	—
Sector structure			
Number of mobile operators	—	3	
Herfindahl-Hirschman Index (HHI) (scale = 0–10,000)	—	3,876	
Sector performance			
Access			
Mobile cellular subscriptions (per 100 people)	80	109[a]	118[a]
Mobile cellular subscriptions (% prepaid)	1.4	1.9[a]	36[a]
Population covered by a mobile-cellular network (%)	99	100	100
Mobile broadband subscriptions (per 100 people)	26.0	97.9[a]	69.6[a]
Mobile broadband (% of total mobile subscriptions)	32.4	89.3[a]	57.6[a]
Usage			
Households with a mobile telephone (%)	88	97	93
Mobile voice usage (minutes per user per month)	274	*312*	339
Population using mobile Internet (%)	27.0	40.8	24.3
Short Message Service (SMS) users (% of mobile users)	—	99.8	78.2[a]
Affordability			
Mobile basket (% of GNI per capita)	*1.2*	0.9	1.0

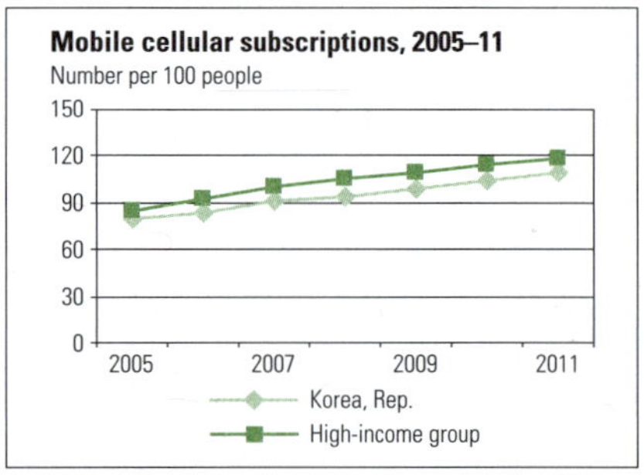

Mobile cellular subscriptions, 2005–11

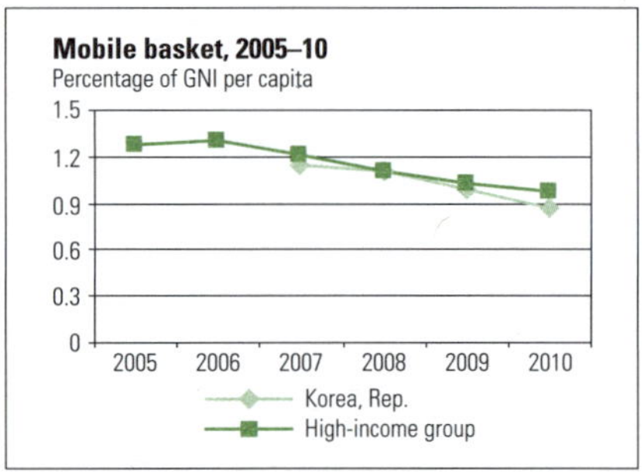

Mobile basket, 2005–10

Sources: Economic and social context: IMF, UIS, UN, WHO and World Bank; Sector structure: ictDATA.org; Sector performance: ictDATA.org, ITU; Wireless Intelligence, and World Bank.
Notes: Use of italics in the column entries indicates years or periods other than those specified. — Not available. GNI = gross national income.
a. Data are for 2011.

Kuwait

	Kuwait 2005	Kuwait 2010	High-income group 2010
Economic and social context			
Population (total, million)	2	3	1,127
GNI per capita, World Bank Atlas method (current US$)	34,650	—	38,746
Rural population (% of total)	2	2	22
Expected years of schooling (years)	*12*	—	*16*
Physicians density (per 1,000 people)	1.8	*1.8*	2.8
Depositors with commercial banks (per 1,000 adults)	—	—	—
Sector structure			
Number of mobile operators	—	3	
Herfindahl-Hirschman Index (HHI) (scale = 0–10,000)	—	3,746	
Sector performance			
Access			
Mobile cellular subscriptions (per 100 people)	101	183[a]	118[a]
Mobile cellular subscriptions (% prepaid)	80	75[a]	36[a]
Population covered by a mobile-cellular network (%)	100	100	100
Mobile broadband subscriptions (per 100 people)	1.1	94.4[a]	69.6[a]
Mobile broadband (% of total mobile subscriptions)	1.0	52.7[a]	57.6[a]
Usage			
Households with a mobile telephone (%)	—	—	93
Mobile voice usage (minutes per user per month)	—	—	339
Population using mobile Internet (%)	—	—	24.3
Short Message Service (SMS) users (% of mobile users)	—	95.0	78.2[a]
Affordability			
Mobile basket (% of GNI per capita)	—	—	1.0

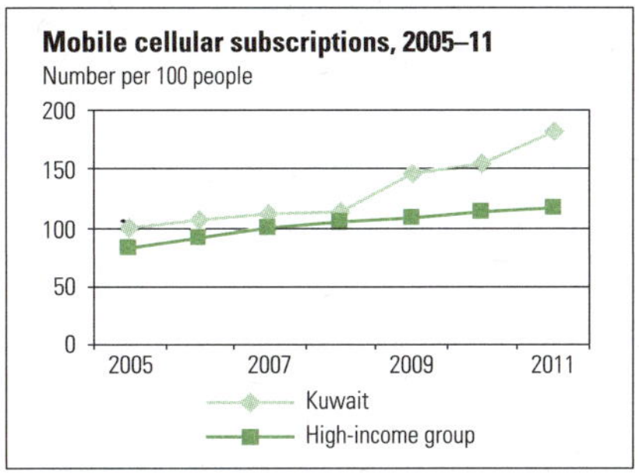

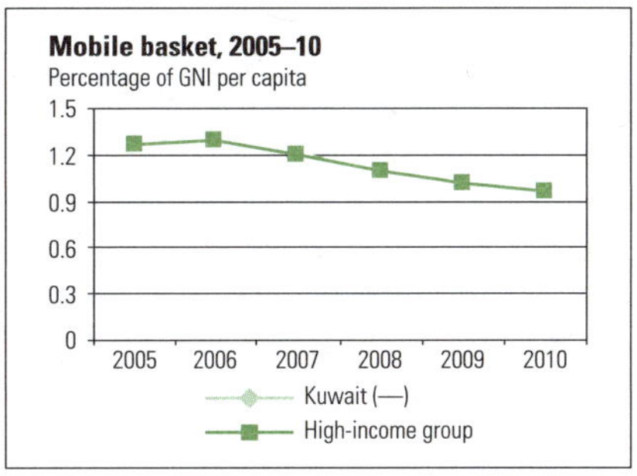

Kyrgyz Republic

	Kyrgyz Republic 2005	Kyrgyz Republic 2010	Low-income group 2010	Europe & Central Asia Region 2010
Economic and social context				
Population (total, million)	5	5	796	405
GNI per capita, World Bank Atlas method (current US$)	450	830	530	7,272
Rural population (% of total)	64	63	72	36
Expected years of schooling (years)	12	*12*	*9*	*13*
Physicians density (per 1,000 people)	*2.4*	—	0.2	3.2
Depositors with commercial banks (per 1,000 adults)	—	181	—	894
Sector structure				
Number of mobile operators	—	4		
Herfindahl-Hirschman Index (HHI) (scale = 0–10,000)	—	4,253		
Sector performance				
Access				
Mobile cellular subscriptions (per 100 people)	10	90	43[a]	125[a]
Mobile cellular subscriptions (% prepaid)	91	94[a]	98[a]	82[a]
Population covered by a mobile-cellular network (%)	—	96	—	96
Mobile broadband subscriptions (per 100 people)	—	3.5[a]	—	22.6[a]
Mobile broadband (% of total mobile subscriptions)	—	3.1[a]	—	18.0[a]
Usage				
Households with a mobile telephone (%)	10	88	43	82
Mobile voice usage (minutes per user per month)	—	—	—	288[a]
Population using mobile Internet (%)	—	2.6	—	8.5
Short Message Service (SMS) users (% of mobile users)	—	*48.0*	—	69.8[a]
Affordability				
Mobile basket (% of GNI per capita)	24.1	5.2	28.8	3.1

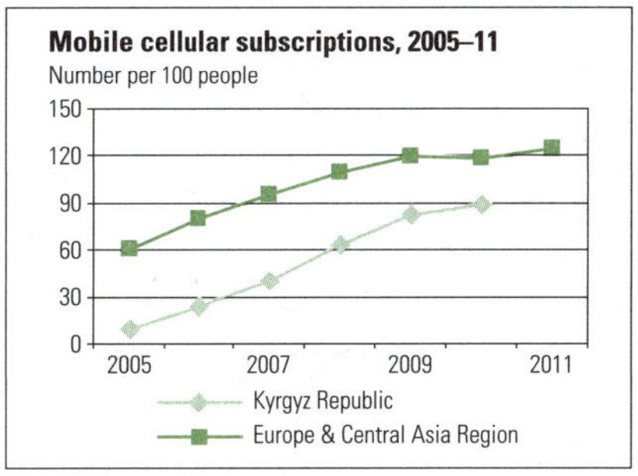

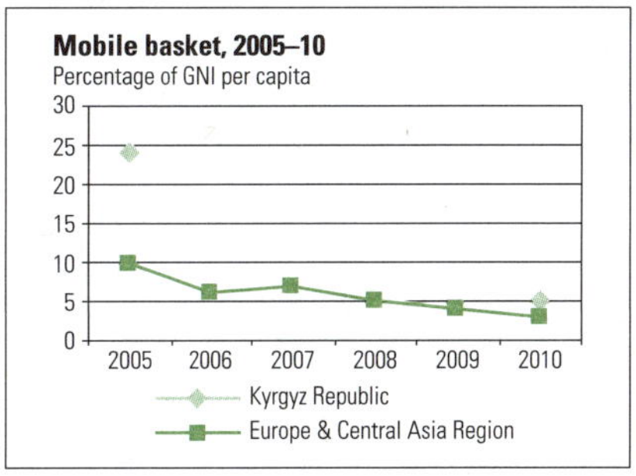

Sources: Economic and social context: IMF, UIS, UN, WHO and World Bank; Sector structure: ictDATA.org; Sector performance: ictDATA.org, ITU; Wireless Intelligence, and World Bank.
Notes: Use of italics in the column entries indicates years or periods other than those specified. — Not available. GNI = gross national income.
a. Data are for 2011.

Lao People's Democratic Republic

	Lao People's Democratic Republic 2005	Lao People's Democratic Republic 2010	Lower-middle-income group 2010	East Asia & Pacific Region 2010
Economic and social context				
Population (total, million)	6	6	2,519	1,962
GNI per capita, World Bank Atlas method (current US$)	470	1,040	1,623	3,696
Rural population (% of total)	73	67	61	54
Expected years of schooling (years)	9	*9*	*10*	*12*
Physicians density (per 1,000 people)	0.3	—	0.8	1.2
Depositors with commercial banks (per 1,000 adults)	—	44	—	—
Sector structure				
Number of mobile operators	—	4		
Herfindahl-Hirschman Index (HHI) (scale = 0–10,000)	—	3,670		
Sector performance				
Access				
Mobile cellular subscriptions (per 100 people)	11	53	78[a]	83[a]
Mobile cellular subscriptions (% prepaid)	98	99[a]	96[a]	85[a]
Population covered by a mobile-cellular network (%)	55	*80*	*86*	*99*
Mobile broadband subscriptions (per 100 people)	—	3.3[a]	7.3[a]	11.6[a]
Mobile broadband (% of total mobile subscriptions)	—	6.8[a]	9.0[a]	14.4[a]
Usage				
Households with a mobile telephone (%)	—	—	77	83
Mobile voice usage (minutes per user per month)	—	—	276[a]	367[a]
Population using mobile Internet (%)	—	—	2.9	22.4[a]
Short Message Service (SMS) users (% of mobile users)	—	—	61.9[a]	84.0[a]
Affordability				
Mobile basket (% of GNI per capita)	—	7.3	7.2	5.7

Mobile cellular subscriptions, 2005–11
Number per 100 people

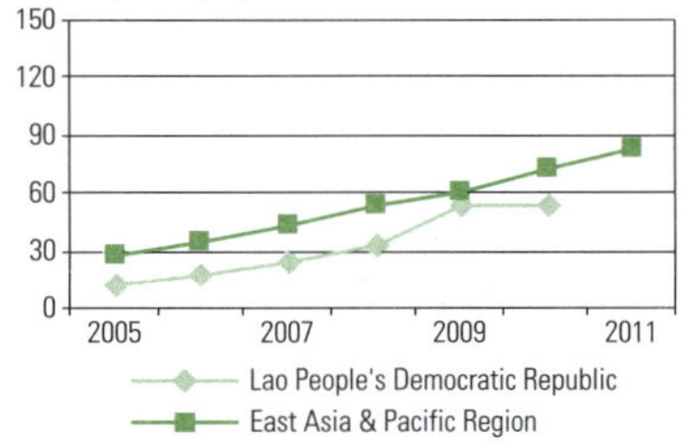

Mobile basket, 2005–10
Percentage of GNI per capita

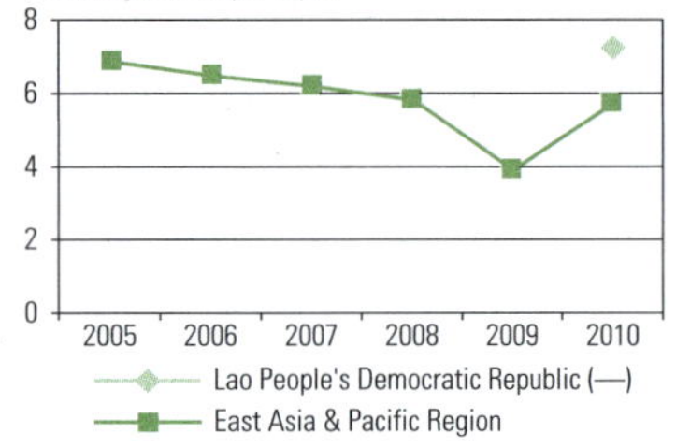

Latvia

	Latvia 2005	Latvia 2010	Upper-middle-income group 2010	Europe & Central Asia Region 2010
Economic and social context				
Population (total, million)	2	2	2,452	405
GNI per capita, World Bank Atlas method (current US$)	6,810	11,640	5,886	7,272
Rural population (% of total)	32	32	43	36
Expected years of schooling (years)	16	*15*	*13*	*13*
Physicians density (per 1,000 people)	*3.1*	*3.0*	1.7	3.2
Depositors with commercial banks (per 1,000 adults)	932	1,286	—	894
Sector structure				
Number of mobile operators	—	3		
Herfindahl-Hirschman Index (HHI) (scale = 0–10,000)	—	3,891		
Sector performance				
Access				
Mobile cellular subscriptions (per 100 people)	81	114[a]	92[a]	125[a]
Mobile cellular subscriptions (% prepaid)	54	51[a]	81[a]	82[a]
Population covered by a mobile-cellular network (%)	98	—	*99*	96
Mobile broadband subscriptions (per 100 people)	0.1	35.3[a]	14.3[a]	22.6[a]
Mobile broadband (% of total mobile subscriptions)	0.2	30.7[a]	15.4[a]	18.0[a]
Usage				
Households with a mobile telephone (%)	75	95	84	82
Mobile voice usage (minutes per user per month)	—	248[a]	325[a]	288[a]
Population using mobile Internet (%)	—	10.3	22.9[a]	8.5
Short Message Service (SMS) users (% of mobile users)	—	—	74.4[a]	69.8[a]
Affordability				
Mobile basket (% of GNI per capita)	4.4	1.0	2.9	3.1

Mobile cellular subscriptions, 2005–11
Number per 100 people

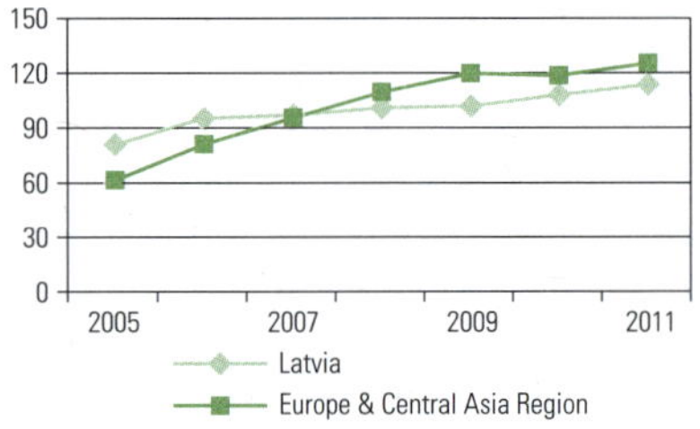

Mobile basket, 2005–10
Percentage of GNI per capita

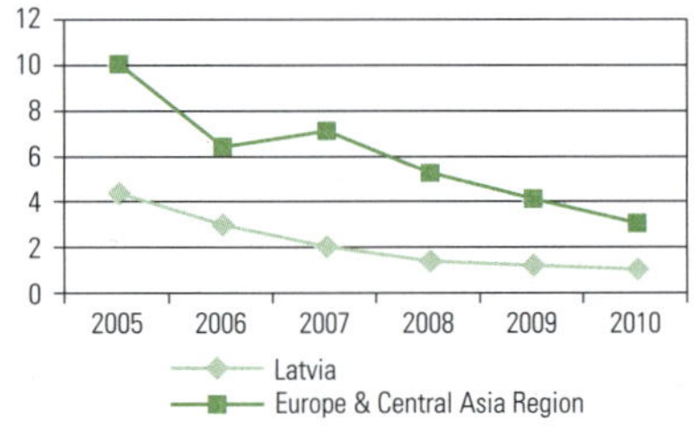

Sources: Economic and social context: IMF, UIS, UN, WHO and World Bank; Sector structure: ictDATA.org; Sector performance: ictDATA.org, ITU; Wireless Intelligence, and World Bank.
Notes: Use of italics in the column entries indicates years or periods other than those specified. — Not available. GNI = gross national income.
a. Data are for 2011.

Lebanon

	Lebanon 2005	Lebanon 2010	Upper-middle-income group 2010	Middle East & North Africa Region 2010
Economic and social context				
Population (total, million)	4	4	2,452	331
GNI per capita, World Bank Atlas method (current US$)	5,710	8,880	5,886	3,874
Rural population (% of total)	13	13	43	42
Expected years of schooling (years)	*13*	*14*	*13*	*12*
Physicians density (per 1,000 people)	2.4	*3.5*	1.7	1.4
Depositors with commercial banks (per 1,000 adults)	742	*873*	—	*443*
Sector structure				
Number of mobile operators	—	2		
Herfindahl-Hirschman Index (HHI) (scale = 0–10,000)	—	5,015		
Sector performance				
Access				
Mobile cellular subscriptions (per 100 people)	25	68	92[a]	89[a]
Mobile cellular subscriptions (% prepaid)	83	84[a]	81[a]	87[a]
Population covered by a mobile-cellular network (%)	100	*95*	*99*	—
Mobile broadband subscriptions (per 100 people)	—	0.8[a]	14.3[a]	—
Mobile broadband (% of total mobile subscriptions)	—	1.0[a]	15.4[a]	—
Usage				
Households with a mobile telephone (%)	50	80	84	—
Mobile voice usage (minutes per user per month)	—	—	325[a]	—
Population using mobile Internet (%)	—	9.4[a]	22.9[a]	4.5
Short Message Service (SMS) users (% of mobile users)	—	87.0[a]	74.4[a]	—
Affordability				
Mobile basket (% of GNI per capita)	6.9	3.7	2.9	3.6

Mobile cellular subscriptions, 2005–11
Number per 100 people

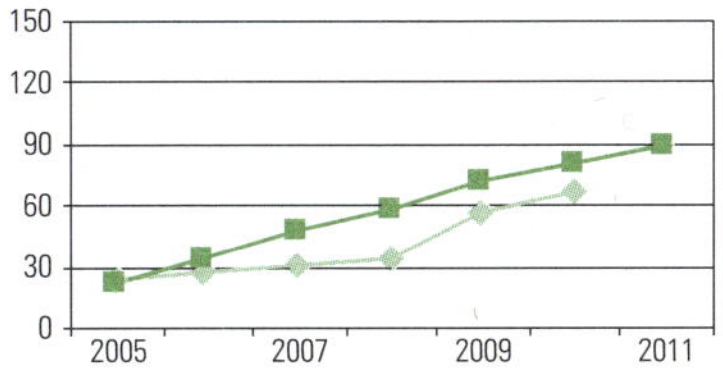

Mobile basket, 2005–10
Percentage of GNI per capita

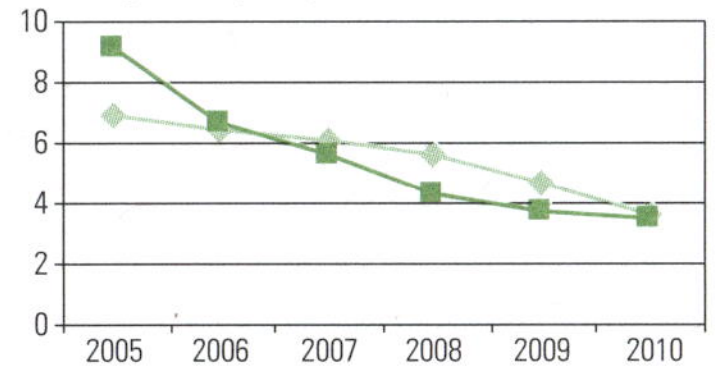

Lesotho

	Lesotho 2005	Lesotho 2010	Lower-middle-income group 2010	Sub-Saharan Africa Region 2010
Economic and social context				
Population (total, million)	2	2	2,519	853
GNI per capita, World Bank Atlas method (current US$)	760	1,090	1,623	1,188
Rural population (% of total)	77	73	61	63
Expected years of schooling (years)	10	—	*10*	*9*
Physicians density (per 1,000 people)	*0.1*	—	0.8	0.2
Depositors with commercial banks (per 1,000 adults)	271	291	—	*167*
Sector structure				
Number of mobile operators	—	2		
Herfindahl-Hirschman Index (HHI) (scale = 0–10,000)	—	6,800		
Sector performance				
Access				
Mobile cellular subscriptions (per 100 people)	12	44	78[a]	57[a]
Mobile cellular subscriptions (% prepaid)	99	99[a]	96[a]	96[a]
Population covered by a mobile-cellular network (%)	29	—	*86*	72
Mobile broadband subscriptions (per 100 people)	—	6.1[a]	7.3[a]	5.6[a]
Mobile broadband (% of total mobile subscriptions)	—	11.1[a]	9.0[a]	10.1[a]
Usage				
Households with a mobile telephone (%)	—	—	77	52
Mobile voice usage (minutes per user per month)	—	29[a]	276[a]	—
Population using mobile Internet (%)	—	—	2.9	—
Short Message Service (SMS) users (% of mobile users)	—	—	61.9[a]	—
Affordability				
Mobile basket (% of GNI per capita)	—	26.5	7.2	19.5

Mobile cellular subscriptions, 2005–11
Number per 100 people

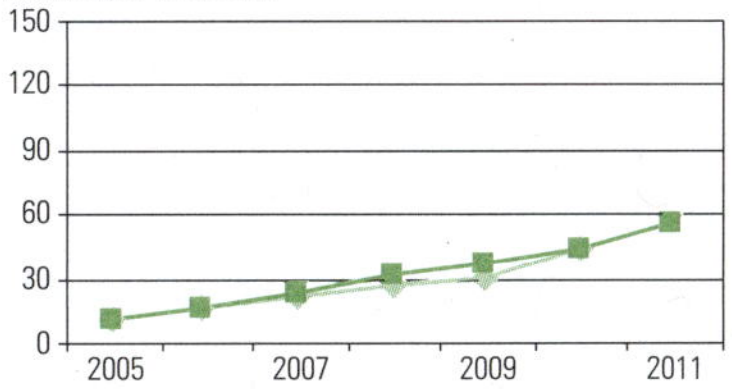

Mobile basket, 2005–10
Percentage of GNI per capita

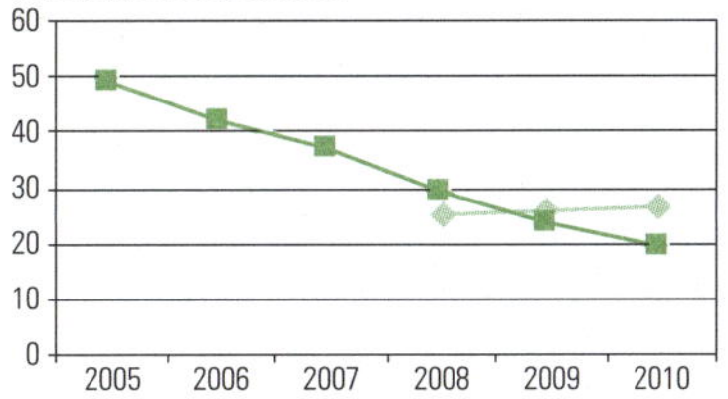

Sources: Economic and social context: IMF, UIS, UN, WHO and World Bank; Sector structure: ictDATA.org; Sector performance: ictDATA.org, ITU; Wireless Intelligence, and World Bank.
Notes: Use of italics in the column entries indicates years or periods other than those specified. — Not available. GNI = gross national income.
a. Data are for 2011.

Libya

	Libya 2005	Libya 2010	Upper-middle-income group 2010	Middle East & North Africa Region 2010
Economic and social context				
Population (total, million)	6	6	2,452	331
GNI per capita, World Bank Atlas method (current US$)	6,460	*12,320*	5,886	3,874
Rural population (% of total)	23	22	43	42
Expected years of schooling (years)	*17*	—	*13*	*12*
Physicians density (per 1,000 people)	*1.3*	*1.9*	1.7	1.4
Depositors with commercial banks (per 1,000 adults)	—	—	—	*443*
Sector structure				
Number of mobile operators	—	2		
Herfindahl-Hirschman Index (HHI) (scale = 0–10,000)	—	—		
Sector performance				
Access				
Mobile cellular subscriptions (per 100 people)	35	172	92[a]	89[a]
Mobile cellular subscriptions (% prepaid)	99	99[a]	81[a]	87[a]
Population covered by a mobile-cellular network (%)	*71*	*98*	*99*	—
Mobile broadband subscriptions (per 100 people)	—	49.8[a]	14.3[a]	—
Mobile broadband (% of total mobile subscriptions)	—	39.9[a]	15.4[a]	—
Usage				
Households with a mobile telephone (%)	—	—	84	—
Mobile voice usage (minutes per user per month)	—	—	325[a]	—
Population using mobile Internet (%)	—	—	22.9[a]	4.5
Short Message Service (SMS) users (% of mobile users)	—	—	74.4[a]	—
Affordability				
Mobile basket (% of GNI per capita)	—	—	2.9	3.6

Mobile cellular subscriptions, 2005–11
Number per 100 people

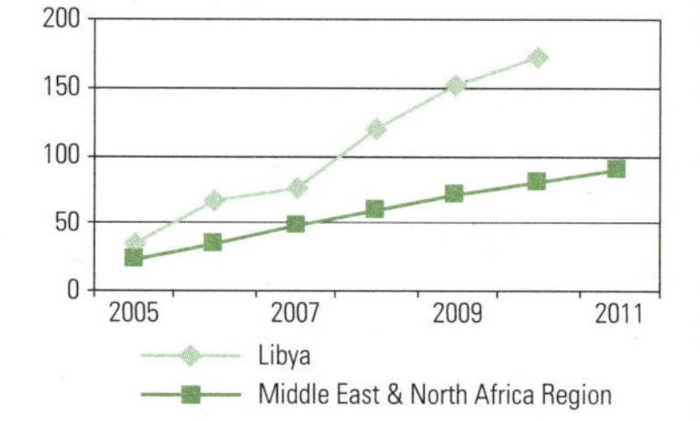

Mobile basket, 2005–10
Percentage of GNI per capita

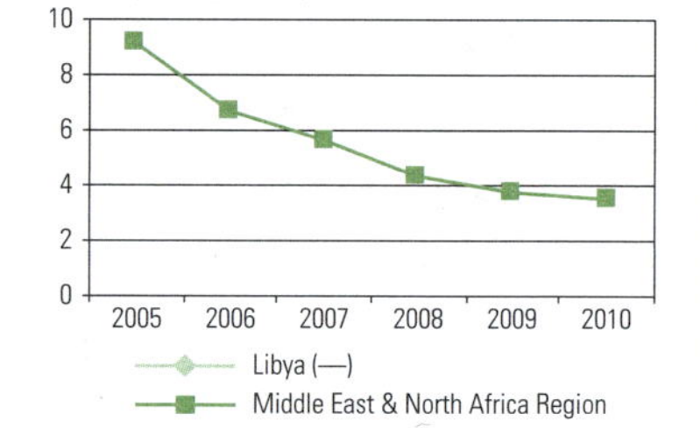

Lithuania

	Lithuania 2005	Lithuania 2010	Upper-middle-income group 2010	Europe & Central Asia Region 2010
Economic and social context				
Population (total, million)	3	3	2,452	405
GNI per capita, World Bank Atlas method (current US$)	7,280	11,510	5,886	7,272
Rural population (% of total)	33	33	43	36
Expected years of schooling (years)	16	*16*	*13*	*13*
Physicians density (per 1,000 people)	*4.0*	*3.6*	1.7	3.2
Depositors with commercial banks (per 1,000 adults)	—	—	—	894
Sector structure				
Number of mobile operators	—	3		
Herfindahl-Hirschman Index (HHI) (scale = 0–10,000)	—	3,396		
Sector performance				
Access				
Mobile cellular subscriptions (per 100 people)	128	151[a]	92[a]	125[a]
Mobile cellular subscriptions (% prepaid)	68	56[a]	81[a]	82[a]
Population covered by a mobile-cellular network (%)	100	100	*99*	96
Mobile broadband subscriptions (per 100 people)	*0.5*	35.3[a]	14.3[a]	22.6[a]
Mobile broadband (% of total mobile subscriptions)	*0.3*	24.1[a]	15.4[a]	18.0[a]
Usage				
Households with a mobile telephone (%)	73	92	84	82
Mobile voice usage (minutes per user per month)	77	174[a]	325[a]	288[a]
Population using mobile Internet (%)	3.2	15.3[a]	22.9[a]	8.5
Short Message Service (SMS) users (% of mobile users)	—	79.0[a]	74.4[a]	69.8[a]
Affordability				
Mobile basket (% of GNI per capita)	2.2	1.0	2.9	3.1

Mobile cellular subscriptions, 2005–11
Number per 100 people

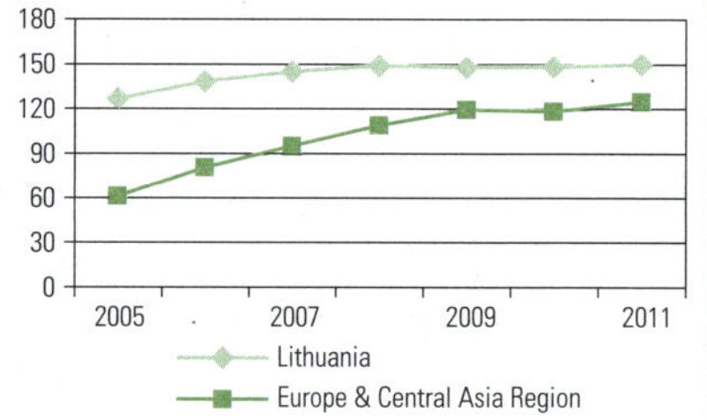

Mobile basket, 2005–10
Percentage of GNI per capita

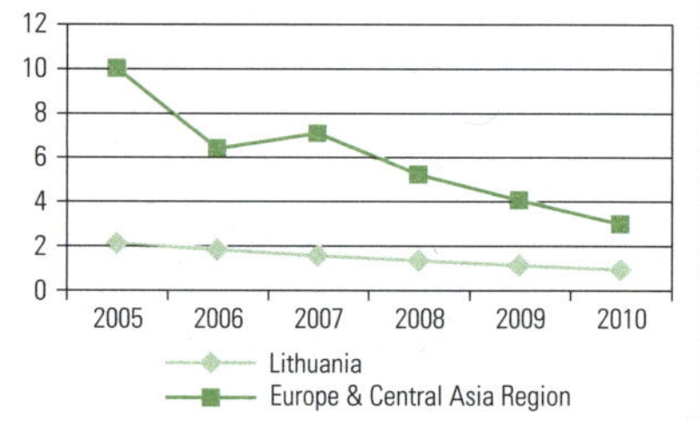

Sources: Economic and social context: IMF, UIS, UN, WHO and World Bank; Sector structure: ictDATA.org; Sector performance: ictDATA.org, ITU; Wireless Intelligence, and World Bank.
Notes: Use of italics in the column entries indicates years or periods other than those specified. — Not available. GNI = gross national income.
a. Data are for 2011.

Macedonia, Former Yugoslav Republic of

	Macedonia, Former Yugoslav Republic of 2005	Macedonia, Former Yugoslav Republic of 2010	Upper-middle-income group 2010	Europe & Central Asia Region 2010
Economic and social context				
Population (total, million)	2	2	2,452	405
GNI per capita, World Bank Atlas method (current US$)	2,900	4,570	5,886	7,272
Rural population (% of total)	35	32	43	36
Expected years of schooling (years)	12	*13*	*13*	*13*
Physicians density (per 1,000 people)	*2.5*	*2.6*	1.7	3.2
Depositors with commercial banks (per 1,000 adults)	—	—	—	894
Sector structure				
Number of mobile operators	—	3		
Herfindahl-Hirschman Index (HHI) (scale = 0–10,000)	—	4,315		
Sector performance				
Access				
Mobile cellular subscriptions (per 100 people)	55	110[a]	92[a]	125[a]
Mobile cellular subscriptions (% prepaid)	86	62[a]	81[a]	82[a]
Population covered by a mobile-cellular network (%)	99	100	*99*	96
Mobile broadband subscriptions (per 100 people)	—	19.0[a]	14.3[a]	22.6[a]
Mobile broadband (% of total mobile subscriptions)	—	16.8[a]	15.4[a]	18.0[a]
Usage				
Households with a mobile telephone (%)	*71*	85	84	82
Mobile voice usage (minutes per user per month)	—	142[a]	325[a]	288[a]
Population using mobile Internet (%)	—	2.9	22.9[a]	8.5
Short Message Service (SMS) users (% of mobile users)	—	*52.0*	74.4[a]	69.8[a]
Affordability				
Mobile basket (% of GNI per capita)	10.4	6.1	2.9	3.1

Mobile cellular subscriptions, 2005–11
Number per 100 people

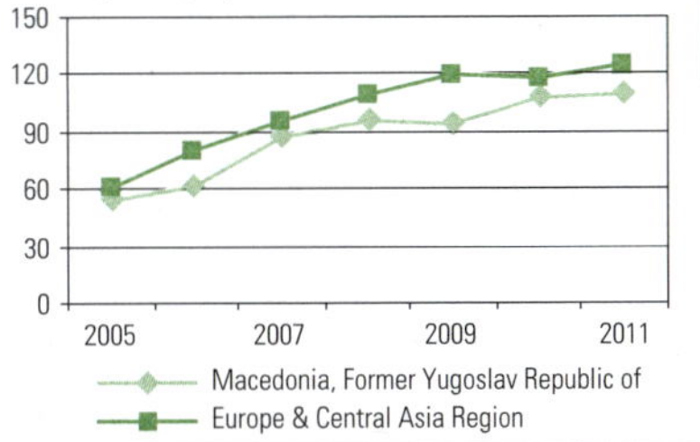

Mobile basket, 2005–10
Percentage of GNI per capita

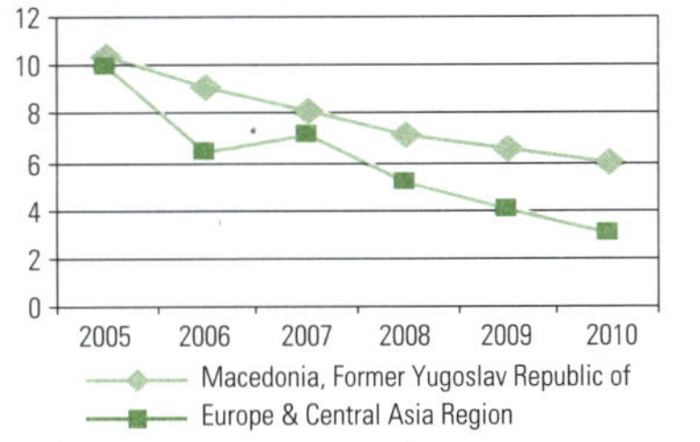

Madagascar

	Madagascar 2005	Madagascar 2010	Low-income group 2010	Sub-Saharan Africa Region 2010
Economic and social context				
Population (total, million)	18	21	796	853
GNI per capita, World Bank Atlas method (current US$)	300	430	530	1,188
Rural population (% of total)	72	70	72	63
Expected years of schooling (years)	9	*11*	*9*	*9*
Physicians density (per 1,000 people)	*0.3*	—	0.2	0.2
Depositors with commercial banks (per 1,000 adults)	18	45	—	*167*
Sector structure				
Number of mobile operators	—	3		
Herfindahl-Hirschman Index (HHI) (scale = 0–10,000)	—	3,528		
Sector performance				
Access				
Mobile cellular subscriptions (per 100 people)	3	37	43[a]	57[a]
Mobile cellular subscriptions (% prepaid)	98	99[a]	98[a]	96[a]
Population covered by a mobile-cellular network (%)	23	—	—	72
Mobile broadband subscriptions (per 100 people)	—	1.6[a]	—	5.6[a]
Mobile broadband (% of total mobile subscriptions)	—	5.4[a]	—	10.1[a]
Usage				
Households with a mobile telephone (%)	4	26	43	52
Mobile voice usage (minutes per user per month)	—	—	—	—
Population using mobile Internet (%)	—	—	—	—
Short Message Service (SMS) users (% of mobile users)	—	—	—	—
Affordability				
Mobile basket (% of GNI per capita)	87.3	43.0	28.8	19.5

Mobile cellular subscriptions, 2005–11
Number per 100 people

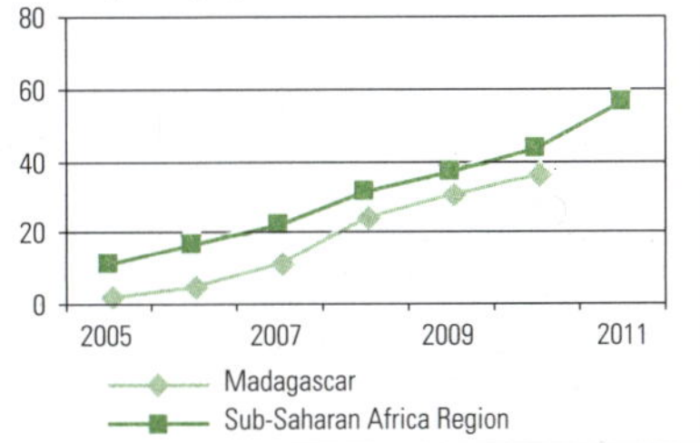

Mobile basket, 2005–10
Percentage of GNI per capita

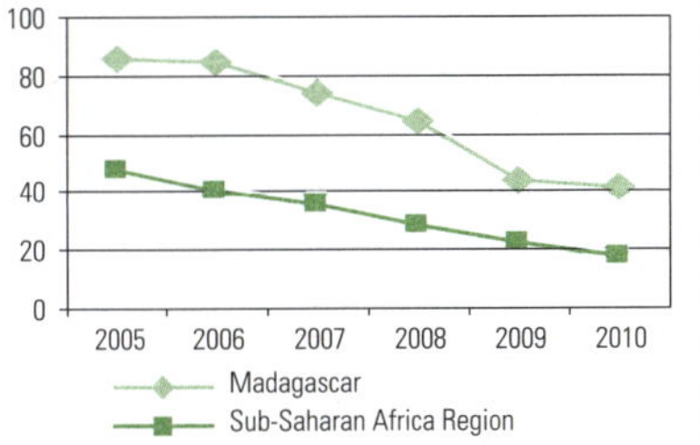

Sources: Economic and social context: IMF, UIS, UN, WHO and World Bank; Sector structure: ictDATA.org; Sector performance: ictDATA.org, ITU; Wireless Intelligence, and World Bank.
Notes: Use of italics in the column entries indicates years or periods other than those specified. — Not available. GNI = gross national income.
a. Data are for 2011.

Malawi

	Malawi 2005	Malawi 2010	Low-income group 2010	Sub-Saharan Africa Region 2010
Economic and social context				
Population (total, million)	13	15	796	853
GNI per capita, World Bank Atlas method (current US$)	220	330	530	1,188
Rural population (% of total)	83	80	72	63
Expected years of schooling (years)	9	—	*9*	*9*
Physicians density (per 1,000 people)	*0.02*	*0.02*	0.2	0.2
Depositors with commercial banks (per 1,000 adults)	—	—	—	*167*
Sector structure				
Number of mobile operators	—	2		
Herfindahl-Hirschman Index (HHI) (scale = 0–10,000)	—	5,626		
Sector performance				
Access				
Mobile cellular subscriptions (per 100 people)	3	26[a]	43[a]	57[a]
Mobile cellular subscriptions (% prepaid)	97	99[a]	98[a]	96[a]
Population covered by a mobile-cellular network (%)	70	85	—	72
Mobile broadband subscriptions (per 100 people)	—	0.6[a]	—	5.6[a]
Mobile broadband (% of total mobile subscriptions)	—	2.2[a]	—	10.1[a]
Usage				
Households with a mobile telephone (%)	3	39	43	52
Mobile voice usage (minutes per user per month)	—	—	—	—
Population using mobile Internet (%)	—	—	—	—
Short Message Service (SMS) users (% of mobile users)	—	—	—	—
Affordability				
Mobile basket (% of GNI per capita)	*109.1*	77.1	28.8	19.5

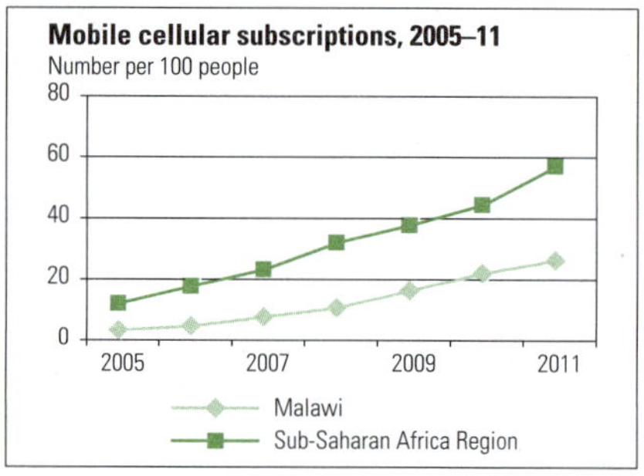

Mobile cellular subscriptions, 2005–11

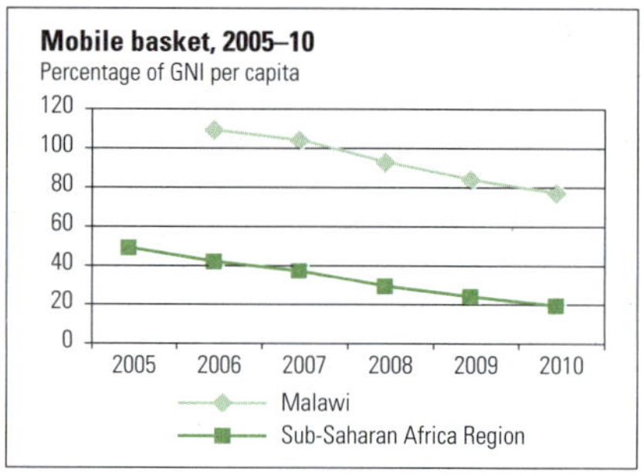

Mobile basket, 2005–10

Malaysia

	Malaysia 2005	Malaysia 2010	Upper-middle-income group 2010	East Asia & Pacific Region 2010
Economic and social context				
Population (total, million)	26	28	2,452	1,962
GNI per capita, World Bank Atlas method (current US$)	5,110	7,760	5,886	3,696
Rural population (% of total)	32	28	43	54
Expected years of schooling (years)	13	*13*	*13*	*12*
Physicians density (per 1,000 people)	—	*0.9*	1.7	1.2
Depositors with commercial banks (per 1,000 adults)	1,892	*1,458*	—	—
Sector structure				
Number of mobile operators	—	4		
Herfindahl-Hirschman Index (HHI) (scale = 0–10,000)	—	3,451		
Sector performance				
Access				
Mobile cellular subscriptions (per 100 people)	75	124[a]	92[a]	83[a]
Mobile cellular subscriptions (% prepaid)	85	77[a]	81[a]	85[a]
Population covered by a mobile-cellular network (%)	—	95	*99*	*99*
Mobile broadband subscriptions (per 100 people)	0.3	38.2[a]	14.3[a]	11.6[a]
Mobile broadband (% of total mobile subscriptions)	0.4	30.5[a]	15.4[a]	14.4[a]
Usage				
Households with a mobile telephone (%)	55	90	84	83
Mobile voice usage (minutes per user per month)	*162*	223[a]	325[a]	367[a]
Population using mobile Internet (%)	3.8	18.0[a]	22.9[a]	22.4[a]
Short Message Service (SMS) users (% of mobile users)	—	*76.0*	74.4[a]	84.0[a]
Affordability				
Mobile basket (% of GNI per capita)	2.0	1.2	2.9	5.7

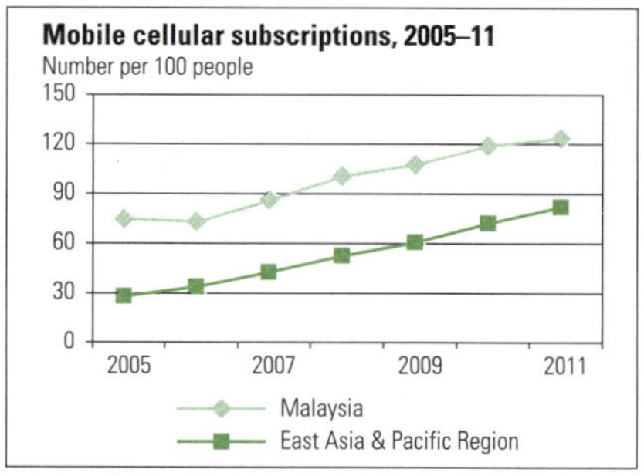

Mobile cellular subscriptions, 2005–11

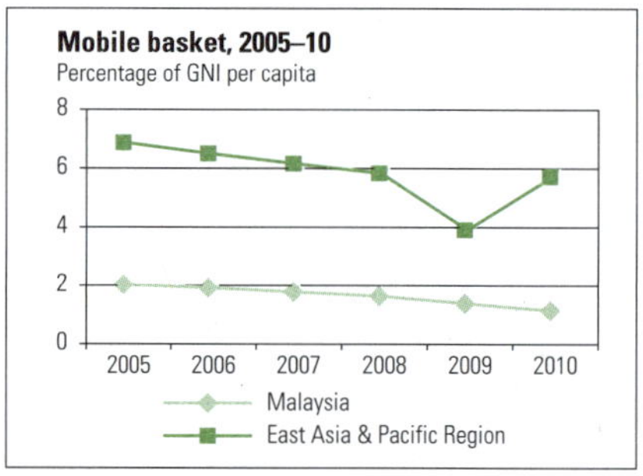

Mobile basket, 2005–10

Sources: Economic and social context: IMF, UIS, UN, WHO and World Bank; Sector structure: ictDATA.org; Sector performance: ictDATA.org, ITU; Wireless Intelligence, and World Bank.
Notes: Use of italics in the column entries indicates years or periods other than those specified. — Not available. GNI = gross national income.
a. Data are for 2011.

Mali

	Mali 2005	Mali 2010	Low-income group 2010	Sub-Saharan Africa Region 2010
Economic and social context				
Population (total, million)	13	15	796	853
GNI per capita, World Bank Atlas method (current US$)	390	600	530	1,188
Rural population (% of total)	70	67	72	63
Expected years of schooling (years)	7	*8*	*9*	*9*
Physicians density (per 1,000 people)	*0.08*	*0.05*	0.2	0.2
Depositors with commercial banks (per 1,000 adults)	—	—	—	*167*
Sector structure				
Number of mobile operators	—	2		
Herfindahl-Hirschman Index (HHI) (scale = 0–10,000)	—	5,690		
Sector performance				
Access				
Mobile cellular subscriptions (per 100 people)	6	69[a]	43[a]	57[a]
Mobile cellular subscriptions (% prepaid)	98	100[a]	98[a]	96[a]
Population covered by a mobile-cellular network (%)	*20*	—	—	72
Mobile broadband subscriptions (per 100 people)	—	3.3[a]	—	5.6[a]
Mobile broadband (% of total mobile subscriptions)	—	4.9[a]	—	10.1[a]
Usage				
Households with a mobile telephone (%)	15	21	43	52
Mobile voice usage (minutes per user per month)	30	—	—	—
Population using mobile Internet (%)	—	—	—	—
Short Message Service (SMS) users (% of mobile users)	—	—	—	—
Affordability				
Mobile basket (% of GNI per capita)	104.3	28.8	28.8	19.5

Mobile cellular subscriptions, 2005–11
Number per 100 people

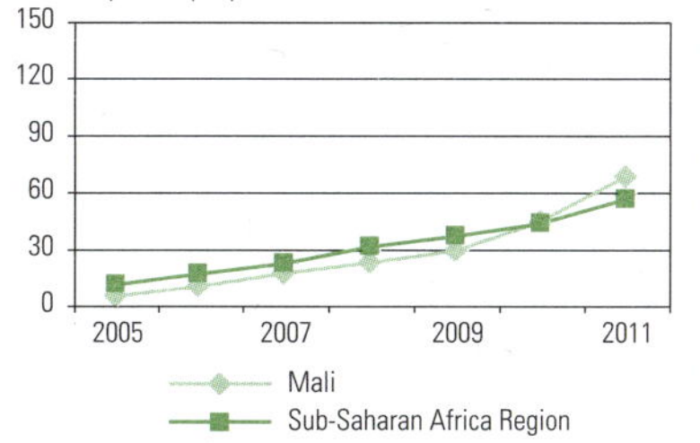

Mobile basket, 2005–10
Percentage of GNI per capita

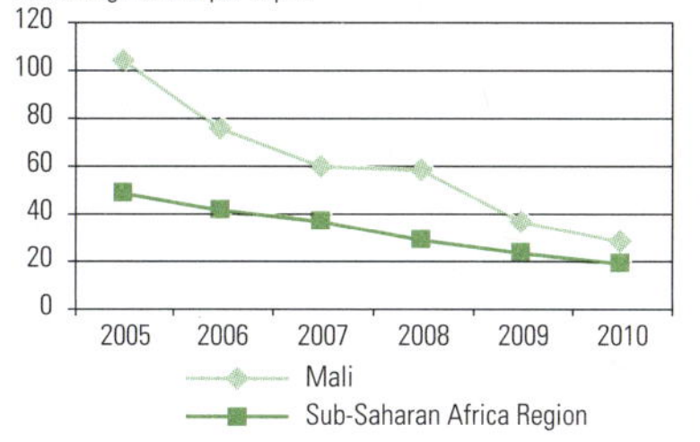

Mauritania

	Mauritania 2005	Mauritania 2010	Lower-middle-income group 2010	Sub-Saharan Africa Region 2010
Economic and social context				
Population (total, million)	3	3	2,519	853
GNI per capita, World Bank Atlas method (current US$)	720	1,000	1,623	1,188
Rural population (% of total)	60	59	61	63
Expected years of schooling (years)	8	—	*10*	*9*
Physicians density (per 1,000 people)	*0.1*	*0.1*	0.8	0.2
Depositors with commercial banks (per 1,000 adults)	—	—	—	*167*
Sector structure				
Number of mobile operators	—	3		
Herfindahl-Hirschman Index (HHI) (scale = 0–10,000)	—	—		
Sector performance				
Access				
Mobile cellular subscriptions (per 100 people)	24	84[a]	78[a]	57[a]
Mobile cellular subscriptions (% prepaid)	98	97[a]	96[a]	96[a]
Population covered by a mobile-cellular network (%)	26	*62*	*86*	72
Mobile broadband subscriptions (per 100 people)	—	—	7.3[a]	5.6[a]
Mobile broadband (% of total mobile subscriptions)	—	—	9.0[a]	10.1[a]
Usage				
Households with a mobile telephone (%)	—	—	77	52
Mobile voice usage (minutes per user per month)	—	—	276[a]	—
Population using mobile Internet (%)	—	—	2.9	—
Short Message Service (SMS) users (% of mobile users)	—	—	61.9[a]	—
Affordability				
Mobile basket (% of GNI per capita)	*19.1*	17.5	7.2	19.5

Mobile cellular subscriptions, 2005–11
Number per 100 people

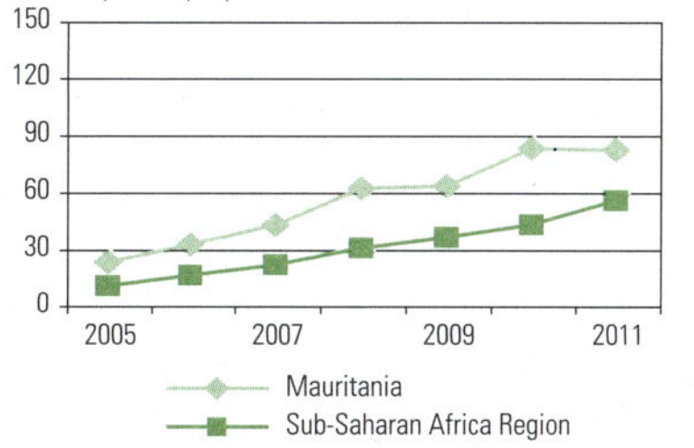

Mobile basket, 2005–10
Percentage of GNI per capita

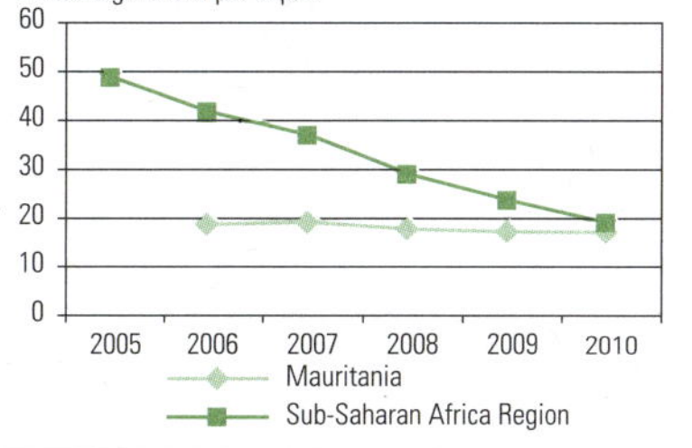

Sources: Economic and social context: IMF, UIS, UN, WHO and World Bank; Sector structure: ictDATA.org; Sector performance: ictDATA.org, ITU; Wireless Intelligence, and World Bank.
Notes: Use of italics in the column entries indicates years or periods other than those specified. — Not available. GNI = gross national income.
a. Data are for 2011.

Mauritius

	Mauritius 2005	Mauritius 2010	Upper-middle-income group 2010	Sub-Saharan Africa Region 2010
Economic and social context				
Population (total, million)	1	1	2,452	853
GNI per capita, World Bank Atlas method (current US$)	5,360	7,850	5,886	1,188
Rural population (% of total)	58	57	43	63
Expected years of schooling (years)	13	*14*	*13*	*9*
Physicians density (per 1,000 people)	*1.1*	—	1.7	0.2
Depositors with commercial banks (per 1,000 adults)	322	465	—	*167*
Sector structure				
Number of mobile operators	—	3		
Herfindahl-Hirschman Index (HHI) (scale = 0–10,000)	—	4,366		
Sector performance				
Access				
Mobile cellular subscriptions (per 100 people)	53	93	92[a]	57[a]
Mobile cellular subscriptions (% prepaid)	93	93[a]	81[a]	96[a]
Population covered by a mobile-cellular network (%)	97	99	*99*	72
Mobile broadband subscriptions (per 100 people)	0.4	21.5[a]	14.3[a]	5.6[a]
Mobile broadband (% of total mobile subscriptions)	0.7	22.2[a]	15.4[a]	10.1[a]
Usage				
Households with a mobile telephone (%)	60	88	84	52
Mobile voice usage (minutes per user per month)	—	—	325[a]	—
Population using mobile Internet (%)	—	—	22.9[a]	—
Short Message Service (SMS) users (% of mobile users)	—	—	74.4[a]	—
Affordability				
Mobile basket (% of GNI per capita)	1.5	1.0	2.9	19.5

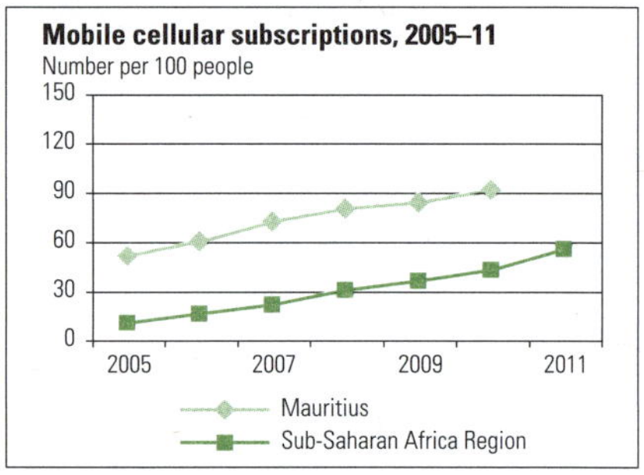

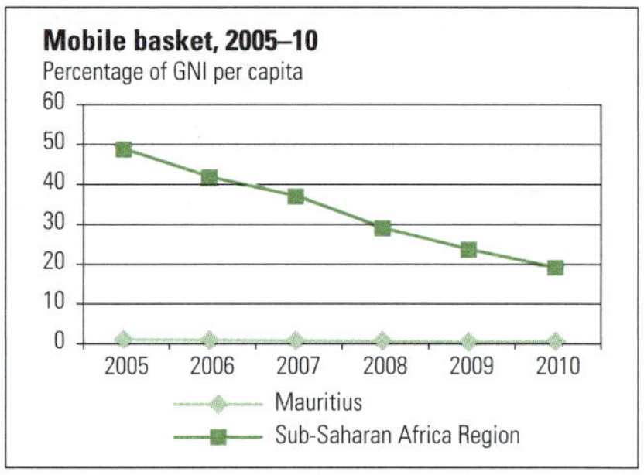

Mexico

	Mexico 2005	Mexico 2010	Upper-middle-income group 2010	Latin America & the Caribbean Region 2010
Economic and social context				
Population (total, million)	106	113	2,452	583
GNI per capita, World Bank Atlas method (current US$)	7,820	8,930	5,886	7,741
Rural population (% of total)	24	22	43	21
Expected years of schooling (years)	13	*14*	*13*	*14*
Physicians density (per 1,000 people)	*2.9*	*2.0*	1.7	1.8
Depositors with commercial banks (per 1,000 adults)	—	1,205	—	—
Sector structure				
Number of mobile operators	—	3		
Herfindahl-Hirschman Index (HHI) (scale = 0–10,000)	—	5,500		
Sector performance				
Access				
Mobile cellular subscriptions (per 100 people)	44	82[a]	92[a]	109[a]
Mobile cellular subscriptions (% prepaid)	91	85[a]	81[a]	81[a]
Population covered by a mobile-cellular network (%)	86	93	*99*	98
Mobile broadband subscriptions (per 100 people)	0.03	17.4[a]	14.3[a]	16.1[a]
Mobile broadband (% of total mobile subscriptions)	0.08	21.3[a]	15.4[a]	15.2[a]
Usage				
Households with a mobile telephone (%)	42	71	84	84
Mobile voice usage (minutes per user per month)	97	203[a]	325[a]	141[a]
Population using mobile Internet (%)	—	6.4[a]	22.9[a]	4.4
Short Message Service (SMS) users (% of mobile users)	—	82.0[a]	74.4[a]	—
Affordability				
Mobile basket (% of GNI per capita)	2.5	2.3	2.9	3.7

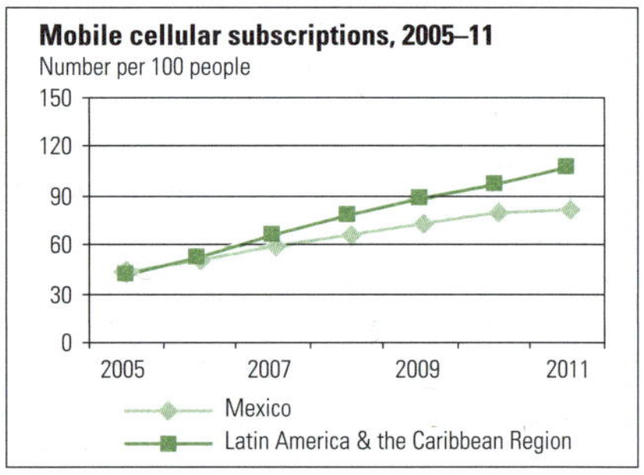

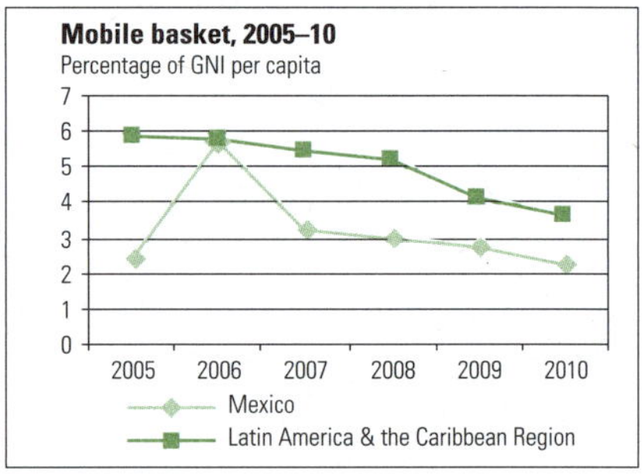

Sources: Economic and social context: IMF, UIS, UN, WHO and World Bank; Sector structure: ictDATA.org; Sector performance: ictDATA.org, ITU; Wireless Intelligence, and World Bank.
Notes: Use of italics in the column entries indicates years or periods other than those specified. — Not available. GNI = gross national income.
a. Data are for 2011.

Moldova

	Moldova 2005	Moldova 2010	Lower-middle-income group 2010	Europe & Central Asia Region 2010
Economic and social context				
Population (total, million)	4	4	2,519	405
GNI per capita, World Bank Atlas method (current US$)	890	1,810	1,623	7,272
Rural population (% of total)	57	59	61	36
Expected years of schooling (years)	12	*12*	*10*	*13*
Physicians density (per 1,000 people)	*2.7*	*2.7*	0.8	3.2
Depositors with commercial banks (per 1,000 adults)	811	1,197	—	894
Sector structure				
Number of mobile operators	—	3		
Herfindahl-Hirschman Index (HHI) (scale = 0–10,000)	—	5,077		
Sector performance				
Access				
Mobile cellular subscriptions (per 100 people)	30	106[a]	78[a]	125[a]
Mobile cellular subscriptions (% prepaid)	84	84[a]	96[a]	82[a]
Population covered by a mobile-cellular network (%)	97	—	*86*	96
Mobile broadband subscriptions (per 100 people)	—	26.6[a]	7.3[a]	22.6[a]
Mobile broadband (% of total mobile subscriptions)	—	27.2[a]	9.0[a]	18.0[a]
Usage				
Households with a mobile telephone (%)	31	68	77	82
Mobile voice usage (minutes per user per month)	—	—	276[a]	288[a]
Population using mobile Internet (%)	0.3	3.4	2.9	8.5
Short Message Service (SMS) users (% of mobile users)	—	*33.0*	61.9[a]	69.8[a]
Affordability				
Mobile basket (% of GNI per capita)	17.4	8.4	7.2	3.1

Mobile cellular subscriptions, 2005–11
Number per 100 people

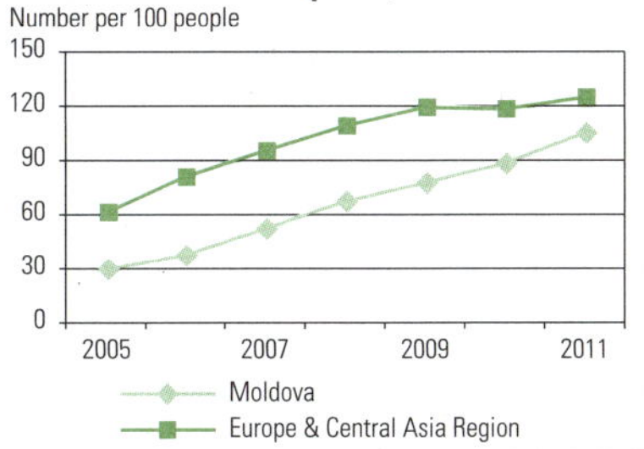

Mobile basket, 2005–10
Percentage of GNI per capita

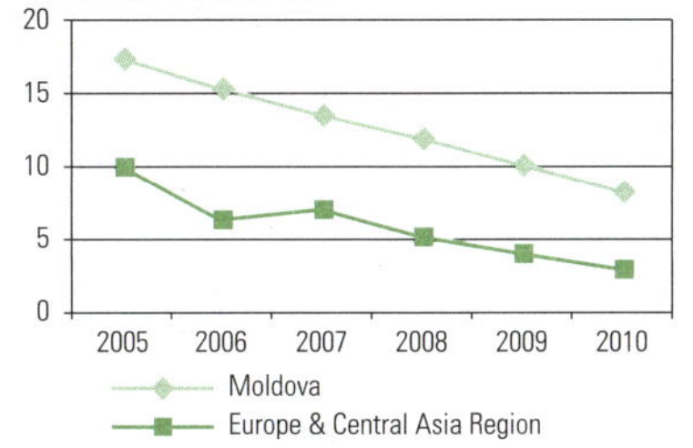

Mongolia

	Mongolia 2005	Mongolia 2010	Lower-middle-income group 2010	East Asia & Pacific Region 2010
Economic and social context				
Population (total, million)	3	3	2,519	1,962
GNI per capita, World Bank Atlas method (current US$)	890	1,870	1,623	3,696
Rural population (% of total)	43	43	61	54
Expected years of schooling (years)	13	*14*	*10*	*12*
Physicians density (per 1,000 people)	—	*2.8*	0.8	1.2
Depositors with commercial banks (per 1,000 adults)	346	*1,339*	—	—
Sector structure				
Number of mobile operators	—	4		
Herfindahl-Hirschman Index (HHI) (scale = 0–10,000)	—	3,102		
Sector performance				
Access				
Mobile cellular subscriptions (per 100 people)	22	92	78[a]	83[a]
Mobile cellular subscriptions (% prepaid)	91	96[a]	96[a]	85[a]
Population covered by a mobile-cellular network (%)	29	85	*86*	*99*
Mobile broadband subscriptions (per 100 people)	—	9.3[a]	7.3[a]	11.6[a]
Mobile broadband (% of total mobile subscriptions)	—	11.1[a]	9.0[a]	14.4[a]
Usage				
Households with a mobile telephone (%)	*28*	86	77	83
Mobile voice usage (minutes per user per month)	—	—	276[a]	367[a]
Population using mobile Internet (%)	—	—	2.9	22.4[a]
Short Message Service (SMS) users (% of mobile users)	—	—	61.9[a]	84.0[a]
Affordability				
Mobile basket (% of GNI per capita)	—	*2.4*	7.2	5.7

Mobile cellular subscriptions, 2005–11
Number per 100 people

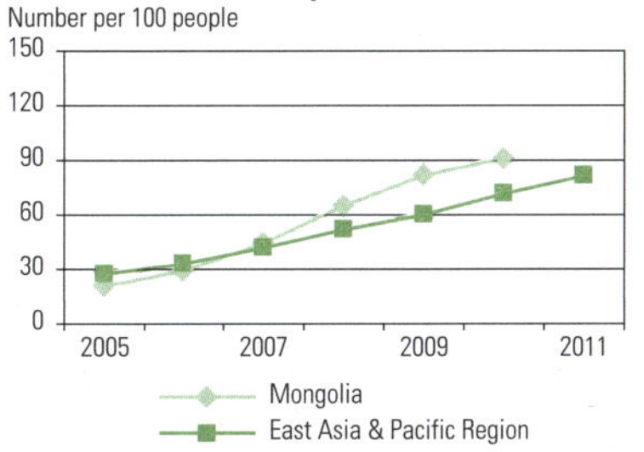

Mobile basket, 2005–10
Percentage of GNI per capita

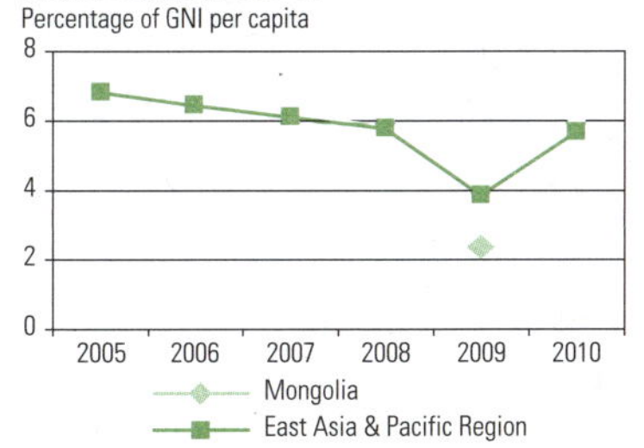

Sources: Economic and social context: IMF, UIS, UN, WHO and World Bank; Sector structure: ictDATA.org; Sector performance: ictDATA.org, ITU; Wireless Intelligence, and World Bank.
Notes: Use of italics in the column entries indicates years or periods other than those specified. — Not available. GNI = gross national income.
a. Data are for 2011.

Morocco

	Morocco 2005	Morocco 2010	Lower-middle-income group 2010	Middle East & North Africa Region 2010
Economic and social context				
Population (total, million)	30	32	2,519	331
GNI per capita, World Bank Atlas method (current US$)	1,960	2,850	1,623	3,874
Rural population (% of total)	45	43	61	42
Expected years of schooling (years)	10	—	*10*	*12*
Physicians density (per 1,000 people)	*0.5*	*0.6*	0.8	1.4
Depositors with commercial banks (per 1,000 adults)	301	694	—	*443*
Sector structure				
Number of mobile operators	—	3		
Herfindahl-Hirschman Index (HHI) (scale = 0–10,000)	—	4,108		
Sector performance				
Access				
Mobile cellular subscriptions (per 100 people)	41	113[a]	78[a]	89[a]
Mobile cellular subscriptions (% prepaid)	96	96[a]	96[a]	87[a]
Population covered by a mobile-cellular network (%)	98	98	*86*	—
Mobile broadband subscriptions (per 100 people)	—	17.5[a]	7.3[a]	—
Mobile broadband (% of total mobile subscriptions)	—	15.4[a]	9.0[a]	—
Usage				
Households with a mobile telephone (%)	59	84	77	—
Mobile voice usage (minutes per user per month)	—	70	276[a]	—
Population using mobile Internet (%)	0.04	3.4	2.9	4.5
Short Message Service (SMS) users (% of mobile users)	69.0	70.0	61.9[a]	—
Affordability				
Mobile basket (% of GNI per capita)	20.2	14.3	7.2	3.6

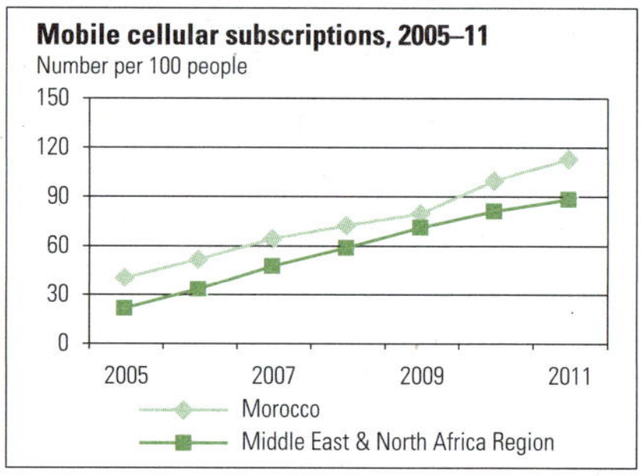

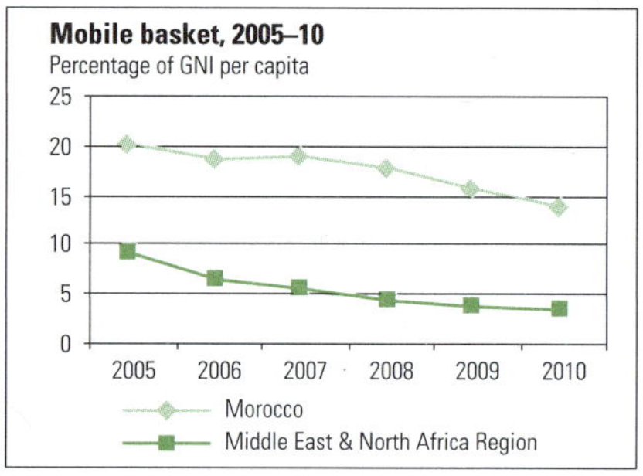

Mozambique

	Mozambique 2005	Mozambique 2010	Low-income group 2010	Sub-Saharan Africa Region 2010
Economic and social context				
Population (total, million)	21	23	796	853
GNI per capita, World Bank Atlas method (current US$)	290	440	530	1,188
Rural population (% of total)	66	62	72	63
Expected years of schooling (years)	8	—	*9*	*9*
Physicians density (per 1,000 people)	*0.03*	*0.03*	0.2	0.2
Depositors with commercial banks (per 1,000 adults)	—	—	—	*167*
Sector structure				
Number of mobile operators	—	2		
Herfindahl-Hirschman Index (HHI) (scale = 0–10,000)	—	5,050		
Sector performance				
Access				
Mobile cellular subscriptions (per 100 people)	7	25	43[a]	57[a]
Mobile cellular subscriptions (% prepaid)	99	98[a]	98[a]	96[a]
Population covered by a mobile-cellular network (%)	—	32	—	72
Mobile broadband subscriptions (per 100 people)	—	3.1[a]	—	5.6[a]
Mobile broadband (% of total mobile subscriptions)	—	8.7[a]	—	10.1[a]
Usage				
Households with a mobile telephone (%)	—	—	43	52
Mobile voice usage (minutes per user per month)	—	—	—	—
Population using mobile Internet (%)	—	—	—	—
Short Message Service (SMS) users (% of mobile users)	—	—	—	—
Affordability				
Mobile basket (% of GNI per capita)	*66.3*	46.4	28.8	19.5

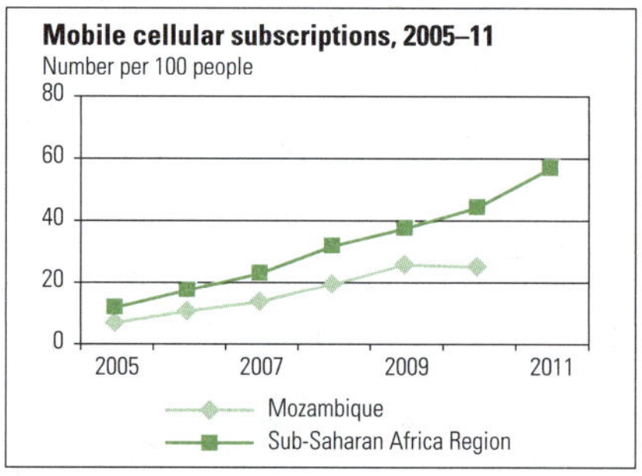

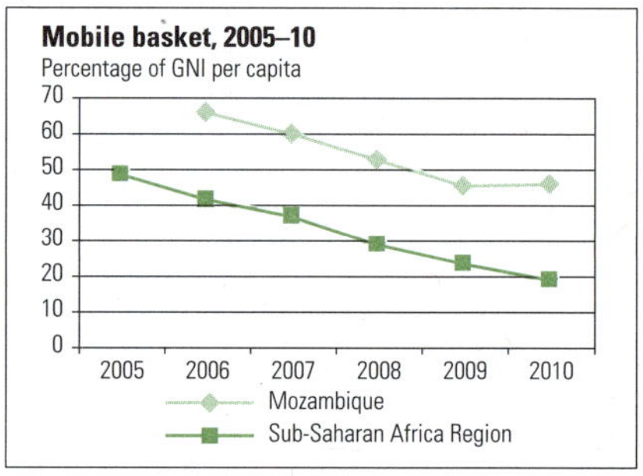

Sources: Economic and social context: IMF, UIS, UN, WHO and World Bank; Sector structure: ictDATA.org; Sector performance: ictDATA.org, ITU; Wireless Intelligence, and World Bank.
Notes: Use of italics in the column entries indicates years or periods other than those specified. — Not available. GNI = gross national income.
a. Data are for 2011.

Myanmar

	Myanmar 2005	Myanmar 2010	Low-income group 2010	East Asia & Pacific Region 2010
Economic and social context				
Population (total, million)	46	48	796	1,962
GNI per capita, World Bank Atlas method (current US$)	—	—	530	3,696
Rural population (% of total)	69	66	72	54
Expected years of schooling (years)	*9*	—	*9*	*12*
Physicians density (per 1,000 people)	*0.4*	*0.5*	0.2	1.2
Depositors with commercial banks (per 1,000 adults)	—	—	—	—
Sector structure				
Number of mobile operators	—	1		
Herfindahl-Hirschman Index (HHI) (scale = 0–10,000)	—	10,000		
Sector performance				
Access				
Mobile cellular subscriptions (per 100 people)	0.3	1	43[a]	83[a]
Mobile cellular subscriptions (% prepaid)	94	99[a]	98[a]	85[a]
Population covered by a mobile-cellular network (%)	*10*	—	—	*99*
Mobile broadband subscriptions (per 100 people)	—	0.05[a]	—	11.6[a]
Mobile broadband (% of total mobile subscriptions)	—	1.1[a]	—	14.4[a]
Usage				
Households with a mobile telephone (%)	—	—	43	83
Mobile voice usage (minutes per user per month)	—	—	—	367[a]
Population using mobile Internet (%)	—	—	—	22.4[a]
Short Message Service (SMS) users (% of mobile users)	—	—	—	84.0[a]
Affordability				
Mobile basket (% of GNI per capita)	—	—	28.8	5.7

Mobile cellular subscriptions, 2005–11
Number per 100 people

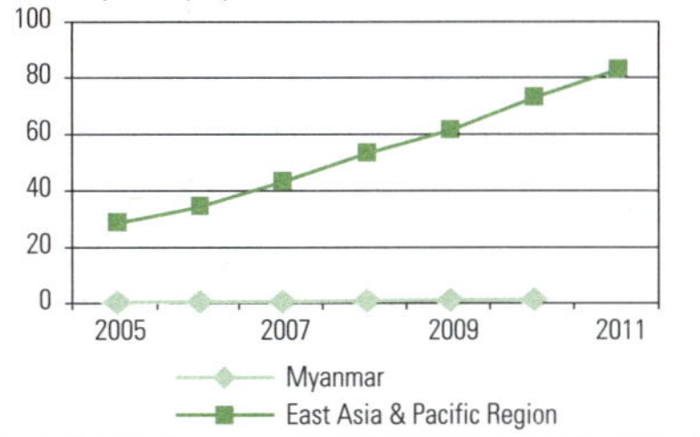

Mobile basket, 2005–10
Percentage of GNI per capita

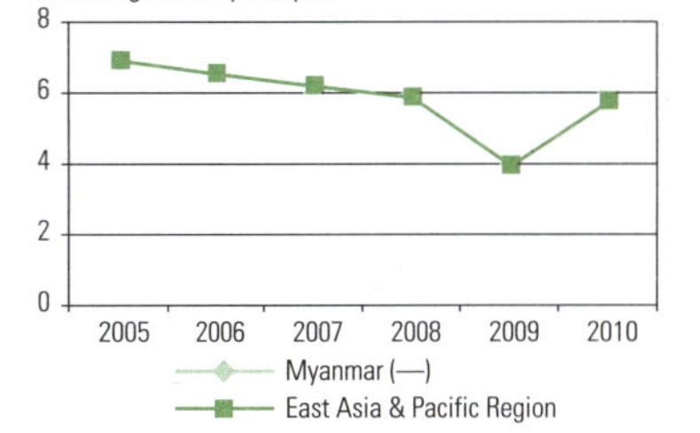

Namibia

	Namibia 2005	Namibia 2010	Upper-middle-income group 2010	Sub-Saharan Africa Region 2010
Economic and social context				
Population (total, million)	2	2	2,452	853
GNI per capita, World Bank Atlas method (current US$)	3,300	4,510	5,886	1,188
Rural population (% of total)	65	62	43	63
Expected years of schooling (years)	12	—	*13*	*9*
Physicians density (per 1,000 people)	*0.3*	—	1.7	0.2
Depositors with commercial banks (per 1,000 adults)	—	624	—	*167*
Sector structure				
Number of mobile operators	—	2		
Herfindahl-Hirschman Index (HHI) (scale = 0–10,000)	—	6,717		
Sector performance				
Access				
Mobile cellular subscriptions (per 100 people)	22	110[a]	92[a]	57[a]
Mobile cellular subscriptions (% prepaid)	91	96[a]	81[a]	96[a]
Population covered by a mobile-cellular network (%)	88	—	*99*	72
Mobile broadband subscriptions (per 100 people)	*0.1*	11.1[a]	14.3[a]	5.6[a]
Mobile broadband (% of total mobile subscriptions)	*0.4*	10.0[a]	15.4[a]	10.1[a]
Usage				
Households with a mobile telephone (%)	40	55	84	52
Mobile voice usage (minutes per user per month)	—	—	325[a]	—
Population using mobile Internet (%)	—	7.0	22.9[a]	—
Short Message Service (SMS) users (% of mobile users)	—	—	74.4[a]	—
Affordability				
Mobile basket (% of GNI per capita)	9.6	4.5	2.9	19.5

Mobile cellular subscriptions, 2005–11
Number per 100 people

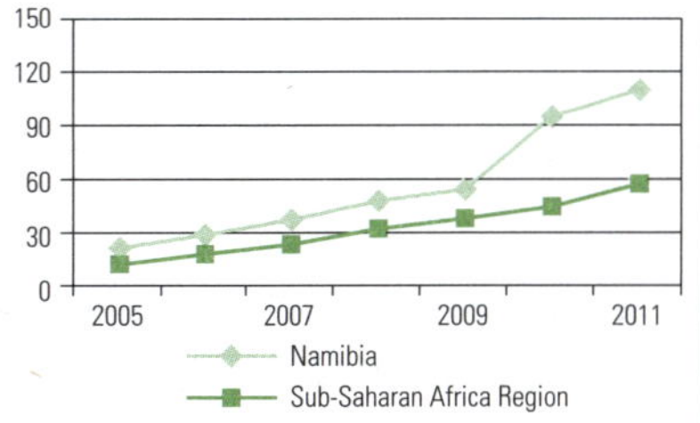

Mobile basket, 2005–10
Percentage of GNI per capita

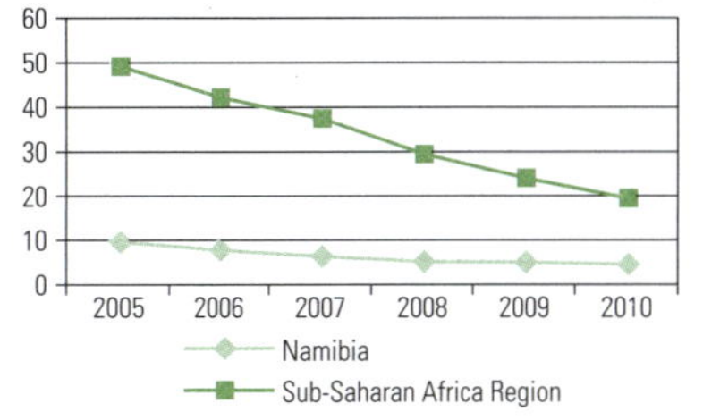

Sources: Economic and social context: IMF, UIS, UN, WHO and World Bank; Sector structure: ictDATA.org; Sector performance: ictDATA.org, ITU; Wireless Intelligence, and World Bank.
Notes: Use of italics in the column entries indicates years or periods other than those specified. — Not available. GNI = gross national income.
a. Data are for 2011.

Nepal

	Nepal 2005	Nepal 2010	Low-income group 2010	South Asia Region 2010
Economic and social context				
Population (total, million)	27	30	796	1,633
GNI per capita, World Bank Atlas method (current US$)	290	490	530	1,176
Rural population (% of total)	84	82	72	70
Expected years of schooling (years)	—	—	*9*	*10*
Physicians density (per 1,000 people)	*0.2*	—	0.2	0.6
Depositors with commercial banks (per 1,000 adults)	—	—	—	249
Sector structure				
Number of mobile operators	—	4		
Herfindahl-Hirschman Index (HHI) (scale = 0–10,000)	—	4,826		
Sector performance				
Access				
Mobile cellular subscriptions (per 100 people)	1	44[a]	43[a]	67[a]
Mobile cellular subscriptions (% prepaid)	71	98[a]	98[a]	96[a]
Population covered by a mobile-cellular network (%)	*10*	35	—	*84*
Mobile broadband subscriptions (per 100 people)	—	0.4[a]	—	3.3[a]
Mobile broadband (% of total mobile subscriptions)	—	0.8[a]	—	4.6[a]
Usage				
Households with a mobile telephone (%)	3	54	43	54
Mobile voice usage (minutes per user per month)	—	123[a]	—	305[a]
Population using mobile Internet (%)	—	—	—	3.3[a]
Short Message Service (SMS) users (% of mobile users)	—	—	—	47.0[a]
Affordability				
Mobile basket (% of GNI per capita)	18.9	6.6	28.8	3.2

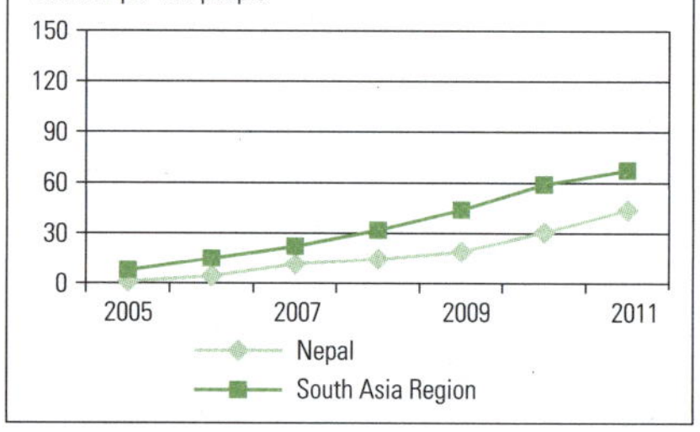

Mobile cellular subscriptions, 2005–11
Number per 100 people

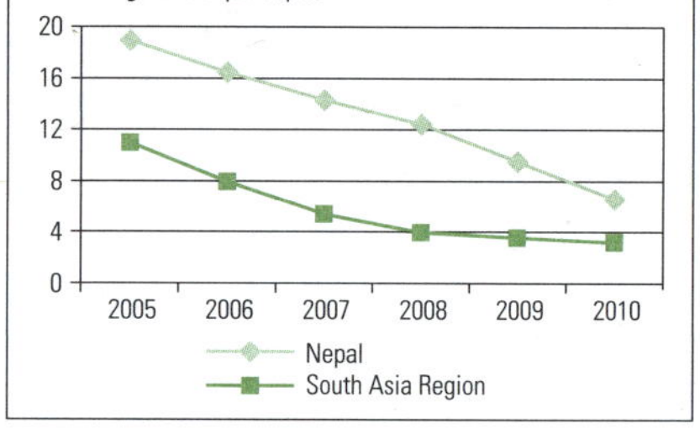

Mobile basket, 2005–10
Percentage of GNI per capita

Netherlands

	Netherlands 2005	Netherlands 2010	High-income group 2010
Economic and social context			
Population (total, million)	16	17	1,127
GNI per capita, World Bank Atlas method (current US$)	39,880	49,030	38,746
Rural population (% of total)	20	17	22
Expected years of schooling (years)	16	*17*	*16*
Physicians density (per 1,000 people)	3.7	*2.9*	2.8
Depositors with commercial banks (per 1,000 adults)	1,769	1,769	—
Sector structure			
Number of mobile operators	—	3	
Herfindahl-Hirschman Index (HHI) (scale = 0–10,000)	—	3,789	
Sector performance			
Access			
Mobile cellular subscriptions (per 100 people)	97	128[a]	118[a]
Mobile cellular subscriptions (% prepaid)	57	41[a]	36[a]
Population covered by a mobile-cellular network (%)	100	—	100
Mobile broadband subscriptions (per 100 people)	1.6	50.7[a]	69.6[a]
Mobile broadband (% of total mobile subscriptions)	1.6	42.3[a]	57.6[a]
Usage			
Households with a mobile telephone (%)	91	94	93
Mobile voice usage (minutes per user per month)	136	159[a]	339
Population using mobile Internet (%)	—	15.6	24.3
Short Message Service (SMS) users (% of mobile users)	—	—	78.2[a]
Affordability			
Mobile basket (% of GNI per capita)	1.5	0.8	1.0

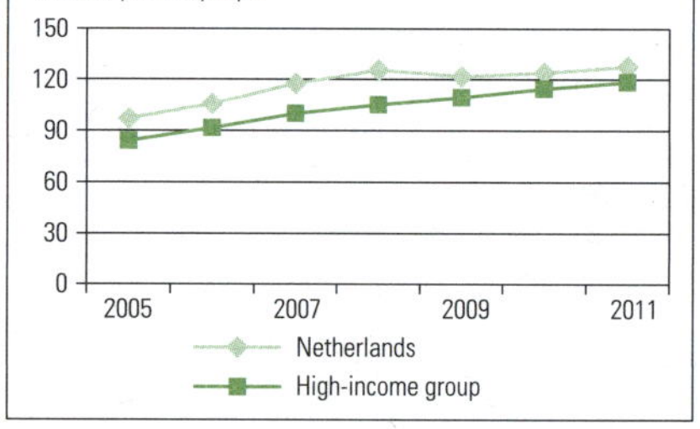

Mobile cellular subscriptions, 2005–11
Number per 100 people

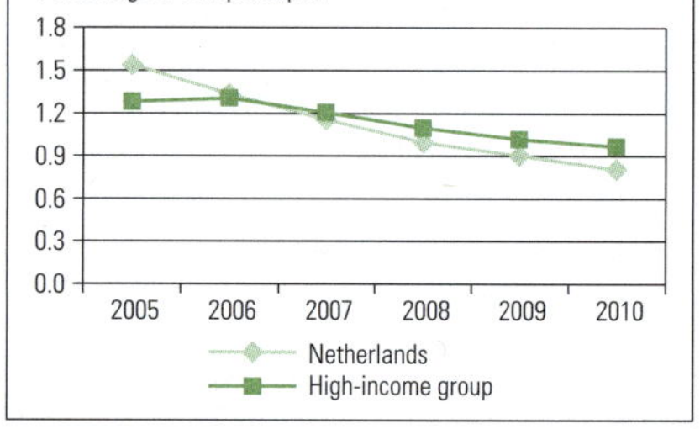

Mobile basket, 2005–10
Percentage of GNI per capita

Sources: Economic and social context: IMF, UIS, UN, WHO and World Bank; Sector structure: ictDATA.org; Sector performance: ictDATA.org, ITU; Wireless Intelligence, and World Bank.
Notes: Use of italics in the column entries indicates years or periods other than those specified. — Not available. GNI = gross national income.
a. Data are for 2011.

New Zealand

	New Zealand 2005	New Zealand 2010	High-income group 2010
Economic and social context			
Population (total, million)	4	4	1,127
GNI per capita, World Bank Atlas method (current US$)	24,840	*28,770*	38,746
Rural population (% of total)	14	13	22
Expected years of schooling (years)	19	*20*	*16*
Physicians density (per 1,000 people)	*2.4*	2.7	2.8
Depositors with commercial banks (per 1,000 adults)	—	—	—
Sector structure			
Number of mobile operators	—	3	
Herfindahl-Hirschman Index (HHI) (scale = 0–10,000)	—	4,229	
Sector performance			
Access			
Mobile cellular subscriptions (per 100 people)	85	108	118[a]
Mobile cellular subscriptions (% prepaid)	70	65[a]	36[a]
Population covered by a mobile-cellular network (%)	98	97	100
Mobile broadband subscriptions (per 100 people)	4.2	77.1[a]	69.6[a]
Mobile broadband (% of total mobile subscriptions)	4.5	64.1[a]	57.6[a]
Usage			
Households with a mobile telephone (%)	*86*	90	93
Mobile voice usage (minutes per user per month)	83	—	339
Population using mobile Internet (%)	—	18.3	24.3
Short Message Service (SMS) users (% of mobile users)	—	—	78.2[a]
Affordability			
Mobile basket (% of GNI per capita)	1.9	*1.7*	1.0

Mobile cellular subscriptions, 2005–11
Number per 100 people

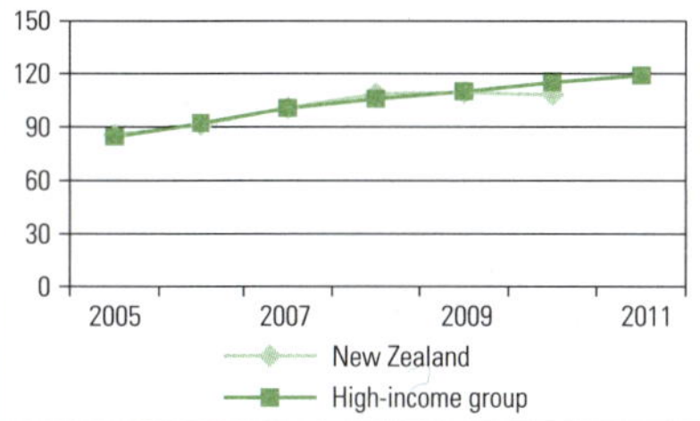

Mobile basket, 2005–10
Percentage of GNI per capita

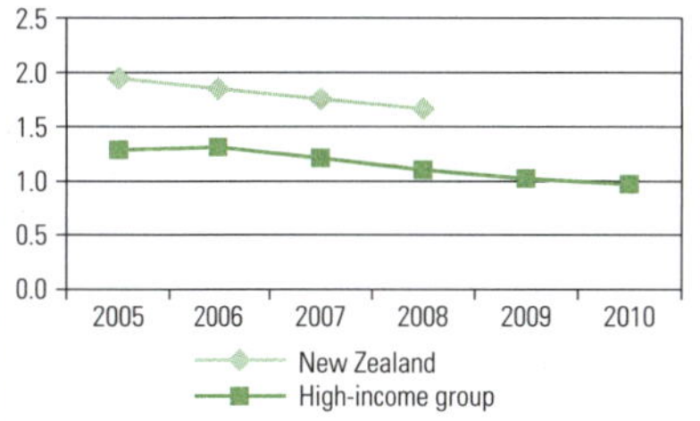

Nicaragua

	Nicaragua 2005	Nicaragua 2010	Lower-middle-income group 2010	Latin America & the Caribbean Region 2010
Economic and social context				
Population (total, million)	5	6	2,519	583
GNI per capita, World Bank Atlas method (current US$)	890	1,110	1,623	7,741
Rural population (% of total)	44	43	61	21
Expected years of schooling (years)	*11*	—	*10*	*14*
Physicians density (per 1,000 people)	*0.4*	—	0.8	1.8
Depositors with commercial banks (per 1,000 adults)	—	—	—	—
Sector structure				
Number of mobile operators	—	2		
Herfindahl-Hirschman Index (HHI) (scale = 0–10,000)	—	5,512		
Sector performance				
Access				
Mobile cellular subscriptions (per 100 people)	21	72[a]	78[a]	109[a]
Mobile cellular subscriptions (% prepaid)	87	92[a]	96[a]	81[a]
Population covered by a mobile-cellular network (%)	70	—	*86*	98
Mobile broadband subscriptions (per 100 people)	—	4.6[a]	7.3[a]	16.1[a]
Mobile broadband (% of total mobile subscriptions)	—	6.4[a]	9.0[a]	15.2[a]
Usage				
Households with a mobile telephone (%)	24	*62*	77	84
Mobile voice usage (minutes per user per month)	—	—	276[a]	141[a]
Population using mobile Internet (%)	—	—	2.9	4.4
Short Message Service (SMS) users (% of mobile users)	—	—	61.9[a]	—
Affordability				
Mobile basket (% of GNI per capita)	*36.4*	14.3	7.2	3.7

Mobile cellular subscriptions, 2005–11
Number per 100 people

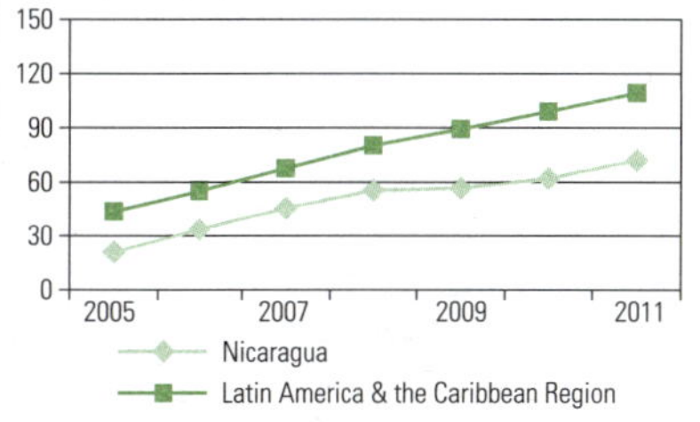

Mobile basket, 2005–10
Percentage of GNI per capita

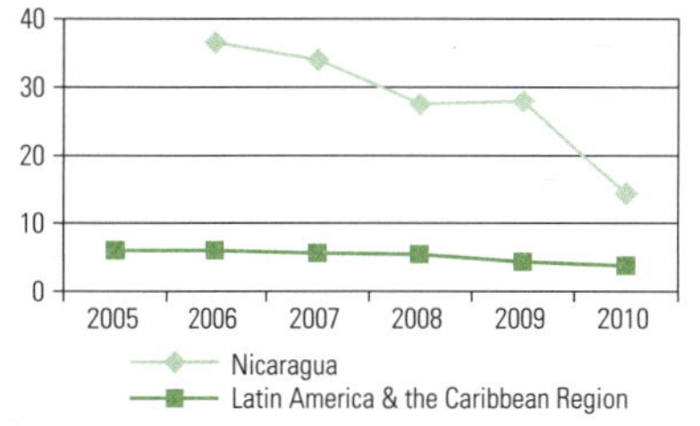

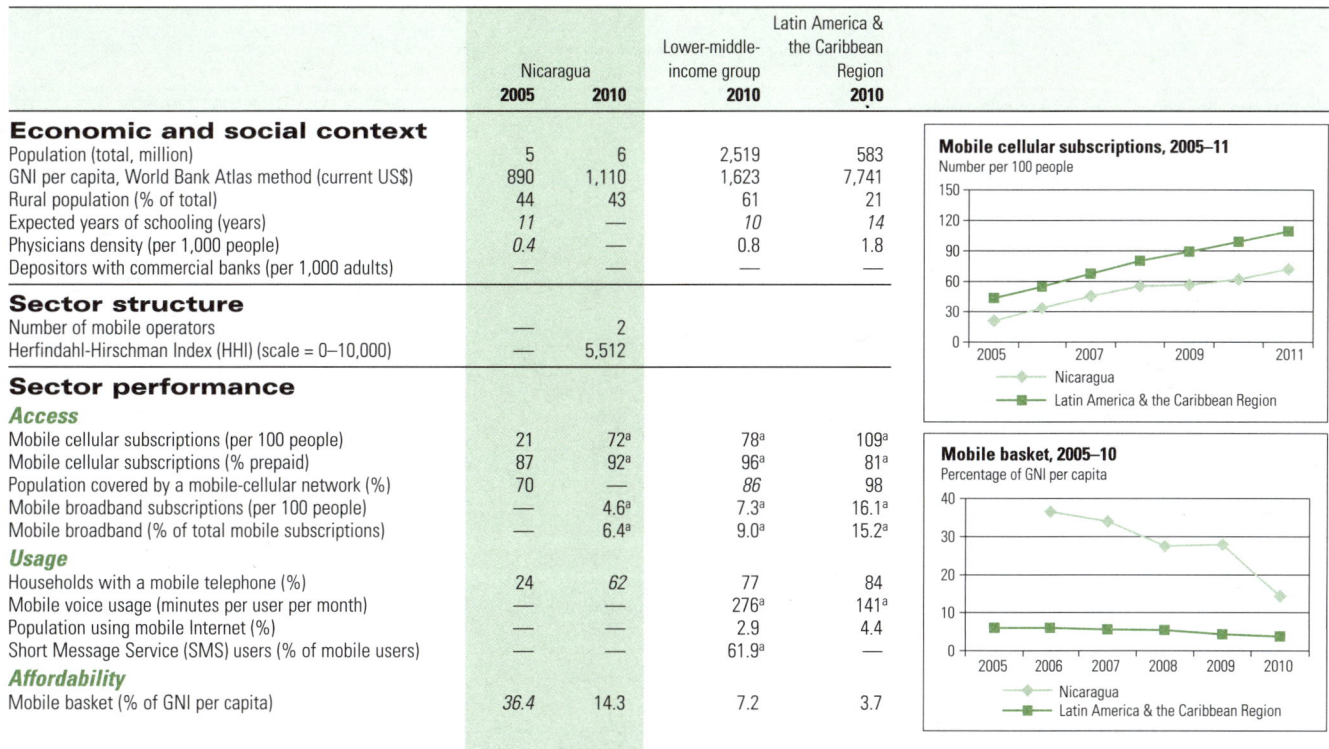

Sources: Economic and social context: IMF, UIS, UN, WHO and World Bank; Sector structure: ictDATA.org; Sector performance: ictDATA.org, ITU; Wireless Intelligence, and World Bank.
Notes: Use of italics in the column entries indicates years or periods other than those specified. — Not available. GNI = gross national income.
a. Data are for 2011.

Niger

	Niger 2005	Niger 2010	Low-income group 2010	Sub-Saharan Africa Region 2010
Economic and social context				
Population (total, million)	13	16	796	853
GNI per capita, World Bank Atlas method (current US$)	260	370	530	1,188
Rural population (% of total)	84	83	72	63
Expected years of schooling (years)	4	5	*9*	*9*
Physicians density (per 1,000 people)	*0.02*	*0.02*	0.2	0.2
Depositors with commercial banks (per 1,000 adults)	—	—	—	*167*
Sector structure				
Number of mobile operators	—	3		
Herfindahl-Hirschman Index (HHI) (scale = 0–10,000)	—	4,890		
Sector performance				
Access				
Mobile cellular subscriptions (per 100 people)	2	24	43[a]	57[a]
Mobile cellular subscriptions (% prepaid)	99	99[a]	98[a]	96[a]
Population covered by a mobile-cellular network (%)	15	—	—	72
Mobile broadband subscriptions (per 100 people)	—	—	—	5.6[a]
Mobile broadband (% of total mobile subscriptions)	—	—	—	10.1[a]
Usage				
Households with a mobile telephone (%)	—	*32*	43	52
Mobile voice usage (minutes per user per month)	—	—	—	—
Population using mobile Internet (%)	—	—	—	—
Short Message Service (SMS) users (% of mobile users)	—	*11.0*	—	—
Affordability				
Mobile basket (% of GNI per capita)	125.3	67.5	28.8	19.5

Mobile cellular subscriptions, 2005–11
Number per 100 people

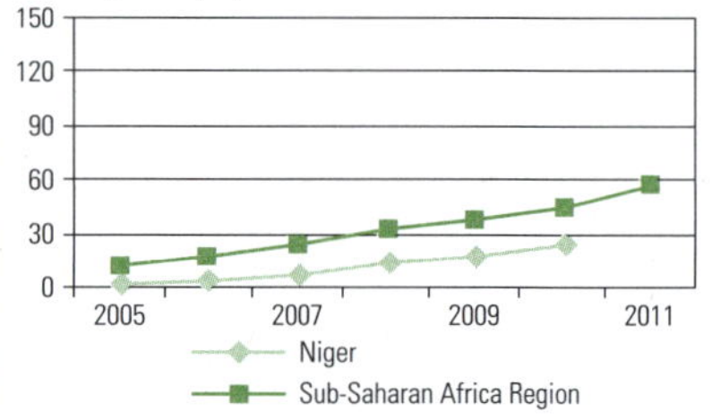

Mobile basket, 2005–10
Percentage of GNI per capita

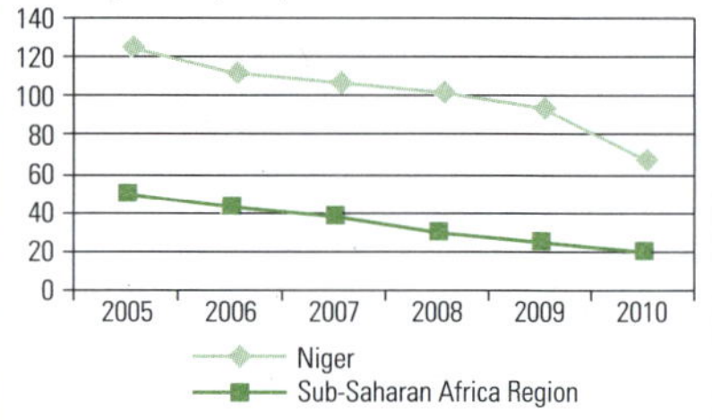

Nigeria

	Nigeria 2005	Nigeria 2010	Lower-middle-income group 2010	Sub-Saharan Africa Region 2010
Economic and social context				
Population (total, million)	140	158	2,519	853
GNI per capita, World Bank Atlas method (current US$)	630	1,230	1,623	1,188
Rural population (% of total)	54	50	61	63
Expected years of schooling (years)	9	—	*10*	*9*
Physicians density (per 1,000 people)	*0.3*	*0.4*	0.8	0.2
Depositors with commercial banks (per 1,000 adults)	—	—	—	*167*
Sector structure				
Number of mobile operators	—	5		
Herfindahl-Hirschman Index (HHI) (scale = 0–10,000)	—	3,305		
Sector performance				
Access				
Mobile cellular subscriptions (per 100 people)	13	59[a]	78[a]	57[a]
Mobile cellular subscriptions (% prepaid)	99	97[a]	96[a]	96[a]
Population covered by a mobile-cellular network (%)	58	90	*86*	72
Mobile broadband subscriptions (per 100 people)	*0.01*	3.9[a]	7.3[a]	5.6[a]
Mobile broadband (% of total mobile subscriptions)	*0.06*	6.6[a]	9.0[a]	10.1[a]
Usage				
Households with a mobile telephone (%)	40	60	77	52
Mobile voice usage (minutes per user per month)	—	—	276[a]	—
Population using mobile Internet (%)	—	1.3	2.9	—
Short Message Service (SMS) users (% of mobile users)	—	*26.0*	61.9[a]	—
Affordability				
Mobile basket (% of GNI per capita)	49.1	13.4	7.2	19.5

Mobile cellular subscriptions, 2005–11
Number per 100 people

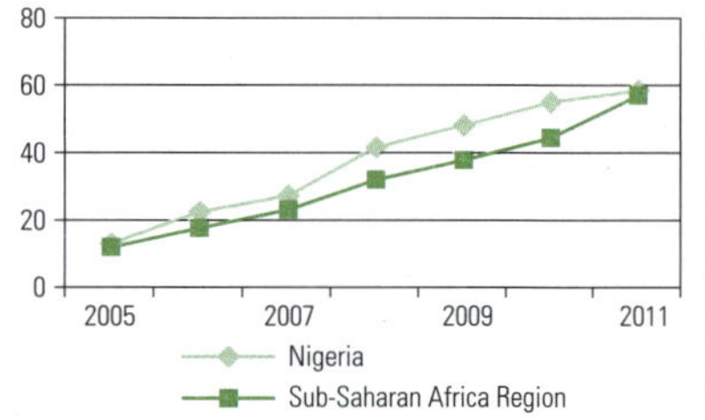

Mobile basket, 2005–10
Percentage of GNI per capita

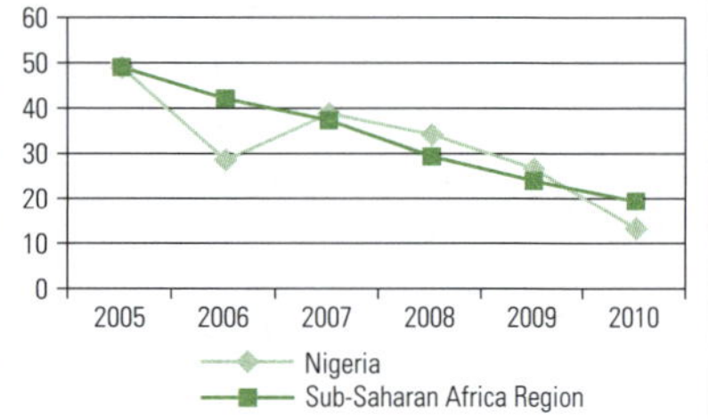

Sources: Economic and social context: IMF, UIS, UN, WHO and World Bank; Sector structure: ictDATA.org; Sector performance: ictDATA.org, ITU; Wireless Intelligence, and World Bank.
Notes: Use of italics in the column entries indicates years or periods other than those specified. — Not available. GNI = gross national income.
a. Data are for 2011.

Norway

	Norway 2005	Norway 2010	High-income group 2010
Economic and social context			
Population (total, million)	5	5	1,127
GNI per capita, World Bank Atlas method (current US$)	62,490	87,350	38,746
Rural population (% of total)	23	22	22
Expected years of schooling (years)	17	*17*	*16*
Physicians density (per 1,000 people)	*3.8*	*4.2*	2.8
Depositors with commercial banks (per 1,000 adults)	422	529	—
Sector structure			
Number of mobile operators	—	3	
Herfindahl-Hirschman Index (HHI) (scale = 0–10,000)	—	4,478	
Sector performance			
Access			
Mobile cellular subscriptions (per 100 people)	103	115[a]	118[a]
Mobile cellular subscriptions (% prepaid)	37	32[a]	36[a]
Population covered by a mobile-cellular network (%)	100	100	100
Mobile broadband subscriptions (per 100 people)	1.9	62.2[a]	69.6[a]
Mobile broadband (% of total mobile subscriptions)	1.8	44.5[a]	57.6[a]
Usage			
Households with a mobile telephone (%)	94	95	93
Mobile voice usage (minutes per user per month)	189	246[a]	339
Population using mobile Internet (%)	6.5	16.4	24.3
Short Message Service (SMS) users (% of mobile users)	—	—	78.2[a]
Affordability			
Mobile basket (% of GNI per capita)	0.5	0.3	1.0

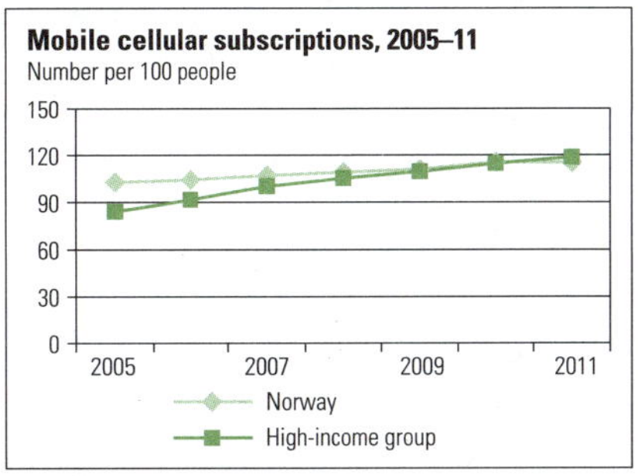

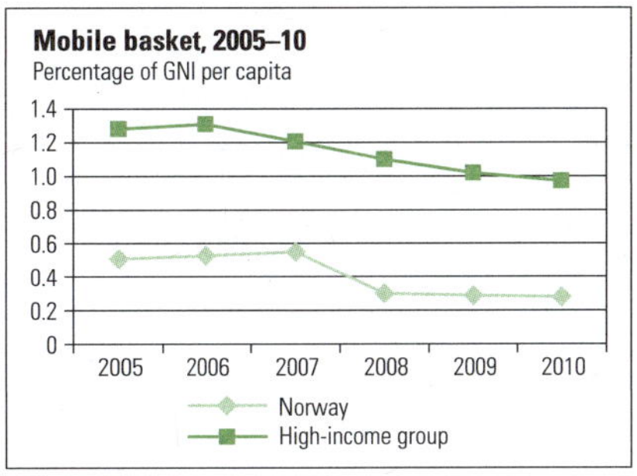

Oman

	Oman 2005	Oman 2010	High-income group 2010
Economic and social context			
Population (total, million)	2	3	1,127
GNI per capita, World Bank Atlas method (current US$)	11,190	*18,260*	38,746
Rural population (% of total)	29	28	22
Expected years of schooling (years)	*11*	*12*	*16*
Physicians density (per 1,000 people)	1.7	*1.9*	2.8
Depositors with commercial banks (per 1,000 adults)	—	*1,012*	—
Sector structure			
Number of mobile operators	—	2	
Herfindahl-Hirschman Index (HHI) (scale = 0–10,000)	—	5,072	
Sector performance			
Access			
Mobile cellular subscriptions (per 100 people)	55	169[a]	118[a]
Mobile cellular subscriptions (% prepaid)	89	80[a]	36[a]
Population covered by a mobile-cellular network (%)	92	98	100
Mobile broadband subscriptions (per 100 people)	—	26.5[a]	69.6[a]
Mobile broadband (% of total mobile subscriptions)	—	15.9[a]	57.6[a]
Usage			
Households with a mobile telephone (%)	80	95	93
Mobile voice usage (minutes per user per month)	—	—	339
Population using mobile Internet (%)	—	—	24.3
Short Message Service (SMS) users (% of mobile users)	—	*82.0*	78.2[a]
Affordability			
Mobile basket (% of GNI per capita)	0.9	*0.6*	1.0

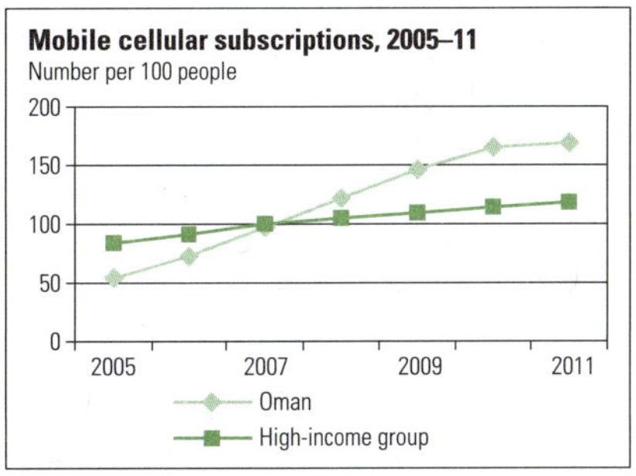

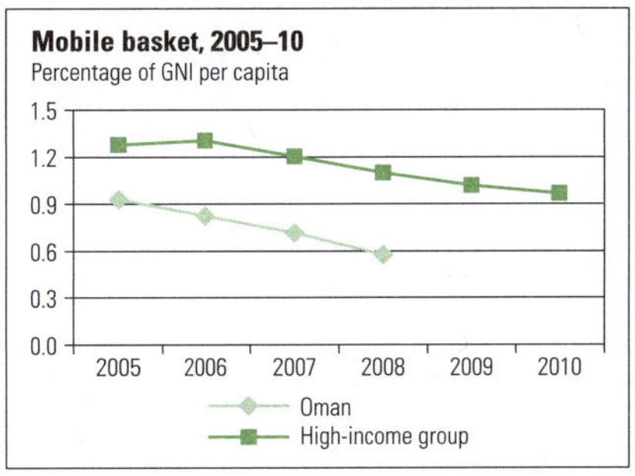

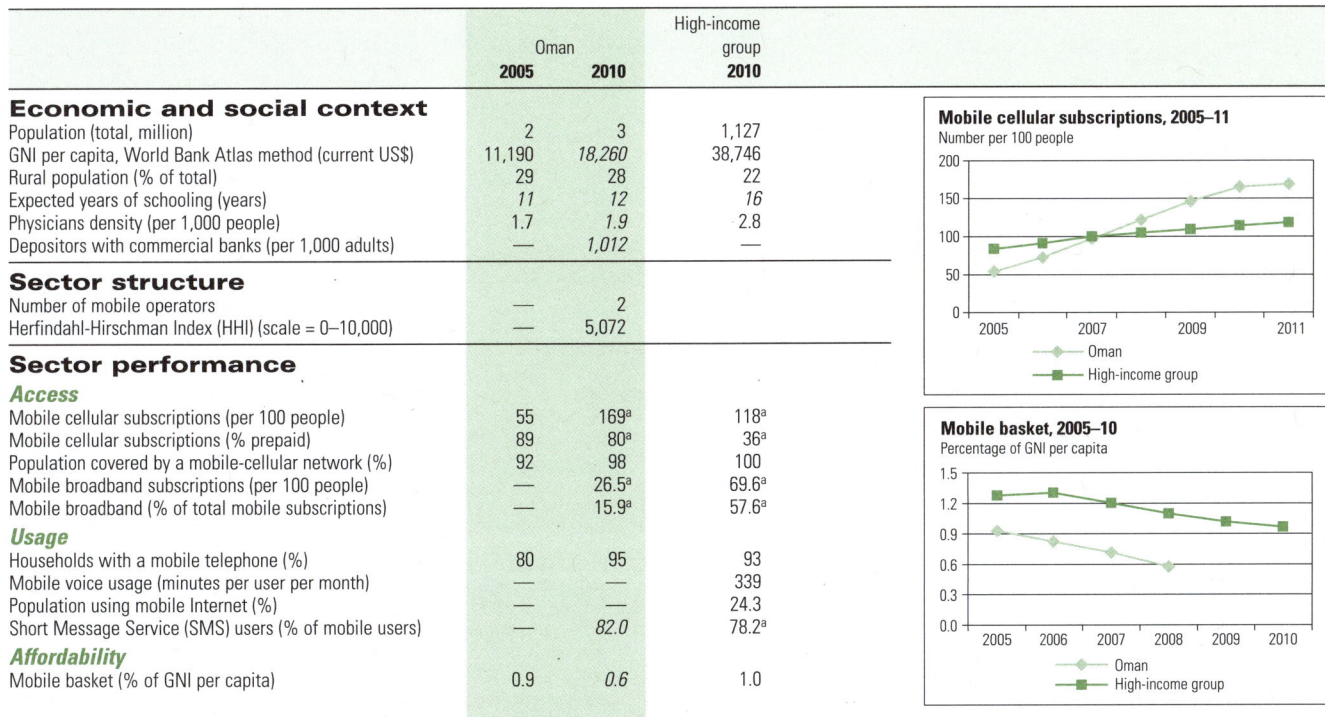

Sources: Economic and social context: IMF, UIS, UN, WHO and World Bank; Sector structure: ictDATA.org; Sector performance: ictDATA.org, ITU; Wireless Intelligence, and World Bank.
Notes: Use of italics in the column entries indicates years or periods other than those specified. — Not available. GNI = gross national income.
a. Data are for 2011.

Pakistan

	Pakistan 2005	Pakistan 2010	Lower-middle-income group 2010	South Asia Region 2010
Economic and social context				
Population (total, million)	159	174	2,519	1,633
GNI per capita, World Bank Atlas method (current US$)	710	1,050	1,623	1,176
Rural population (% of total)	65	63	61	70
Expected years of schooling (years)	6	*7*	*10*	*10*
Physicians density (per 1,000 people)	0.8	*0.8*	0.8	0.6
Depositors with commercial banks (per 1,000 adults)	131	249	—	249
Sector structure				
Number of mobile operators	—	5		
Herfindahl-Hirschman Index (HHI) (scale = 0–10,000)	—	2,282		
Sector performance				
Access				
Mobile cellular subscriptions (per 100 people)	8	64[a]	78[a]	67[a]
Mobile cellular subscriptions (% prepaid)	97	98[a]	96[a]	96[a]
Population covered by a mobile-cellular network (%)	36	92	*86*	*84*
Mobile broadband subscriptions (per 100 people)	—	—	7.3[a]	3.3[a]
Mobile broadband (% of total mobile subscriptions)	—	—	9.0[a]	4.6[a]
Usage				
Households with a mobile telephone (%)	33	48	77	54
Mobile voice usage (minutes per user per month)	151	205[a]	276[a]	305[a]
Population using mobile Internet (%)	—	1.7	2.9	3.3[a]
Short Message Service (SMS) users (% of mobile users)	—	44.0[a]	61.9[a]	47.0[a]
Affordability				
Mobile basket (% of GNI per capita)	8.5	2.9	7.2	3.2

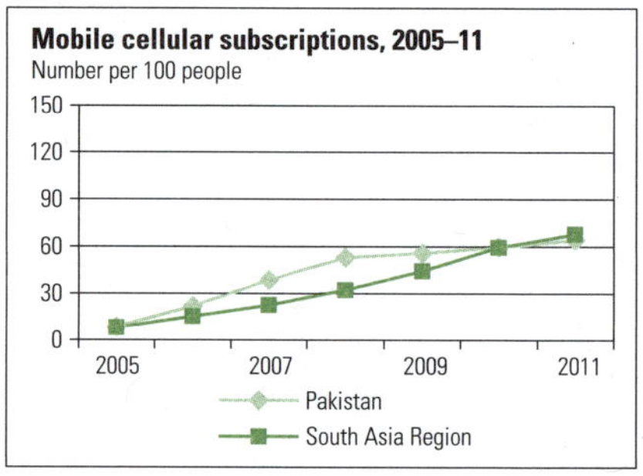

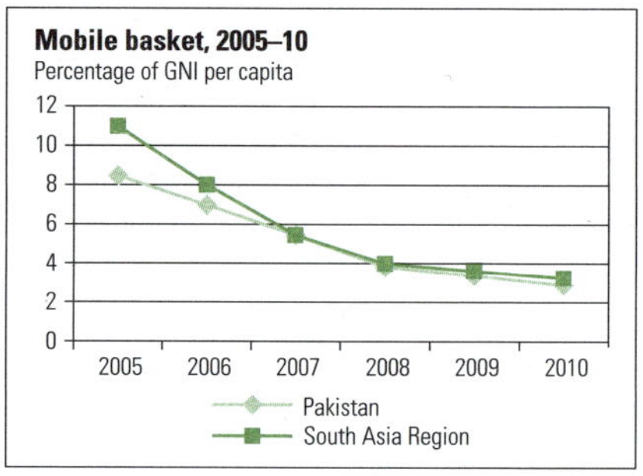

Panama

	Panama 2005	Panama 2010	Upper-middle-income group 2010	Latin America & the Caribbean Region 2010
Economic and social context				
Population (total, million)	3	4	2,452	583
GNI per capita, World Bank Atlas method (current US$)	4,640	6,970	5,886	7,741
Rural population (% of total)	29	25	43	21
Expected years of schooling (years)	13	*13*	*13*	*14*
Physicians density (per 1,000 people)	—	—	1.7	1.8
Depositors with commercial banks (per 1,000 adults)	—	—	—	—
Sector structure				
Number of mobile operators	—	4		
Herfindahl-Hirschman Index (HHI) (scale = 0–10,000)	—	—		
Sector performance				
Access				
Mobile cellular subscriptions (per 100 people)	54	204[a]	92[a]	109[a]
Mobile cellular subscriptions (% prepaid)	92	94[a]	81[a]	81[a]
Population covered by a mobile-cellular network (%)	75	91	*99*	98
Mobile broadband subscriptions (per 100 people)	—	5.5[a]	14.3[a]	16.1[a]
Mobile broadband (% of total mobile subscriptions)	—	3.8[a]	15.4[a]	15.2[a]
Usage				
Households with a mobile telephone (%)	*64*	84	84	84
Mobile voice usage (minutes per user per month)	—	—	325[a]	141[a]
Population using mobile Internet (%)	—	—	22.9[a]	4.4
Short Message Service (SMS) users (% of mobile users)	—	—	74.4[a]	—
Affordability				
Mobile basket (% of GNI per capita)	*9.6*	1.5	2.9	3.7

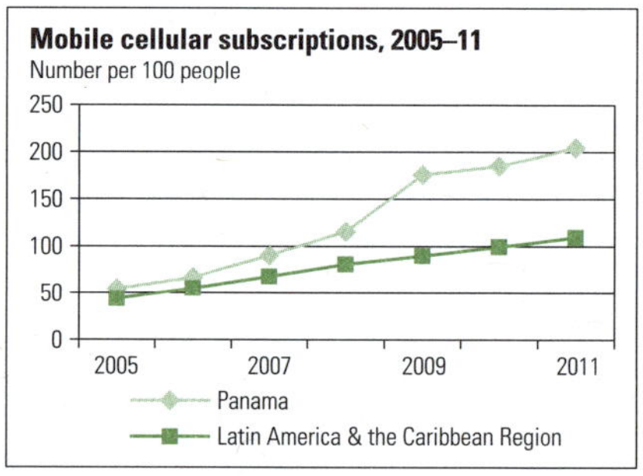

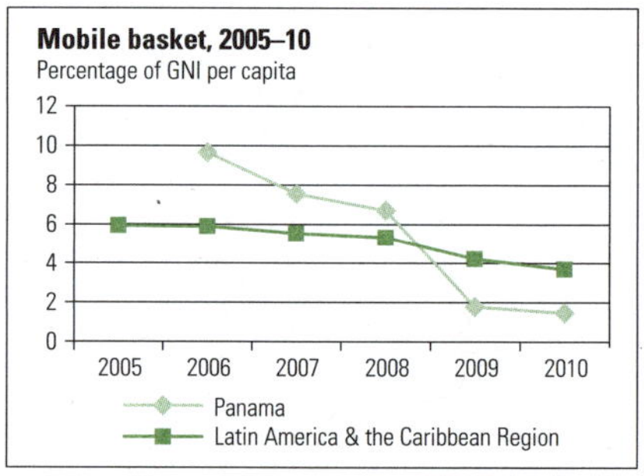

Sources: Economic and social context: IMF, UIS, UN, WHO and World Bank; Sector structure: ictDATA.org; Sector performance: ictDATA.org, ITU; Wireless Intelligence, and World Bank.
Notes: Use of italics in the column entries indicates years or periods other than those specified. — Not available. GNI = gross national income.
a. Data are for 2011.

Papua New Guinea

	Papua New Guinea 2005	Papua New Guinea 2010	Lower-middle-income group 2010	East Asia & Pacific Region 2010
Economic and social context				
Population (total, million)	6	7	2,519	1,962
GNI per capita, World Bank Atlas method (current US$)	680	1,300	1,623	3,696
Rural population (% of total)	87	88	61	54
Expected years of schooling (years)	—	—	*10*	*12*
Physicians density (per 1,000 people)	—	*0.1*	0.8	1.2
Depositors with commercial banks (per 1,000 adults)	157	*178*	—	—
Sector structure				
Number of mobile operators	—	2		
Herfindahl-Hirschman Index (HHI) (scale = 0–10,000)	—	—		
Sector performance				
Access				
Mobile cellular subscriptions (per 100 people)	1	28	78[a]	83[a]
Mobile cellular subscriptions (% prepaid)	98	99[a]	96[a]	85[a]
Population covered by a mobile-cellular network (%)	—	—	*86*	*99*
Mobile broadband subscriptions (per 100 people)	—	—	7.3[a]	11.6[a]
Mobile broadband (% of total mobile subscriptions)	—	—	9.0[a]	14.4[a]
Usage				
Households with a mobile telephone (%)	—	—	77	83
Mobile voice usage (minutes per user per month)	—	—	276[a]	367[a]
Population using mobile Internet (%)	—	—	2.9	22.4[a]
Short Message Service (SMS) users (% of mobile users)	—	—	61.9[a]	84.0[a]
Affordability				
Mobile basket (% of GNI per capita)	—	21.5	7.2	5.7

Mobile cellular subscriptions, 2005–11
Number per 100 people

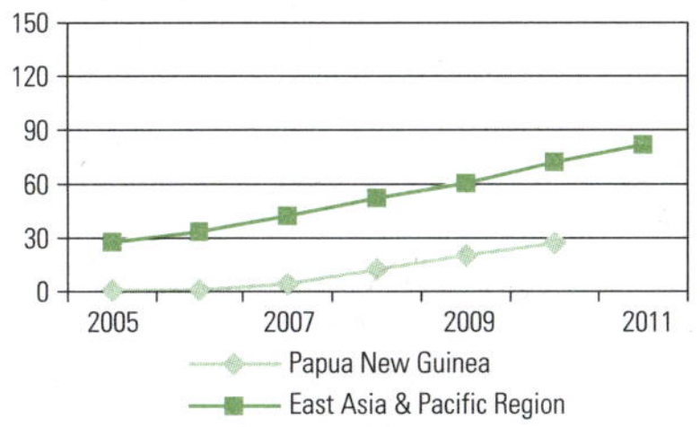

Mobile basket, 2005–10
Percentage of GNI per capita

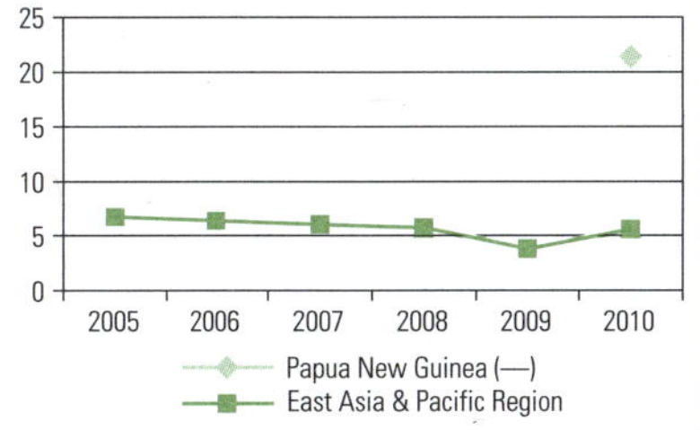

Paraguay

	Paraguay 2005	Paraguay 2010	Lower-middle-income group 2010	Latin America & the Caribbean Region 2010
Economic and social context				
Population (total, million)	6	6	2,519	583
GNI per capita, World Bank Atlas method (current US$)	1,220	2,720	1,623	7,741
Rural population (% of total)	42	39	61	21
Expected years of schooling (years)	12	*12*	*10*	*14*
Physicians density (per 1,000 people)	—	—	0.8	1.8
Depositors with commercial banks (per 1,000 adults)	—	—	—	—
Sector structure				
Number of mobile operators	—	4		
Herfindahl-Hirschman Index (HHI) (scale = 0–10,000)	—	3,655		
Sector performance				
Access				
Mobile cellular subscriptions (per 100 people)	32	96	78[a]	109[a]
Mobile cellular subscriptions (% prepaid)	87	84[a]	96[a]	81[a]
Population covered by a mobile-cellular network (%)	—	94	*86*	98
Mobile broadband subscriptions (per 100 people)	—	4.5[a]	7.3[a]	16.1[a]
Mobile broadband (% of total mobile subscriptions)	—	4.4[a]	9.0[a]	15.2[a]
Usage				
Households with a mobile telephone (%)	49	85	77	84
Mobile voice usage (minutes per user per month)	—	—	276[a]	141[a]
Population using mobile Internet (%)	—	—	2.9	4.4
Short Message Service (SMS) users (% of mobile users)	—	—	61.9[a]	—
Affordability				
Mobile basket (% of GNI per capita)	7.7	3.8	7.2	3.7

Mobile cellular subscriptions, 2005–11
Number per 100 people

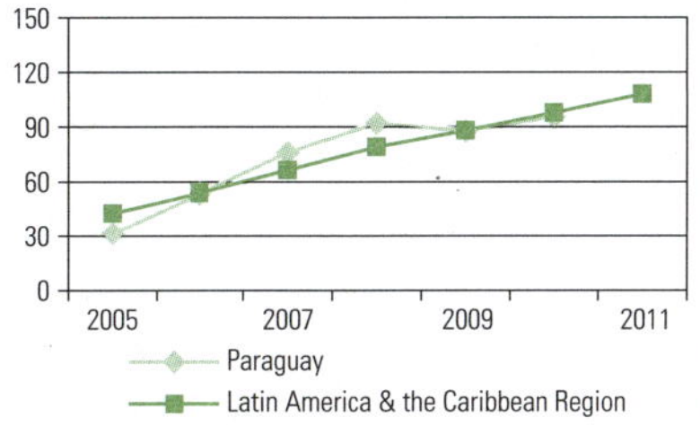

Mobile basket, 2005–10
Percentage of GNI per capita

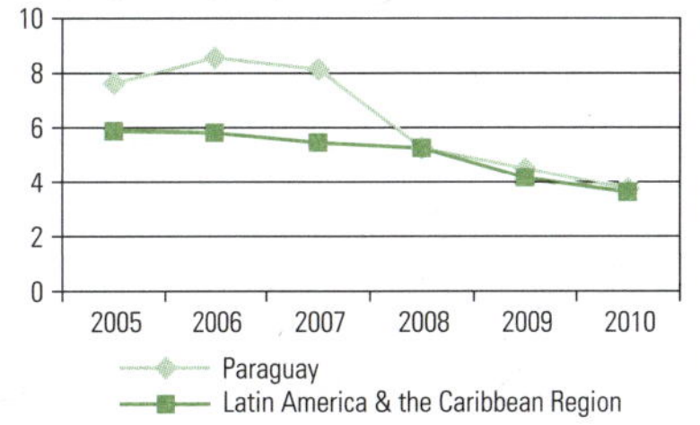

Sources: Economic and social context: IMF, UIS, UN, WHO and World Bank; Sector structure: ictDATA.org; Sector performance: ictDATA.org, ITU; Wireless Intelligence, and World Bank.
Notes: Use of italics in the column entries indicates years or periods other than those specified. — Not available. GNI = gross national income.
a. Data are for 2011.

Peru

	Peru 2005	Peru 2010	Upper-middle-income group 2010	Latin America & the Caribbean Region 2010
Economic and social context				
Population (total, million)	28	29	2,452	583
GNI per capita, World Bank Atlas method (current US$)	2,680	4,700	5,886	7,741
Rural population (% of total)	29	28	43	21
Expected years of schooling (years)	13	—	*13*	*14*
Physicians density (per 1,000 people)	—	*0.9*	1.7	1.8
Depositors with commercial banks (per 1,000 adults)	237	436	—	—
Sector structure				
Number of mobile operators	—	3		
Herfindahl-Hirschman Index (HHI) (scale = 0–10,000)	—	5,115		
Sector performance				
Access				
Mobile cellular subscriptions (per 100 people)	20	101[a]	92[a]	109[a]
Mobile cellular subscriptions (% prepaid)	82	79[a]	81[a]	81[a]
Population covered by a mobile-cellular network (%)	*87*	97	*99*	98
Mobile broadband subscriptions (per 100 people)	—	9.1[a]	14.3[a]	16.1[a]
Mobile broadband (% of total mobile subscriptions)	—	10.0[a]	15.4[a]	15.2[a]
Usage				
Households with a mobile telephone (%)	21	73	84	84
Mobile voice usage (minutes per user per month)	74	109[a]	325[a]	141[a]
Population using mobile Internet (%)	—	5.8	22.9[a]	4.4
Short Message Service (SMS) users (% of mobile users)	—	—	74.4[a]	—
Affordability				
Mobile basket (% of GNI per capita)	14.9	11.0	2.9	3.7

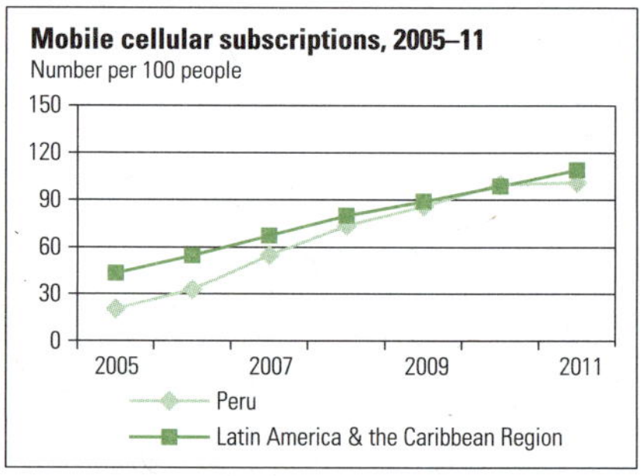

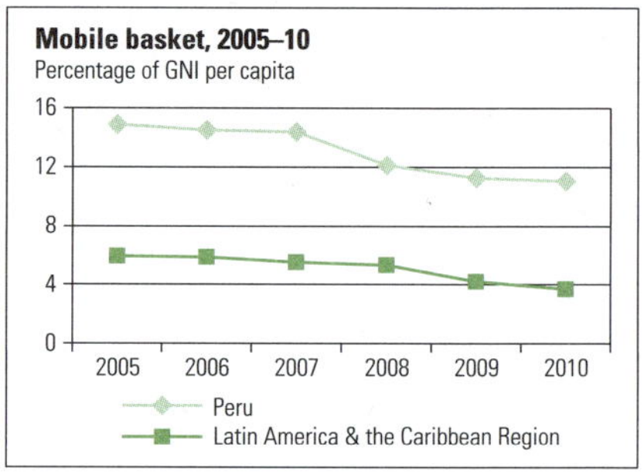

Philippines

	Philippines 2005	Philippines 2010	Lower-middle-income group 2010	East Asia & Pacific Region 2010
Economic and social context				
Population (total, million)	86	93	2,519	1,962
GNI per capita, World Bank Atlas method (current US$)	1,210	2,060	1,623	3,696
Rural population (% of total)	37	34	61	54
Expected years of schooling (years)	12	*12*	*10*	*12*
Physicians density (per 1,000 people)	*1.2*	—	0.8	1.2
Depositors with commercial banks (per 1,000 adults)	370	488	—	—
Sector structure				
Number of mobile operators	—	3		
Herfindahl-Hirschman Index (HHI) (scale = 0–10,000)	—	3,931		
Sector performance				
Access				
Mobile cellular subscriptions (per 100 people)	41	101[a]	78[a]	83[a]
Mobile cellular subscriptions (% prepaid)	97	96[a]	96[a]	85[a]
Population covered by a mobile-cellular network (%)	99	99	*86*	*99*
Mobile broadband subscriptions (per 100 people)	0.0	23.1[a]	7.3[a]	11.6[a]
Mobile broadband (% of total mobile subscriptions)	0.0	23.2[a]	9.0[a]	14.4[a]
Usage				
Households with a mobile telephone (%)	47	80	77	83
Mobile voice usage (minutes per user per month)	—	69[a]	276[a]	367[a]
Population using mobile Internet (%)	0.5	9.8[a]	2.9	22.4[a]
Short Message Service (SMS) users (% of mobile users)	—	97.0	61.9[a]	84.0[a]
Affordability				
Mobile basket (% of GNI per capita)	6.9	5.9	7.2	5.7

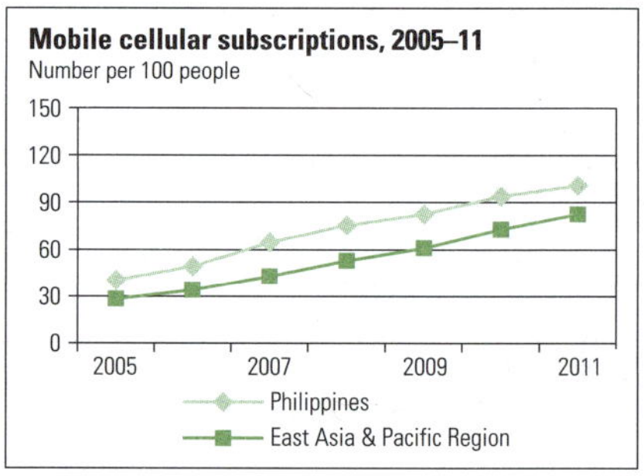

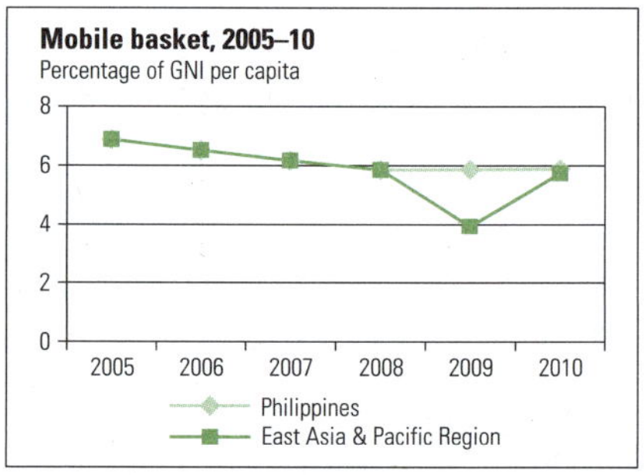

Sources: Economic and social context: IMF, UIS, UN, WHO and World Bank; Sector structure: ictDATA.org; Sector performance: ictDATA.org, ITU; Wireless Intelligence, and World Bank.
Notes: Use of italics in the column entries indicates years or periods other than those specified. — Not available. GNI = gross national income.
a. Data are for 2011.

Poland

	Poland 2005	Poland 2010	High-income group 2010
Economic and social context			
Population (total, million)	38	38	1,127
GNI per capita, World Bank Atlas method (current US$)	7,290	12,440	38,746
Rural population (% of total)	39	39	22
Expected years of schooling (years)	15	*15*	*16*
Physicians density (per 1,000 people)	2.0	*2.2*	2.8
Depositors with commercial banks (per 1,000 adults)	—	—	—
Sector structure			
Number of mobile operators	—	5	
Herfindahl-Hirschman Index (HHI) (scale = 0–10,000)	—	2,692	
Sector performance			
Access			
Mobile cellular subscriptions (per 100 people)	76	111[a]	118[a]
Mobile cellular subscriptions (% prepaid)	62	51[a]	36[a]
Population covered by a mobile-cellular network (%)	99	99	100
Mobile broadband subscriptions (per 100 people)	0.05	43.4[a]	69.6[a]
Mobile broadband (% of total mobile subscriptions)	0.06	33.2[a]	57.6[a]
Usage			
Households with a mobile telephone (%)	62	88	93
Mobile voice usage (minutes per user per month)	*67*	148[a]	339
Population using mobile Internet (%)	—	3.7	24.3
Short Message Service (SMS) users (% of mobile users)	—	85.0[a]	78.2[a]
Affordability			
Mobile basket (% of GNI per capita)	2.3	1.5	1.0

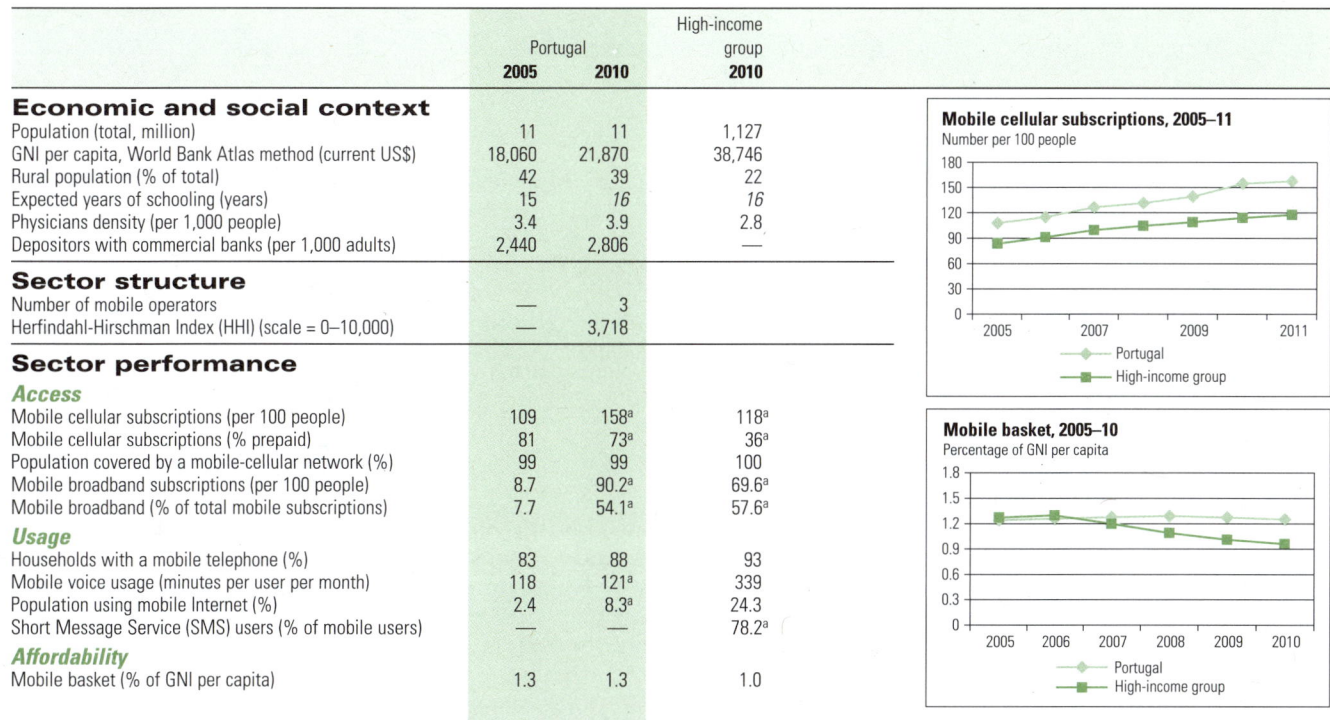

Mobile cellular subscriptions, 2005–11
Number per 100 people

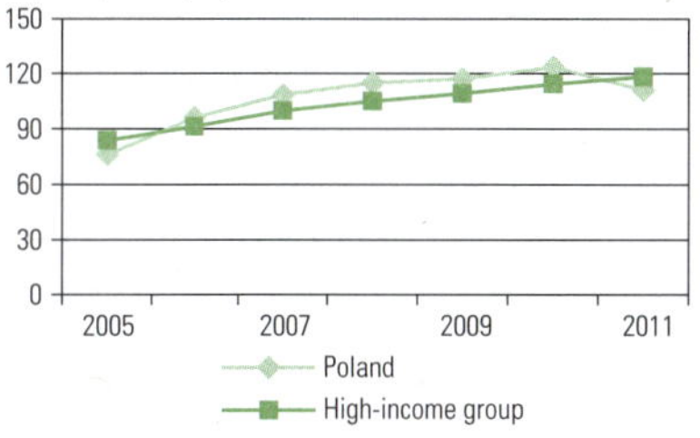

Mobile basket, 2005–10
Percentage of GNI per capita

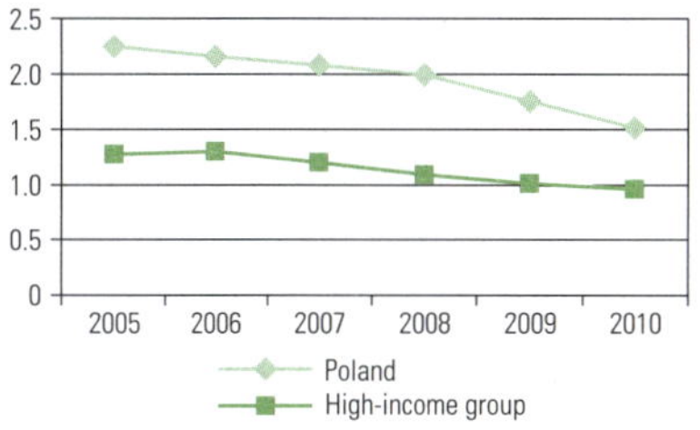

Portugal

	Portugal 2005	Portugal 2010	High-income group 2010
Economic and social context			
Population (total, million)	11	11	1,127
GNI per capita, World Bank Atlas method (current US$)	18,060	21,870	38,746
Rural population (% of total)	42	39	22
Expected years of schooling (years)	15	*16*	*16*
Physicians density (per 1,000 people)	3.4	3.9	2.8
Depositors with commercial banks (per 1,000 adults)	2,440	2,806	—
Sector structure			
Number of mobile operators	—	3	
Herfindahl-Hirschman Index (HHI) (scale = 0–10,000)	—	3,718	
Sector performance			
Access			
Mobile cellular subscriptions (per 100 people)	109	158[a]	118[a]
Mobile cellular subscriptions (% prepaid)	81	73[a]	36[a]
Population covered by a mobile-cellular network (%)	99	99	100
Mobile broadband subscriptions (per 100 people)	8.7	90.2[a]	69.6[a]
Mobile broadband (% of total mobile subscriptions)	7.7	54.1[a]	57.6[a]
Usage			
Households with a mobile telephone (%)	83	88	93
Mobile voice usage (minutes per user per month)	118	121[a]	339
Population using mobile Internet (%)	2.4	8.3[a]	24.3
Short Message Service (SMS) users (% of mobile users)	—	—	78.2[a]
Affordability			
Mobile basket (% of GNI per capita)	1.3	1.3	1.0

Mobile cellular subscriptions, 2005–11
Number per 100 people

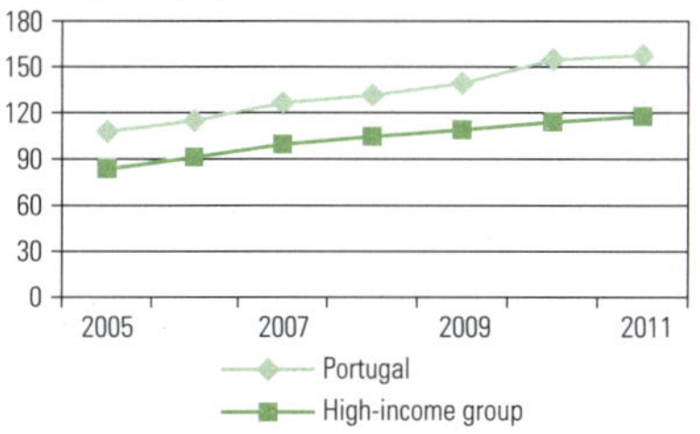

Mobile basket, 2005–10
Percentage of GNI per capita

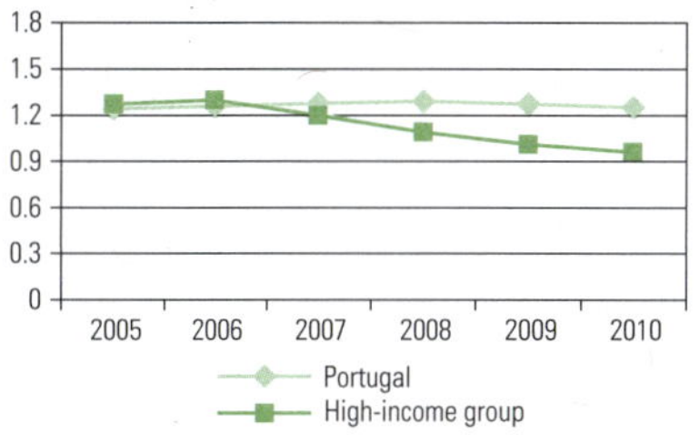

Sources: Economic and social context: IMF, UIS, UN, WHO and World Bank; Sector structure: ictDATA.org; Sector performance: ictDATA.org, ITU; Wireless Intelligence, and World Bank.
Notes: Use of italics in the column entries indicates years or periods other than those specified. — Not available. GNI = gross national income.
a. Data are for 2011.

Puerto Rico

	Puerto Rico 2005	Puerto Rico 2010	High-income group 2010
Economic and social context			
Population (total, million)	4	4	1,127
GNI per capita, World Bank Atlas method (current US$)	14,190	15,500	38,746
Rural population (% of total)	2	1	22
Expected years of schooling (years)	—	—	*16*
Physicians density (per 1,000 people)	—	—	2.8
Depositors with commercial banks (per 1,000 adults)	—	—	—
Sector structure			
Number of mobile operators	—	—	
Herfindahl-Hirschman Index (HHI) (scale = 0–10,000)	—	—	
Sector performance			
Access			
Mobile cellular subscriptions (per 100 people)	51	74	118[a]
Mobile cellular subscriptions (% prepaid)	10	17[a]	36[a]
Population covered by a mobile-cellular network (%)	—	—	100
Mobile broadband subscriptions (per 100 people)	0.4	11.7[a]	69.6[a]
Mobile broadband (% of total mobile subscriptions)	0.8	15.8[a]	57.6[a]
Usage			
Households with a mobile telephone (%)	—	—	93
Mobile voice usage (minutes per user per month)	—	—	339
Population using mobile Internet (%)	—	—	24.3
Short Message Service (SMS) users (% of mobile users)	—	—	78.2[a]
Affordability			
Mobile basket (% of GNI per capita)	—	—	1.0

Mobile cellular subscriptions, 2005–11
Number per 100 people

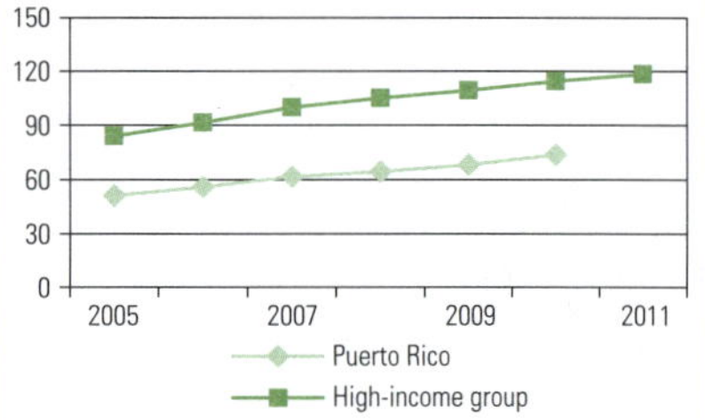

Mobile basket, 2005–10
Percentage of GNI per capita

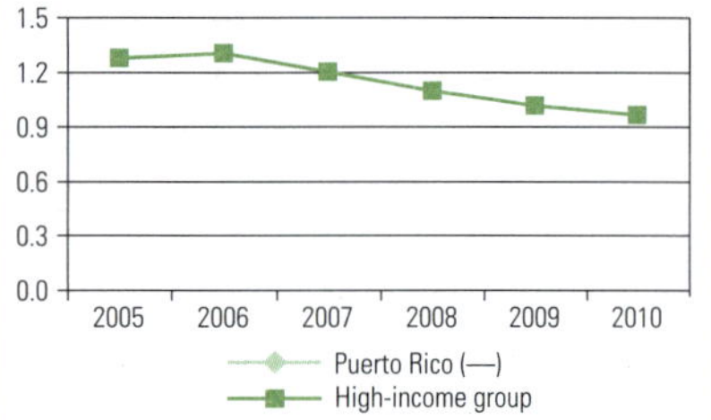

Qatar

	Qatar 2005	Qatar 2010	High-income group 2010
Economic and social context			
Population (total, million)	0.82	2	1,127
GNI per capita, World Bank Atlas method (current US$)	—	—	38,746
Rural population (% of total)	5	4	22
Expected years of schooling (years)	14	*12*	*16*
Physicians density (per 1,000 people)	2.6	—	2.8
Depositors with commercial banks (per 1,000 adults)	672	770	—
Sector structure			
Number of mobile operators	—	2	
Herfindahl-Hirschman Index (HHI) (scale = 0–10,000)	—	6,250	
Sector performance			
Access			
Mobile cellular subscriptions (per 100 people)	87	153[a]	118[a]
Mobile cellular subscriptions (% prepaid)	81	87[a]	36[a]
Population covered by a mobile-cellular network (%)	99	100	100
Mobile broadband subscriptions (per 100 people)	*0.7*	43.5[a]	69.6[a]
Mobile broadband (% of total mobile subscriptions)	*0.7*	28.4[a]	57.6[a]
Usage			
Households with a mobile telephone (%)	97	99	93
Mobile voice usage (minutes per user per month)	—	—	339
Population using mobile Internet (%)	—	32.4	24.3
Short Message Service (SMS) users (% of mobile users)	—	—	78.2[a]
Affordability			
Mobile basket (% of GNI per capita)	—	—	1.0

Mobile cellular subscriptions, 2005–11
Number per 100 people

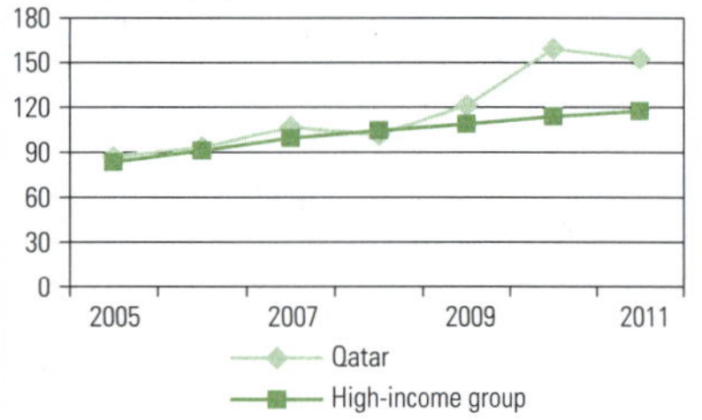

Mobile basket, 2005–10
Percentage of GNI per capita

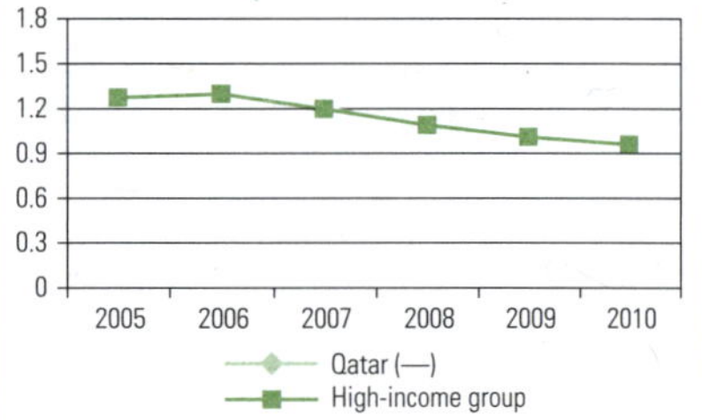

Sources: Economic and social context: IMF, UIS, UN, WHO and World Bank; Sector structure: ictDATA.org; Sector performance: ictDATA.org, ITU; Wireless Intelligence, and World Bank.
Notes: Use of italics in the column entries indicates years or periods other than those specified. — Not available. GNI = gross national income.
a. Data are for 2011.

Romania

	Romania 2005	Romania 2010	Upper-middle-income group 2010	Europe & Central Asia Region 2010
Economic and social context				
Population (total, million)	22	21	2,452	405
GNI per capita, World Bank Atlas method (current US$)	3,920	7,850	5,886	7,272
Rural population (% of total)	46	45	43	36
Expected years of schooling (years)	14	*15*	*13*	*13*
Physicians density (per 1,000 people)	*1.9*	*2.3*	1.7	3.2
Depositors with commercial banks (per 1,000 adults)	—	—	—	894
Sector structure				
Number of mobile operators	—	4		
Herfindahl-Hirschman Index (HHI) (scale = 0–10,000)	—	3,130		
Sector performance				
Access				
Mobile cellular subscriptions (per 100 people)	62	110[a]	92[a]	125[a]
Mobile cellular subscriptions (% prepaid)	67	68[a]	81[a]	82[a]
Population covered by a mobile-cellular network (%)	*97*	100	*99*	96
Mobile broadband subscriptions (per 100 people)	0.7	38.8[a]	14.3[a]	22.6[a]
Mobile broadband (% of total mobile subscriptions)	1.1	30.0[a]	15.4[a]	18.0[a]
Usage				
Households with a mobile telephone (%)	50	77	84	82
Mobile voice usage (minutes per user per month)	—	213[a]	325[a]	288[a]
Population using mobile Internet (%)	0.4	8.0[a]	22.9[a]	8.5
Short Message Service (SMS) users (% of mobile users)	—	—	74.4[a]	69.8[a]
Affordability				
Mobile basket (% of GNI per capita)	4.5	3.1	2.9	3.1

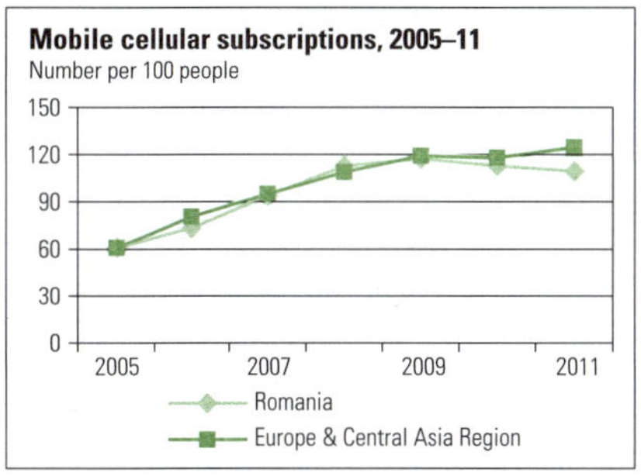

Mobile cellular subscriptions, 2005–11

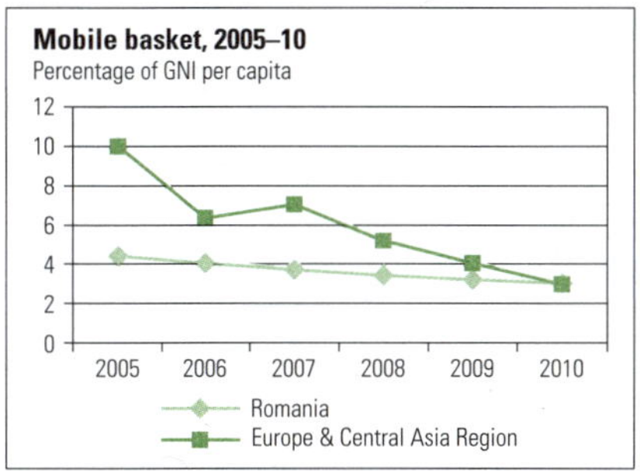

Mobile basket, 2005–10

Russian Federation

	Russian Federation 2005	Russian Federation 2010	Upper-middle-income group 2010	Europe & Central Asia Region 2010
Economic and social context				
Population (total, million)	143	142	2,452	405
GNI per capita, World Bank Atlas method (current US$)	4,460	9,900	5,886	7,272
Rural population (% of total)	27	27	43	36
Expected years of schooling (years)	14	*14*	*13*	*13*
Physicians density (per 1,000 people)	4.0	—	1.7	3.2
Depositors with commercial banks (per 1,000 adults)	—	—	—	894
Sector structure				
Number of mobile operators	—	6		
Herfindahl-Hirschman Index (HHI) (scale = 0–10,000)	—	2,570		
Sector performance				
Access				
Mobile cellular subscriptions (per 100 people)	84	160[a]	92[a]	125[a]
Mobile cellular subscriptions (% prepaid)	91	88[a]	81[a]	82[a]
Population covered by a mobile-cellular network (%)	*95*	—	*99*	96
Mobile broadband subscriptions (per 100 people)	*0.03*	25.0[a]	14.3[a]	22.6[a]
Mobile broadband (% of total mobile subscriptions)	*0.03*	15.5[a]	15.4[a]	18.0[a]
Usage				
Households with a mobile telephone (%)	32	*90*	84	82
Mobile voice usage (minutes per user per month)	136	275[a]	325[a]	288[a]
Population using mobile Internet (%)	0.3	17.0[a]	22.9[a]	8.5
Short Message Service (SMS) users (% of mobile users)	—	75.0[a]	74.4[a]	69.8[a]
Affordability				
Mobile basket (% of GNI per capita)	2.9	1.1	2.9	3.1

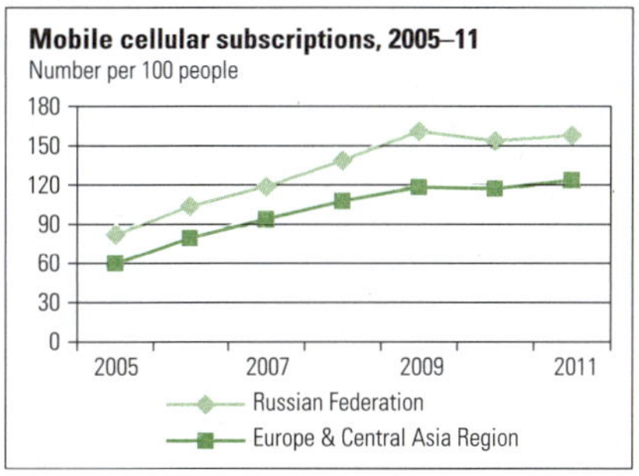

Mobile cellular subscriptions, 2005–11

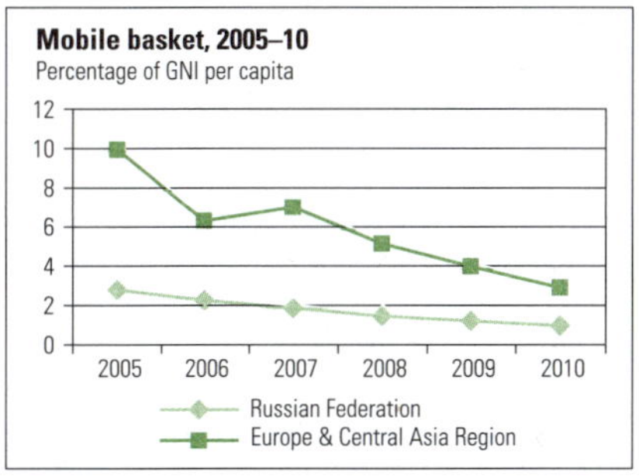

Mobile basket, 2005–10

Sources: Economic and social context: IMF, UIS, UN, WHO and World Bank; Sector structure: ictDATA.org; Sector performance: ictDATA.org, ITU; Wireless Intelligence, and World Bank.
Notes: Use of italics in the column entries indicates years or periods other than those specified. — Not available. GNI = gross national income.
a. Data are for 2011.

Rwanda

	Rwanda 2005	Rwanda 2010	Low-income group 2010	Sub-Saharan Africa Region 2010
Economic and social context				
Population (total, million)	9	11	796	853
GNI per capita, World Bank Atlas method (current US$)	270	520	530	1,188
Rural population (% of total)	83	81	72	63
Expected years of schooling (years)	9	*11*	*9*	*9*
Physicians density (per 1,000 people)	0.02	—	0.2	0.2
Depositors with commercial banks (per 1,000 adults)	9	*218*	—	*167*
Sector structure				
Number of mobile operators	—	3		
Herfindahl-Hirschman Index (HHI) (scale = 0–10,000)	—	5,609		
Sector performance				
Access				
Mobile cellular subscriptions (per 100 people)	2	39[a]	43[a]	57[a]
Mobile cellular subscriptions (% prepaid)	99	100[a]	98[a]	96[a]
Population covered by a mobile-cellular network (%)	75	96	—	72
Mobile broadband subscriptions (per 100 people)	—	6.2[a]	—	5.6[a]
Mobile broadband (% of total mobile subscriptions)	—	16.7[a]	—	10.1[a]
Usage				
Households with a mobile telephone (%)	5	40	43	52
Mobile voice usage (minutes per user per month)	—	96[a]	—	—
Population using mobile Internet (%)	—	0.5	—	—
Short Message Service (SMS) users (% of mobile users)	—	*35.0*	—	—
Affordability				
Mobile basket (% of GNI per capita)	97.0	32.1	28.8	19.5

Mobile cellular subscriptions, 2005–11
Number per 100 people

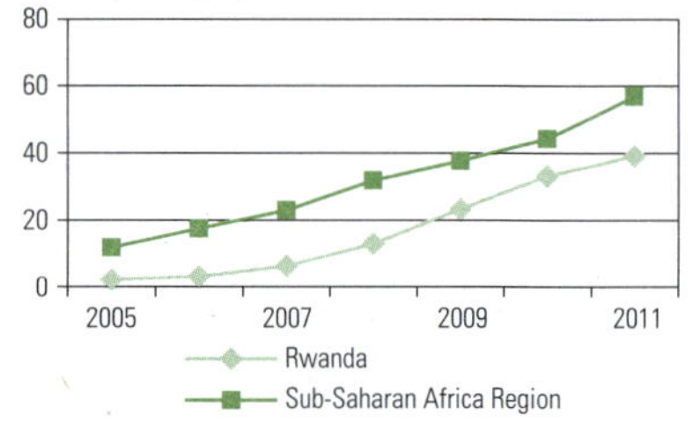

Mobile basket, 2005–10
Percentage of GNI per capita

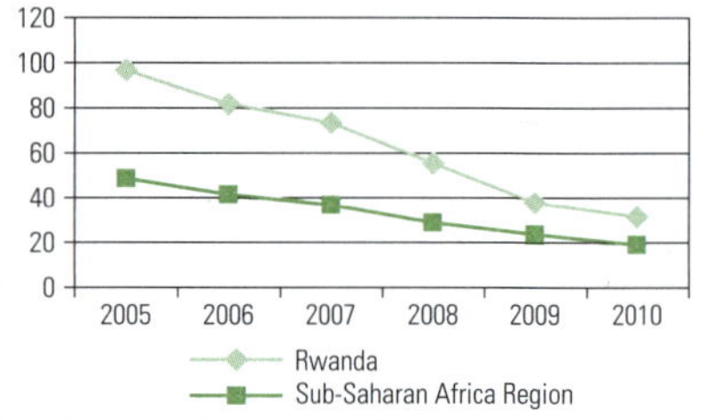

Saudi Arabia

	Saudi Arabia 2005	Saudi Arabia 2010	High-income group 2010
Economic and social context			
Population (total, million)	24	27	1,127
GNI per capita, World Bank Atlas method (current US$)	12,230	*16,190*	38,746
Rural population (% of total)	19	16	22
Expected years of schooling (years)	13	*14*	*16*
Physicians density (per 1,000 people)	*1.4*	*0.9*	2.8
Depositors with commercial banks (per 1,000 adults)	480	*780*	—
Sector structure			
Number of mobile operators	—	3	
Herfindahl-Hirschman Index (HHI) (scale = 0–10,000)	—	3,802	
Sector performance			
Access			
Mobile cellular subscriptions (per 100 people)	59	200[a]	118[a]
Mobile cellular subscriptions (% prepaid)	85	81[a]	36[a]
Population covered by a mobile-cellular network (%)	96	99	100
Mobile broadband subscriptions (per 100 people)	*0.6*	86.2[a]	69.6[a]
Mobile broadband (% of total mobile subscriptions)	*0.7*	42.8[a]	57.6[a]
Usage			
Households with a mobile telephone (%)	95	99	93
Mobile voice usage (minutes per user per month)	—	—	339
Population using mobile Internet (%)	—	7.3	24.3
Short Message Service (SMS) users (% of mobile users)	—	—	78.2[a]
Affordability			
Mobile basket (% of GNI per capita)	1.7	*1.0*	1.0

Mobile cellular subscriptions, 2005–11
Number per 100 people

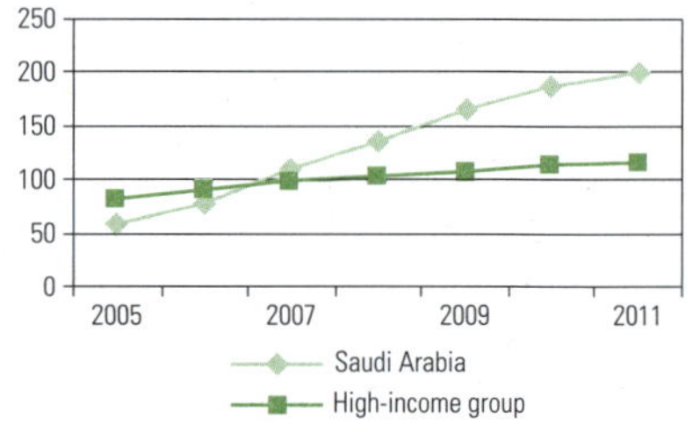

Mobile basket, 2005–10
Percentage of GNI per capita

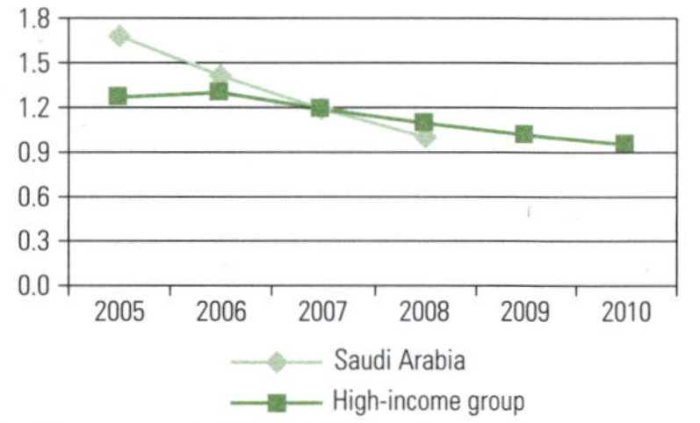

Sources: Economic and social context: IMF, UIS, UN, WHO and World Bank; Sector structure: ictDATA.org; Sector performance: ictDATA.org, ITU; Wireless Intelligence, and World Bank.
Notes: Use of italics in the column entries indicates years or periods other than those specified. — Not available. GNI = gross national income.
a. Data are for 2011.

Senegal

	Senegal 2005	Senegal 2010	Lower-middle-income group 2010	Sub-Saharan Africa Region 2010
Economic and social context				
Population (total, million)	11	12	2,519	853
GNI per capita, World Bank Atlas method (current US$)	800	1,080	1,623	1,188
Rural population (% of total)	58	57	61	63
Expected years of schooling (years)	7	*7*	*10*	*9*
Physicians density (per 1,000 people)	*0.1*	*0.1*	0.8	0.2
Depositors with commercial banks (per 1,000 adults)	—	—	—	*167*
Sector structure				
Number of mobile operators	—	3		
Herfindahl-Hirschman Index (HHI) (scale = 0–10,000)	—	4,893		
Sector performance				
Access				
Mobile cellular subscriptions (per 100 people)	16	74[a]	78[a]	57[a]
Mobile cellular subscriptions (% prepaid)	98	99[a]	96[a]	96[a]
Population covered by a mobile-cellular network (%)	85	90	*86*	72
Mobile broadband subscriptions (per 100 people)	—	6.9[a]	7.3[a]	5.6[a]
Mobile broadband (% of total mobile subscriptions)	—	8.8[a]	9.0[a]	10.1[a]
Usage				
Households with a mobile telephone (%)	30	86	77	52
Mobile voice usage (minutes per user per month)	—	—	276[a]	—
Population using mobile Internet (%)	—	0.3	2.9	—
Short Message Service (SMS) users (% of mobile users)	—	—	61.9[a]	—
Affordability				
Mobile basket (% of GNI per capita)	38.7	14.1	7.2	19.5

Mobile cellular subscriptions, 2005–11
Number per 100 people

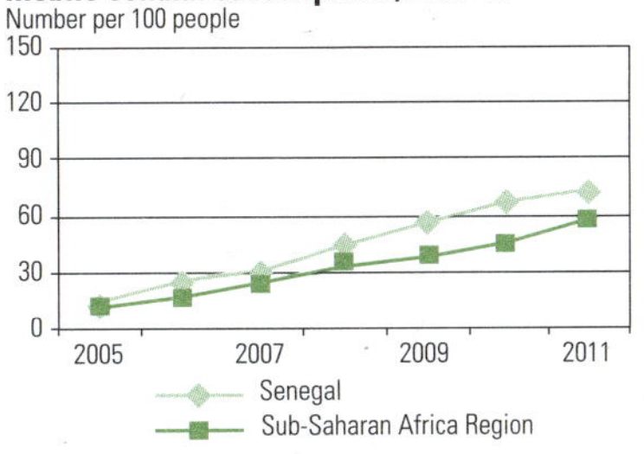

Mobile basket, 2005–10
Percentage of GNI per capita

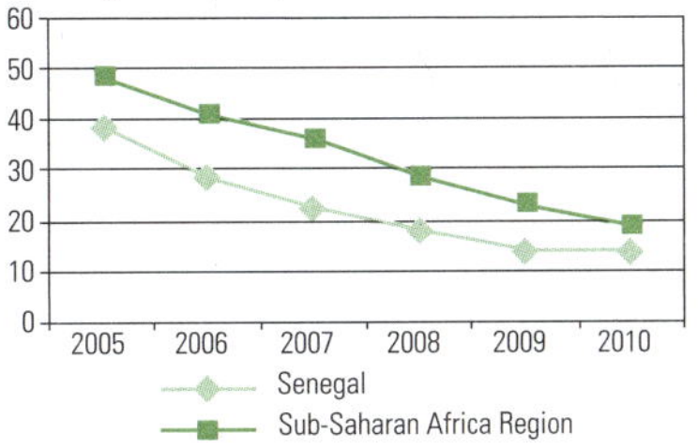

Serbia

	Serbia 2005	Serbia 2010	Upper-middle-income group 2010	Europe & Central Asia Region 2010
Economic and social context				
Population (total, million)	7	7	2,452	405
GNI per capita, World Bank Atlas method (current US$)	3,430	5,630	5,886	7,272
Rural population (% of total)	49	48	43	36
Expected years of schooling (years)	*14*	*14*	*13*	*13*
Physicians density (per 1,000 people)	*2.0*	*2.1*	1.7	3.2
Depositors with commercial banks (per 1,000 adults)	—	—	—	894
Sector structure				
Number of mobile operators	—	—		
Herfindahl-Hirschman Index (HHI) (scale = 0–10,000)	—	—		
Sector performance				
Access				
Mobile cellular subscriptions (per 100 people)	74	143[a]	92[a]	125[a]
Mobile cellular subscriptions (% prepaid)	87	70[a]	81[a]	82[a]
Population covered by a mobile-cellular network (%)	96	97	*99*	96
Mobile broadband subscriptions (per 100 people)	*0.3*	19.5[a]	14.3[a]	22.6[a]
Mobile broadband (% of total mobile subscriptions)	*0.3*	13.6[a]	15.4[a]	18.0[a]
Usage				
Households with a mobile telephone (%)	70	82	84	82
Mobile voice usage (minutes per user per month)	—	—	325[a]	288[a]
Population using mobile Internet (%)	—	4.1	22.9[a]	8.5
Short Message Service (SMS) users (% of mobile users)	—	*64.0*	74.4[a]	69.8[a]
Affordability				
Mobile basket (% of GNI per capita)	4.0	2.5	2.9	3.1

Mobile cellular subscriptions, 2005–11
Number per 100 people

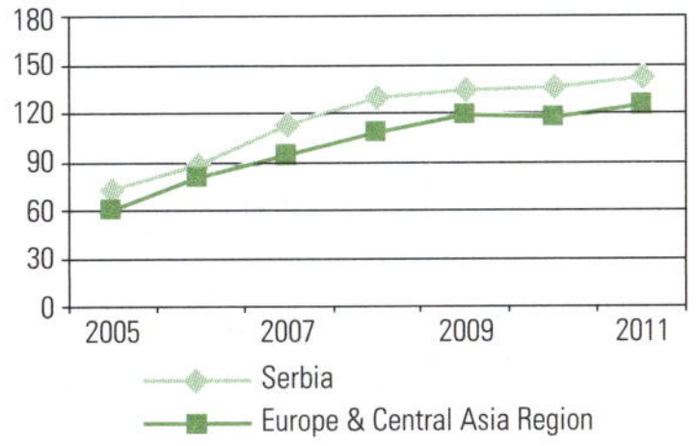

Mobile basket, 2005–10
Percentage of GNI per capita

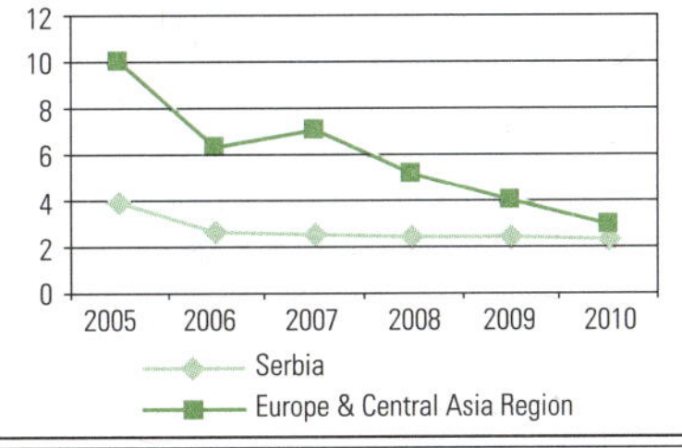

Sources: Economic and social context: IMF, UIS, UN, WHO and World Bank; Sector structure: ictDATA.org; Sector performance: ictDATA.org, ITU; Wireless Intelligence, and World Bank.
Notes: Use of italics in the column entries indicates years or periods other than those specified. — Not available. GNI = gross national income.
a. Data are for 2011.

Sierra Leone

	Sierra Leone 2005	Sierra Leone 2010	Low-income group 2010	Sub-Saharan Africa Region 2010
Economic and social context				
Population (total, million)	5	6	796	853
GNI per capita, World Bank Atlas method (current US$)	230	340	530	1,188
Rural population (% of total)	63	62	72	63
Expected years of schooling (years)	—	—	*9*	*9*
Physicians density (per 1,000 people)	*0.03*	*0.02*	0.2	0.2
Depositors with commercial banks (per 1,000 adults)	*61*	190	—	*167*
Sector structure				
Number of mobile operators	—	3		
Herfindahl-Hirschman Index (HHI) (scale = 0–10,000)	—	3,522		
Sector performance				
Access				
Mobile cellular subscriptions (per 100 people)	*14*	34	43[a]	57[a]
Mobile cellular subscriptions (% prepaid)	99	99[a]	98[a]	96[a]
Population covered by a mobile-cellular network (%)	*70*	—	—	72
Mobile broadband subscriptions (per 100 people)	—	0.8[a]	—	5.6[a]
Mobile broadband (% of total mobile subscriptions)	—	1.7[a]	—	10.1[a]
Usage				
Households with a mobile telephone (%)	—	37	43	52
Mobile voice usage (minutes per user per month)	—	—	—	—
Population using mobile Internet (%)	—	—	—	—
Short Message Service (SMS) users (% of mobile users)	—	—	—	—
Affordability				
Mobile basket (% of GNI per capita)	82.6	—	28.8	19.5

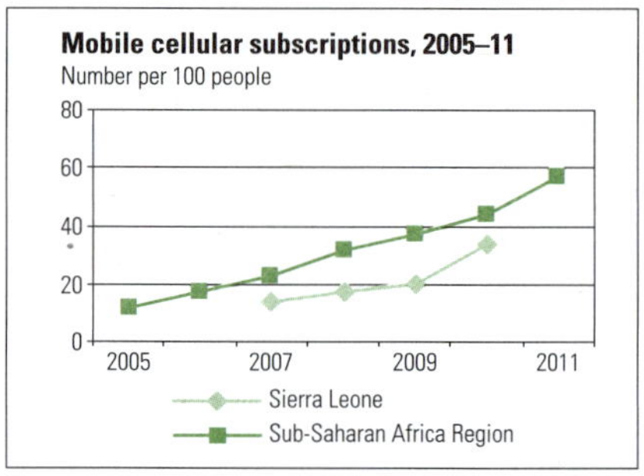

Mobile cellular subscriptions, 2005–11

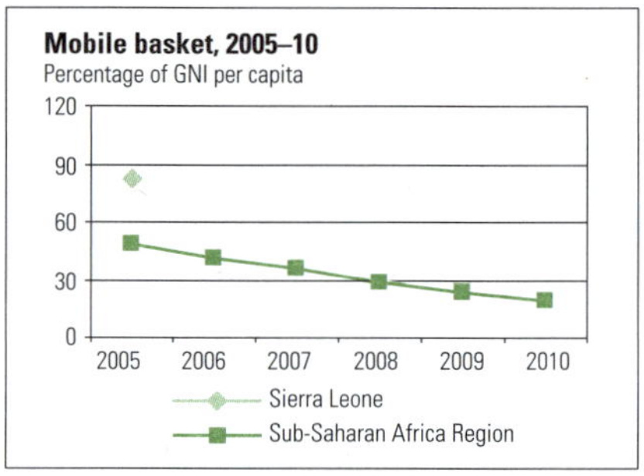

Mobile basket, 2005–10

Singapore

	Singapore 2005	Singapore 2010	High-income group 2010
Economic and social context			
Population (total, million)	4	5	1,127
GNI per capita, World Bank Atlas method (current US$)	27,180	40,070	38,746
Rural population (% of total)	0	0	22
Expected years of schooling (years)	—	—	*16*
Physicians density (per 1,000 people)	*1.5*	*1.8*	2.8
Depositors with commercial banks (per 1,000 adults)	2,031	2,134	—
Sector structure			
Number of mobile operators	—	3	
Herfindahl-Hirschman Index (HHI) (scale = 0–10,000)	—	3,520	
Sector performance			
Access			
Mobile cellular subscriptions (per 100 people)	103	150[a]	118[a]
Mobile cellular subscriptions (% prepaid)	35	48[a]	36[a]
Population covered by a mobile-cellular network (%)	100	100	100
Mobile broadband subscriptions (per 100 people)	4.1	81.9[a]	69.6[a]
Mobile broadband (% of total mobile subscriptions)	4.0	54.4[a]	57.6[a]
Usage			
Households with a mobile telephone (%)	91	96	93
Mobile voice usage (minutes per user per month)	312	366[a]	339
Population using mobile Internet (%)	—	25.6	24.3
Short Message Service (SMS) users (% of mobile users)	—	—	78.2[a]
Affordability			
Mobile basket (% of GNI per capita)	0.4	0.2	1.0

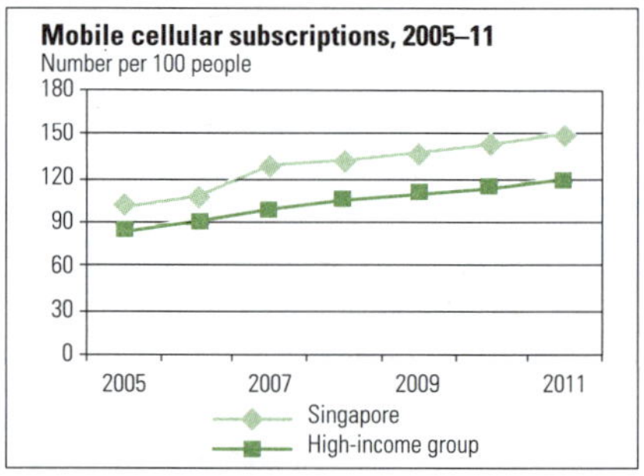

Mobile cellular subscriptions, 2005–11

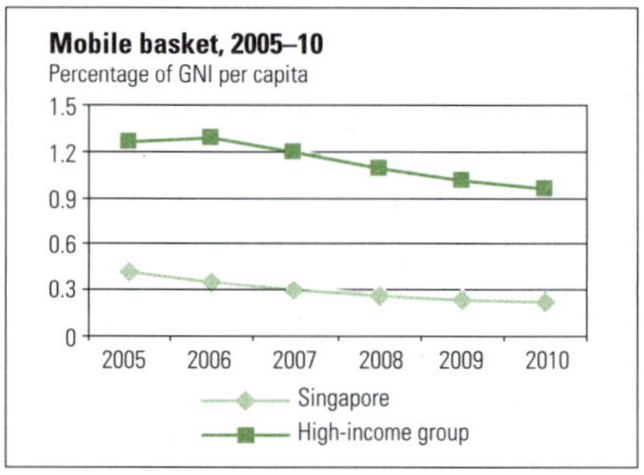

Mobile basket, 2005–10

Sources: Economic and social context: IMF, UIS, UN, WHO and World Bank; Sector structure: ictDATA.org; Sector performance: ictDATA.org, ITU; Wireless Intelligence, and World Bank.
Notes: Use of italics in the column entries indicates years or periods other than those specified. — Not available. GNI = gross national income.
a. Data are for 2011.

Slovak Republic

	Slovak Republic 2005	Slovak Republic 2010	High-income group 2010
Economic and social context			
Population (total, million)	5	5	1,127
GNI per capita, World Bank Atlas method (current US$)	11,040	16,840	38,746
Rural population (% of total)	44	43	22
Expected years of schooling (years)	14	*15*	*16*
Physicians density (per 1,000 people)	*3.1*	—	2.8
Depositors with commercial banks (per 1,000 adults)	—	—	—
Sector structure			
Number of mobile operators	—	3	
Herfindahl-Hirschman Index (HHI) (scale = 0–10,000)	—	3,918	
Sector performance			
Access			
Mobile cellular subscriptions (per 100 people)	84	110[a]	118[a]
Mobile cellular subscriptions (% prepaid)	58	49[a]	36[a]
Population covered by a mobile-cellular network (%)	100	100	100
Mobile broadband subscriptions (per 100 people)	*2.2*	40.9[a]	69.6[a]
Mobile broadband (% of total mobile subscriptions)	*2.4*	34.6[a]	57.6[a]
Usage			
Households with a mobile telephone (%)	85	88	93
Mobile voice usage (minutes per user per month)	—	—	339
Population using mobile Internet (%)	—	17.9	24.3
Short Message Service (SMS) users (% of mobile users)	—	—	78.2[a]
Affordability			
Mobile basket (% of GNI per capita)	1.4	2.7	1.0

Mobile cellular subscriptions, 2005–11
Number per 100 people

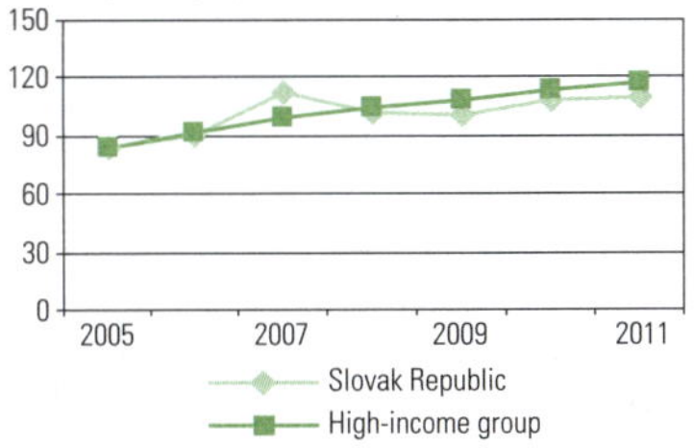

Mobile basket, 2005–10
Percentage of GNI per capita

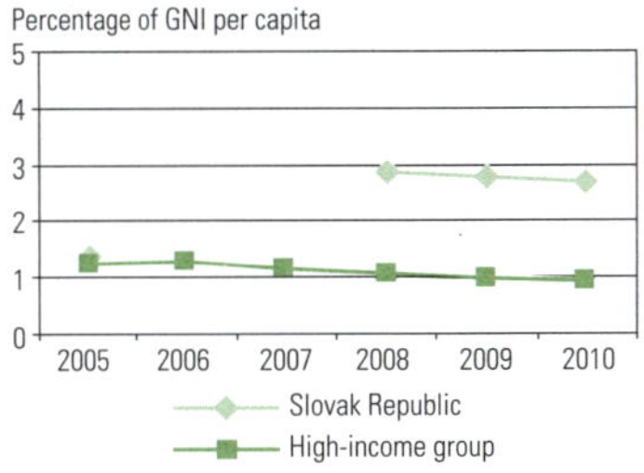

Slovenia

	Slovenia 2005	Slovenia 2010	High-income group 2010
Economic and social context			
Population (total, million)	2	2	1,127
GNI per capita, World Bank Atlas method (current US$)	18,070	23,900	38,746
Rural population (% of total)	51	52	22
Expected years of schooling (years)	16	*17*	*16*
Physicians density (per 1,000 people)	2.4	*2.5*	2.8
Depositors with commercial banks (per 1,000 adults)	—	—	—
Sector structure			
Number of mobile operators	—	4	
Herfindahl-Hirschman Index (HHI) (scale = 0–10,000)	—	4,100	
Sector performance			
Access			
Mobile cellular subscriptions (per 100 people)	88	105[a]	118[a]
Mobile cellular subscriptions (% prepaid)	47	33[a]	36[a]
Population covered by a mobile-cellular network (%)	99	100	100
Mobile broadband subscriptions (per 100 people)	1.3	43.7[a]	69.6[a]
Mobile broadband (% of total mobile subscriptions)	1.5	44.0[a]	57.6[a]
Usage			
Households with a mobile telephone (%)	87	94	93
Mobile voice usage (minutes per user per month)	*134*	*151*	339
Population using mobile Internet (%)	2.5	13.9	24.3
Short Message Service (SMS) users (% of mobile users)	—	—	78.2[a]
Affordability			
Mobile basket (% of GNI per capita)	1.3	1.0	1.0

Mobile cellular subscriptions, 2005–11
Number per 100 people

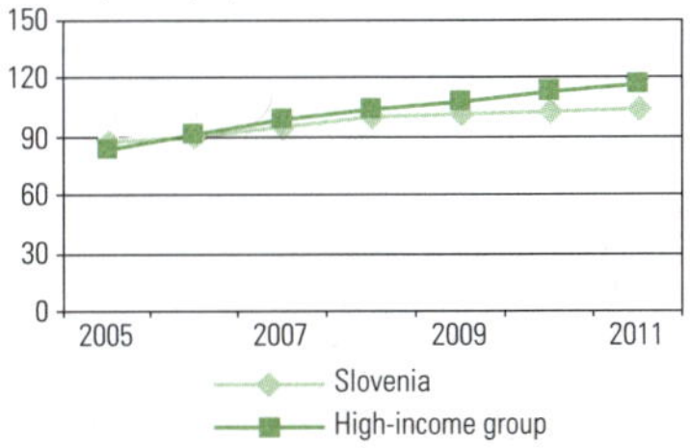

Mobile basket, 2005–10
Percentage of GNI per capita

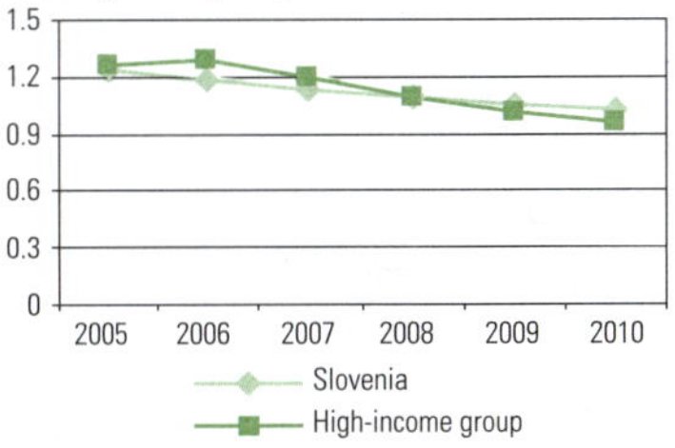

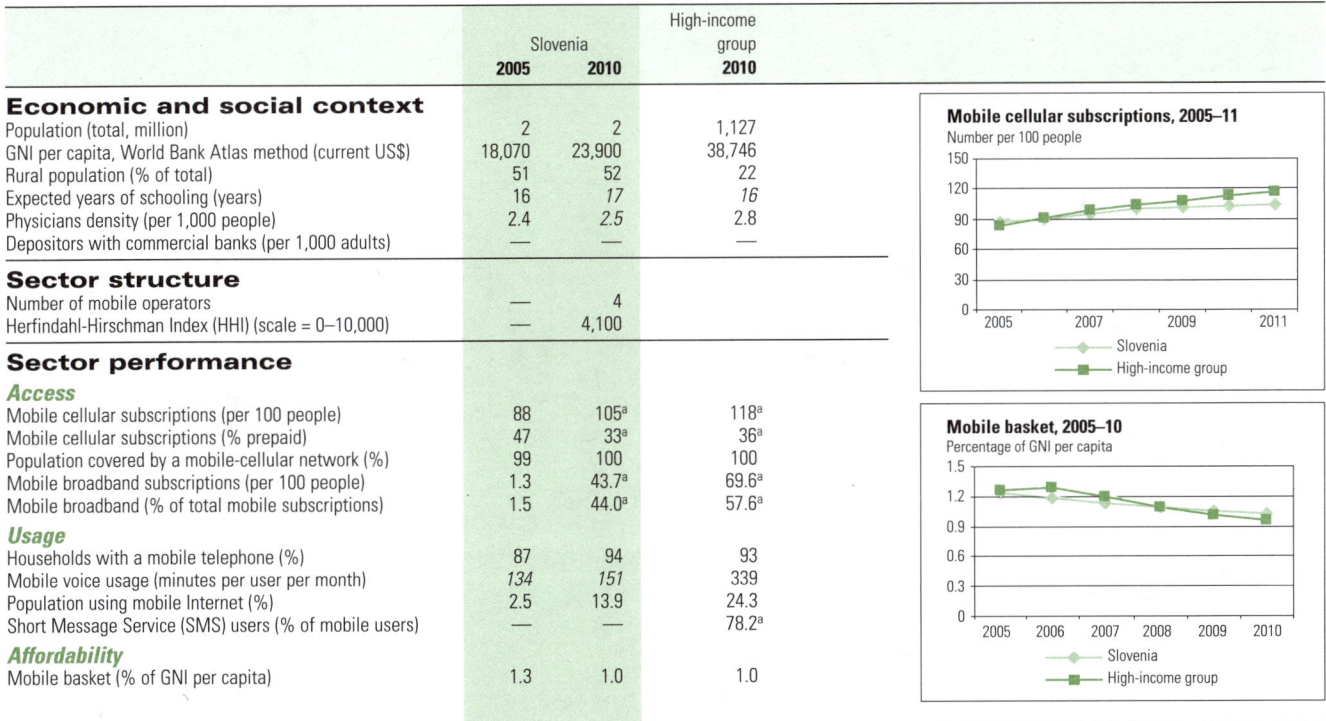

Sources: Economic and social context: IMF, UIS, UN, WHO and World Bank; Sector structure: ictDATA.org; Sector performance: ictDATA.org, ITU; Wireless Intelligence, and World Bank.
Notes: Use of italics in the column entries indicates years or periods other than those specified. — Not available. GNI = gross national income.
a. Data are for 2011.

South Africa

	South Africa 2005	South Africa 2010	Upper-middle-income group 2010	Sub-Saharan Africa Region 2010
Economic and social context				
Population (total, million)	47	50	2,452	853
GNI per capita, World Bank Atlas method (current US$)	4,850	6,090	5,886	1,188
Rural population (% of total)	41	38	43	63
Expected years of schooling (years)	—	—	*13*	*9*
Physicians density (per 1,000 people)	*0.8*	—	1.7	0.2
Depositors with commercial banks (per 1,000 adults)	522	978	—	*167*
Sector structure				
Number of mobile operators	—	4		
Herfindahl-Hirschman Index (HHI) (scale = 0–10,000)	—	3,850		
Sector performance				
Access				
Mobile cellular subscriptions (per 100 people)	72	128[a]	92[a]	57[a]
Mobile cellular subscriptions (% prepaid)	85	82[a]	81[a]	96[a]
Population covered by a mobile-cellular network (%)	96	—	*99*	72
Mobile broadband subscriptions (per 100 people)	0.4	27.8[a]	14.3[a]	5.6[a]
Mobile broadband (% of total mobile subscriptions)	0.6	21.9[a]	15.4[a]	10.1[a]
Usage				
Households with a mobile telephone (%)	62	86	84	52
Mobile voice usage (minutes per user per month)	98	110	325[a]	—
Population using mobile Internet (%)	0.2	6.2	22.9[a]	—
Short Message Service (SMS) users (% of mobile users)	—	*50.0*	74.4[a]	—
Affordability				
Mobile basket (% of GNI per capita)	6.1	4.6	2.9	19.5

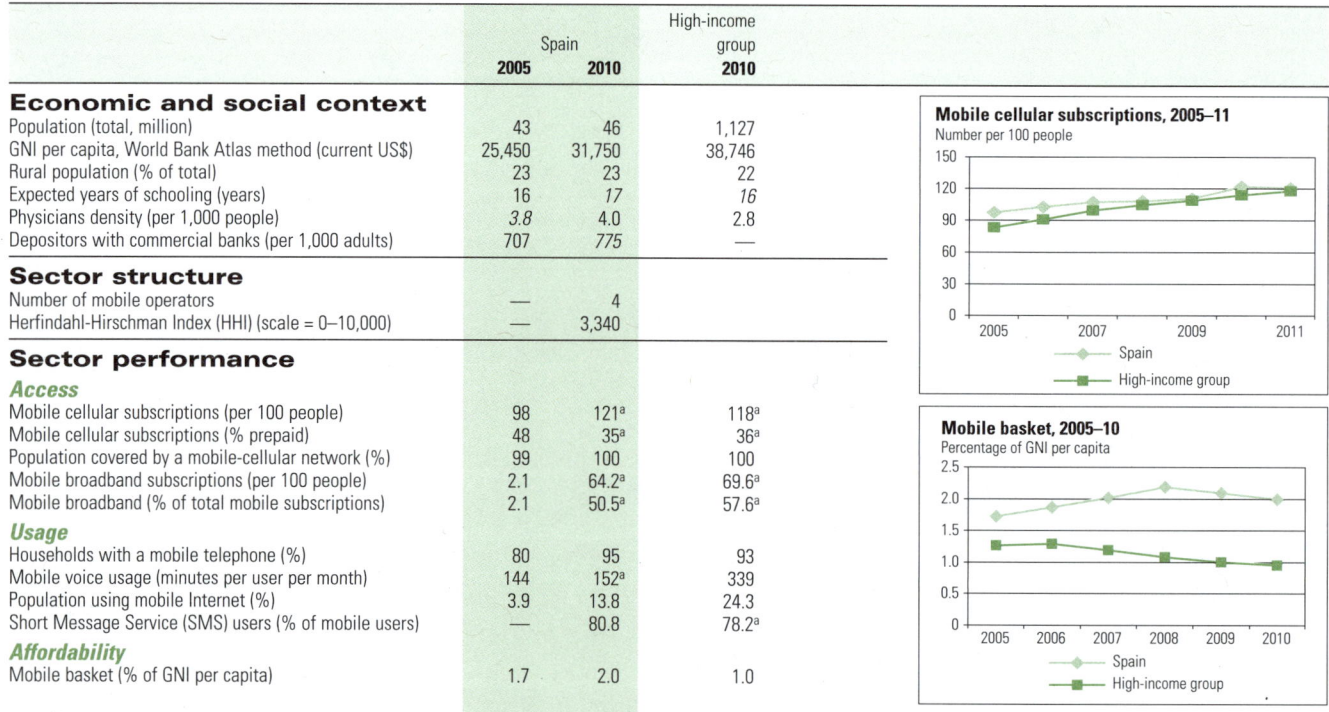

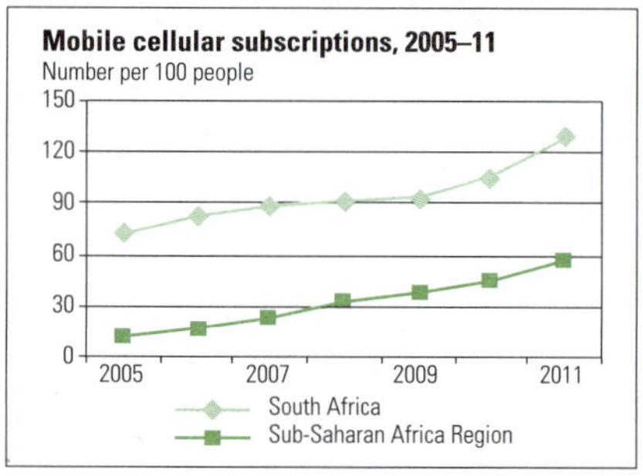

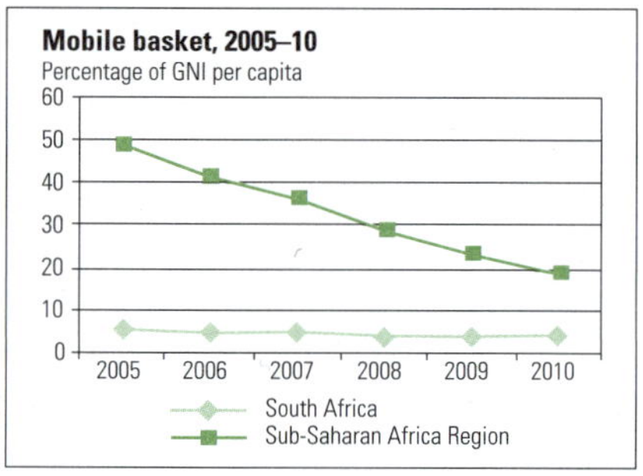

Spain

	Spain 2005	Spain 2010	High-income group 2010
Economic and social context			
Population (total, million)	43	46	1,127
GNI per capita, World Bank Atlas method (current US$)	25,450	31,750	38,746
Rural population (% of total)	23	23	22
Expected years of schooling (years)	16	*17*	*16*
Physicians density (per 1,000 people)	*3.8*	4.0	2.8
Depositors with commercial banks (per 1,000 adults)	707	*775*	—
Sector structure			
Number of mobile operators	—	4	
Herfindahl-Hirschman Index (HHI) (scale = 0–10,000)	—	3,340	
Sector performance			
Access			
Mobile cellular subscriptions (per 100 people)	98	121[a]	118[a]
Mobile cellular subscriptions (% prepaid)	48	35[a]	36[a]
Population covered by a mobile-cellular network (%)	99	100	100
Mobile broadband subscriptions (per 100 people)	2.1	64.2[a]	69.6[a]
Mobile broadband (% of total mobile subscriptions)	2.1	50.5[a]	57.6[a]
Usage			
Households with a mobile telephone (%)	80	95	93
Mobile voice usage (minutes per user per month)	144	152[a]	339
Population using mobile Internet (%)	3.9	13.8	24.3
Short Message Service (SMS) users (% of mobile users)	—	80.8	78.2[a]
Affordability			
Mobile basket (% of GNI per capita)	1.7	2.0	1.0

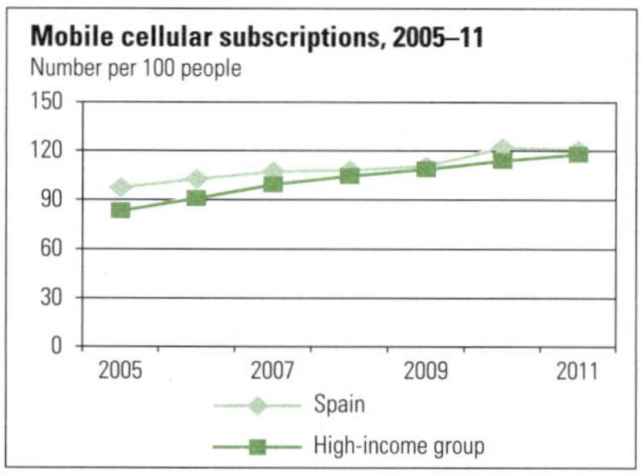

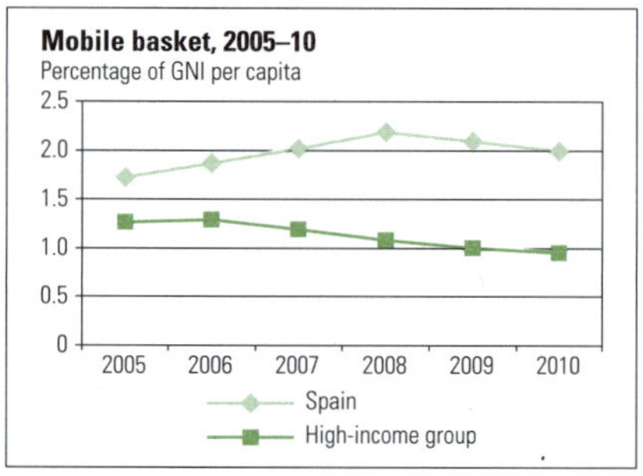

Sources: Economic and social context: IMF, UIS, UN, WHO and World Bank; Sector structure: ictDATA.org; Sector performance: ictDATA.org, ITU; Wireless Intelligence, and World Bank.
Notes: Use of italics in the column entries indicates years or periods other than those specified. — Not available. GNI = gross national income.
a. Data are for 2011.

Sri Lanka

	Sri Lanka 2005	Sri Lanka 2010	Lower-middle-income group 2010	South Asia Region 2010
Economic and social context				
Population (total, million)	20	21	2,519	1,633
GNI per capita, World Bank Atlas method (current US$)	1,190	2,240	1,623	1,176
Rural population (% of total)	85	85	61	70
Expected years of schooling (years)	*13*	—	*10*	*10*
Physicians density (per 1,000 people)	*0.5*	—	0.8	0.6
Depositors with commercial banks (per 1,000 adults)	—	—	—	249
Sector structure				
Number of mobile operators	—	5		
Herfindahl-Hirschman Index (HHI) (scale = 0–10,000)	—	2,810		
Sector performance				
Access				
Mobile cellular subscriptions (per 100 people)	17	87[a]	78[a]	67[a]
Mobile cellular subscriptions (% prepaid)	85	94[a]	96[a]	96[a]
Population covered by a mobile-cellular network (%)	85	98	*86*	*84*
Mobile broadband subscriptions (per 100 people)	*0.03*	9.8[a]	7.3[a]	3.3[a]
Mobile broadband (% of total mobile subscriptions)	*0.10*	10.5[a]	9.0[a]	4.6[a]
Usage				
Households with a mobile telephone (%)	20	60	77	54
Mobile voice usage (minutes per user per month)	—	*121*	276[a]	305[a]
Population using mobile Internet (%)	—	4.4	2.9	3.3[a]
Short Message Service (SMS) users (% of mobile users)	—	—	61.9[a]	47.0[a]
Affordability				
Mobile basket (% of GNI per capita)	6.5	1.0	7.2	3.2

Mobile cellular subscriptions, 2005–11
Number per 100 people

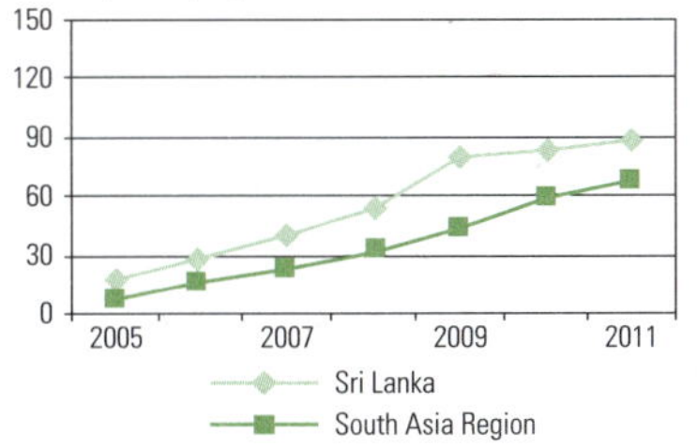

Mobile basket, 2005–10
Percentage of GNI per capita

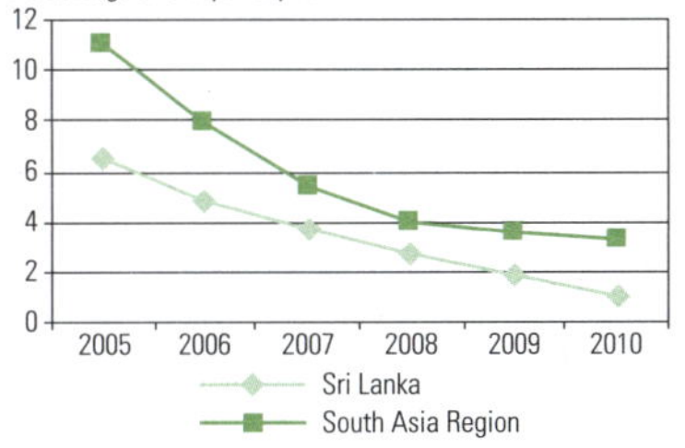

Sudan

	Sudan[b] 2005	Sudan[b] 2010	Lower-middle-income group 2010	Sub-Saharan Africa Region 2010
Economic and social context				
Population (total, million)	38	44	2,519	853
GNI per capita, World Bank Atlas method (current US$)	610	1,270	1,623	1,188
Rural population (% of total)	59	55	61	63
Expected years of schooling (years)	—	—	*10*	*9*
Physicians density (per 1,000 people)	*0.3*	*0.3*	0.8	0.2
Depositors with commercial banks (per 1,000 adults)	—	—	—	*167*
Sector structure				
Number of mobile operators	—	3		
Herfindahl-Hirschman Index (HHI) (scale = 0–10,000)	—	4,402		
Sector performance				
Access				
Mobile cellular subscriptions (per 100 people)	5	50[a]	78[a]	57[a]
Mobile cellular subscriptions (% prepaid)	97	99[a]	96[a]	96[a]
Population covered by a mobile-cellular network (%)	34	*66*	*86*	72
Mobile broadband subscriptions (per 100 people)	*0.03*	8.6[a]	7.3[a]	5.6[a]
Mobile broadband (% of total mobile subscriptions)	*0.23*	15.7[a]	9.0[a]	10.1[a]
Usage				
Households with a mobile telephone (%)	—	—	77	52
Mobile voice usage (minutes per user per month)	—	—	276[a]	—
Population using mobile Internet (%)	—	—	2.9	—
Short Message Service (SMS) users (% of mobile users)	—	—	61.9[a]	—
Affordability				
Mobile basket (% of GNI per capita)	17.7	*3.3*	7.2	19.5

Mobile cellular subscriptions, 2005–11
Number per 100 people

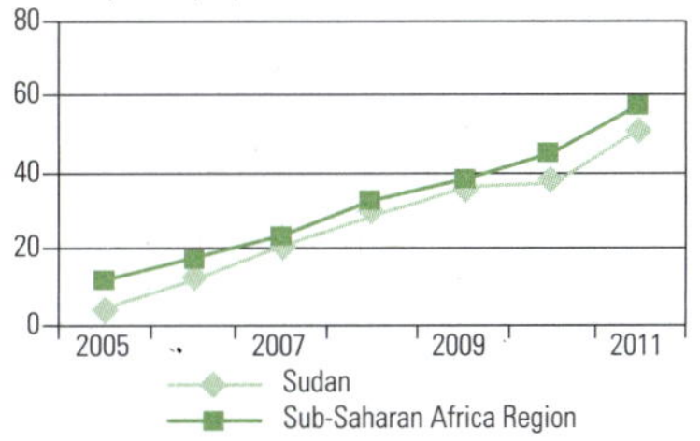

Mobile basket, 2005–10
Percentage of GNI per capita

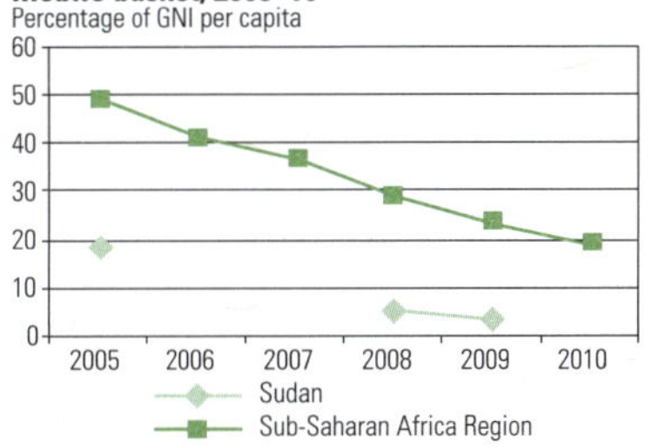

Sources: Economic and social context: IMF, UIS, UN, WHO and World Bank; Sector structure: ictDATA.org; Sector performance: ictDATA.org, ITU; Wireless Intelligence, and World Bank.
Notes: Use of italics in the column entries indicates years or periods other than those specified. — Not available. GNI = gross national income.
a. Data are for 2011.
b. Data for Sudan include South Sudan.

Swaziland

	Swaziland 2005	Swaziland 2010	Lower-middle-income group 2010	Sub-Saharan Africa Region 2010
Economic and social context				
Population (total, million)	1	1	2,519	853
GNI per capita, World Bank Atlas method (current US$)	2,600	2,930	1,623	1,188
Rural population (% of total)	76	75	61	63
Expected years of schooling (years)	10	—	*10*	*9*
Physicians density (per 1,000 people)	*0.2*	—	0.8	0.2
Depositors with commercial banks (per 1,000 adults)	352	455	—	*167*
Sector structure				
Number of mobile operators	—	1		
Herfindahl-Hirschman Index (HHI) (scale = 0–10,000)	—	10,000		
Sector performance				
Access				
Mobile cellular subscriptions (per 100 people)	20	78[a]	78[a]	57[a]
Mobile cellular subscriptions (% prepaid)	98	98[a]	96[a]	96[a]
Population covered by a mobile-cellular network (%)	*90*	*91*	*86*	72
Mobile broadband subscriptions (per 100 people)	—	0.4[a]	7.3[a]	5.6[a]
Mobile broadband (% of total mobile subscriptions)	—	0.5[a]	9.0[a]	10.1[a]
Usage				
Households with a mobile telephone (%)	*60*	—	77	52
Mobile voice usage (minutes per user per month)	305	79	276[a]	—
Population using mobile Internet (%)	—	—	2.9	—
Short Message Service (SMS) users (% of mobile users)	—	—	61.9[a]	—
Affordability				
Mobile basket (% of GNI per capita)	14.2	9.9	7.2	19.5

Mobile cellular subscriptions, 2005–11
Number per 100 people

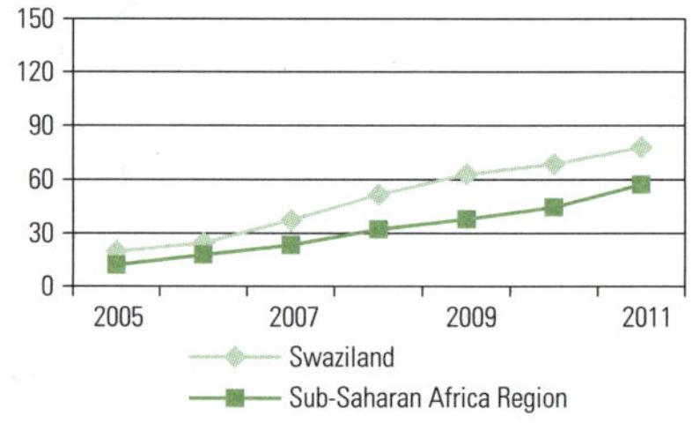

Mobile basket, 2005–10
Percentage of GNI per capita

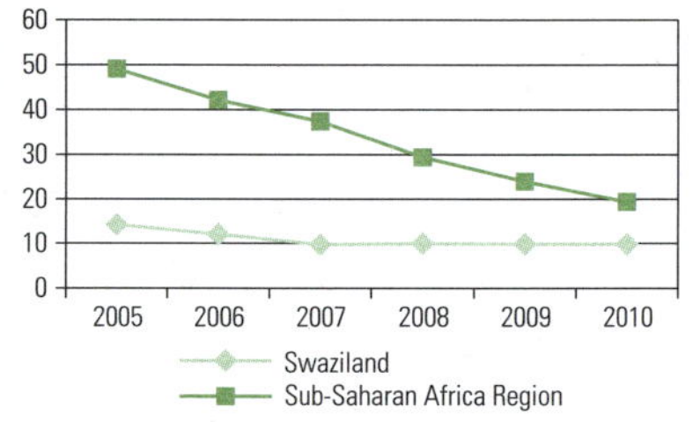

Sweden

	Sweden 2005	Sweden 2010	High-income group 2010
Economic and social context			
Population (total, million)	9	9	1,127
GNI per capita, World Bank Atlas method (current US$)	42,920	50,100	38,746
Rural population (% of total)	16	15	22
Expected years of schooling (years)	16	*16*	*16*
Physicians density (per 1,000 people)	*3.6*	*3.8*	2.8
Depositors with commercial banks (per 1,000 adults)	—	—	—
Sector structure			
Number of mobile operators	—	4	
Herfindahl-Hirschman Index (HHI) (scale = 0–10,000)	—	2,990	
Sector performance			
Access			
Mobile cellular subscriptions (per 100 people)	101	139[a]	118[a]
Mobile cellular subscriptions (% prepaid)	56	38[a]	36[a]
Population covered by a mobile-cellular network (%)	99	99	100
Mobile broadband subscriptions (per 100 people)	7.2	114.2[a]	69.6[a]
Mobile broadband (% of total mobile subscriptions)	6.5	77.1[a]	57.6[a]
Usage			
Households with a mobile telephone (%)	95	97	93
Mobile voice usage (minutes per user per month)	140	242[a]	339
Population using mobile Internet (%)	5.4	19.9	24.3
Short Message Service (SMS) users (% of mobile users)	—	91.0[a]	78.2[a]
Affordability			
Mobile basket (% of GNI per capita)	0.9	0.4	1.0

Mobile cellular subscriptions, 2005–11
Number per 100 people

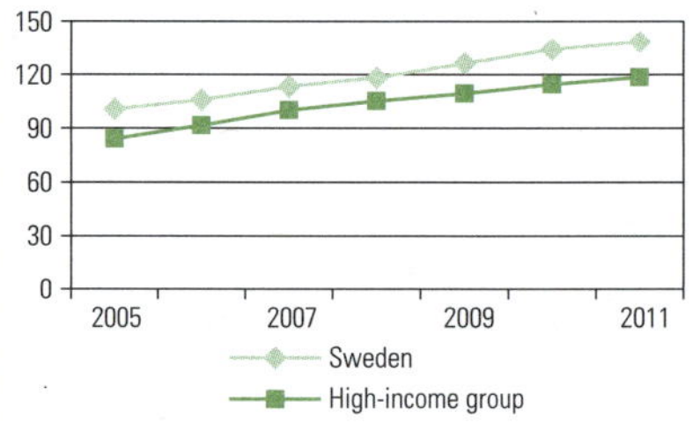

Mobile basket, 2005–10
Percentage of GNI per capita

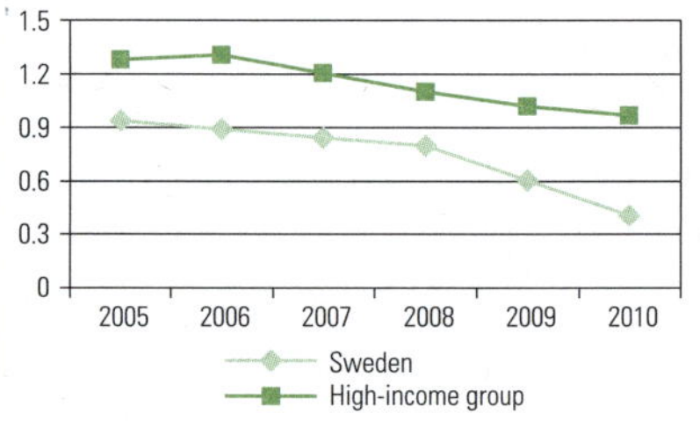

Sources: Economic and social context: IMF, UIS, UN, WHO and World Bank; Sector structure: ictDATA.org; Sector performance: ictDATA.org, ITU; Wireless Intelligence, and World Bank.
Notes: Use of italics in the column entries indicates years or periods other than those specified. — Not available. GNI = gross national income.
a. Data are for 2011.

Switzerland

	Switzerland 2005	Switzerland 2010	High-income group 2010
Economic and social context			
Population (total, million)	7	8	1,127
GNI per capita, World Bank Atlas method (current US$)	56,870	71,520	38,746
Rural population (% of total)	27	26	22
Expected years of schooling (years)	15	*16*	*16*
Physicians density (per 1,000 people)	*4.0*	*4.1*	2.8
Depositors with commercial banks (per 1,000 adults)	—	—	—
Sector structure			
Number of mobile operators	—	3	
Herfindahl-Hirschman Index (HHI) (scale = 0–10,000)	—	4,371	
Sector performance			
Access			
Mobile cellular subscriptions (per 100 people)	92	123	118[a]
Mobile cellular subscriptions (% prepaid)	39	39[a]	36[a]
Population covered by a mobile-cellular network (%)	100	100	100
Mobile broadband subscriptions (per 100 people)	1.4	57.1[a]	69.6[a]
Mobile broadband (% of total mobile subscriptions)	1.5	45.6[a]	57.6[a]
Usage			
Households with a mobile telephone (%)	84	92	93
Mobile voice usage (minutes per user per month)	124	130[a]	339
Population using mobile Internet (%)	—	19.2	24.3
Short Message Service (SMS) users (% of mobile users)	—	—	78.2[a]
Affordability			
Mobile basket (% of GNI per capita)	1.2	1.0	1.0

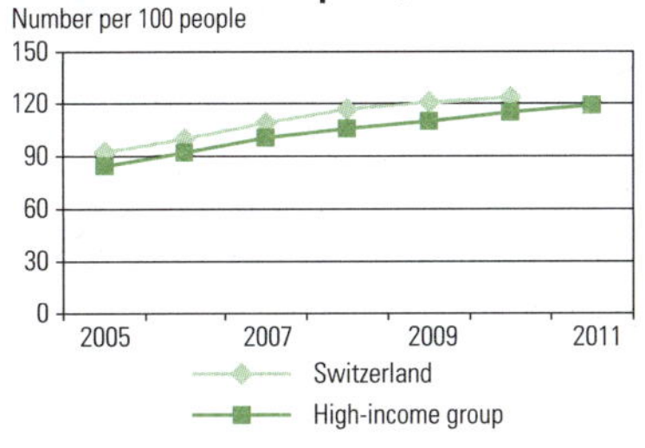

Mobile cellular subscriptions, 2005–11

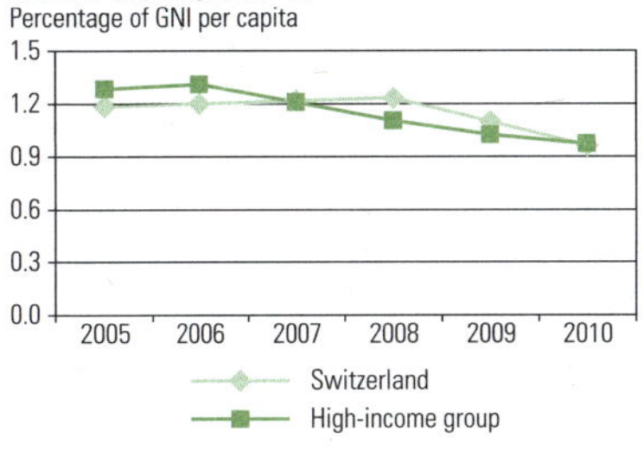

Mobile basket, 2005–10

Syrian Arab Republic

	Syrian Arab Republic 2005	Syrian Arab Republic 2010	Lower-middle-income group 2010	Middle East & North Africa Region 2010
Economic and social context				
Population (total, million)	18	20	2,519	331
GNI per capita, World Bank Atlas method (current US$)	1,500	2,750	1,623	3,874
Rural population (% of total)	47	45	61	42
Expected years of schooling (years)	11	—	*10*	*12*
Physicians density (per 1,000 people)	*0.5*	*1.5*	0.8	1.4
Depositors with commercial banks (per 1,000 adults)	—	220	—	*443*
Sector structure				
Number of mobile operators	—	2		
Herfindahl-Hirschman Index (HHI) (scale = 0–10,000)	—	5,050		
Sector performance				
Access				
Mobile cellular subscriptions (per 100 people)	16	60[a]	78[a]	89[a]
Mobile cellular subscriptions (% prepaid)	62	84[a]	96[a]	87[a]
Population covered by a mobile-cellular network (%)	92	98	*86*	—
Mobile broadband subscriptions (per 100 people)	—	2.2[a]	7.3[a]	—
Mobile broadband (% of total mobile subscriptions)	—	3.7[a]	9.0[a]	—
Usage				
Households with a mobile telephone (%)	—	—	77	—
Mobile voice usage (minutes per user per month)	—	—	276[a]	—
Population using mobile Internet (%)	—	—	2.9	4.5
Short Message Service (SMS) users (% of mobile users)	—	93.0	61.9[a]	—
Affordability				
Mobile basket (% of GNI per capita)	—	8.7	7.2	3.6

Mobile cellular subscriptions, 2005–11

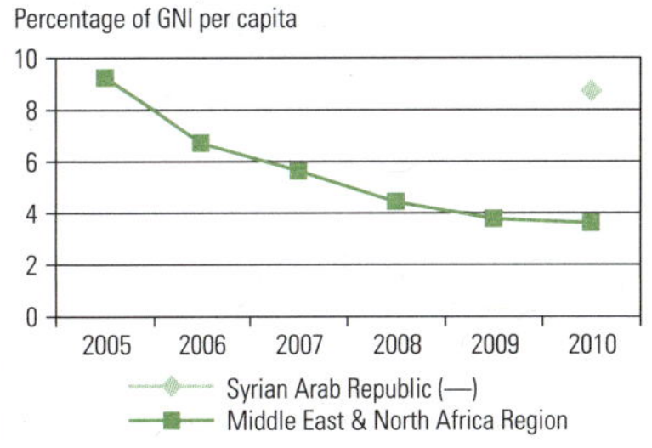

Mobile basket, 2005–10

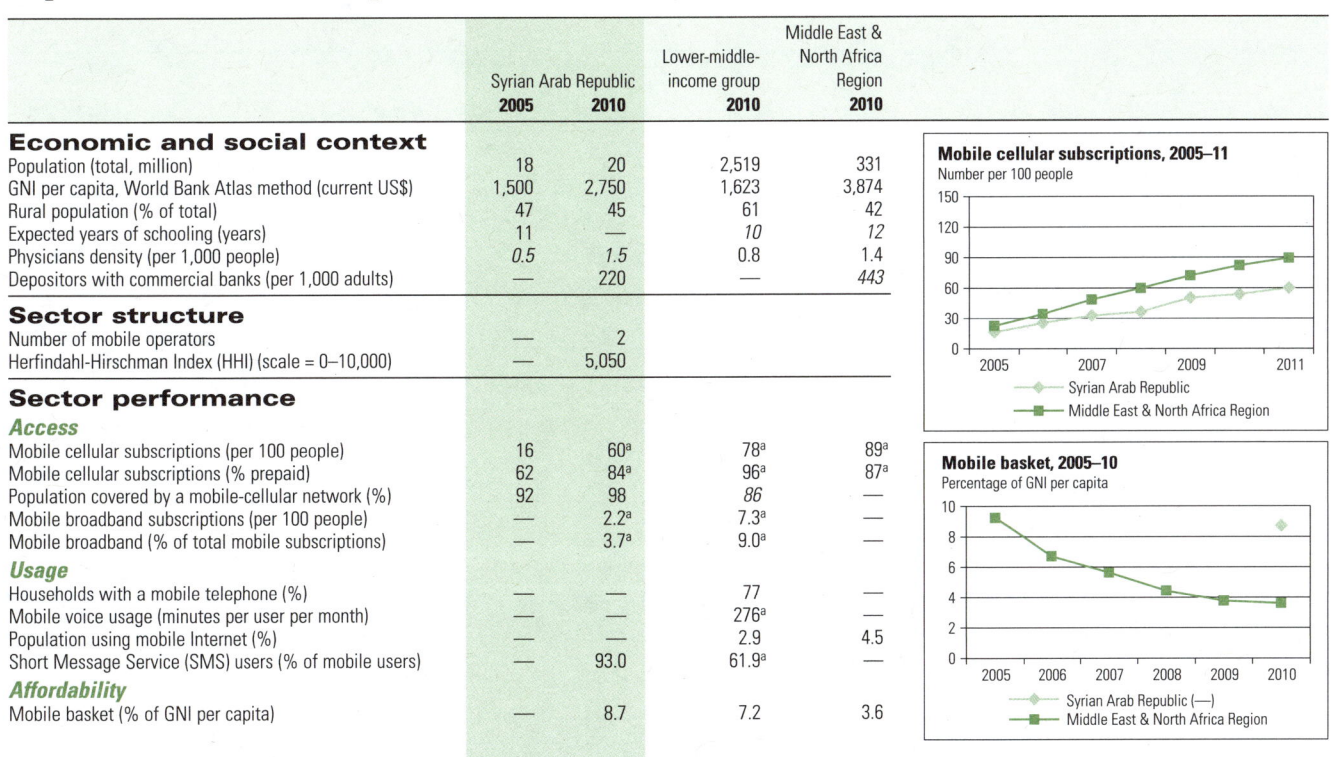

Sources: Economic and social context: IMF, UIS, UN, WHO and World Bank; Sector structure: ictDATA.org; Sector performance: ictDATA.org, ITU; Wireless Intelligence, and World Bank.
Notes: Use of italics in the column entries indicates years or periods other than those specified. — Not available. GNI = gross national income.
a. Data are for 2011.

Tajikistan

	Tajikistan 2005	Tajikistan 2010	Low-income group 2010	Europe & Central Asia Region 2010
Economic and social context				
Population (total, million)	6	7	796	405
GNI per capita, World Bank Atlas method (current US$)	340	800	530	7,272
Rural population (% of total)	74	74	72	36
Expected years of schooling (years)	11	*11*	*9*	*13*
Physicians density (per 1,000 people)	*2.0*	*2.1*	0.2	3.2
Depositors with commercial banks (per 1,000 adults)	—	—	—	894
Sector structure				
Number of mobile operators	—	5		
Herfindahl-Hirschman Index (HHI) (scale = 0–10,000)	—	2,545		
Sector performance				
Access				
Mobile cellular subscriptions (per 100 people)	4	85[a]	43[a]	125[a]
Mobile cellular subscriptions (% prepaid)	87	95[a]	98[a]	82[a]
Population covered by a mobile-cellular network (%)	—	—	—	96
Mobile broadband subscriptions (per 100 people)	0.01	11.2[a]	—	22.6[a]
Mobile broadband (% of total mobile subscriptions)	0.16	11.9[a]	—	18.0[a]
Usage				
Households with a mobile telephone (%)	11	80	43	82
Mobile voice usage (minutes per user per month)	*216*	182	—	288[a]
Population using mobile Internet (%)	—	2.0	—	8.5
Short Message Service (SMS) users (% of mobile users)	—	*22.0*	—	69.8[a]
Affordability				
Mobile basket (% of GNI per capita)	90.3	2.7	28.8	3.1

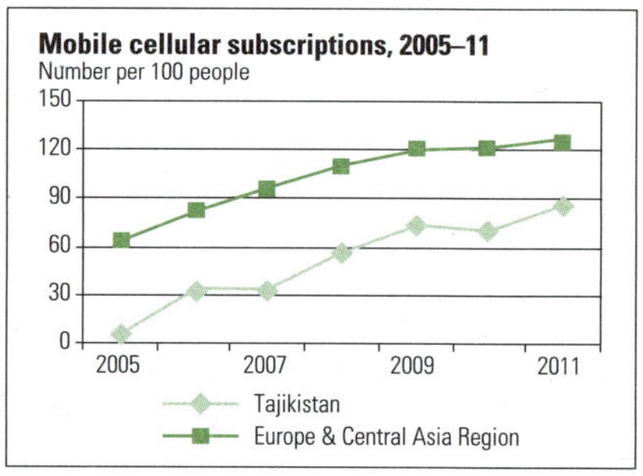

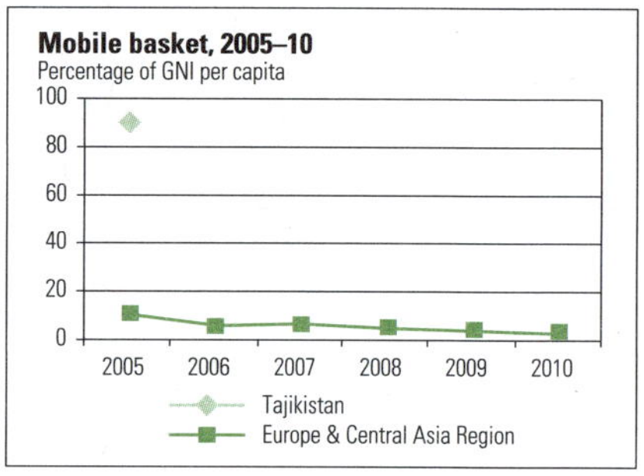

Tanzania

	Tanzania 2005	Tanzania 2010	Low-income group 2010	Sub-Saharan Africa Region 2010
Economic and social context				
Population (total, million)	39	45	796	853
GNI per capita, World Bank Atlas method (current US$)	390	540	530	1,188
Rural population (% of total)	76	74	72	63
Expected years of schooling (years)	8	—	*9*	*9*
Physicians density (per 1,000 people)	*0.01*	—	0.2	0.2
Depositors with commercial banks (per 1,000 adults)	83	*131*	—	*167*
Sector structure				
Number of mobile operators	—	4		
Herfindahl-Hirschman Index (HHI) (scale = 0–10,000)	—	3,082		
Sector performance				
Access				
Mobile cellular subscriptions (per 100 people)	8	56[a]	43[a]	57[a]
Mobile cellular subscriptions (% prepaid)	100	100[a]	98[a]	96[a]
Population covered by a mobile-cellular network (%)	45	85	—	72
Mobile broadband subscriptions (per 100 people)	*0.1*	3.8[a]	—	5.6[a]
Mobile broadband (% of total mobile subscriptions)	*0.4*	6.9[a]	—	10.1[a]
Usage				
Households with a mobile telephone (%)	9	45	43	52
Mobile voice usage (minutes per user per month)	—	68[a]	—	—
Population using mobile Internet (%)	—	—	—	—
Short Message Service (SMS) users (% of mobile users)	—	51.0	—	—
Affordability				
Mobile basket (% of GNI per capita)	52.4	21.6	28.8	19.5

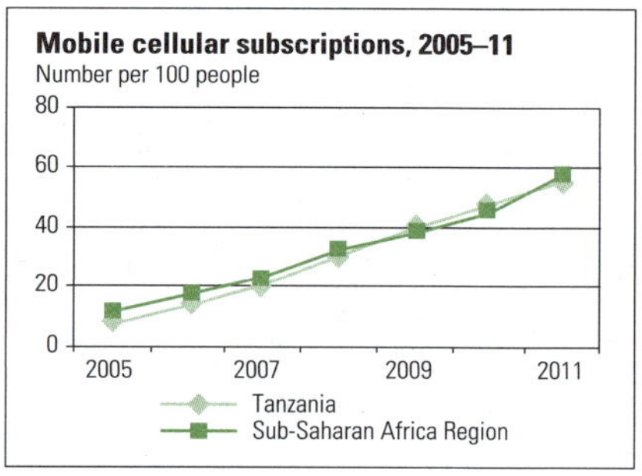

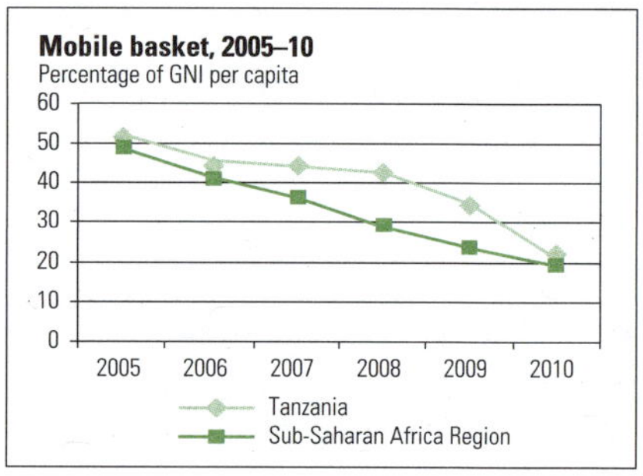

Sources: Economic and social context: IMF, UIS, UN, WHO and World Bank; Sector structure: ictDATA.org; Sector performance: ictDATA.org, ITU; Wireless Intelligence, and World Bank.
Notes: Use of italics in the column entries indicates years or periods other than those specified. — Not available. GNI = gross national income.
a. Data are for 2011.

Thailand

	Thailand 2005	Thailand 2010	Upper-middle-income group 2010	East Asia & Pacific Region 2010
Economic and social context				
Population (total, million)	67	69	2,452	1,962
GNI per capita, World Bank Atlas method (current US$)	2,560	4,150	5,886	3,696
Rural population (% of total)	68	66	43	54
Expected years of schooling (years)	12	*12*	*13*	*12*
Physicians density (per 1,000 people)	0.3	0.3	1.7	1.2
Depositors with commercial banks (per 1,000 adults)	*984*	1,120	—	—
Sector structure				
Number of mobile operators	—	5		
Herfindahl-Hirschman Index (HHI) (scale = 0–10,000)	—	3,409		
Sector performance				
Access				
Mobile cellular subscriptions (per 100 people)	47	109[a]	92[a]	83[a]
Mobile cellular subscriptions (% prepaid)	85	90[a]	81[a]	85[a]
Population covered by a mobile-cellular network (%)	—	—	*99*	*99*
Mobile broadband subscriptions (per 100 people)	—	5.6[a]	14.3[a]	11.6[a]
Mobile broadband (% of total mobile subscriptions)	—	5.1[a]	15.4[a]	14.4[a]
Usage				
Households with a mobile telephone (%)	70	90	84	83
Mobile voice usage (minutes per user per month)	*493*	321[a]	325[a]	367[a]
Population using mobile Internet (%)	2.1	13.7[a]	22.9[a]	22.4[a]
Short Message Service (SMS) users (% of mobile users)	—	—	74.4[a]	84.0[a]
Affordability				
Mobile basket (% of GNI per capita)	6.5	2.5	2.9	5.7

Mobile cellular subscriptions, 2005–11
Number per 100 people

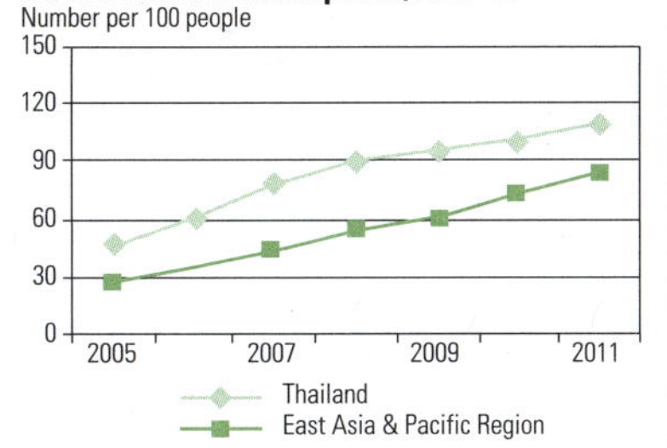

Mobile basket, 2005–10
Percentage of GNI per capita

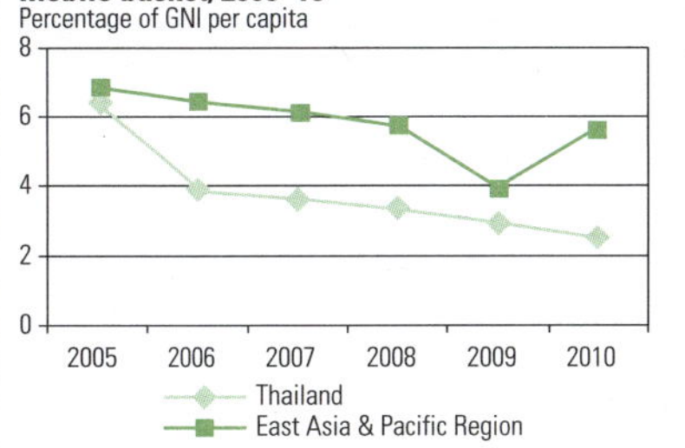

Timor-Leste

	Timor-Leste 2005	Timor-Leste 2010	Lower-middle-income group 2010	East Asia & Pacific Region 2010
Economic and social context				
Population (total, million)	1	1	2,519	1,962
GNI per capita, World Bank Atlas method (current US$)	730	2,220	1,623	3,696
Rural population (% of total)	74	72	61	54
Expected years of schooling (years)	*11*	*11*	*10*	*12*
Physicians density (per 1,000 people)	*0.1*	—	0.8	1.2
Depositors with commercial banks (per 1,000 adults)	—	—	—	—
Sector structure				
Number of mobile operators	—	—		
Herfindahl-Hirschman Index (HHI) (scale = 0–10,000)	—	—		
Sector performance				
Access				
Mobile cellular subscriptions (per 100 people)	3	*32*	78[a]	83[a]
Mobile cellular subscriptions (% prepaid)	96	98[a]	96[a]	85[a]
Population covered by a mobile-cellular network (%)	50	*69*	*86*	*99*
Mobile broadband subscriptions (per 100 people)	—	—	7.3[a]	11.6[a]
Mobile broadband (% of total mobile subscriptions)	—	—	9.0[a]	14.4[a]
Usage				
Households with a mobile telephone (%)	—	—	77	83
Mobile voice usage (minutes per user per month)	97	*87*	276[a]	367[a]
Population using mobile Internet (%)	—	—	2.9	22.4[a]
Short Message Service (SMS) users (% of mobile users)	—	—	61.9[a]	84.0[a]
Affordability				
Mobile basket (% of GNI per capita)	—	8.7	7.2	5.7

Mobile cellular subscriptions, 2005–11
Number per 100 people

Mobile basket, 2005–10
Percentage of GNI per capita

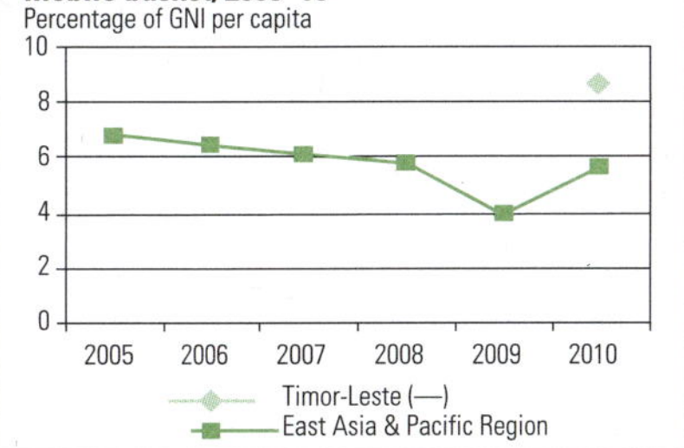

Sources: Economic and social context: IMF, UIS, UN, WHO and World Bank; Sector structure: ictDATA.org; Sector performance: ictDATA.org, ITU; Wireless Intelligence, and World Bank.
Notes: Use of italics in the column entries indicates years or periods other than those specified. — Not available. GNI = gross national income.
a. Data are for 2011.

Togo

	Togo 2005	Togo 2010	Low-income group 2010	Sub-Saharan Africa Region 2010
Economic and social context				
Population (total, million)	5	6	796	853
GNI per capita, World Bank Atlas method (current US$)	340	490	530	1,188
Rural population (% of total)	60	57	72	63
Expected years of schooling (years)	10	—	*9*	*9*
Physicians density (per 1,000 people)	*0.04*	*0.05*	0.2	0.2
Depositors with commercial banks (per 1,000 adults)	53	*181*	—	*167*
Sector structure				
Number of mobile operators	—	2		
Herfindahl-Hirschman Index (HHI) (scale = 0–10,000)	—	6,100		
Sector performance				
Access				
Mobile cellular subscriptions (per 100 people)	8	41	43[a]	57[a]
Mobile cellular subscriptions (% prepaid)	99	99[a]	98[a]	96[a]
Population covered by a mobile-cellular network (%)	85	—	—	72
Mobile broadband subscriptions (per 100 people)	—	—	—	5.6[a]
Mobile broadband (% of total mobile subscriptions)	—	—	—	10.1[a]
Usage				
Households with a mobile telephone (%)	*22*	—	43	52
Mobile voice usage (minutes per user per month)	—	33	—	—
Population using mobile Internet (%)	—	—	—	—
Short Message Service (SMS) users (% of mobile users)	—	—	—	—
Affordability				
Mobile basket (% of GNI per capita)	103.4	48.7	28.8	19.5

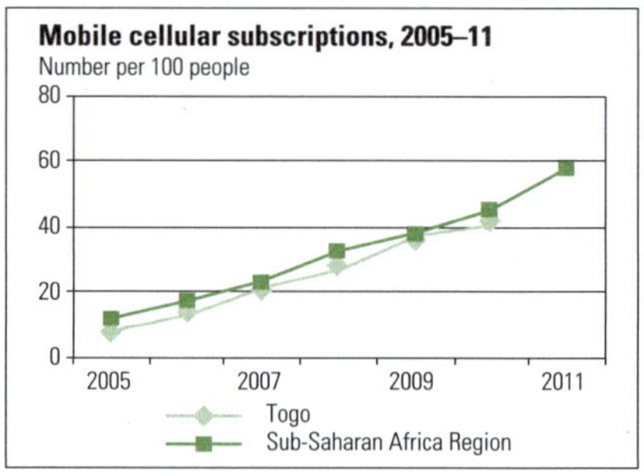

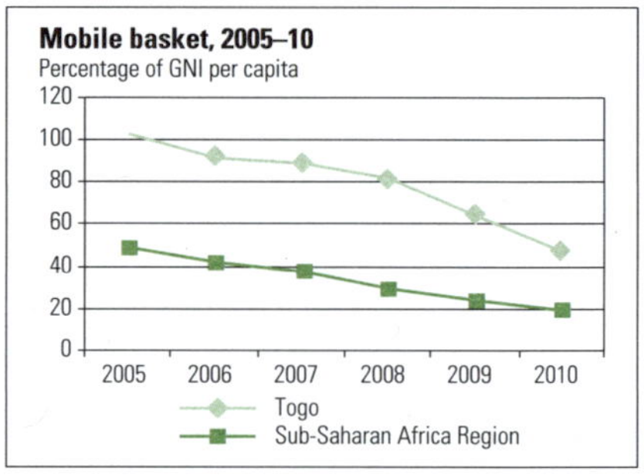

Trinidad and Tobago

	Trinidad and Tobago 2005	Trinidad and Tobago 2010	High-income group 2010
Economic and social context			
Population (total, million)	1	1	1,127
GNI per capita, World Bank Atlas method (current US$)	10,880	15,380	38,746
Rural population (% of total)	88	86	22
Expected years of schooling (years)	11	—	*16*
Physicians density (per 1,000 people)	*1.2*	—	2.8
Depositors with commercial banks (per 1,000 adults)	—	—	—
Sector structure			
Number of mobile operators	—	2	
Herfindahl-Hirschman Index (HHI) (scale = 0–10,000)	—	5,003	
Sector performance			
Access			
Mobile cellular subscriptions (per 100 people)	70	134[a]	118[a]
Mobile cellular subscriptions (% prepaid)	88	89[a]	36[a]
Population covered by a mobile-cellular network (%)	*100*	100	100
Mobile broadband subscriptions (per 100 people)	—	—	69.6[a]
Mobile broadband (% of total mobile subscriptions)	—	—	57.6[a]
Usage			
Households with a mobile telephone (%)	60	—	93
Mobile voice usage (minutes per user per month)	—	—	339
Population using mobile Internet (%)	—	—	24.3
Short Message Service (SMS) users (% of mobile users)	—	—	78.2[a]
Affordability			
Mobile basket (% of GNI per capita)	1.6	0.9	1.0

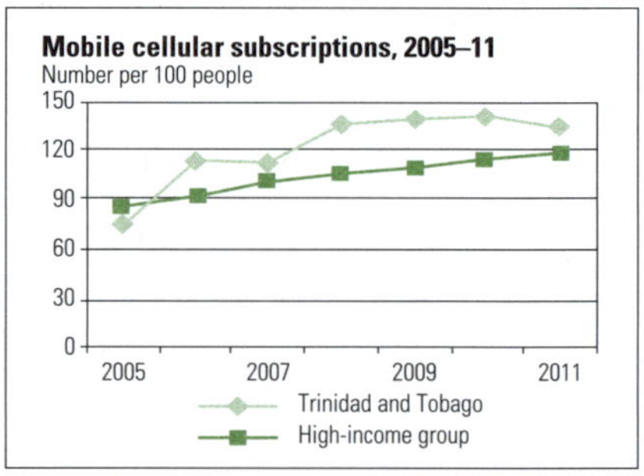

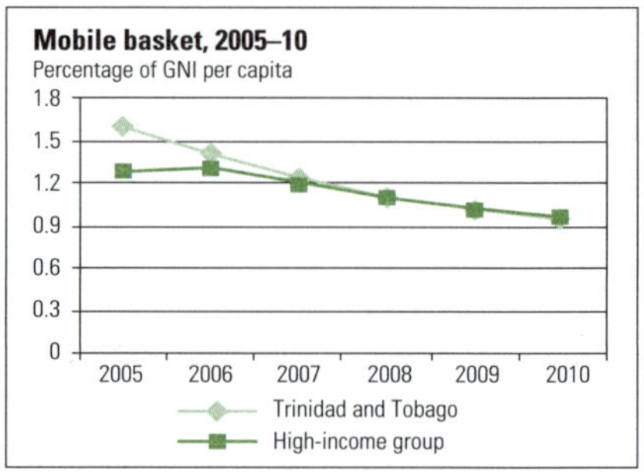

Sources: Economic and social context: IMF, UIS, UN, WHO and World Bank; Sector structure: ictDATA.org; Sector performance: ictDATA.org, ITU; Wireless Intelligence, and World Bank.
Notes: Use of italics in the column entries indicates years or periods other than those specified. — Not available. GNI = gross national income.
a. Data are for 2011.

Tunisia

	Tunisia 2005	Tunisia 2010	Upper-middle-income group 2010	Middle East & North Africa Region 2010
Economic and social context				
Population (total, million)	10	11	2,452	331
GNI per capita, World Bank Atlas method (current US$)	3,200	4,160	5,886	3,874
Rural population (% of total)	35	33	43	42
Expected years of schooling (years)	14	*14*	*13*	*12*
Physicians density (per 1,000 people)	*1.3*	*1.2*	1.7	1.4
Depositors with commercial banks (per 1,000 adults)	—	—	—	*443*
Sector structure				
Number of mobile operators	—	3		
Herfindahl-Hirschman Index (HHI) (scale = 0–10,000)	—	4,497		
Sector performance				
Access				
Mobile cellular subscriptions (per 100 people)	57	106[a]	92[a]	89[a]
Mobile cellular subscriptions (% prepaid)	99	98[a]	81[a]	87[a]
Population covered by a mobile-cellular network (%)	98	100	*99*	—
Mobile broadband subscriptions (per 100 people)	—	2.3[a]	14.3[a]	—
Mobile broadband (% of total mobile subscriptions)	—	1.9[a]	15.4[a]	—
Usage				
Households with a mobile telephone (%)	—	—	84	—
Mobile voice usage (minutes per user per month)	—	*171*	325[a]	—
Population using mobile Internet (%)	—	—	22.9[a]	4.5
Short Message Service (SMS) users (% of mobile users)	—	—	4.4[a]	—
Affordability				
Mobile basket (% of GNI per capita)	4.3	2.9	2.9	3.6

Mobile cellular subscriptions, 2005–11
Number per 100 people

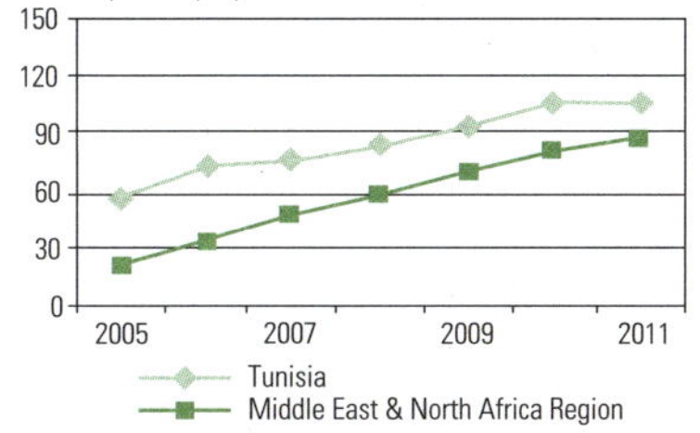

Mobile basket, 2005–10
Percentage of GNI per capita

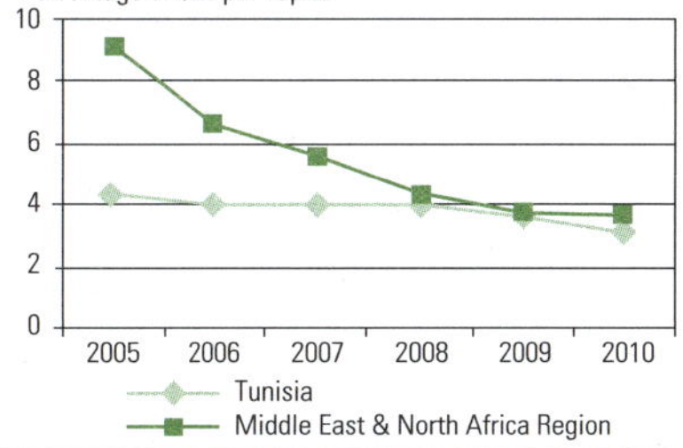

Turkey

	Turkey 2005	Turkey 2010	Upper-middle-income group 2010	Europe & Central Asia Region 2010
Economic and social context				
Population (total, million)	68	73	2,452	405
GNI per capita, World Bank Atlas method (current US$)	6,480	9,890	5,886	7,272
Rural population (% of total)	33	30	43	36
Expected years of schooling (years)	11	*12*	*13*	*13*
Physicians density (per 1,000 people)	1.3	*1.5*	1.7	3.2
Depositors with commercial banks (per 1,000 adults)	1,362	*1,265*	—	894
Sector structure				
Number of mobile operators	—	3		
Herfindahl-Hirschman Index (HHI) (scale = 0–10,000)	—	4,020		
Sector performance				
Access				
Mobile cellular subscriptions (per 100 people)	64	88[a]	92[a]	125[a]
Mobile cellular subscriptions (% prepaid)	80	65[a]	81[a]	82[a]
Population covered by a mobile-cellular network (%)	96	100	*99*	96
Mobile broadband subscriptions (per 100 people)	—	38.4[a]	14.3[a]	22.6[a]
Mobile broadband (% of total mobile subscriptions)	—	43.3[a]	15.4[a]	18.0[a]
Usage				
Households with a mobile telephone (%)	73	91	84	82
Mobile voice usage (minutes per user per month)	70	261[a]	325[a]	288[a]
Population using mobile Internet (%)	0.1	12.2[a]	22.9[a]	8.5
Short Message Service (SMS) users (% of mobile users)	—	64.0[a]	74.4[a]	69.8[a]
Affordability				
Mobile basket (% of GNI per capita)	7.3	5.3	2.9	3.1

Mobile cellular subscriptions, 2005–11
Number per 100 people

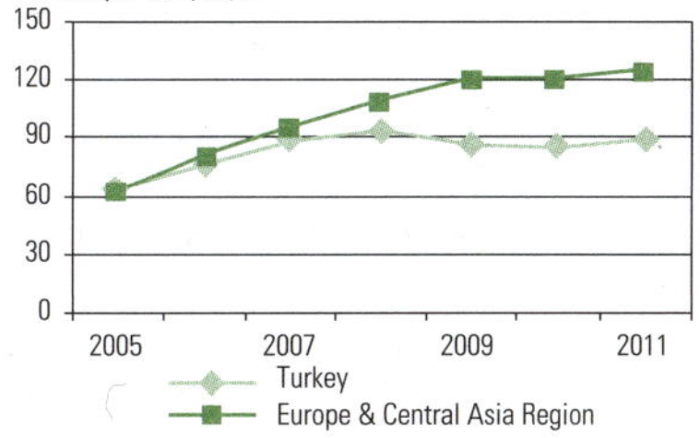

Mobile basket, 2005–10
Percentage of GNI per capita

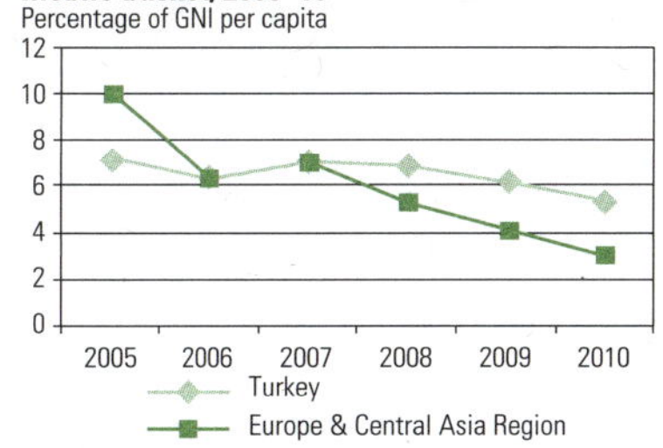

Sources: Economic and social context: IMF, UIS, UN, WHO and World Bank; Sector structure: ictDATA.org; Sector performance: ictDATA.org, ITU; Wireless Intelligence, and World Bank.
Notes: Use of italics in the column entries indicates years or periods other than those specified. — Not available. GNI = gross national income.
a. Data are for 2011.

Turkmenistan

	Turkmenistan 2005	Turkmenistan 2010	Lower-middle-income group 2010	Europe & Central Asia Region 2010
Economic and social context				
Population (total, million)	5	5	2,519	405
GNI per capita, World Bank Atlas method (current US$)	1,650	3,790	1,623	7,272
Rural population (% of total)	53	51	61	36
Expected years of schooling (years)	—	—	*10*	*13*
Physicians density (per 1,000 people)	*2.5*	*2.4*	0.8	3.2
Depositors with commercial banks (per 1,000 adults)	—	—	—	894
Sector structure				
Number of mobile operators	—	2		
Herfindahl-Hirschman Index (HHI) (scale = 0–10,000)	—	6,622		
Sector performance				
Access				
Mobile cellular subscriptions (per 100 people)	2	62	78[a]	125[a]
Mobile cellular subscriptions (% prepaid)	89	95[a]	96[a]	82[a]
Population covered by a mobile-cellular network (%)	14	—	*86*	96
Mobile broadband subscriptions (per 100 people)	—	—	7.3[a]	22.6[a]
Mobile broadband (% of total mobile subscriptions)	—	—	9.0[a]	18.0[a]
Usage				
Households with a mobile telephone (%)	—	—	77	82
Mobile voice usage (minutes per user per month)	256	292	276[a]	288[a]
Population using mobile Internet (%)	—	—	2.9	8.5
Short Message Service (SMS) users (% of mobile users)	—	—	61.9[a]	69.8[a]
Affordability				
Mobile basket (% of GNI per capita)	—	—	7.2	3.1

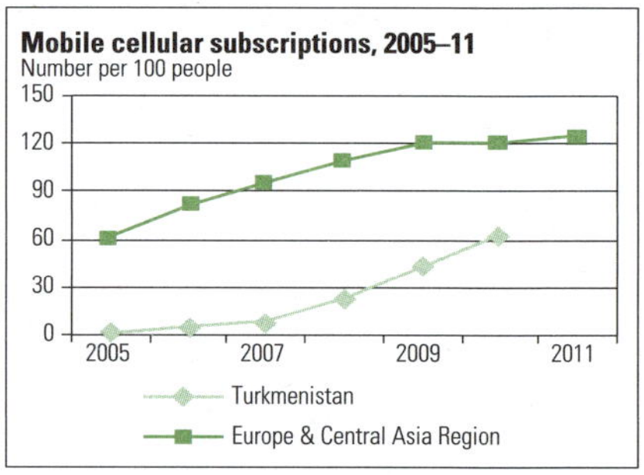

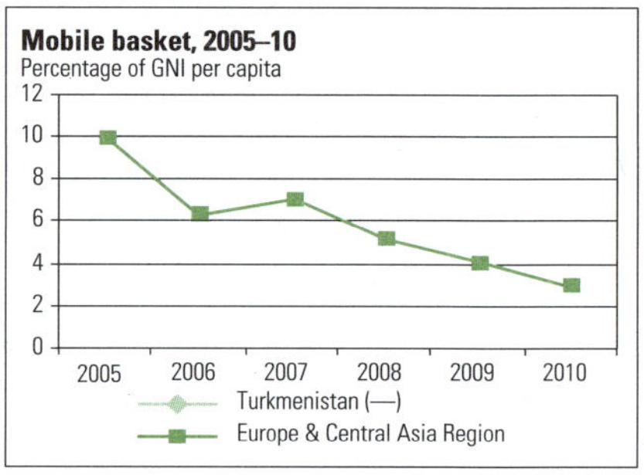

Uganda

	Uganda 2005	Uganda 2010	Low-income group 2010	Sub-Saharan Africa Region 2010
Economic and social context				
Population (total, million)	28	33	796	853
GNI per capita, World Bank Atlas method (current US$)	300	500	530	1,188
Rural population (% of total)	88	87	72	63
Expected years of schooling (years)	10	*11*	*9*	*9*
Physicians density (per 1,000 people)	0.1	—	0.2	0.2
Depositors with commercial banks (per 1,000 adults)	97	192	—	*167*
Sector structure				
Number of mobile operators	—	5		
Herfindahl-Hirschman Index (HHI) (scale = 0–10,000)	—	4,384		
Sector performance				
Access				
Mobile cellular subscriptions (per 100 people)	5	43[a]	43[a]	57[a]
Mobile cellular subscriptions (% prepaid)	99	99[a]	98[a]	96[a]
Population covered by a mobile-cellular network (%)	70	*100*	—	72
Mobile broadband subscriptions (per 100 people)	—	1.1[a]	—	5.6[a]
Mobile broadband (% of total mobile subscriptions)	—	2.7[a]	—	10.1[a]
Usage				
Households with a mobile telephone (%)	10	52	43	52
Mobile voice usage (minutes per user per month)	—	67[a]	—	—
Population using mobile Internet (%)	—	—	—	—
Short Message Service (SMS) users (% of mobile users)	—	—	—	—
Affordability				
Mobile basket (% of GNI per capita)	57.6	29.3	28.8	19.5

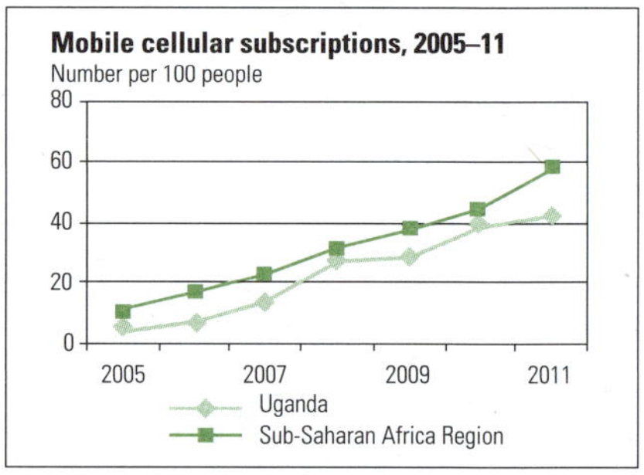

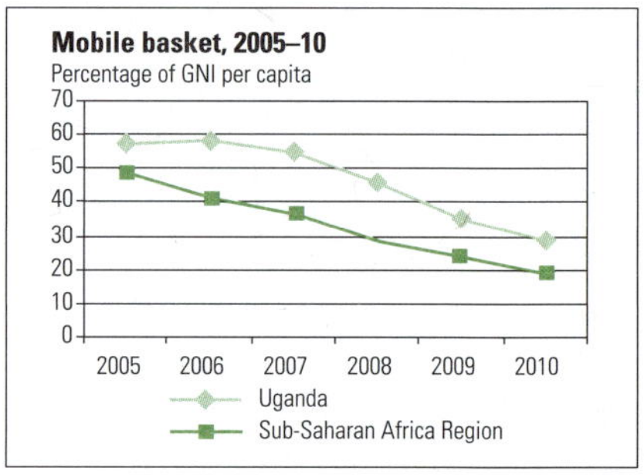

Sources: Economic and social context: IMF, UIS, UN, WHO and World Bank; Sector structure: ictDATA.org; Sector performance: ictDATA.org, ITU; Wireless Intelligence, and World Bank.
Notes: Use of italics in the column entries indicates years or periods other than those specified. — Not available. GNI = gross national income.
a. Data are for 2011.

Ukraine

	Ukraine 2005	Ukraine 2010	Lower-middle-income group 2010	Europe & Central Asia Region 2010
Economic and social context				
Population (total, million)	47	46	2,519	405
GNI per capita, World Bank Atlas method (current US$)	1,540	3,000	1,623	7,272
Rural population (% of total)	32	32	61	36
Expected years of schooling (years)	14	*15*	*10*	*13*
Physicians density (per 1,000 people)	*3.1*	*3.2*	0.8	3.2
Depositors with commercial banks (per 1,000 adults)	2,708	3,220	—	894
Sector structure				
Number of mobile operators	—	4		
Herfindahl-Hirschman Index (HHI) (scale = 0–10,000)	—	4,063		
Sector performance				
Access				
Mobile cellular subscriptions (per 100 people)	64	118[a]	78[a]	125[a]
Mobile cellular subscriptions (% prepaid)	92	92[a]	96[a]	82[a]
Population covered by a mobile-cellular network (%)	96	100	*86*	96
Mobile broadband subscriptions (per 100 people)	0.0	5.8[a]	7.3[a]	22.6[a]
Mobile broadband (% of total mobile subscriptions)	0.1	4.7[a]	9.0[a]	18.0[a]
Usage				
Households with a mobile telephone (%)	44	84	77	82
Mobile voice usage (minutes per user per month)	113	469[a]	276[a]	288[a]
Population using mobile Internet (%)	—	2.4	2.9	8.5
Short Message Service (SMS) users (% of mobile users)	—	72.0[a]	61.9[a]	69.8[a]
Affordability				
Mobile basket (% of GNI per capita)	13.5	3.0	7.2	3.1

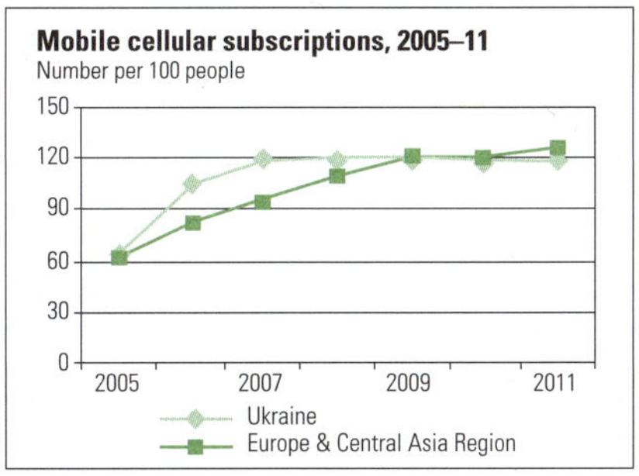

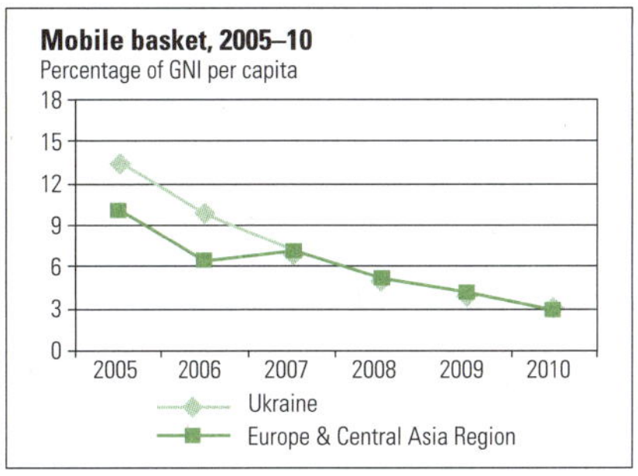

United Arab Emirates

	United Arab Emirates 2005	United Arab Emirates 2010	High-income group 2010
Economic and social context			
Population (total, million)	4	8	1,127
GNI per capita, World Bank Atlas method (current US$)	42,280	*41,930*	38,746
Rural population (% of total)	22	22	22
Expected years of schooling (years)	*11*	*13*	*16*
Physicians density (per 1,000 people)	1.5	—	2.8
Depositors with commercial banks (per 1,000 adults)	—	—	—
Sector structure			
Number of mobile operators	—	2	
Herfindahl-Hirschman Index (HHI) (scale = 0–10,000)	—	5,887	
Sector performance			
Access			
Mobile cellular subscriptions (per 100 people)	111	149[a]	118[a]
Mobile cellular subscriptions (% prepaid)	89	89[a]	36[a]
Population covered by a mobile-cellular network (%)	100	100	100
Mobile broadband subscriptions (per 100 people)	4.1	74.8[a]	69.6[a]
Mobile broadband (% of total mobile subscriptions)	3.7	45.4[a]	57.6[a]
Usage			
Households with a mobile telephone (%)	95	97	93
Mobile voice usage (minutes per user per month)	—	—	339
Population using mobile Internet (%)	4.9	9.3	24.3
Short Message Service (SMS) users (% of mobile users)	—	—	78.2[a]
Affordability			
Mobile basket (% of GNI per capita)	0.2	*0.3*	1.0

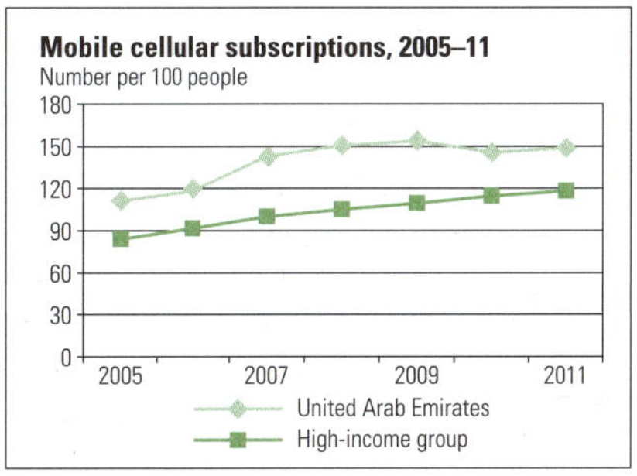

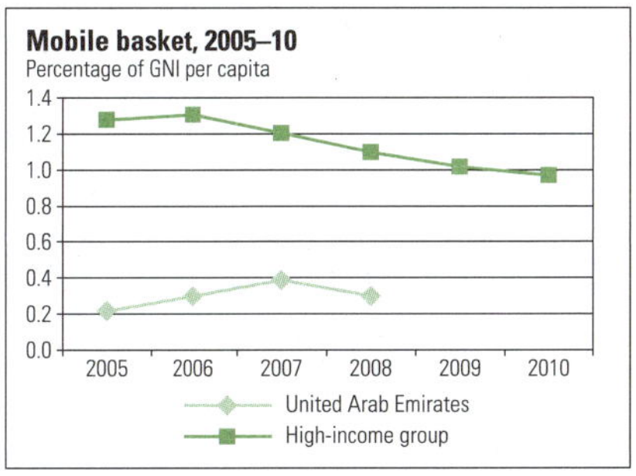

Sources: Economic and social context: IMF, UIS, UN, WHO and World Bank; Sector structure: ictDATA.org; Sector performance: ictDATA.org, ITU; Wireless Intelligence, and World Bank.
Notes: Use of italics in the column entries indicates years or periods other than those specified. — Not available. GNI = gross national income.
a. Data are for 2011.

United Kingdom

	United Kingdom 2005	United Kingdom 2010	High-income group 2010
Economic and social context			
Population (total, million)	60	62	1,127
GNI per capita, World Bank Atlas method (current US$)	38,850	38,200	38,746
Rural population (% of total)	10	10	22
Expected years of schooling (years)	17	*16*	*16*
Physicians density (per 1,000 people)	*2.2*	2.7	2.8
Depositors with commercial banks (per 1,000 adults)	—	—	—
Sector structure			
Number of mobile operators	—	4	
Herfindahl-Hirschman Index (HHI) (scale = 0–10,000)	—	2,495	
Sector performance			
Access			
Mobile cellular subscriptions (per 100 people)	109	130[a]	118[a]
Mobile cellular subscriptions (% prepaid)	67	50[a]	36[a]
Population covered by a mobile-cellular network (%)	99	100	100
Mobile broadband subscriptions (per 100 people)	7.7	67.5[a]	69.6[a]
Mobile broadband (% of total mobile subscriptions)	6.9	55.2[a]	57.6[a]
Usage			
Households with a mobile telephone (%)	88	93	93
Mobile voice usage (minutes per user per month)	151	192[a]	339
Population using mobile Internet (%)	9.3	20.2	24.3
Short Message Service (SMS) users (% of mobile users)	83.5	90.3	78.2[a]
Affordability			
Mobile basket (% of GNI per capita)	1.1	1.0	1.0

Mobile cellular subscriptions, 2005–11
Number per 100 people

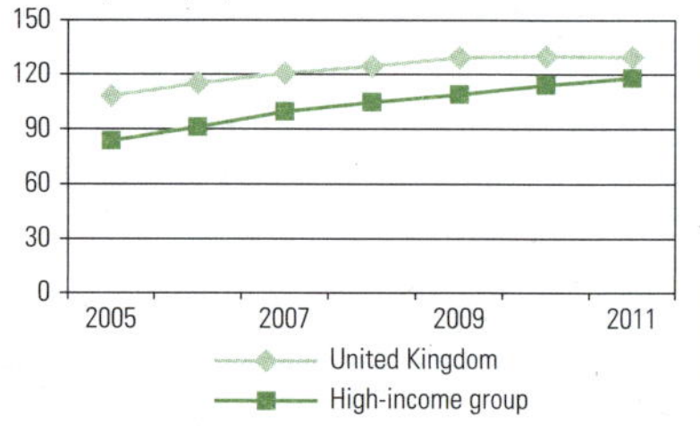

Mobile basket, 2005–10
Percentage of GNI per capita

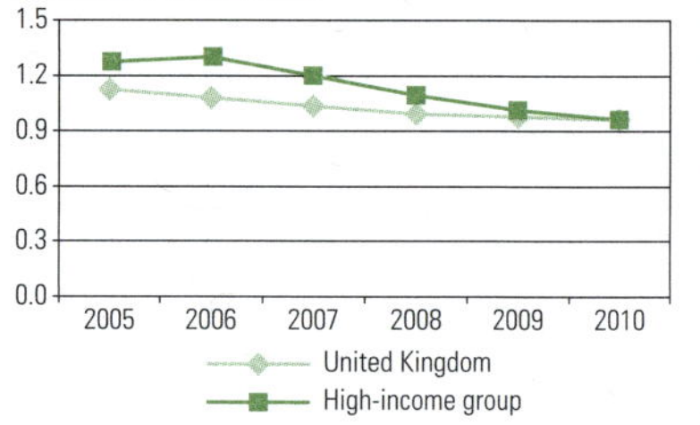

United States

	United States 2005	United States 2010	High-income group 2010
Economic and social context			
Population (total, million)	296	309	1,127
GNI per capita, World Bank Atlas method (current US$)	44,660	47,340	38,746
Rural population (% of total)	19	18	22
Expected years of schooling (years)	16	*16*	*16*
Physicians density (per 1,000 people)	*2.7*	*2.4*	2.8
Depositors with commercial banks (per 1,000 adults)	*337*	—	—
Sector structure			
Number of mobile operators	—	4	
Herfindahl-Hirschman Index (HHI) (scale = 0–10,000)	—	2,848	
Sector performance			
Access			
Mobile cellular subscriptions (per 100 people)	70	106[a]	118[a]
Mobile cellular subscriptions (% prepaid)	11	16[a]	36[a]
Population covered by a mobile-cellular network (%)	99	100	100
Mobile broadband subscriptions (per 100 people)	2.1	72.8[a]	69.6[a]
Mobile broadband (% of total mobile subscriptions)	3.0	67.0[a]	57.6[a]
Usage			
Households with a mobile telephone (%)	51	85	93
Mobile voice usage (minutes per user per month)	*683*	772	339
Population using mobile Internet (%)	6.6	35.6[a]	24.3
Short Message Service (SMS) users (% of mobile users)	—	68.0	78.2[a]
Affordability			
Mobile basket (% of GNI per capita)	0.5	0.8	1.0

Mobile cellular subscriptions, 2005–11
Number per 100 people

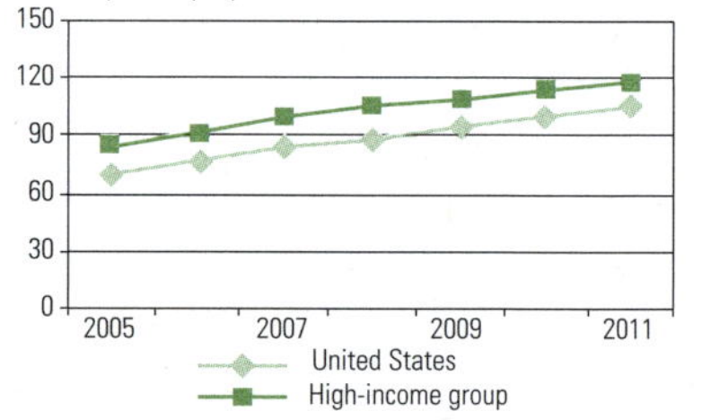

Mobile basket, 2005–10
Percentage of GNI per capita

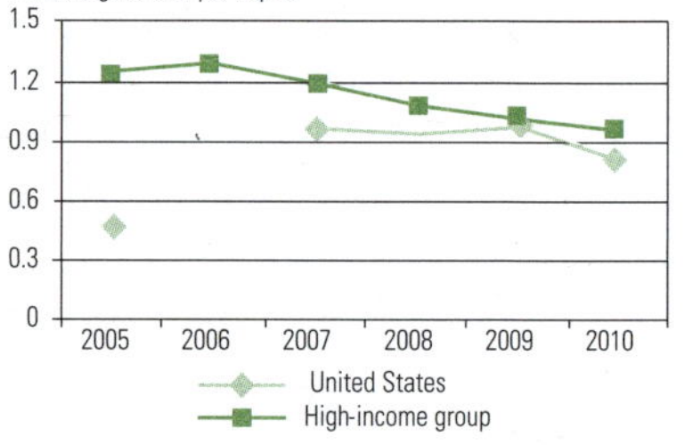

Sources: Economic and social context: IMF, UIS, UN, WHO and World Bank; Sector structure: ictDATA.org; Sector performance: ictDATA.org, ITU; Wireless Intelligence, and World Bank.
Notes: Use of italics in the column entries indicates years or periods other than those specified. — Not available. GNI = gross national income.
a. Data are for 2011.

Uruguay

	Uruguay 2005	Uruguay 2010	Upper-middle-income group 2010	Latin America & the Caribbean Region 2010
Economic and social context				
Population (total, million)	3	3	2,452	583
GNI per capita, World Bank Atlas method (current US$)	4,740	10,230	5,886	7,741
Rural population (% of total)	8	8	43	21
Expected years of schooling (years)	*15*	*16*	*13*	*14*
Physicians density (per 1,000 people)	*4.2*	*3.7*	1.7	1.8
Depositors with commercial banks (per 1,000 adults)	341	*538*	—	—
Sector structure				
Number of mobile operators	—	3		
Herfindahl-Hirschman Index (HHI) (scale = 0–10,000)	—	3,746		
Sector performance				
Access				
Mobile cellular subscriptions (per 100 people)	35	136[a]	92[a]	109[a]
Mobile cellular subscriptions (% prepaid)	85	71[a]	81[a]	81[a]
Population covered by a mobile-cellular network (%)	100	100	*99*	98
Mobile broadband subscriptions (per 100 people)	—	21.7[a]	14.3[a]	16.1[a]
Mobile broadband (% of total mobile subscriptions)	—	15.4[a]	15.4[a]	15.2[a]
Usage				
Households with a mobile telephone (%)	35	*83*	84	84
Mobile voice usage (minutes per user per month)	—	—	325[a]	141[a]
Population using mobile Internet (%)	—	4.8	22.9[a]	4.4
Short Message Service (SMS) users (% of mobile users)	—	—	74.4[a]	—
Affordability				
Mobile basket (% of GNI per capita)	6.3	2.1	2.9	3.7

Mobile cellular subscriptions, 2005–11
Number per 100 people

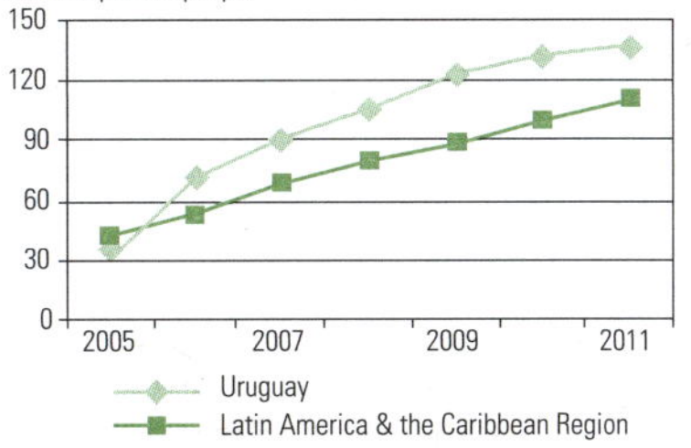

Mobile basket, 2005–10
Percentage of GNI per capita

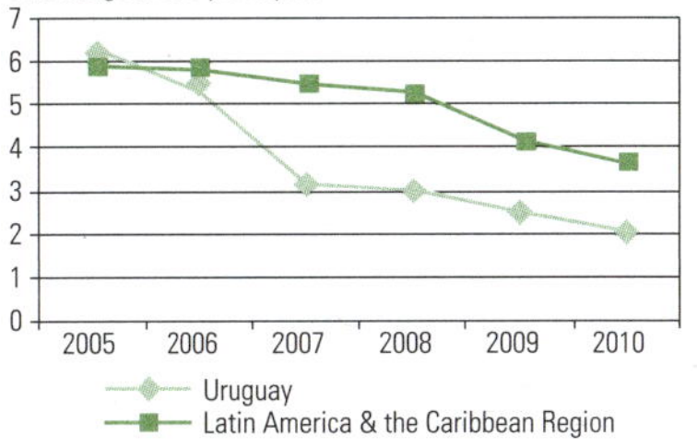

Uzbekistan

	Uzbekistan 2005	Uzbekistan 2010	Lower-middle-income group 2010	Europe & Central Asia Region 2010
Economic and social context				
Population (total, million)	26	28	2,519	405
GNI per capita, World Bank Atlas method (current US$)	530	1,280	1,623	7,272
Rural population (% of total)	63	63	61	36
Expected years of schooling (years)	12	*11*	*10*	*13*
Physicians density (per 1,000 people)	2.7	*2.6*	0.8	3.2
Depositors with commercial banks (per 1,000 adults)	676	957	—	894
Sector structure				
Number of mobile operators	—	5		
Herfindahl-Hirschman Index (HHI) (scale = 0–10,000)	—	3,339		
Sector performance				
Access				
Mobile cellular subscriptions (per 100 people)	3	84[a]	78[a]	125[a]
Mobile cellular subscriptions (% prepaid)	90	95[a]	96[a]	82[a]
Population covered by a mobile-cellular network (%)	75	*93*	*86*	96
Mobile broadband subscriptions (per 100 people)	—	6.6[a]	7.3[a]	22.6[a]
Mobile broadband (% of total mobile subscriptions)	—	8.1[a]	9.0[a]	18.0[a]
Usage				
Households with a mobile telephone (%)	50	87	77	82
Mobile voice usage (minutes per user per month)	450	389[a]	276[a]	288[a]
Population using mobile Internet (%)	—	0.7	2.9	8.5
Short Message Service (SMS) users (% of mobile users)	—	*25.0*	61.9[a]	69.8[a]
Affordability				
Mobile basket (% of GNI per capita)	18.3	2.8	7.2	3.1

Mobile cellular subscriptions, 2005–11
Number per 100 people

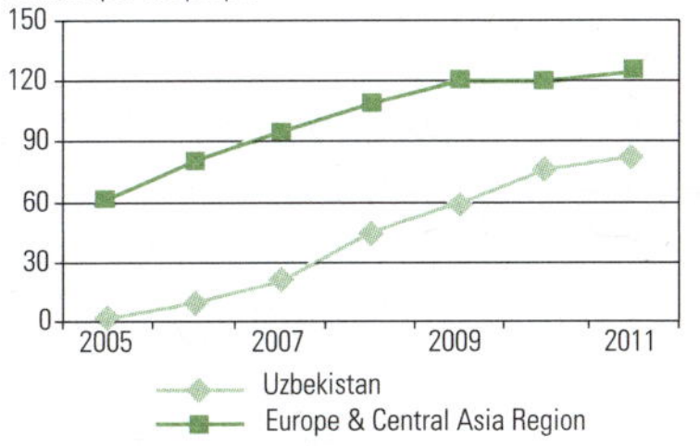

Mobile basket, 2005–10
Percentage of GNI per capita

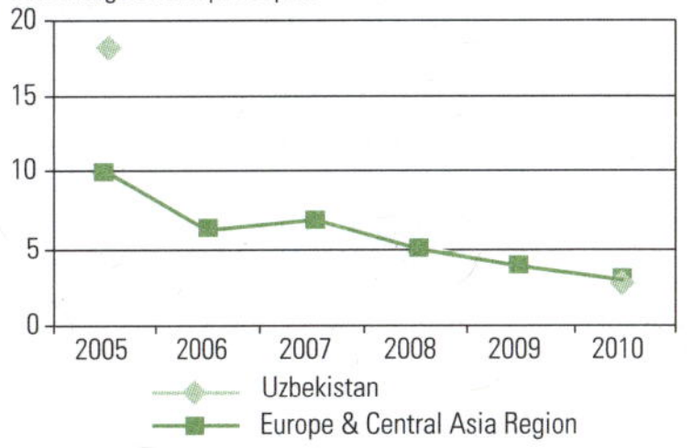

Sources: Economic and social context: IMF, UIS, UN, WHO and World Bank; Sector structure: ictDATA.org; Sector performance: ictDATA.org, ITU; Wireless Intelligence, and World Bank.
Notes: Use of italics in the column entries indicates years or periods other than those specified. — Not available. GNI = gross national income.
a. Data are for 2011.

Venezuela, RB

	Venezuela, RB 2005	Venezuela, RB 2010	Upper-middle-income group 2010	Latin America & the Caribbean Region 2010
Economic and social context				
Population (total, million)	27	29	2,452	583
GNI per capita, World Bank Atlas method (current US$)	4,950	11,590	5,886	7,741
Rural population (% of total)	8	6	43	21
Expected years of schooling (years)	*12*	*14*	*13*	*14*
Physicians density (per 1,000 people)	—	—	1.7	1.8
Depositors with commercial banks (per 1,000 adults)	—	—	—	—
Sector structure				
Number of mobile operators	—	3		
Herfindahl-Hirschman Index (HHI) (scale = 0–10,000)	—	—		
Sector performance				
Access				
Mobile cellular subscriptions (per 100 people)	47	98[a]	92[a]	109[a]
Mobile cellular subscriptions (% prepaid)	95	94[a]	81[a]	81[a]
Population covered by a mobile-cellular network (%)	*85*	—	*99*	98
Mobile broadband subscriptions (per 100 people)	0.3	26.1[a]	14.3[a]	16.1[a]
Mobile broadband (% of total mobile subscriptions)	0.5	24.7[a]	15.4[a]	15.2[a]
Usage				
Households with a mobile telephone (%)	26	46	84	84
Mobile voice usage (minutes per user per month)	*116*	—	325[a]	141[a]
Population using mobile Internet (%)	—	5.5	22.9[a]	4.4
Short Message Service (SMS) users (% of mobile users)	—	—	74.4[a]	—
Affordability				
Mobile basket (% of GNI per capita)	5.5	2.3	2.9	3.7

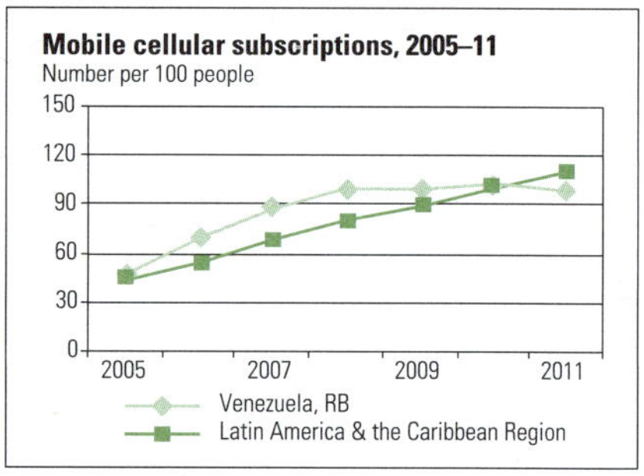

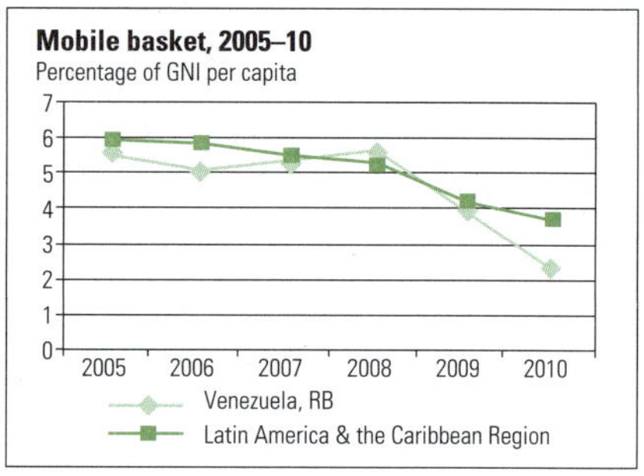

Vietnam

	Vietnam 2005	Vietnam 2010	Lower-middle-income group 2010	East Asia & Pacific Region 2010
Economic and social context				
Population (total, million)	82	87	2,519	1,962
GNI per capita, World Bank Atlas method (current US$)	630	1,160	1,623	3,696
Rural population (% of total)	74	71	61	54
Expected years of schooling (years)	—	—	*10*	*12*
Physicians density (per 1,000 people)	—	*1.2*	0.8	1.2
Depositors with commercial banks (per 1,000 adults)	—	—	—	—
Sector structure				
Number of mobile operators	—	7		
Herfindahl-Hirschman Index (HHI) (scale = 0–10,000)	—	2,664		
Sector performance				
Access				
Mobile cellular subscriptions (per 100 people)	12	134[a]	78[a]	83[a]
Mobile cellular subscriptions (% prepaid)	92	88[a]	96[a]	85[a]
Population covered by a mobile-cellular network (%)	*70*	—	*86*	*99*
Mobile broadband subscriptions (per 100 people)	*0.01*	25.6[a]	7.3[a]	11.6[a]
Mobile broadband (% of total mobile subscriptions)	*0.03*	16.4[a]	9.0[a]	14.4[a]
Usage				
Households with a mobile telephone (%)	30	50	77	83
Mobile voice usage (minutes per user per month)	—	—	276[a]	367[a]
Population using mobile Internet (%)	—	8.2	2.9	22.4[a]
Short Message Service (SMS) users (% of mobile users)	—	49.0	61.9[a]	84.0[a]
Affordability				
Mobile basket (% of GNI per capita)	19.1	5.6	7.2	5.7

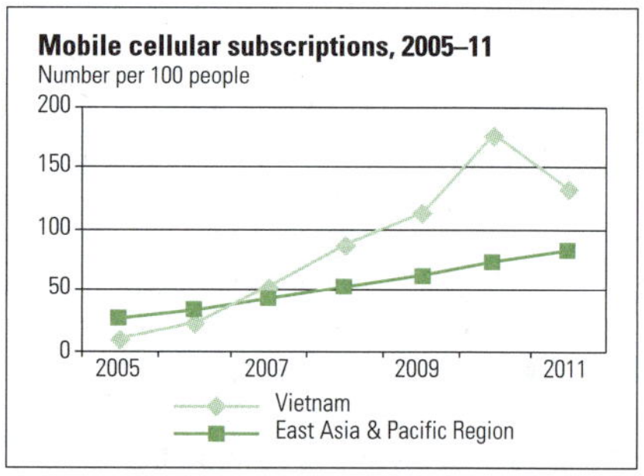

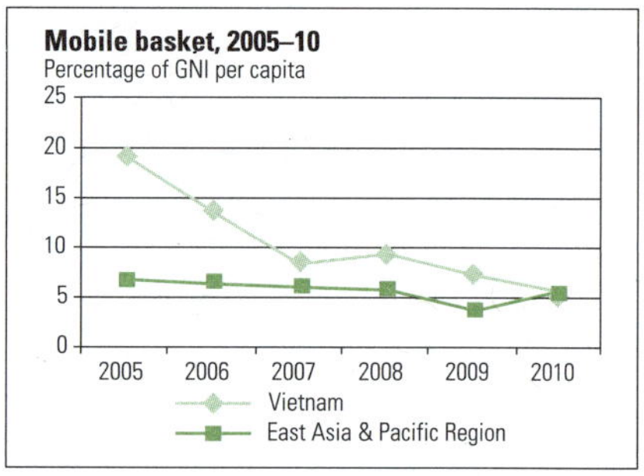

Sources: Economic and social context: IMF, UIS, UN, WHO and World Bank; Sector structure: ictDATA.org; Sector performance: ictDATA.org, ITU; Wireless Intelligence, and World Bank.
Notes: Use of italics in the column entries indicates years or periods other than those specified. — Not available. GNI = gross national income.
a. Data are for 2011.

West Bank and Gaza

	West Bank and Gaza 2005	West Bank and Gaza 2010	Lower-middle-income group 2010	Middle East & North Africa Region 2010
Economic and social context				
Population (total, million)	4	4	2,519	331
GNI per capita, World Bank Atlas method (current US$)	1,250	—	1,623	3,874
Rural population (% of total)	28	28	61	42
Expected years of schooling (years)	13	*13*	*10*	*12*
Physicians density (per 1,000 people)	—	—	0.8	1.4
Depositors with commercial banks (per 1,000 adults)	—	*543*	—	*443*
Sector structure				
Number of mobile operators	—	2		
Herfindahl-Hirschman Index (HHI) (scale = 0–10,000)	—	6,800		
Sector performance				
Access				
Mobile cellular subscriptions (per 100 people)	16	*45*	78[a]	89[a]
Mobile cellular subscriptions (% prepaid)	90	90[a]	96[a]	87[a]
Population covered by a mobile-cellular network (%)	95	—	*86*	—
Mobile broadband subscriptions (per 100 people)	—	—	7.3[a]	—
Mobile broadband (% of total mobile subscriptions)	—	—	9.0[a]	—
Usage				
Households with a mobile telephone (%)	*37*	*92*	77	—
Mobile voice usage (minutes per user per month)	—	—	276[a]	—
Population using mobile Internet (%)	—	—	2.9	4.5
Short Message Service (SMS) users (% of mobile users)	—	94.0	61.9[a]	—
Affordability				
Mobile basket (% of GNI per capita)	23.2	—	7.2	3.6

Mobile cellular subscriptions, 2005–11
Number per 100 people

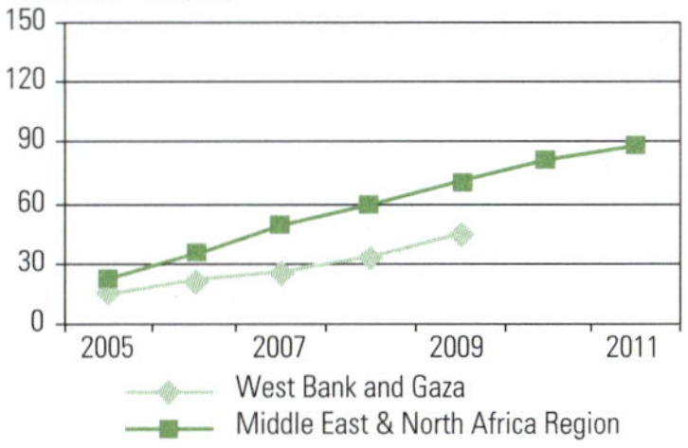

Mobile basket, 2005–10
Percentage of GNI per capita

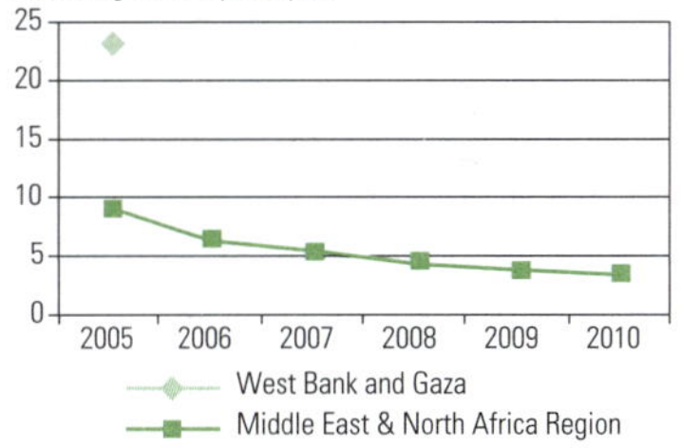

Yemen, Rep.

	Yemen, Rep. 2005	Yemen, Rep. 2010	Lower-middle-income group 2010	Middle East & North Africa Region 2010
Economic and social context				
Population (total, million)	21	24	2,519	331
GNI per capita, World Bank Atlas method (current US$)	690	1,170	1,623	3,874
Rural population (% of total)	71	68	61	42
Expected years of schooling (years)	9	—	*10*	*12*
Physicians density (per 1,000 people)	*0.3*	*0.3*	0.8	1.4
Depositors with commercial banks (per 1,000 adults)	54	*101*	—	*443*
Sector structure				
Number of mobile operators	—	3		
Herfindahl-Hirschman Index (HHI) (scale = 0–10,000)	—	3,450		
Sector performance				
Access				
Mobile cellular subscriptions (per 100 people)	11	36[a]	78[a]	89[a]
Mobile cellular subscriptions (% prepaid)	92	87[a]	96[a]	87[a]
Population covered by a mobile-cellular network (%)	68	84	*86*	—
Mobile broadband subscriptions (per 100 people)	—	—	7.3[a]	—
Mobile broadband (% of total mobile subscriptions)	—	—	9.0[a]	—
Usage				
Households with a mobile telephone (%)	—	—	77	—
Mobile voice usage (minutes per user per month)	—	—	276[a]	—
Population using mobile Internet (%)	—	—	2.9	4.5
Short Message Service (SMS) users (% of mobile users)	—	—	61.9[a]	—
Affordability				
Mobile basket (% of GNI per capita)	20.1	8.3	7.2	3.6

Mobile cellular subscriptions, 2005–11
Number per 100 people

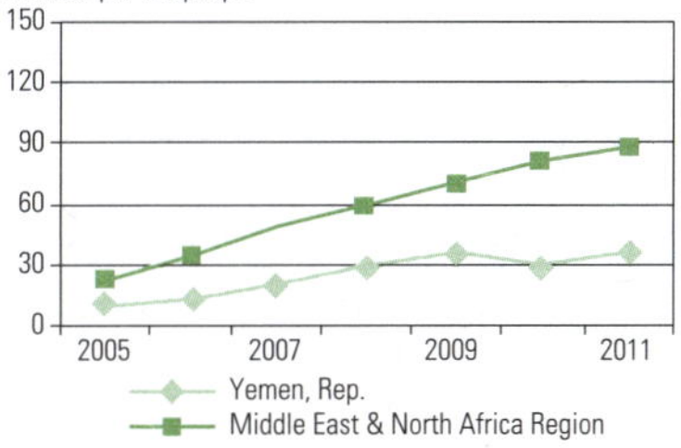

Mobile basket, 2005–10
Percentage of GNI per capita

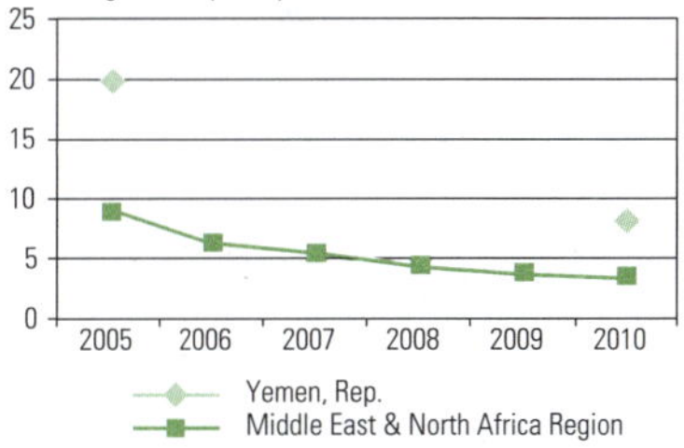

Sources: Economic and social context: IMF, UIS, UN, WHO and World Bank; Sector structure: ictDATA.org; Sector performance: ictDATA.org, ITU; Wireless Intelligence, and World Bank.
Notes: Use of italics in the column entries indicates years or periods other than those specified. — Not available. GNI = gross national income.
a. Data are for 2011.

Zambia

	Zambia 2005	Zambia 2010	Lower-middle-income group 2010	Sub-Saharan Africa Region 2010
Economic and social context				
Population (total, million)	11	13	2,519	853
GNI per capita, World Bank Atlas method (current US$)	500	1,070	1,623	1,188
Rural population (% of total)	65	64	61	63
Expected years of schooling (years)	—	—	*10*	*9*
Physicians density (per 1,000 people)	*0.1*	—	0.8	0.2
Depositors with commercial banks (per 1,000 adults)	—	—	—	*167*
Sector structure				
Number of mobile operators	—	3		
Herfindahl-Hirschman Index (HHI) (scale = 0–10,000)	—	5,478		
Sector performance				
Access				
Mobile cellular subscriptions (per 100 people)	8	54[a]	78[a]	57[a]
Mobile cellular subscriptions (% prepaid)	99	99[a]	96[a]	96[a]
Population covered by a mobile-cellular network (%)	65	90	*86*	72
Mobile broadband subscriptions (per 100 people)	—	0.3[a]	7.3[a]	5.6[a]
Mobile broadband (% of total mobile subscriptions)	—	0.5[a]	9.0[a]	10.1[a]
Usage				
Households with a mobile telephone (%)	15	58	77	52
Mobile voice usage (minutes per user per month)	—	—	276[a]	—
Population using mobile Internet (%)	—	2.8	2.9	—
Short Message Service (SMS) users (% of mobile users)	—	—	61.9[a]	—
Affordability				
Mobile basket (% of GNI per capita)	51.0	19.0	7.2	19.5

Mobile cellular subscriptions, 2005–11
Number per 100 people

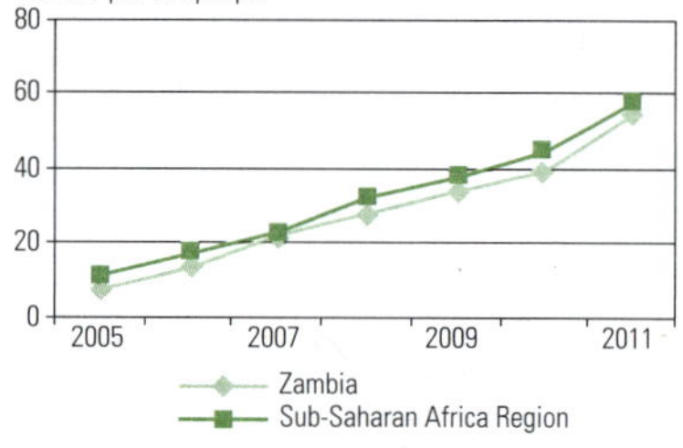

Mobile basket, 2005–10
Percentage of GNI per capita

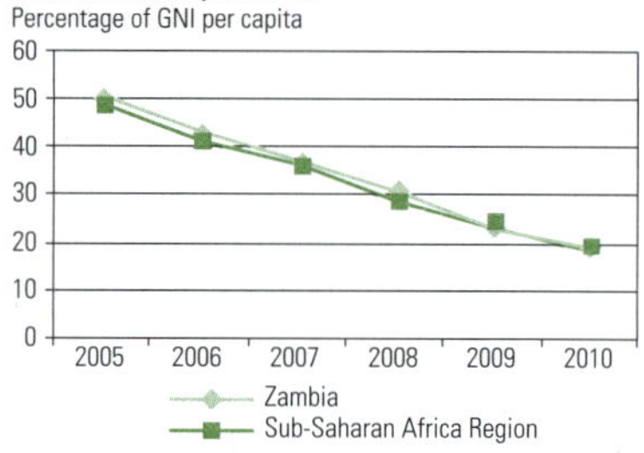

Zimbabwe

	Zimbabwe 2005	Zimbabwe 2010	Low-income group 2010	Sub-Saharan Africa Region 2010
Economic and social context				
Population (total, million)	13	13	796	853
GNI per capita, World Bank Atlas method (current US$)	440	460	530	1,188
Rural population (% of total)	64	62	72	63
Expected years of schooling (years)	—	—	*9*	*9*
Physicians density (per 1,000 people)	*0.2*	—	0.2	0.2
Depositors with commercial banks (per 1,000 adults)	—	—	—	*167*
Sector structure				
Number of mobile operators	—	3		
Herfindahl-Hirschman Index (HHI) (scale = 0–10,000)	—	—		
Sector performance				
Access				
Mobile cellular subscriptions (per 100 people)	5	60	43[a]	57[a]
Mobile cellular subscriptions (% prepaid)	87	93[a]	98[a]	96[a]
Population covered by a mobile-cellular network (%)	70	80	—	72
Mobile broadband subscriptions (per 100 people)	—	8.6[a]	—	5.6[a]
Mobile broadband (% of total mobile subscriptions)	—	12.4[a]	—	0.1[a]
Usage				
Households with a mobile telephone (%)	10	54	43	52
Mobile voice usage (minutes per user per month)	*119*	*98*	—	—
Population using mobile Internet (%)	—	0.6	—	—
Short Message Service (SMS) users (% of mobile users)	—	47.0	—	—
Affordability				
Mobile basket (% of GNI per capita)	16.4	53.5	28.8	19.5

Mobile cellular subscriptions, 2005–11
Number per 100 people

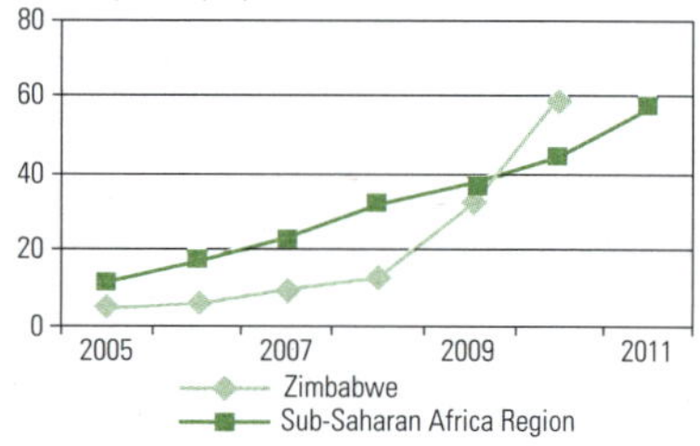

Mobile basket, 2005–10
Percentage of GNI per capita

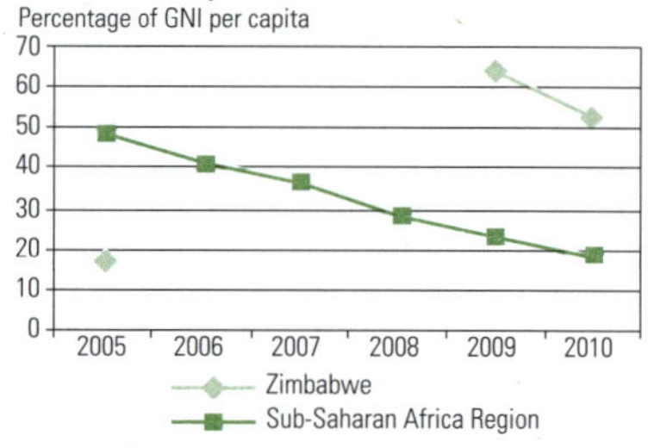

Sources: Economic and social context: IMF, UIS, UN, WHO and World Bank; Sector structure: ictDATA.org; Sector performance: ictDATA.org, ITU; Wireless Intelligence, and World Bank.
Notes: Use of italics in the column entries indicates years or periods other than those specified. — Not available. GNI = gross national income.
a. Data are for 2011.

Key mobile indicators for other economies, 2010

	Population (total, thousand) 2010	GNI per capita, World Bank Atlas method (current US$) 2010	Mobile cellular subscriptions (per 100 people) 2010	Population covered by a mobile-cellular network (%) 2010	Number of mobile operators 2010	Mobile cellular basket (% of GNI per capita) 2010
Afghanistan	34,385	410	39[a]	*75*	4	—
American Samoa	68	—[b]	—	—	—	—
Andorra	85	*41,750*	77	99	—	—
Antigua and Barbuda	88	13,280	186	100	3	2.0
Aruba	108	—[c]	122	*99*	—	—
Bahamas, The	343	22,240	125	100	1	0.9
Barbados	274	*12,660*	128	*100*	2	*1.0*
Belize	345	3,810	56	—	1	9.8
Bermuda	65	—[c]	137	—	3	—
Bhutan	726	1,870	52	100	2	2.9
Brunei Darussalam	399	*31,800*	109[a]	—	2	—
Cape Verde	496	3,270	75	85	2	15.3
Cayman Islands	56	—[c]	178	100	—	—
Channel Islands	153	*67,960*	—	—	—	—
Comoros	735	750	32[a]	—	1	38.8
Curaçao	143	—[c]	—	—	—	—
Djibouti	889	*1,270*	23[a]	90	1	*6.2*
Dominica	68	6,740	147	90	2	2.6
Equatorial Guinea	700	14,550	58	—	2	—
Faeroe Islands	49	—[c]	122	100	—	—
Fiji	860	3,630	81	*65*	2	6.2
French Polynesia	271	—[c]	80	80	1	—
Gibraltar	29	—[c]	103	—	—	—
Greenland	57	*26,020*	101	100	—	—
Grenada	104	6,960	112	—	2	2.5
Guam	179	—[c]	—	—	—	—
Guyana	755	2,870	74	95	2	3.9
Iceland	318	32,640	118	99	4	0.6
Isle of Man	83	*48,910*	—	—	—	—
Kiribati	100	2,000	10	—	1	10.4
Korea, Dem. People's Rep.	24,346	—[d]	3[a]	—	1	—
Kosovo	1,815	3,290	86	—	2	—
Liberia	3,994	200	41[a]	—	4	—
Liechtenstein	36	*137,070*	80	*95*	—	—
Luxembourg	507	76,980	143	100	3	0.4
Macao SAR, China	544	*34,880*	206	100	3	*0.1*
Maldives	316	5,750	156	100	2	1.2

(continued next page)

Key mobile indicators *continued*

	Population (total, thousand) 2010	GNI per capita, World Bank Atlas method (current US$) 2010	Mobile cellular subscriptions (per 100 people) 2010	Population covered by a mobile-cellular network 2010	Number of mobile operators 2010	Mobile cellular basket (% of GNI per capita) 2010
Malta	416	19,130	109	100	3	1.4
Marshall Islands	54	3,640	7	—	—	—
Mayotte	204	—[b]	—	—	—	—
Micronesia, Fed. Sts.	111	2,740	25	—	1	4.0
Monaco	35	*183,150*	—	—	—	—
Montenegro	632	6,740	183[a]	100	—	2.9
New Caledonia	247	—[c]	89	*89*	—	—
Northern Mariana Islands	61	—[c]	—	—	2	—
Palau	20	6,560	71	95	—	—
Samoa	184	2,980	91	—	2	7.1
San Marino	32	*50,400*	76	98	—	—
São Tomé and Principe	165	1,200	62	88	1	12.7
Seychelles	87	9,710	146	*98*	3	2.0
Sint Maarten (Dutch part)	38	—[c]	—	—	—	—
Solomon Islands	538	1,030	33[a]	—	2	—
Somalia	9,331	—[d]	34[a]	—	—	—
South Sudan	9,948[f]	—[e]	24	—	5	—
St. Kitts and Nevis	52	11,830	154	—	2	1.5
St. Lucia	174	6,560	113	*100*	2	4.1
St. Martin (French part)	30	—[c]	—	—	—	—
St. Vincent and the Grenadines	109	6,320	113	100	2	2.8
Suriname	525	*5,920*	170	—	3	*1.9*
Tonga	104	3,290	52	*90*	2	4.0
Turks and Caicos Islands	38	—[c]	—	—	—	—
Tuvalu	10	4,760	25	—	—	—
Vanuatu	240	2,640	27	—	2	10.6
Virgin Islands (U.S.)	110	—[c]	—	—	—	—

Sources: Economic and social context: IMF, UIS, UN, WHO and World Bank; Sector structure: ictDATA.org; Sector performance: ictDATA.org, ITU; Wireless Intelligence, and World Bank.

Notes: Use of italics in the column entries indicates years other than those specified. — Not available. GNI = gross national income.

a. Data are for 2011.
b. Estimated to be upper middle income ($3,976–$12,275).
c. Estimated to be high income ($12,276 or more).
d. Estimated to be low income ($1,005 or less).
e. Estimated to be lower middle income ($1,006–$3,975).
f. 2010 estimate.

Contributors

Maja Andjelkovic is interested in the potential of the mobile industry to create opportunities in emerging markets and in the role of mobile technology in human development. With the World Bank's Information for Development Program (*info*Dev), she works on supporting entrepreneurs to establish businesses in Africa, Asia, and Europe. During LLM studies at the University of Kent, she examined public-private governance of the internet. As a PhD student at Oxford University's Internet Institute, she is focusing on social aspects of innovation in the context of mobile entrepreneurship.

Kevin Donovan is a research associate in the World Bank's *info*Dev. He graduated from Georgetown University's School of Foreign Service with a degree in science, technology, and international affairs and spent part of 2012 as a Fulbright recipient studying the intersection of digital technology and democratic engagement in South Africa. He previously served on the board of directors of Students for Free Culture, an international NGO working to reform intellectual property rights, and managed Georgetown's OpenCourseWare initiative.

Neil Fantom is a manager in the Development Economics Data Group of the World Bank. He leads the team that provides open access to the World Bank's databases on development and manages the compilation and publication of the World Development Indicators and Global Development Finance. Before joining the World Bank, he worked for the U.K. Department for International Development (DFID) and the European Commission. Mr. Fantom studied statistics and mathematics in the United Kingdom, at University College London, the University of Oxford, and the University of Durham.

Nicolas Friederici is a consultant at *info*Dev at the World Bank. His research interests cover ICT for development and mobile innovation. He has authored academic publications in the fields of broadband economics and policy, social online behavior, and knowledge management. He is also involved in *info*Dev's operational work in mobile innovation and incubation. He was a Fulbright scholar at Michigan State University where he received a master's degree in telecommunication, information studies, and media. He also holds a Diplom in media studies and media management from the University of Cologne.

Naomi J. Halewood is an ICT policy specialist with the ICT Unit of the World Bank, where she works on projects involving telecommunications sector development and modernizing government through the use of ICT, mainly in the East Asia and Pacific Region. Her contributions to publications examine the role of ICT and, more recently, the mobile phone in various development contexts such as agriculture, public service delivery, and small and medium enterprise development. Before joining the ICT Unit, she worked in the

Development Economics Data Group of the World Bank. She holds master's degrees in international development and business administration from American University and a bachelor's degree in political science and sociology from Brandeis University.

Carol Hullin specializes in health informatics. She works as a professor of health informatics in developing countries such as Argentina, Bolivia, and Chile. As a consultant to the World Bank's ICT Sector Unit, she works on the eHealth strategy. Her areas of expertise include ICT for health, education, and social development, with an emphasis in policy-making and utilization of mobile technologies. She has a PhD in health informatics from Melbourne University, Australia. Currently, she is the vice president of the Latino American and Caribbean Federation of Medical Informatics, and a founder of the Chilean and Peruvian Health Informatics Association, an NGO within the International Medical Informatics Association in Geneva, Switzerland.

Saori Imaizumi is a consultant in the World Bank's education team in the South Asia Region. She specializes in skills development, engineering education, and ICT and education, conducting operational and analytical work in India and Pakistan. With her background as an IT/management consultant in the private sector, she actively leverages business acumen and technologies to help solve issues in an innovative way, especially in the education sector. Her work includes the use of mobile phone for a tracer study and use of ICT in teacher education and monitoring and evaluation. She holds a master's degree in development economics and international business from the Fletcher School at Tufts University and a bachelor's degree in comparative politics and international relations from Wesleyan University.

Tim Kelly acted as Task Team Leader and led the research and drafting of this report. He is a lead ICT policy specialist working with the ICT Sector Unit and *info*Dev within the World Bank Group. He previously worked at the International Telecommunication Union (ITU) and Organisation for Economic Co-operation and Development (OECD). He is the author or co-author of more than 30 books in the field of ICT4D, including the OECD *Communications Outlook* and the ITU *Internet Reports*. At the Bank, he is co-author of the *Broadband Strategies Handbook*. He holds a PhD in geography from the University of Cambridge, U.K.

Buyant Erdene Khaltarkhuu is a statistical analyst in the Development Economics Data Group of the World Bank. She is an author and producer of the States and Markets section of the World Development Indicators database and publication, responsible for data and statistics on topics such as the private sector, the financial system, governance, transport, ICT, and science and technology. She has a master's degree in economics from Northern Illinois University.

Kaoru Kimura is an operations analyst with the ICT Sector Unit at the World Bank Group. She has worked on several operational and analytical projects in Sub-Saharan Africa and East Asia. In addition, she has been actively involved in monitoring and evaluation activities in the ICT sector. She has worked on the ICT at-a-glance tables, Core Sector Indicators, and the *Little Data Book on Information and Communication Technology* series. Before joining the ICT Sector Unit, she worked at Nippon Telegraph and Telecommunication in Japan. She has a master's degree in international development studies from the National Graduate Institute for Policy Studies in Japan.

Soong Sup Lee is a senior information officer in the Development Economics Data Group of the World Bank. He is a member of the team that provides open access to the World Bank's databases and leads the team that produces the World Development Indicators database and publication. Mr. Lee has worked in various roles at the World Bank for over 25 years. His current focus is improving the access and usefulness of the World Bank's information for a broad audience. Mr. Lee has a master's degree in business administration from George Washington University and a bachelor's degree in engineering from McGill University.

Samia Melhem is the chair of the e-Development Thematic Group. Her current operational and analytical responsibilities include technical assistance, planning, and supervision of eGoverment operations. In her 20 years of experience in development at the World Bank Group, Ms. Melhem has worked on ICT4D in several sectors: telecoms policy regulation, ICT for public sector reform (taxes, customs, trade), education, the knowledge economy, and private sector development. She has held several positions in different regions such as Africa, the Middle East, and Europe and Central Asia. She is the sector coordinator for governance and accountability, and gender. She holds degrees in electrical engineering (BS), computer sciences (MS), and finance (MBA).

Michael Minges is an independent consultant with more than 20 years of experience advising governments and the private sector on ICT issues in developing countries. He previously worked for Telecommunications Management Group (TMG) where he was senior market analyst. Before joining TMG, he served as head of the Markets, Economics and Finance Unit at the International Telecommunication Union (ITU). While at the ITU he launched the *World Telecommunications Development Report*, a principal industry publication, and designed the Digital Access Index for measuring ICT progress. He also worked at the International Monetary Fund as an information technology specialist. Mr. Minges holds an MBA in information systems from George Washington University.

Victor Mulas is an ICT policy specialist in the World Bank's ICT Sector Unit. His expertise lies in policy analysis and advisory work on sector reform, policy strategy, regulatory frameworks, ICT-led innovation and transformation, and institutional capacity building. Before coming to the World Bank, he worked for Telecommunications Management Group, a global consulting firm, for an affiliate of the Tiscali group in Spain, and as an associate lawyer for a telecommunications law firm in Spain. Mr. Mulas holds an MBA with an International Business Diplomacy certificate from the McDonough School of Business at Georgetown University, an LLM in telecommunications law from Universidad de Comillas, and a law degree from Universidad Autonoma de Madrid.

William Prince is a senior information officer in the Development Economics Data Group of the World Bank, where he leads the Data Administration and Quality team responsible for production and content management of electronic data products, including the online Open Data versions of *World Development Indicators* and *Global Development Finance*. He also provides overall data management and support for the Data Group's clients. Before joining the Bank, he was a statistical consultant for British Telecom. He has a master's degree in business administration in decision analysis from Arizona State University.

Siddhartha Raja is a policy specialist with the ICT Sector Unit of the World Bank Group. He works with governments in South Asia, Eastern Europe, and Central Asia on ICT sector strategy and telecommunications policy development and provides advisory services on using mobile tools to support service delivery and good governance. He has published books on media convergence and broadband telecommunications during his time with the World Bank. Mr. Raja has a bachelor's degree in telecommunications engineering from the University of Bombay, a master's degree in infrastructure policy studies from Stanford University, and a doctorate in telecommunications policy from the University of Illinois.

Priya Surya is a technology consultant in the World Bank's South Asia Rural Livelihoods Unit, where she works on strategy and implementation of mobile, smartcard, big data, and multimedia technologies for improving livelihood outcomes and public service delivery. Her work focuses on driving customer adoption and improving the user experience. Her areas of expertise include financial inclusion, branchless banking, mobile-based data collection, agricultural value chain, innovative business models, and community-driven development. She has an MPA in international development from the Harvard Kennedy School and a BA in economics from Macaulay Honors College at the City University of New York.

Masatake Yamamichi is a consultant in the World Bank's ICT Sector Unit. His expertise lies in ICT policies, telecommunications reform, and eGovernment, and their relevant areas, such as ICT-enabled social development and employment. He is also involved in operational work with client countries in the Middle East and North Africa and the unit's global analytical work and portfolio review. He has contributed to a number of ICT-related publications as an author, researcher, and reviewer. He holds a bachelor's degree in economics from the University of Tokyo and a master's degree in international relations from the Maxwell School of Syracuse University.

ECO-AUDIT

Environmental Benefits Statement

The World Bank is committed to preserving endangered forests and natural resources. The Office of the Publisher has chosen to print ***Information and Communications for Development 2012: Maximizing Mobile*** on recycled paper with 50 percent postconsumer fiber in accordance with the recommended standards for paper usage set by the Green Press Initiative, a nonprofit program supporting publishers in using fiber that is not sourced from endangered forests. For more information, visit www.greenpressinitiative.org.

Saved:

- 31 trees
- 13 million BTU of total energy
- 3.153 pounds of net greenhouse gases (CO_2 equivalent)
- 14,216 gallons of waste water
- 901 pounds of solid waste

green press INITIATIVE